THE BIOLOGY OF THE INTERFERON SYSTEM 1986

The Biology of the Interferon System 1986

Proceedings of the 1986 ISIR-TNO meeting on the interferon system,
7–12 September 1986, Dipoli Congress Center, Espoo, Finland

Edited by

K. CANTELL
National Public Health Institute, Helsinki, Finland

H. SCHELLEKENS
Primate Center, TNO, Rijswijk, The Netherlands

1987 **MARTINUS NIJHOFF PUBLISHERS**
a member of the KLUWER ACADEMIC PUBLISHERS GROUP
DORDRECHT / BOSTON / LANCASTER

Distributors

for the United States and Canada: Kluwer Academic Publishers, P.O. Box 358, Accord Station, Hingham, MA 02018-0358, USA
for the UK and Ireland: Kluwer Academic Publishers, MTP Press Limited, Falcon House, Queen Square, Lancaster LA1 1RN, UK
for all other countries: Kluwer Academic Publishers Group, Distribution Center, P.O. Box 322, 3300 AH Dordrecht, The Netherlands

Library of Congress Cataloging in Publication Data

Biology of the interferon system 1986.

  Includes bibliographies and indexes.
  1. Interferon--Congresses. 1. Cantell, K. (Kari)
II. Schellekens, Hubb. III. International Society for
Interferon Research. IV. Nederlandse Centrale
Organisatie voor Toegepast-Natuurwetenschappelijk
Onderzoek. [DNLM: 1. Interferons--congresses.
QW 800 86156]
QR187.5.B56 1987   616.07'9   86-31269

ISBN 978-94-010-8084-2   ISBN 978-94-009-3543-3 (eBook)
DOI 10.1007/978-94-009-3543-3

Copyright

ACKNOWLEDGEMENTS

Sponsors (in alphabetical order):

Academy of Finland

Amgen

Bioferon biochemische Substanzen GmbH and Co.

Biogen SA

Boehringer Ingelheim Zentrale GmbH

Bristol-Myers Company

Ciba-Geigy AG

Finnair

Finnish Cancer Society

Finnish Ministry of Education

Finnish Red Cross Blood Transfusion Service

Hayashibara Biochemical Laboratories, Inc.

F. Hoffmann-La Roche and Co. AG

Interferon Sciences, Inc.

Sigrid Jusélius Foundation

KLM Royal Dutch Airlines

Mary Ann Liebert, Inc. Publishers

Institut Mérieux

Martinus Nijhoff Publishers

Orion Corporation Limited

E.I. du Pont de Nemours and Company

Roche Laboratories, a division of Hoffmann-La Roche Inc.

Roussel Uclaf

Schering Corporation, USA, and its Essex Companies, and
    AESCA, UNICET, Kirby-Warrick

Shionogi Research Laboratories/Shionogi and Co., Ltd.

Tecnomara AG

Triton Biosciences Inc.

Wagons-lits Reizen

Wellcome Biotech

ACKNOWLEDGMENTS

CONTENTS

GENES AND PROTEINS

MODES OF ACTION

RECEPTORS

INDUCTION

CELL BIOLOGY

IMMUNOLOGY AND INTERACTION WITH OTHER LYMPHOKINES

ANIMAL MODELS

CLINICAL STUDIES

Antitumor effects

GENES AND PROTEINS

AMINO ACID RESIDUES AFFECTING ANTIVIRAL AND ANTIPROLIFERATIVE ACTIVITY IN INTERFERON-α's

M.J. TYMMS, M.W. BEILHARZ, I.T. NISBET, P.J. CHAMBERS, B. McINNES, J.C. TURTON, P.J. HERTZOG AND A.W. LINNANE

CENTRE FOR MOLECULAR BIOLOGY & MEDICINE, CLAYTON, VICTORIA 3168, AUSTRALIA

INTRODUCTION

Human interferon-α's (IFN-α's) are a family of more than 13 subtypes which, although highly homologous, exhibit wide variations in the magnitudes of their biological activities (1,2).

Site-directed in vitro mutagenesis is being used in our laboratory to study the relationship between structure and function in human IFN-α's. We have concentrated on making amino acid substitutions in hydrophilic regions of IFN-α's which are highly conserved between IFN-α subtypes. Two such regions include the amino acid sequences from position 29 to 43 and from 116 to 136. In the Sternberg-Cohen model for the three dimensional structure of human interferon-α these regions are juxtaposed as anti-parallel β-sheets (3).

In this report we examine IFN-α analogues with single and double amino acid substitutions concentrated at and around tyrosine 123. The analysis of these analogues suggests that this region of IFN-α is important for antiviral and antiproliferative activities on human cells. We also report on substitutions elsewhere in the IFN-α molecule. This additional data indicates the importance of the region encompassing residues 29 to 43 and the possibility of differentially altering antiviral and antiproliferative activities.

MATERIALS AND METHODS

Construction of IFN analogues and their expression in yeast. Synthesis of appropriate oligonucleotides, construction of IFN-α templates in M13 phage, site-directed in vitro mutagenesis, transfer of IFN-α sequences from M13 phage to yeast vectors, expression, purification and quantification of met-IFN-α's have all been described elsewhere (4,5,6).

Expression In vitro of IFN-α's. In addition to yeast expression, IFN-α proteins were synthesised in cell-free extracts of rabbit reticulocytes using IFN-α mRNA synthesised in vitro using SP6 polymerase. IFN-α1 and modified IFN-α1 genes were cloned as BamHI fragments into the BglII site of pSP64T (7). This vector has an SP6 polymerase promoter upstream of the BglII cloning site. Linearisation of the vector with BamHI results in a template suitable for run-off transcription by SP6 polymerase.

Transcription reactions contained 50 μg.ml^{-1} linearized template in a solution containing 40 mM Tris-HCl pH 7.5, 6 mM MgCl$_2$, 10 mM dithiothreitol, 2 mM spermidine, 100 μg.ml^{-1} bovine serum albumin (nuclease-free), 500 μM each of ATP, CTP, GTP and UTP, 5 mM diguanosine triphosphate (Pharmacia), 1000 U.ml^{-1} ribonuclease inhibitor (RN'asin, Promega Biotec) and 5000 U.ml^{-1} SP6 RNA polymerase (Promega Biotec). Reactions were performed at 40°C for 60 min. DNA template was removed by treatment of the transcription mix with 50 U.ml^{-1} of DNase I (RQ1 RN'ase free, Promega Biotech) for 15 min at 40°C. Protein-free RNA was recovered by phenol extraction followed by precipitation with a 0.1 volume of 2.5 M potassium acetate pH 5.5 and 2.5 volumes of ethanol at -20°C overnight. The RNA precipitate was recovered by centrifugation at 10,000 g for 10 min and was washed twice with cold 80% ethanol. The pellet was lyophilyzed and resuspended at a concentration of 500 μg.ml^{-1} in water containing 1000 U ml^{-1} RN'asin. RNA prepared in this way maintains full biological activity for at least 4 months when stored at -80°C.

Synthetic mRNA was translated in commercially available, nuclease-treated, rabbit reticulocyte lysate using the protocol provided (Promega-Biotech) with the addition of ^{35}S-methionine (>1000 Ci/mmol, Amersham) and 1000 U.ml^{-1} RN'asin. Incorporation of ^{35}S-methionine into TCA insoluble material was measured as described by Pratt (8).

Biological Assays. Monoclonal antibody-purified IFN-α1 and IFN-α4A (α4A gene isolated in our laboratory and previously referred to as αM1, see ref. 9) or rabbit reticulocyte lysate preparations were assayed for antiviral activity on human HEp-2 cells and bovine MDBK cells challenged with Semliki Forest Virus as previously described (5). Antiproliferative activities were assayed on human Daudi cells as described by Evinger and Pestka (10).

RESULTS AND DISCUSSION

Expression in vitro of IFN-α's. The generation of mutated IFN-α genes, their expression in yeast or E. coli, and purification is a process with many steps. We have found that the critical step is expression of IFN-α analogues. Some analogues are expressed in insufficient quantities for biological testing. To circumvent this problem and speed up the analysis of IFN analogues we have developed an in vitro expression system as an alternative to expression in yeast or E. coli.

IFN-α1 and modified IFN-α genes were cloned into the BglII site of plasmid pSP64T for transcription. IFN-α mRNA's synthesised from this vector using SP6 RNA polymerase were efficiently translated in a cell-free protein synthesizing system from rabbit reticulocytes. All IFN-α1 mRNA's were translated with equal efficiency in rabbit reticulocyte lysate.

An example is given in Figure 1 which shows the level of in vitro expression of IFN-α1 and an IFN-α1 analogue which is expressed at very low levels in both yeast and E. coli. In vitro, the IFN-α1 analogue is expressed with an equal efficiency to unmodified IFN-α1. The yield of IFN-α1 in this translation was 10^6 IU.ml^{-1} (antiviral activity on bovine cells). Gel analysis shows that the predominant ^{35}S-labelled product has the expected apparent molecular weight for IFN-α1 of about 18 kD (Figure 2). A small proportion of this is present as dimer with an apparent molecular weight of 32 kD.

Antiviral and antiproliferative activities of IFN-α's synthesised in vitro can be determined by directly assaying the rabbit reticulocyte lysate. Activities can be calculated relative to unmodified IFN-α using ^{35}S-methionine incorporation data to correct for small variations in IFN yield. We have tested the biological activity of a number of IFN-α's and analogues prepared in this way and their properties are indistinguishable from material prepared from yeast (data not shown).

Substitutions at position 123 of IFN-α1. The tyrosine at 123 in IFN-α1 is conserved in all mammalian IFN-α's suggesting a functional role for this residue. We have previously substituted tyrosine 123 with serine and glycine, both of which resulted in a loss of antiviral activity on human cells (2). To further explore the requirements at this position in IFN-α1 we made five additional substitutions which cover a wider range of hydrophobicities and sizes (K, D, A, W, F). Table 1 gives the relative antiviral specific activities of these analogues and an analogue in which the tyrosine at 123 was deleted. All of the substitutions except for phenylalanine showed a greater than 80% loss of antiviral activity on human cells. The antiviral activity on bovine cells for all substitutions was not significantly altered from that of unmodified IFN-α1. This data emphasises that this residue is important for human cell activity and not bovine activity. Only the analogue with tyrosine deleted showed significantly reduced antiviral activity on bovine cells. Phenylalanine is the most conservative amino acid substitution since it differs from tyrosine by only an hydroxyl group. This result demonstrates that a conservative amino acid replacement at the highly conserved tyrosine 123 does not significantly change the ability of IFN-α1 to elicit an antiviral response on human cells. It is therefore unlikely that the hydroxyl group on tyrosine plays a vital functional role.

6

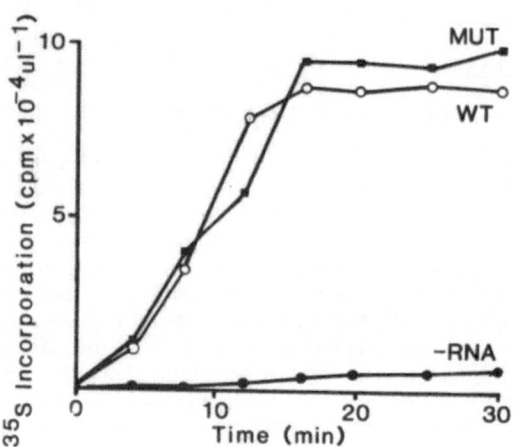

Fig. 1. Time course of incorporation of ^{35}S-methionine during translation of 5' capped IFNα1-mRNA in rabbit reticulocyte lysate. Translations contained unmodified IFN-α1 mRNA (WT), IFN-α1 mRNA directing $Y_{123} \rightarrow S_{123}$ substitution (MUT) and a no RNA control (-RNA).

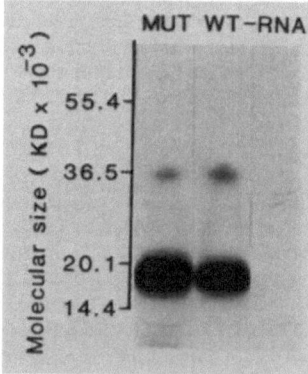

Fig. 2. Products from rabbit reticulocyte lysate translation of IFN-mRNA. The ^{35}S-methionine labelled products of translation were separated on a gradient (9-15%) polyacrylamide gel in the presence of SDS and a borate buffer (11) and were visualized by fluorography. Samples are as described for Fig. 1.

TABLE 1. RELATIVE ANTIVIRAL SPECIFIC ACTIVITIES OF IFN-α1 AND ANALOGUES EXPRESSED <u>IN VITRO</u> IN RABBIT RETICULOCYTE LYSATE.

	Relative antiviral activity[a]	
IFN-α Protein	Human cells (HEp-2)	Bovine cells (MDBK)
IFN-α1	100	100
$Y_{123} \rightarrow S_{123}$	15	150
G_{123}	<10	80
K_{123}	<11	140
D_{123}	<4	40
A_{123}	<8	150
W_{123}	<20	150
F_{123}	60	150
Deletion	<17%	<0.4%

(a) Values calculated from the specific antiviral activities for human and bovine cell assays and expressed as the percentage of unmodified IFN-α activites.

<u>Substitutions adjacent to position 123 in IFN-α1.</u> To examine the biological importance of the region around tyrosine 123 we have made selective amino acid replacements on either side of this residue (Table 2). At position 121 and 122 in human IFN-α's are two positively charged residues which in IFN-α1 are both lysines. These two residues contribute significantly to the local hydrophilicity. An analogue of IFN-α1 in which both of these lysines are substituted by hydrophobic leucine residues shows a 97% lower antiviral and antiproliferative activity on human cells.

The substitution of serine for phenylalanine at position 124 does not show marked change in antiviral activity on human cells. This suggests that the specific amino acid at this position is not critical for activity. The substitution of an uncharged glutamine for arginine at 125 which changes the IFN-α1 sequence to the human IFN-α consensus at this position results in approximately a 70% loss of antiviral and antiproliferative activity on human cells.

This series of substitutions in conjunction with the preceding data on position 123 suggests that the conserved residues 121, 122 and 123 in particular are important for human cell activity. It is noteworthy that none of the substitutions affect bovine antiviral activity.

TABLE 2.   RELATIVE   ANTIVIRAL   SPECIFIC   ACTIVITIES   AND ANTIPROLIFERATIVE ACTIVITIES OF IFN-α1 AND IFN-α4A PROTEINS AND THEIR ANALOGUES.

| Purified IFN-α Protein | Relative activities[a] | | |
	Human Antiviral (HEp2)	Bovine Antiviral (MDBK)	Human anti- proliferative (Daudi)[c]
IFN-α1	100	100	100
$K_{121}, K_{122} \rightarrow L_{121}, L_{122}$	3	109	2
$Y_{123} \rightarrow S_{123}$	15	150	19
$F_{124} \rightarrow S_{124}$	75	207	82
$R_{125} \rightarrow Q_{125}$	27	72	25
IFN-α4A	100	100	100
$K_{31}, R_{33}, \rightarrow E_{31}, E_{33}$	<0.003	<0.001	<0.001
IFN-α1[b]	100	100	100
$C_{86} \rightarrow S_{86}$	39	120	530

(a)   Values calculated from specific antiviral activities for human and bovine cell assays and expressed as the percentage of the unmodified IFN activity, IFN-α1 specific antiviral activities were 1.6 x 10^7 IU/mg on human HEp-2 cells and 5.3 x 10^8 IU/mg on the bovine MDBK cells.   IFN-α4A specific antiviral activities were 5.0 x 10^8 IU/mg on HEp-2 cells and 5.0 x 10^8 IU/mg on MDBK cells.

(b)   Expressed in E. coli with 8 amino acid leader.   All other proteins expressed in yeast as leaderless met-IFN.

(c)   The IFN concentration required to inhibit the growth of exponentially growing Daudi cells by 50% was 100 picograms per ml for IFN-α1 and 10 picograms per ml for IFN α4A.

Substitutions elsewhere in IFN-α's. The above results illustrate the importance of conserved hydrophilic amino acids around tyrosine 123. The region spanning residues 29 to 43 also contain hydrophilic amino acids which are conserved in human and other mammalian IFN-α's. We have made analogues with amino acid substitutions in this region which confirm that it is also important for biological activity (data not shown). For example the substitution of lysine at position 31 and arginine at position 33 in IFN-α4A to glutamic acids results in an analogue with no detectable antiviral activity on human and bovine cells and no detectable antiproliferative activity on human cells (Table 2).

Analogues with amino acid substitutions in the regions around residue 33 or residue 123 exhibit concomitant decreases in human antiviral and human antiproliferative activities. However, some substitutions of other regions of the molecule that we have examined exhibit an apparent separation in these activities. For example, a substitution of serine for cysteine at residue 86 results in a slight decrease in antiviral activity and a five-fold increase in antiproliferative activity on human cells (Table 2). One interpretation of this result is that antiviral and antiproliferative activities of IFN-α's may be mediated via different mechanisms. Alternatively, since the antiviral and antiproliferative activities are assayed on different cell types (HEp2 and Daudi), this result may demonstrate a differential alteration in target cell specificity of the IFN-α1 analogue.

REFERENCES

1. Weck PK, Apperson S, May L, Stebbing N (1981) J Gen Virol 57:233-237
2. Ortaldo JR, Herberman RB, Harvey C, Osheroff P, Pan Y-CE, Kelder B, Pestka S (1984) Proc Natl Acad Sci USA 81:4926-4929
3. Sternberg MJE, Cohen FE (1982) Int J Biol Macromol 4:137-144
4. Nisbet IT, Beilharz MW (1985) Gene Anal Techn 2:23-29
5. Nisbet IT, Beilharz MW, Hertzog PJ, Tymms MJ, Linnane AW (1985) Biochem Int 11:301-309
6. Devenish RJ, Maxwell RJ, Beilharz MW, Nagley P, Linnane AW In: Developments in Biological Standardization (in press).
7. Krieg PA and Melton DA (1974) Nucleic Acid Res 12:7057-7070
8. Pratt JM: Coupled transcription-translation in prokaryotic cell-free systems. In: Transcription and Translation (Eds. Hames BD, Higgins SJ) Oxford: IRL Press, 1984
9. Linnane AW, Beilharz MW, McMullen GL, Macreadie IG, Murphy M, Nisbet IT, Novitski CE, Woodrow GC (1984) Biochem Int 8:725-732
10. Evinger M, Pestka S (1981) Methods Enzymol 79:362-368
11. Shoeman RL, Schweiger HG (1982) J Cell Sci 58:23-33

DIFFERENTIAL STRUCTURES AND ACTIVITIES OF THREE IFN α_2 VARIANTS DEPARTING ONLY AT TWO AMINO ACID POSITIONS.

v. GABAIN[†*], A., OHLSSON[*], M., HOLMGREN[††], E., JOSEPHSSON[††], S., ALKAN[o], S.S. and LUNDGREN[*], E.

† Dept of Bacteriology, Karolinska Institute, S-104 01 Stockholm, Sweden.

* Inst. of Appl. Cell and Molec. Biology, University of Umeå, S-901 87 Umeå, Sweden.

†† KabiGen AB, S-112 87 Stockholm, Sweden.

o Ciba Geigy Ltd., CH-4002 Basle, Switzerland.

INTRODUCTION

Almost 40 human interferon (IFN) α sequences have been derived from various c-DNA and chromosomal libraries (1). Cataloguing DNA segments containing IFN sequences, made it possible to distinguish 18 IFN α related loci in the chromosome (1,2,3). A genetic study based on 25 families, has restricted the number of different IFN α loci to about 17 (4) and has revealed a high degree of polymorphism in the human IFN α loci. For many of the known IFN α sequences it is not possible to decide whether they represent alleles or different loci, when they are compared to each other.

Here we describe the differential properties of three IFN α_2 variants, IFN α_2, IFN α_A and IFN α_2 Umeå (1,3,5), which were isolated from various c-DNA libraries. The three IFN variants differ by only one or two amino acid(s) residing in the N-terminal part.

MATERIALS AND METHODS

For simplicity the three IFN α variants used in this study were designated Var 1 (IFN α_2), Var 2 (IFN α_A) and Var 3 (IFN α_2 Umeå).

Isolation and characterization of the c-DNA clone encoding the Var 3, were described previously (3,5). Coding regions of Var 1 and Var 2 were derived from Var 3 c-DNA by site directed mutagenesis. The nucleotide exchanges at codon positions 23 and 34 of the mature protein are shown in Figure 1. Prior to mutagenesis, the coding region of the c-DNA was manipulated in such a way that the mature IFN protein begins with an extra - inframe- methionine start codon. That coding region was inserted in the production plasmid under the control of a synthetic ribosomal binding site and the trp promotor (6). For site directed mutagenesis, coding region of Var 3 was excised from the production plasmid at flanking restriction sites, and inserted in a suitable M13 vector. The desired mutants were obtained by primer mismatch DNA synthesis (7) and the mutagenized coding regions were reinserted in the original production plasmid. The coding regions of all production plasmids encoding either of the three IFN $_2$ variants were confirmed by DNA sequencing.

The three variants were parallely isolated from the respective E.coli production strains. Cultivation of cells and purification of IFN followed a previously described protocol based on the extraction of inclusion bodies (8). In addition, the enriched IFN α_2 preparations were furhter purified by applying them to a monoclonal antibody column. The purity of the preprations was verified by detection of only one single band for each IFN variant on silver-stained SDS PAGE. The IFN concentrations and the antiviral activities were determined as described previously (8). Maintenance of biological activities of all the preparations was confirmed by reiterating the antiviral assay prior to any of the tests described.

Monoclonal antibodies (MAbs) 2K2 (9), 144BS (10), NK2 (11) and 1/24 (12) were described previously. ELISA, the combined immunoprecipitation-bioassay (CI-BA) (9) and neutralization assay (10) followed the published protocol or the instruction of the manufacturer. Each of the three assays was performed at least in two independent experiments.

The antiproliferative effect of the three IFN α_2 variants was tested on the human Burkitt's lymphoma line Daudi by monitoring the thymidine uptake and viable cell counts in at least three independent experiments.

RESULTS

The sequence of the three recombinant IFN α_2 variants prepared for this study are displayed in Figure 1. With exception of the extra N-terminal methionine (Materials and Methods), the coding regions of the three variants were identical to those of IFN α_2 (Var 1), IFN α_A (Var 2) and IFN (Umeå) (Var 3).

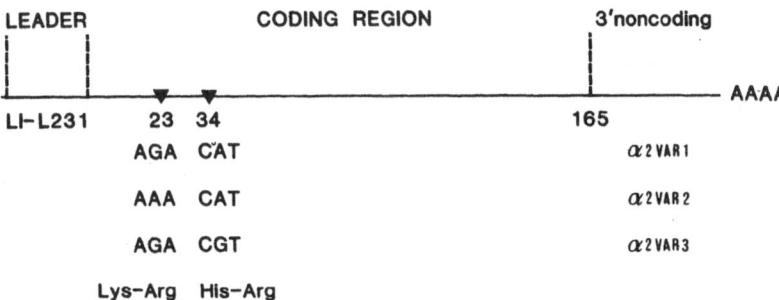

Figure 1 displays the three sequences representing IFN α variants, on which the study is based. L1 – L23 and 1 – 165 mark the amino acid positions of the leader peptide and the mature proteins. The triangles indicate the codon position found to be different in the respective IFN:s.

MONOCLONAL ANTIBODIES

IFN α_2 VARIANTS		1 2K2	2 144	3 NK2	4 I/24
Var 1	ELISA	++	+	+++++	++++
(α_2)	CI-BA	+++	(+)	+++	ND
Arg_{23}-His_{34}	NEUTR	+++	+	++++	++++
Var 2	ELISA	–	–	+++++	++++
(α_A)	CI-BA	+++	+++	+++	ND
	**	+++	+++	+++	ND
Lys_{23}-His_{34}	NEUTR	–	–	++++	++++
	**	–	–	++++	++++
Var 3	ELISA	+	–	+++++	++++
(α_2 Umeå)	CI-BA	++	+	+++	ND
Arg_{23}-Arg_{34}	NEUTR	++	–	++++	++++

Table. The abbreviations for each MAb is given at the top. Each symbol represents a unique MAb reactivity pattern. ++++, 100%; +++, 75%; ++, 50%; +, 25%; –, 0% precipitation of IFN using the CI-BA assay. ++++, 100%; +++, 75%; etc.; 0% recovery of the IFN activity in presence of the antibodies using the neutralization assay. +++++, 10^{-5}; ++++, 10^{-4}; +++, 10^{-3}; ++, 10^{-2}; +, 10^{-1}; –, no detection at max concentration, indicate the maximal MAb-dilutions that gave a detectable signal above background (see text).

** assays were parallely performed using the recombinant IFN α_A from Hoffmann LaRoche.

The variants were submitted to three assays employing four different MAbs directed against Var 1 and Var 2. In the first "CI-BA", IFN variants and MAbs were allowed to react with each other in fluid phase; the IFN-MAb complexes were then precipitated by an anti-Ab and the precipitates were further analyzed by determining the antiviral activity of the redissolved IFN molecules. In the second "neutralization assay", the IFN variants were combined with excess of monoclonal antibodies and thereafter the reduction of the antiviral activities was determined. In the last "ELISA", the wells were coated with the three IFN variants; serially diluted MAbs and, relevant controls were added. Bound MAbs were measured by a second biotinylated antibody, followed by a peroxidase specific, colorimetric reaction. The results of all three assays are displayed in Table.

MAbs 3 and 4 disclosed almost identical affinities to all three IFNα variants in both the "ELISA" and the "CI-BA". In the parallel neutralization assay all three IFN variants were inactivated with identical efficiencies. Therefore the epitope recognized by MAbs 3 and 4 is not affected by the amino acid exchanges, but it is important for their biological function.

In contrast, the affinities of MAbs 1 and 2 to the IFN variants revealed marked differences in all three assays. For example, in the "CI-BA" the three IFN variants are recognized with affinities following a gradient, such that Var 2 disclosed the highest, and Var 3 the lowest affinities to those MAbs. On the other hand, MAbs 1 and 2 did not react with Var 2 in the "ELISA" and in the "neutralization assay". Interestingly, Var 2 disclosed the highest signal strength in the liquid phase assay (CI-BA) employing MAbs 1 and 2, whereas Var 2 failed to react with the same MAbs in the "ELISA" and in the "neutralization assay". However, the other two IFN variants reacted to some degree with MAbs 1 and 2 in all three assays. Finally, the the result of the neutralization assay shows the fact that the three IFN variants are inhibited to a different degree in presence of MAbs.

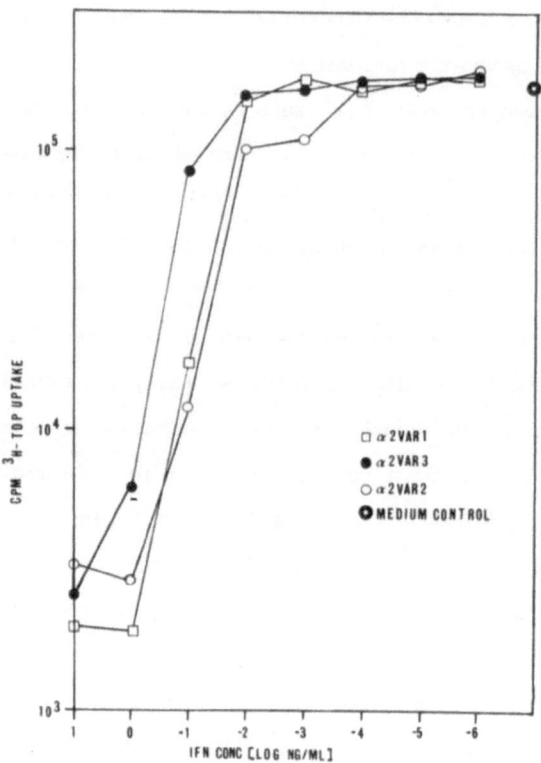

Figure 2: Effect of different IFN α_2 variants on proliferation of human Daudi cells. Exponentially growing cells (2 x 10^5 cells/ml) were exposed to indicated concentrations of the three variants. Proliferation determined as thymidine uptake after 72 hours.

The IFN variants were also assayed in a biological activity not based on MAbs, namely, the known antiproliferative effect of IFN α_2 against human Daudi cells. Figure 2 presents the result of one of several experiments (see Materials and Methods). The result revealed that Var 1 and Var 3 inhibit cell growth between 5 and 10 times more efficiently than Var 3. Identical results were obtained by monitoring "cell counts" in presence of the IFN variants during five days.

DISCUSSION

In the present paper we show that three closely related IFN α_2 variants – departing from each other by only one amino acid at position 23 or 34 – exhibit significant structural and functional differences.

Further investigations are necessary to determine whether the present variants reflect alleles or different loci in the chromosome. However, if the tested variants represent alleles, the results would point to the importance of individual differences in one particular IFN α locus. On the other hand, if they represent different loci the results would support a model that different IFN loci differ in their function. However, if the tested variants have originated from different chromosomal loci, other known IFN α variants, – possibly differing by even more than two amino acids – must be considered as alleles, because the number of IFN loci is restricted.

The amino acid exchange at positions 23, specific for Var 2, is apparently less important for the two biological function tested – the antiproliferative activity. On the other hand, the same amino acid exchange makes that IFN variant more accessable to those critical MAbs. Moreover, the epitopes recognized by those MAbs, ceases, when Var 2 is exposed to target cells or bound to the solid phase of the microtiter well. Therefore binding to receptor or solid phase of that IFN variant might force it in a configuration, which antagonizes MAb epitope binding.

The arginine at position 34, specific for Var 3, seems more important for the biological activity, but less influential concerning the affinities to the antibodies.

Collectively, the results point to the importance of the very N-terminal part of IFN α for its functional structure and agree with previous conclusions (13). It is interesting to note that the analysis of natural IFN variants with minimal differences revealed traits concerning structure and function, while the comparison of variants founded by artificial exchanges at the most conserved sites among IFN α s, did not show significant biological effects (8). This might be interpreted such that the amino acid exchanges among natural variants are the result of a selection of which we do not yet understand the mechanism. Finally, the fact that the two major IFN variants used for clinical trials, Var 1 and Var 2, exhibit serological differences should be considered for evaluation of clinical trials.

ACKNOWLEDGEMENTS

This work was supported by grants from KabiVitrum AB, Stockholm, the Swedish Cancer Society and the Swedish Medical Research Council. We thank Kristina Malmqvist for her excellent support in preparing the manuscript and Håkan Borg for his competent advice concerning the purification of IFN.

REFERENCES
1. Henco, K., Brosius, J., Fujisawa, J., Fujisawa, J.-I., Haynes, J.R., Hochstadt, J., Kavacic, T., Pasek, M., Schamböck, A., Schmid, J., Tokodoro, K., Wächli, M., Nagata, J. & Weissmann, C. J. Mol. Biol. 185, 227-260 (1985)
2. Lund, B., Edlund, T., Lindenmaier, W., Ny, T., Collins, J., Lundgren, E. & von Gabain, A. Proc. Natl. Acad. Sci. USA. 81, 2435-2439 (1984)
3. von Gabain, A., Ohlsson, M., Lindström, E., Lundström, M. & Lundgren, E. Chemica Scripta, in Jörnvall, H. ed. Molecular evolution of life. (1986) in press.
4. Ohlsson, M., Feder, J., Cavalli-Storza, L. & von Gabain, A. Proc. Natl. Acad. Sci. USA. 82, 4473-4476 (1985)
5. Lund, B., von Gabain, A., Edlund, T., Ny, T. & Lundgren, E. J. IFN Res. 5, 229-238 (1985)
6. Grundström, T., von Gabain, A., Nilsson, G., Lundström, M., Andersson, M., Lund, B. & Lundgren, E. Submitted to DNA.
7. Lathe, R.F. Lequoq, J.P. & Evererr, R. in Williamson ed. Genetic engineering 4, 1-56 (1983)
8. Valenzuela, D., Weber, H. & Weissmann, C. Nature 313, 698-790 (1985)
9. Alkan, S.S., Weideli, M.J., Schürch, A.R. in Protides Biol. Fluids Proc. Collogn. 30, 495-498 (1983)
10. Alkan, S.S., Hochkeppel, H.K. & Kuettel, L. in Kirchner & Schelleken eds: The Biology of the IFN System 1984, 91-98 (1985)
11. Secher, D.S. & Burke, D.C. Nature 285, 446-448 (1980)
12. Aguet, M., Gröbke, M. & Dreidning, P. Virology 132, 211-216 (1984)
13. Alkan, S.S. & Braun, D.G. in Synthetic peptides as antigens, eds. Wiley, Chichester (Ciba Foundation Symposium 119) 264-278 (1985)

SELECTIVE ASSOCIATION OF HIGH DENSITY LIPOPROTEIN WITH INTERFERON BETA SERINE

D.A. WIEBE, I.B. ROSENZWEIG, F.J. RUZICKA, E.S. SHRAGO, and E.C. BORDEN

INTRODUCTION

To determine if interferon beta serine (IFN βser) associated with lipoproteins, we studied its distribution in plasma by two independent procedures--ultracentrifugation and agarose gel electrophoresis. In addition, the effects of lipoproteins on IFN βser receptor binding were investigated to determine the influences of IFN βser and lipoprotein association. To investigate the distribution of IFN βser in serum, we focused on the hydrophobic nature of IFN βser, a property it shares with apolipoproteins. Cyclosporin, another pharmacologically-active protein, also has hydrophobic characteristics and as a result was found to be closely associated with lipoproteins [1].

MATERIALS AND METHODS

Recombinant IFN βser stabilized with albumin was supplied by Triton Biosciences Inc. (Alameda, CA 94501). IFN βser labeled with I-125 by the Bolton-Hunter procedure was used for all of our studies [2,3].

Ultracentrifugation. Serum was mixed with trace amounts of I-125 IFN βser. Unlabeled IFN βser was added to obtain the following final concentrations: 250, 500, 1000, and 2000 U/mL. Aliquots were fractionated by sequential ultracentrifugation [4]. Centrifugation was performed at 105,000 x g for 20 hr at 15°C, fractions were separated using tube slicing techniques and subsequent density adjustments made with the addition of solid KBr. Thus, very low density (VLDL), low density (LDL), high density lipoprotein (HDL), and serum proteins were obtained at densities <1.006, 1.006-1.063, 1.063-1.21, and >1.21 g/mL, respectively. Gamma counting was used to determine the amount of labeled IFN βser present in each fraction and results reported as a percentage of the total CPM recovered. In addition, interferon of each fraction was titered by the EMC virus hemagglutination assay [5].

Electrophoresis. Serum and lipoprotein (VLDL, LDL, and HDL) fractions were mixed with trace amounts of I-125 labeled IFN βser. Since only 1.0 μL was applied to the agarose gel for electrophoresis, the required amount of labeled IFN βser was 200 CPM/μL for autoradiography. Electrophoresis slides with 8 sample wells were used and duplicate specimens were placed on each half of the slide. Following electrophoresis in barbital buffer for 30 min, the slide was cut in half and one section was stained for protein with amido black and the other stained for lipids with fat red 7B. The slides were then used for autoradiography with Kodak X-Omat AR film with 3-day exposure prior to development. Mobility of the labeled IFN βser was compared to the protein and lipoprotein mobilities for each fraction.

IFN βser Cell Receptor Assay. Daudi cells were harvested in the exponential growth phase by centrifugation at 1500 x g for 20 min and resuspended in RPMI. Cells were mixed with 0.3 ng/mL I-125 labeled IFN βser in a total of 4.0 mL. Nonspecific binding was determined by the addition of 1000 ng unlabeled IFN βser. To study the effects of various protein mixtures on IFN βser binding, IFN βser was preincubated with the protein material or serum components for 24 hr prior to the addition of Daudi cells. Final receptor assay mixtures were incubated for 16 hr at 4 C. Cells were separated by centrifugation and the resulting pellets were counted for I-125 IFN βser content.

RESULTS AND DISCUSSION

Ultracentrifugation has been used extensively for the separation and isolation of the various lipoprotein classes to quantitate their lipid or protein moieties. We added IFN βser in various concentrations to serum and proceeded to isolate lipoproteins by ultracentrifugation. Interferon content of the isolated fractions was determined on the basis of radioactivity for those samples in which I-125 IFN βser was incorporated into the serum or by measurement of antiviral activity with the same lipoprotein fractions (Table 1). Based on both radioactivity and antiviral activity data, there is evidence for association between IFN βser and HDL.

TABLE 1
PLASMA DISTRIBUTION OF INTERFERON BETA SERINE

IFN Added	Based on Antiviral Activity*				Based on IFN I-125[+]			
(units/mL)	250	500	1000	2000	250	500	1000	2000
VLDL	0	0	0	0	4.0	4.3	3.7	3.6
LDL	0	0	0	0	5.0	4.9	4.6	4.8
HDL	62.1	56.7	63.8	53.4	28.4	27.1	28.3	28.9
Proteins	37.9	43.3	36.2	46.6	62.5	63.7	63.4	62.7

*Antiviral activity data are presented as % of total activity.
+I-125 data are presented as % of total radioactivity found.

Electrophoresis provided an alternative approach to assess whether any association existed between IFN βser and the three lipoprotein classes. Electrophoretic mobility of whole Serum, VLDL, LDL, and HDL with trace quantities of I-125 IFN βser in each sample was assessed (Figure 1). The protein and lipid stained slides demonstrated the integrity of the samples. Autoradiography provided a clear differentiation between the effect of HDL versus other proteins/lipoproteins on the mobility of IFN βser. The presence of HDL had a marked influence on IFN βser mobility, probably through a close association between the two entities.

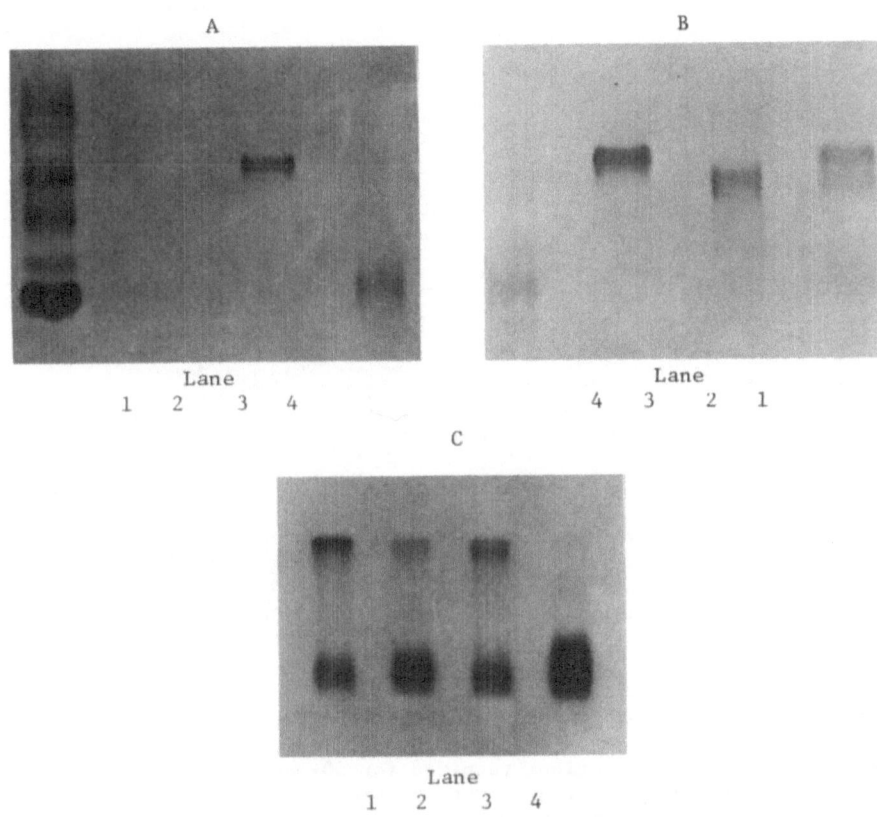

FIGURE 1. Slides A, B, and C represent electrophoretograms produced under identical conditions and developed for protein (amido black), lipid (fat red 7B), and I-125, respectively. Lanes 1-4 correspond to whole serum, VLDL, LDL, and HDL.

The IFN βser cell receptor experiments provided a methodological approach with cells quite different from the previous two procedures. Measurement of specific I-125 IFN βser binding to Daudi cells in the presence of proteins and lipoproteins at various concentrations suggested a possible interaction between IFN βser and the protein moiety (Figure 2). It is equally possible that altered IFN βser binding could be the end result of interactions between the cell receptor and the protein moieties. The lipoproteins were quite effective in inhibiting I-125 IFN βser binding to its cell receptor.

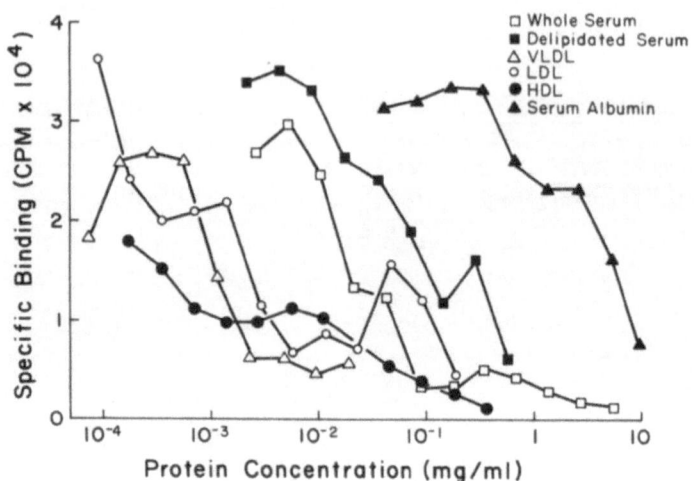

FIGURE 2. IFN βser was preincubated with serum components 24 hours prior to the addition of Daudi cells. Specific binding of I-125 IFN βser to Daudi cells was then determined in the presence of the following protein fractions: whole serum, delipidated serum, albumin, VLDL, LDL, and HDL.

The protein concentration required to suppress specific binding at the 50% level was 370, 6.4, 1.45, 0.165, 0.085, and 0.045 mg/dL for albumin, delipidated serum, whole serum, LDL, VLDL, and HDL, respectively. Concentrations required for 50% suppression of IFN βser were less than the quantities found in normal serum. Therefore, the serum must be diluted to achieve 50% suppression. Content of each of these protein moieties in whole serum was estimated and each protein's contribution to suppression of I-125 IFN βser receptor binding was calculated versus whole serum on the basis of how large a dilution would be necessary to achieve 50% suppression. Whole serum requires a 4000-fold dilution to achieve 50% suppression using a concentration of 6.0 g/dL as the typical protein content of serum. Albumin at a physiologic concentration of 2.5 g/dL in whole serum would require a 7-fold dilution to achieve the concentration necessary for 50% suppression. Thus, albumin as a component of serum has an insignificant role in the suppression of IFN βser receptor binding relative to other proteins in serum. However, HDL required a 3300-fold dilution by the same comparison and, therefore, contributed approximately 80% of the suppression of IFN βser receptor binding found in serum. Other lipoproteins appeared to have a less significant effect, confirming a strong association between IFN βser and HDL.

CONCLUSION

Using three separate and unrelated experimental techniques, we documented that HDL and IFN βser selectively associate with each other. Other proteins have been noted to associate with lipoproteins such as cyclosporin. Other proteins, however, do not display the specificity of IFN βser for a single lipoprotein class. Physiologic concentrations of HDL effectively suppress binding of IFN βser to its receptor which may have significance in therapeutic use of interferons.

REFERENCES

1. Sgoutas D, Love A, and Jerkunica I: Clin. Chem. 31:936, 1985.
2. Bolton AE, and Hunter WM: Biochem. J. 133:529-539, 1973.
3. Ruzicka FJ, Groveman DS, Schmid SM, Cummings KB, and Borden ED: Fred. Proc. 44:1937, 1985.
4. Havel RJ, Eder HA, and Bragdon JH: J. Clin. Invest. 34:1345-1353, 1955.
5. Jameson P, Dixon MA, and Grossberg SE: Proc. Soc. Exp. Biol. Med. 155:173-178, 1977.

STRUCTURE, GENETICS AND FUNCTION OF HUMAN β_2 INTERFERON

PRAVINKUMAR B. SEHGAL AND LESTER T. MAY

The Rockefeller University, New York, NY 10021, USA

It was reported in 1980 that polyadenylated cytoplasmic RNA extracted from human diploid fibroblasts (FS-4 strain) induced with poly(I).poly(C) and cycloheximide could be resolved by electrophoresis through an agarose-methylmercury hydroxide gel into two translationally active (in the Xenopus laevis oocyte assay) mRNA species that coded for human interferon (IFN) of the β serotype (1). The translation product of the 0.9 kb mRNA species, which hybridized a cDNA probe corresponding to the IFN-β gene on chromosome 9, was designated as IFN-β_1 while that of the 1.3 kb mRNA species, which did not cross-hybridize the IFN-β_1 cDNA probe, was designated IFN-β_2 (1). The two species of translationally active human IFN-β mRNA were also resolved by sucrose-gradient sedimentation of polyadenylated RNA extracted from poly(I).poly(C) and cycloheximide-induced SV40-transformed human fibroblasts (SV80 cell line) and partial cDNA clones corresponding to the 1.3 kb IFN-β_2 mRNA were described (2). IFN-β_2 cDNA probes did not cross-hybridize the 0.9 kb IFN-β_1 mRNA (2). Translationally active 1.3 kb long human IFN-β mRNA was detected in appropriately induced human-rodent somatic cell hybrids that lacked human chromosome 9 strongly suggesting that the IFN-β_2 gene was not syntenic with the IFN-β_1 gene (3). Human IFN-β_2 genomic DNA and cDNA clones isolated using the partial cDNA clones described in 1980 (2) have been transfected into rodent cells to yield biologically active human interferon of the β serotype (4). The antiviral activity of recombinant IFN-β_2 is neutralized by polyclonal and monoclonal antibodies to IFN-β_1 (4).

The discovery that IFN-β_2, but not IFN-β_1, is induced by tumor necrosis factor (TNF) in FS-4 cells has generated new excitement in this area of interferon research (5). TNF has a mitogenic effect in FS-4 cell cultures. The induction of IFN-β_2 by TNF appears to represent a feedback mechanism regulating cell proliferation (5). The antiviral effect of TNF and the increased expression of class I HLA gene expression observed in TNF-treated human fibroblasts appears to be mediated by the endogenous production of IFN-β_2 (5, 6). Other cytokines that affect cell proliferation, such as platelet-derived growth factor (PDGF), interleukin-1 (IL-1) and IFN-β_1, also increase the expression of IFN-β_2 in human fibroblasts (7).

STRUCTURE OF HUMAN IFN-β_2

A cDNA library was prepared from polyadenylated mRNA extracted from TNF-induced FS-4 cells and eight IFN-β_2 cDNA clones (out of 8,000 in the library) were isolated using a 21 residue-long synthetic oligo-

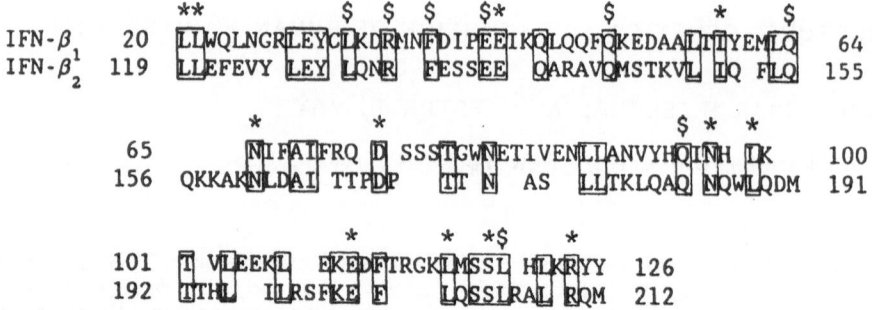

Figure 1. Similarity between IFN-β₁ and β₂ amino acid sequences. The "best" common pattern between the amino acid sequences of IFN-β₁ and β₂ was determined using the Sellers TT algorithm and the pattern score method as mentioned in the text. The symbol " $ " identifies residues that are aligned and identical in IFN-β₁ and all IFN-α species sequenced to date (10), whereas the symbol "*" identifies residues in IFN-β₁ that are also part of an IFN-α consensus sequence (10). From ref. 6.

nucleotide probe (6). Five of these had cDNA inserts ≧1 kb and represent cDNA clones containing the entire coding region of IFN-β₂, the entire 3' noncoding region and most of the 5' noncoding region. The longest of these (p35) appears to contain the entire 5' noncoding region. The 1126-nucleotide long sequence of IFN-β₂ mRNA and the 212-amino acid long sequence of the IFN-β₂ protein were deduced from these cDNA clones. The protein contains a 30-33 residue-long amino-terminal hydrophobic stretch that could serve as the secretory signal sequence and two potential N-glycosylation sites. This structure is identical to that deduced for IFN-β₂ mRNA extracted from poly(I).poly(C)-induced fibroblasts (4).

SIMILARITY BETWEEN HUMAN IFN-β₁ and β₂

At least two different murine monoclonal antibodies that neutralize recombinant IFN-β₁ also neutralize recombinant IFN-β₂ (4, J. Vilček, personal communication). The amino acid sequences of the serologically related human IFN-β₁ and β₂ were compared (6) using the Sellers TT metric alogorithm for locating similarities (8, 9) and the pattern scoring method for evaluating the observed similarities (9). The "best" pattern common to the two amino acid sequences is illustrated in Fig. 1. It aligns a 105-residue long protein segment at the amino-terminal end of IFN-β₁ with a 92-residue long segment at the carboxy-terminal end of IFN-β₂. The pattern contains 39 residues that are aligned and identical in the two sequences. The 42-residue segment of the alignment illustrated in the upper section in Fig. 1 is remarkable because it spans a region that is highly conserved between IFN-β₁ and all the IFN-α species sequenced. This 42-residue alignment has a 'distance' score 3.85 standard deviations

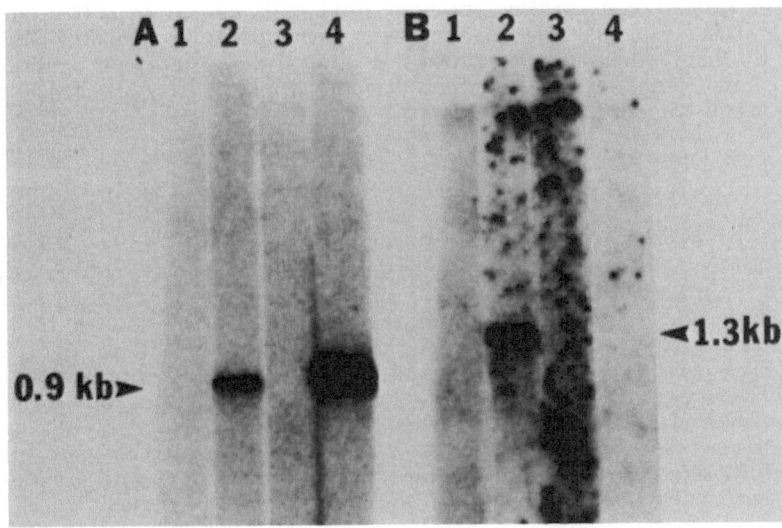

Figure 2. Tissue-specific expression of IFN-β_2. Polyadenylated RNA was extracted from uninduced FS-4 cells (lane 1), from FS-4 cells induced with poly(I).poly(C) and cycloheximide for 4 hours (lane 2), from uninduced Namalwa cells (lane 3) and from Namalwa cells treated with bromodeoxyuridine for 48 hours and then induced with Sendai virus for approximately 9 hours (lane 4). The RNA was analysed using the Northern blot procedure. Lanes illustrated in panel A were probed using an IFN-β_1 cDNA probe (pD19), while those illustrated in panel B were probed using an IFN-β_2 oligonucleotide probe as described in ref. 5.

better than the mean distribution of scores obtained when 400 permuted pairs of sequences having the same amino acid compositions and lengths as the IFN-β protein segments are compared (8). This is likely to represent an underestimate of the statistical significance of the observed alignment because it does not take into account the finding that 10 of the 16 identities in this alignment are also identical in all or most of the 18 or more IFN-α sequences determined (10). The hydropathic index plots across the segments illustrated in Fig. 1 are strikingly similar (6). Thus, there appears to be sufficient structural similarity between IFN-β_1 and β_2 to account for the serological cross-reaction. Whether the alignment depicted in Fig. 1 identifies the common serologic epitopes remains to be experimentally verified.

CHROMOSOMAL ASSIGNMENT OF THE HUMAN IFN-β_2 GENE

The IFN-β_2 gene has been assigned to human chromosome 7 by blot-hybridization analyses of DNA from a panel of human-rodent somatic cell hybrids using a full-length IFN-β_2 cDNA probe (11). This assignment is consistent with previous experimental data in which the expression of the translationally active 1.3 kb IFN-β_2 mRNA was assayed in some of these same somatic cell hybrids (3, 11). Blot-hybridization using DNA from different human cell strains and cell lines and characterization of IFN-β_2 genomic DNA clones isolated from a human DNA library in Charon 4Å reveal at least three polymorphisms of the IFN-β_2 gene (11, unpublished data). The detailed relationships among these different forms of the IFN-β_2 gene remain to be elucidated.

TISSUE-SPECIFIC EXPRESSION OF IFN-β_2

The expression of IFN-β_1 and β_2 can be dissociated in a tissue-specific manner (Fig. 2). Human lymphoblastoid cells (Namalwa), which appear to contain an intact IFN-β_2 gene as judged by restriction fragment analysis (11), does not express detectable IFN-β_2 mRNA in response to inducers like Sendai virus which cause a marked induction of IFN-β_1 gene expression (Fig. 2). On the other hand, poly(I).poly(C) and cycloheximide readily induce both IFN-β_1 and β_2 in human fibroblasts (FS-4) (Fig. 2). At the present time, we have been unable to detect IFN-β_2 mRNA in Namalwa cells treated with a variety of induction protocols. We anticipate that additional cell-type and tissue-specific differences will be uncovered in the future in the expression of the IFN-β_1 and β_2 genes.

SELECTIVE EXPRESSION OF IFN-β_2 in HUMAN FIBROBLASTS

The expression of IFN-β_1 and β_2 in human fibroblasts can be dissociated in a stimulus-specific manner (5-7). Cytokines that affect cell proliferation in human fibroblast cultures enhance the expression of IFN-β_2 but not that of IFN-β_1 (5-7). It was discovered that human TNF induces IFN-β_2 mRNA in human diploid fibroblasts without the induction of any detectable IFN-β_1 mRNA (5). IFN-β_2, but not IFN-β_1, was also found to be involved in a complex network of interactions triggered in human fibroblast cultures by growth modulatory cytokines such as PDGF, IL-1 and IFN-β_1 (7). Endogenously produced IFN-β_2 appears to mediate the antiviral effect of TNF in FS-4 cells since this effect can be blocked by anti-IFN-β serum (5). Polyclonal and monoclonal anti-IFN-β antibodies inhibit the increase in class I HLA gene expression (HLA-B7 mRNA) in TNF-treated FS-4 cells suggesting that TNF-induced IFN-β_2 mediates the enhancing effect of TNF on HLA gene expression in human fibroblasts (6). Endogenously produced IFN-β_2 appears also to modulate (suppress) the mitogenic effect of TNF in FS-4 cells. Anti-IFN-β serum amplifies this effect suggesting that the endogenous production of IFN-β_2 represents a negative feedback mechanism in the regulation of cell proliferation (5). PDGF and IL-1 also lead to the increased expression of IFN-β_2 (7). However, PDGF and IL-1 can override the antiviral effect of endogenously produced IFN-β_2 in FS-4 cell cultures (7). Finally, IFN-β_1 can also directly increase the expression of IFN-β_2 (7). The available data suggest that a complex network of interactions that involves the endogenous production of IFN-β_2 is triggered by several growth modulatory cytokines in human fibroblasts. It is IFN-β_2, not IFN-β_1, that appears to be the quintissential human fibroblast interferon.

The observation that IFN-β_1, but not IFN-α, increase the expression of IFN-β_2 in FS-4 cells in the presence of cycloheximide raises interesting questions about the interactions between these exogenously added interferons and their cell surface receptors (7).

TRANSCRIPTIONAL REGULATION OF IFN-β_2 GENE EXPRESSION BY DIFFERENT CYTOKINES

The increased expression of IFN-β_2 mRNA in human fibroblasts exposed to TNF, IL-1, PDGF and IFN-β_1 occurs largely at the transcriptional level as determined by nuclear runoff assays (Z. Walther, L.T. May and P.B. Sehgal, unpublished data). Transcription of the IFN-β_2 gene occurs constitutively at a low level in FS-4 cells. Exposure of cells to cycloheximide alone causes, at best, only a minimal increase in the rate of IFN-β_2 transcription. Thus the large increase in IFN-β_2 mRNA levels in cycloheximide-treated FS-4 cells described earlier (5) is likely to be mainly a post-transcriptional phenomenon. TNF, PDGF, IFN-β_1 and poly(I).poly(C) cause an increase in IFN-β_2 transcription in the presence of cycloheximide suggesting that new protein synthesis is not required for these cytokines to turn on the IFN-β_2 gene. On the other hand, IL-1α produces a rapid and marked increase in IFN-β_2 gene transcription which, in preliminary experiments, appears to be blocked by cycloheximide. The involvement of new protein synthesis in the regulation of IFN-β_2 gene expression by IL-1α is an unexpected observation that requires further exploration.

CONCLUSIONS

Research carried out during the last six years has led to the elucidation of the structure, genetics and some of the functions of the IFN-β_2 gene located on human chromosome 7. This novel interferon appears to have autocrine biological functions distinct from those of the other interferons. The key distinguishing feature of IFN-β_2 is not that its biological effects per se appear to be qualitatively different from those of the other interferons, but that the stimuli that cause its expression are intimately involved in important physiological processes such as the regulation of cell proliferation and of the expression of cell-surface antigens. Cellular homeostasis is likely to represent a balance between the induction of IFN-β_2 by different cytokines and their ability to override the inhibitory actions of IFN-β_2.

ACKNOWLEDGEMENTS

We thank Drs. Igor Tamm, Frank Ruddle, Michel Revel, Jan Vilček and Masayoshi Kohase for their collaboration, enthusiastic support and encouragement. This work was supported by Research Grant AI-16262 from the National Institutes of Health, a contract from the National Foundation for Cancer Research, a grant from Enzo Biochem, Inc., an Irma T. Hirschl Award (P. B. S.) and an Established Investigatorship from the American Heart Association (P. B. S.).

REFERENCES

1. Sehgal PB and Sagar AD. Nature 288:95-97, 1980.
2. Weissenbach J, Chernajovsky Y, Zeevi M, Shulman L, Soreq H, Nir U, Wallach D, Perricaudet M, Tiollais P and Revel M. Proc. Natl. Acad. Sci. USA 77:7152-7156, 1980.

3. Sagar AD, Sehgal PB, Slate DL and Ruddle FH. J. Exp. Med. 156:744-755, 1982.
4. Zilberstein A, Ruggieri R, Korn JN and Revel M. EMBO J., in press.
5. Kohase M, Henriksen-DeStefano D, May LT, Vilček J and Sehgal PB. Cell 45:659-666, 1986.
6. May LT, Helfgott DC and Sehgal PB. Proc. Natl. Acad. Sci. USA, in press.
7. Kohase M, May LT, Tamm I, Vilček J and Sehgal PB. Mol. Cell. Biol., submitted.
8. Erickson BW, May LT and Sehgal PB. Proc. Natl. Acad. Sci. USA 81:7171-7175, 1984.
9. May LT, Landsberger FR, Inouye I and Sehgal PB. Proc. Natl. Acad. Sci. USA 82:4090-4094, 1985.
10. Henco K, Brosius J, Fujisawa A, Fujisawa J-I, Haynes JR, Hochstadt J, Kovacic T, Pasek M, Schambock A, Schmid J, Todokoro K, Walschli M, Nagata S and Weissmann C. J. Mol. Biol. 185:227-260, 1985.
11. Sehgal PB, Zilberstein A, Ruggieri R, May LT, Ferguson-Smith A, Slate DL, Revel M and Ruddle FH. Proc. Natl. Acad. Sci. USA 83:5219-5222, 1986.

TAILORING OF ALPHA INTERFERONS VIA IN VIVO RECOMBINATION OF THE GENES

E.C. Zwarthoff, A.M.C. Gennissen, M. van Heuvel, I.J. Bosveld and J. Trapman

Department of Pathology, Erasmus University, P.O. Box 1738, 3000 DR Rotterdam, The Netherlands.

ABSTRACT

MuIFN-α1 has a specific antiviral activity that is only 15% of the activity of MuIFN-α4. We have constructed hybrids between these two IFNs, with the objective to localize area(s) within the proteins that are involved in the determination of their biological activity. To obtain as much α1α4 hybrids as possible, we made use of the recombination system of E. coli. This approach resulted in hybrid α1α4 genes with cross-over points scattered along the whole sequence. The specific antiviral activity of the corresponding proteins was determined. Interestingly, the activity of some of these hybrids was 10 to 100 times higher than the activity of the parent IFNs. The very active hybrids contain at least the N-terminal 10 and at most the N-terminal 54 amino acids of the parent α1 protein. The structural difference between IFN-α4 and a hybrid in which the first 10 amino acids are from α1 is only 3 amino acids. The activity of the hybrid is 5- to 10-fold higher. Hybrids with the first 54 amino acids from α1 have activities that are 50 to 100 times higher than the activity of a hybrid in which the first 66 amino acids are derived from α1. Here the structural difference between the hybrids is at most 4 amino acids. We presume that the amino acid residues thus identified are important modulators of the biological activity of α-IFNs.

INTRODUCTION

In all mammals studied so far interferon-α (IFN-α) consists of a group of closely related proteins (see ref. 1 for a review). The mean homology between the proteins of one species is about 80%. Also between species, the IFN-α proteins seem to be reasonably well conserved. The genes for these proteins are clustered in the genome and form the IFN-α gene family. Despite the high degree of homology between the IFN-α subspecies they show striking differences in their antiviral, antiproliferative and other activities (2,3,4). It is thought that IFNs elicit these biological activities only after binding to a surface receptor on the target cells. Thus, it is possible that differences in biological activity can be reduced to differences in affinity for the receptor. Binding studies between human IFNs (HuIFNs) and target cells suggest that this could indeed explain the disparity in activity between HuIFN-α1 and α2 (5).

32

The murine IFN-α4 (MuIFN-α4) gene differs from all other IFN
-α genes characterized so far by the fact that it contains
an in frame deletion of 15 nucleotides (6). This results in the
absence of amino acids 103 through 107 in the corresponding protein. However, the antiviral activity of this protein on mouse
L cells is much higher than that of other MuIFN-α subspecies
(7). The specific activity of MuIFN-α1, for instance, is 6-fold
lower. MuIFN-α1 and MuIFN-α4 differ in 34 (including the deletion) out of 166 residues. To identify the region in the proteins, that is responsible for the high or low specific activity,
we have constructed hybrids between the MuIFN-α1 and -α4 genes
and analyzed the specific activity of the corresponding proteins.

RESULTS

To obtain a large number of hybrids, which differ from each
other in only a few amino acids, we made use of the recombination system of E. coli. We first arranged the MuIFN-α1 and
-α4 genes tandemly in a plasmid, as is shown in Figure 1. This
plasmid was maintained in a RecA$^-$ host. The plasmid was subsequently isolated, linearized with restriction enzyme ClaI and
introduced into an E. coli strain containing a wild type RecA
protein. Recombinant plasmids were identified by hybridization
with a fragment from pAGB that is located near the ClaI site.

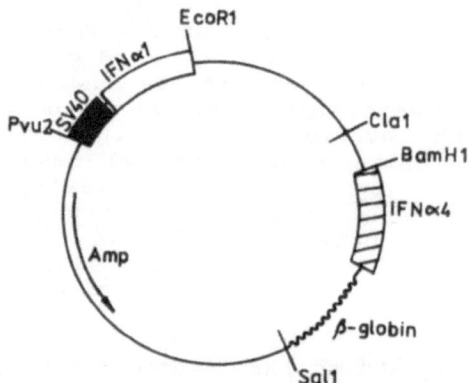

FIGURE 1. Outline of plasmid pAGB. The PvuII-EcoRI fragment is
from plasmid pSVα1 (7). It contains the SV40 origin of replica-
tion/early promoter and the coding sequence of the MuIFN-α1
gene. The BamHI-SalI fragment is from pSVα4 (7). It contains
the coding sequence of the MuIFN-α4 gene, followed by the 3'-
noncoding portion of the rabbit β-globin gene. The remainder
of the plasmid is from pBR328.

A total of 48 recombinant plasmids were isolated and characterized with restriction enzymes specific for either the α1 or the
α4 gene. This resulted in a classification of the recombinants
into 5 groups as is depicted in Figure 2. This figure also shows
that the cross-over points of the hybrid genes are relatively
well distributed throughout the sequence.

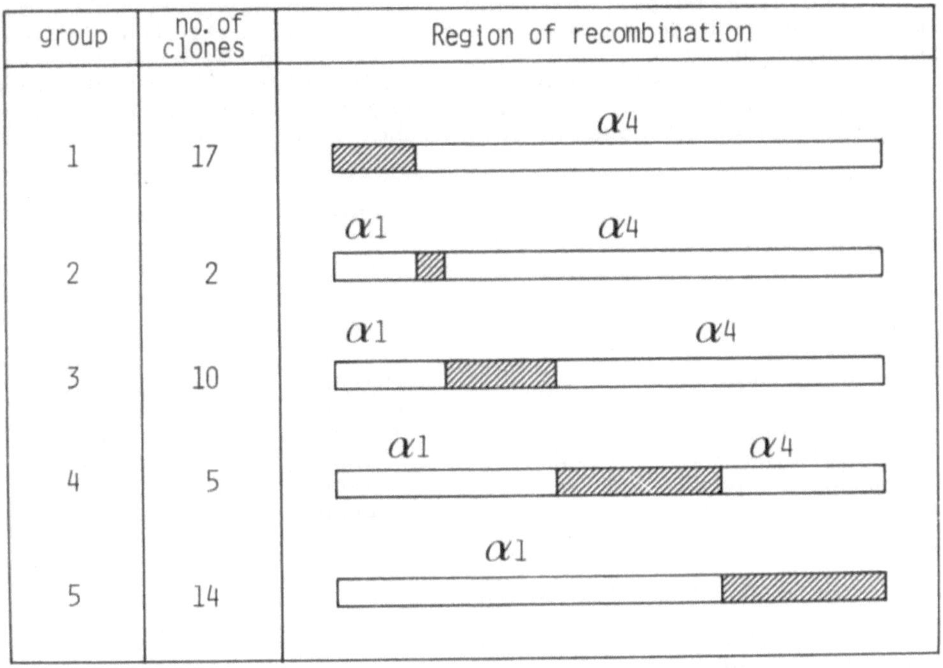

group	no. of clones	Region of recombination
1	17	α4
2	2	α1 α4
3	10	α1 α4
4	5	α1 α4
5	14	α1

Figure 2. Outline of the hybrid MuIFN-α1α4 genes. The second column shows the number of clones obtained within each area. In the third column the genes are shown as boxes. The area in which recombination has taken place is hatched.

Specific antiviral activity of hybrid proteins.

After recombination between the α1 and α4 genes in plasmid pAGB, the hybrid gene is preceded by the SV40 origin of replication/early promoter. This allows an efficient transient expression of the genes in monkey COS cells. The 3'-noncoding region of the β-globin gene, that follows the hybrid gene, provides the polyadenylation signal. COS cells were transfected with these plasmids as described before (7). Two days after transfection the medium was replaced by medium containg ^{35}S-methionine. The next day the antiviral titre of this medium was determined in a cytopathic-effect reduction assay and another sample was used for polyacrylamide gelelectrophoresis. The ^{35}S-labeled proteins were visualized by fluorography. The amount of protein in each IFN band was calculated from densitometric scans of the fluorogram and the incorporation of ^{35}S-methionine into protein. The specific antiviral activities of a representative selection of hybrid proteins, obtained from these calculations, are in Table 1, together with the specific activities of α1 and α4. The hybrid α1α4, which is also included in the Table, was constructed using a common XmnI site (7). Using the classification as introduced in Figure 2, this hybrid belongs in group 4.

The results obtained show that hybrids in group 1, which are almost identical to $\alpha4$ had activities similar to that of $\alpha4$, whereas hybrids from groups 4 (and 5, not shown here) have activities similar to that of $\alpha1$. A surprising result was obtained with the group 2 and 3 hybrids: they were all found to have activities which are considerably higher than those of the parent proteins.

TABLE 1. Antiviral activity of $\alpha1\,\alpha4$ hybrid IFNs on mouse cells

IFN	Group	IU/ml	IUx10^{-7}/mg
$\alpha1$	-	3200	5
$\alpha4$	-	25600	30
14-17[a]	1	25600	40
14-06	2	153600	200
14-12	3	128000	300
14-15	3	307200	800
14-16	3	204800	500
$\alpha1\,\alpha4$	4	3200	9
14-04	4	4800	6
14-11	4	8000	10

[a]14-17 means $\alpha1\,\alpha4$ hybrid clone 17 etc.

Analysis of the structural differences between the hybrid IFNs

In the preceding section we showed that hybrids in groups 2 and 3 have very high specific activities. Hybrids from group 2 can differ from hybrids in group 1 in only a few amino acids, yet their activity is 5 to 10 times higher. To investigate this in more detail, we determined the exact cross-over point in a number of hybrids by nucleotide sequencing. It then appeared that hybrid 14-17 (group 1) differs from hybrids 14-06 and 14-55 (group 2) by only 3 amino acids. In Figure 3 the first 40 amino acids of these hybrids are compared. Note that hybrid 14-17 contains only the signal peptide (not included in the Figure) and the first 4 amino acids of $\alpha1$ and thus is identical to $\alpha4$. A very large difference in activity was found between hybrids from group 3 and hybrids from group 4. The activities of group 3 hybrids are 50 to 100 times higher than the activity of group 4 hybrids. Nucleotide sequencing showed that the minimal difference between these two groups is only 4 amino acids. This is shown in the lower part of Figure 3, where the very active hybrids 10,33 and 64 are compared with the $\alpha1\,\alpha4$ hybrid.

DISCUSSION

This study was started with the objective to explain differences in biological activities of IFN-α proteins in terms of differences in primary structure of the proteins. Using in vivo recombination between two MuIFN-α genes, a large number of overlapping hybrid genes was obtained. When the specific activity of the corresponding proteins was determined, some of them displayed an extremely high activity. Comparing the nucleotide sequences of active and less active hybrids it appeared that

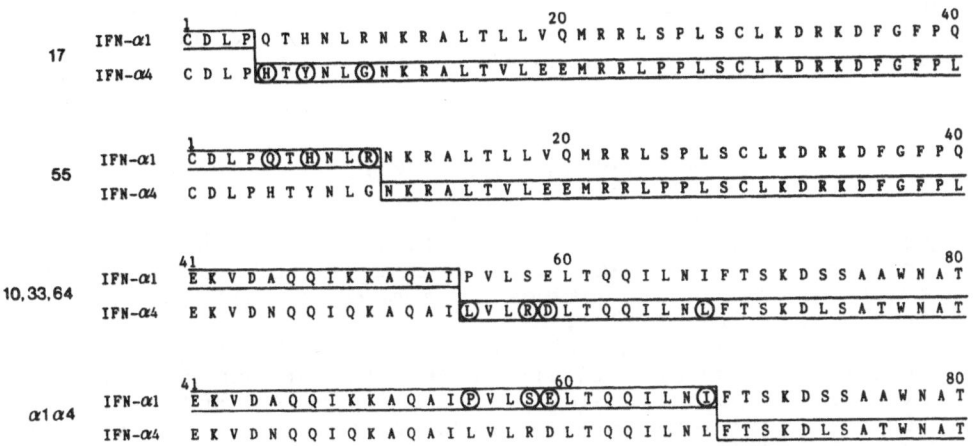

FIGURE 3. Amino acid differences between hybrids in groups 2 and 3 (top-half) and hybrids in groups 3 and 4 (bottom-half). Hybrid 17 contains only the first 4 amino acids of α1, the remainder is from α4. Hybrids 10, 33 and 64 contain the first 54 amino acids of α1, the remainder is derived from α4.

the very active hybrids contain at least the N-terminal 10 and at most the N-terminal 54 amino acids of MuIFN-α1. The less active hybrids differ in as little as 3 amino acids from the more active proteins. The most striking difference is that between hybrids in group 3, with activities of 2-8 x 10^9 IU/mg and hybrids in group 4, with activities of 6-10 x 10^7 IU/mg. The more active proteins contain leu55, arg58 and leu67 compared to pro55, ser58, glu59 and ile67 in the less active proteins. It is possible that the difference is even smaller than these four amino acids. Especially the leu instead of pro and arg instead of ser mutations look promising for a further investigation by site-directed mutagenesis. When the amino acid differences between group 1 and 2 are compared, the exchange of gly at position 10 for arg can cause a considerable effect on the charge of the protein.

The question remains to be answered how these amino acid changes affect the biological activity of the IFN proteins. If these differences are caused by a change in affinity for the cellular receptor, these amino acids may well be part of the receptor binding site. It is also possible that these amino acids influence the tertiary structure of the proteins and as a conquence the stereochemical interaction between ligand and receptor. In several secondary structure models the area between amino acids 50 and 70 is part of an -helix structure (see ref. 8). A proline residue at position 55 could disturb the continuity of this helix. Besides an influence on ligand-receptor interaction, other explanations are also possible. For instance the question whether internalization of IFNs plays a role in the observed biological effects, is still open to debate. However, it seems that with the hybrid IFNs described here, with such large differences in specific activity and such small structural differences, it should be possible to gain a better insight into the way in which IFNs act.

REFERENCES

1. Weissmann, C. and Weber, H. (1986), The interferon genes. In: Progress in nucleic acids research and molecular biology, vol. 33, in press.
2. Weck, P.K., Apperson, S., May, L. and Stebbing, N. (1981), Comparison of the antiviral activities of various cloned human interferon-α subtypes in mammalian cell cultures. J. Gen. Virol. 57, 233-237.
3. Rehberg, E., Kelder, B., Hoal, E.G. and Pestka, S. (1982), Specific molecular activity of recombinant and hybrid leucocyte interferons. J. Biol. Chem. 257, 11497-11502.
4. Lemson, P.J., Vonk, W.P., Van der Korput, J.A.G.M. and Trapman, J. (1984), Isolation and characterization of subspecies of murine interferon alpha. J. Gen. Virol. 65, 1365-1372.
5. Uzé, G., Mogensen, K.E. and Aguet, M. (1985), Receptor dynamics of closely related ligands: 'fast' and 'slow' interferons. EMBO J. 4, 65-70.
6. Zwarthoff, E.C., Mooren, A.T.A. and Trapman, J. (1985), Organization, structure and expression of murine interferon alpha genes. Nucleic Acids Res. 13, 791-804.
7. Van Heuvel, M., Bosveld, I.J., Mooren, A.T.A., Trapman, J. and Zwarthoff, E.C. (1986), Properties of natural and hybrid murine alpha interferons. J. Gen. Virol. 67, in press.
8. Ptitsyn, O.B., Finkelstein, A.V. and Murzin, A.G. (1985), Structural model for interferons. FEBS lett. 186, 143-147.

A NEW TYPE OF INTERFERON-LIKE PROTEIN PRODUCED BY PERIPHERAL BLOOD MONONUCLEAR CELLS TREATED WITH INHIBITORS OF TRANSCRIPTION

F. DIANZANI*, A. DOLEI*° and M. GENTILE*

* Istituto di Virologia, Università La Sapienza, Roma
° Istituto di Microbiologia e Virologia, Univ. di Sassari, Italy.

1. INTRODUCTION

We have previously shown that unstimulated peripheral blood mononuclear cells (PBMC) produce an interferon- (IFN)- like substance when treated with compounds capable of inhibiting transcription, such as actinomycin D, diribofuranosylbenzimidazole, and alpha amanitine (1-3). Since the production did not occur in the presence of inhibitors of protein synthesis, it was inferred that the IFN-like substance was a newly synthesized protein whose mRNA was preformed in at least some of the cells (2).

Characterization experiments showed that this IFN-like protein (ILP) was species specific and active against a wide spectrum of unrelated viruses, induced 2'-5' oligo(A) synthetase activity, and shared several properties with IFN gamma, although it was not neutralized by antibody to IFN gamma or by antibodies to the other types of IFN (2-3). Gel filtration experiments on Sephadex G-50 and G-75 showed that ILP migrates as a single peak at 6000-8000 MW. We suggested that ILP may indeed be a new type of IFN, with much smaller molecular weight and biological properties similar, but not identical, to those of IFN gamma.

In the present report we provide additional data on ILP effects on cellular functions, and a partial biochemical characterization.

2. MATERIAL AND METHODS

2.1 Cells. Human PBMC were obtained from buffy coats of healthy volunteers by Ficoll Hypaque sedimentation and were suspended at 5×10^6 cells/ml in RPMI 1640 medium supplemented with 2% fetal calf serum (FCS), and 20 mM HEPES pH 7.2. Human amnion WISH, melanoma M14, T Molt4, B Namalva lymphoblastoid cell lines, and monocytic U-937 cell line were cultivated in RPMI 1640 supplemented with 10% FCS. Monocyte-enriched preparations were obtained by incubating PBMC on plastic dishes (Falcon) for 1 h at 37° C; adherent cells were gently washed four times with RPMI 1640 plus 5% FCS, then induced as for PBMC. T+B cell-enriched preparations were obtained by depleting adherent cells by two separate runs of adherence on plastic dishes for 1 h at 37° C.

2.2 <u>Interferons</u>. Human recombinant IFN alpha2 was provided by Schering Co (Bloomfield, N.J.). Human native IFN beta was produced in normal diploid fibroblasts and had a specific activity of 10^7 IU/mg of protein. <u>E.coli</u>-derived human IFN gamma was produced by Genentech Inc. IFNs were titrated on WISH cells using Sindbis virus as the challenge virus, as previously described (2); IFN activity were standardized by using the NIH International Reference Standards: IFN alpha (Ga 23-902-530), IFN beta (G-023-902-527) IFN gamma (Gg 23-901-530) supplied by the National Institute of Allergy and Infectious Diseases.

2.3. <u>Reagents</u>. Actinomycin D and cycloheximide were purchased by Sigma Chemical Co (St. Louis, MO).

2.4. <u>ILP induction</u>. Induction was performed as previously described (1-3) by treating the cultures with cycloheximide (20 µg/ml) and actinomycin D (5 µg/ml). One hour later cells were centrifuged and washed three times (2 min at 2,500 rpm) in the cold to remove the metabolic inhibitors. The presence of cycloheximide was not critical for the induction process, but given the very rapid kinetics of production (1), this protein synthesis inhibitor was added to prevent any ILP production before removal of transcription inhibitors and washing of the cultures. The cells were resuspended in prewarmed medium at 10^6 cells/ml and incubated overnight at 37° C. The supernates of the cultures were then titrated for IFN activity by conventional assays.

2.5. <u>ILP characterization</u>. ILP was characterized as IFN according to the current criteria (4). Crude ILP preparations were lyophilized and chromatographed on the following columns (Pharmacia, Uppsala): Sephadex G-75 (elution buffer (EB): PBS); DEAE-Sephadex A-25 (EB: PBS, 0-1 M NaCl, pH 5.5); Blue-Sepharose 4B (EB: 20 mM $Na_2 HPO_4$, 1 M NaCl, containing ethylene glycol 0-30-50 % (v/v).

2.6. <u>MHC Class II determination</u>. Class II antigens were determined on cell lysates by radioimmunoassay as described (5).

2.7. <u>E.coli phagocytosis determination</u>. The monocytic U-937 cell line was treated with ILP or IFN for 48 h, then extensively washed and incubated at 37° C for 2 h under gentle shaking, in a 1:100 ratio with <u>E.coli</u>, with/without a non-agglutinating dilution of inactivated human antiserum against <u>E.coli</u>. Phagocytosis was determined as described in (6).

3. RESULTS.

3.1. <u>Production of ILP</u>.

Human PBMC treated with cycloheximide and actinomycin D for 1 h at 37° C, as described in Methods, produced ILP at titers ranging from 10^2 to $10^{2.5}$. As shown on Table 1, PBMC from healthy donors, as well as monocyte-enriched and T+B cell-enriched preparations are capable to produce ILP. Also human cell lines of the B, T and monocyte lineages, i.e. Namalva, Molt4 and U-937 cell lines produce ILP when stimulated with actinomycin D. No IFN activity was recovered in untreated concomitant cell cultures and without metabolic inhibitors, or treated with cycloheximide alone and used as control.

Molecular weight determination of ILP produced by normal PBMC and by the cell lines by gel filtration through Sephadex G-75 columns (2) gave the following values: PBMC 6700, Namalva 6500, Molt4 6200, and U-937 6700, respectively.

TABLE 1. ILP production by PBMC and by cells lines of B, T and monocyte lineages.

Type of culture		ILP titer ($U/10^6$ cells/ml)	
		Induced	Uninduced
NORMAL CELLS:	Total PBMC	128	<4
	Monocytes	64	<4
	T+B cells	16	<4
CELL LINES:	U-937 (monocytic)	192	<4
	Namalva (B)	94	<4
	Molt4 (T)	192	<4

3.2. ILP purification steps.

The various crude preparations were concentrated by lyophilization or precipitation with ammonium sulphate (50-90 percent of saturation), then gel-filtrated on a Sephadex G-75 column equilibrated with PBS. Eighty % of the activity was recovered in a single peak, with a MW ranging from 6000 to 8000 (mean MW 6760+850); the peak fractions were stabilized with bovine serum albumin (BSA, 100 µg/ml) and rechromatographed on a Blue-Sepharose column equilibrated with 20mM Na_2HPO_4, 1 M NaCl. The activity was eluted by adding ethylene glycol (EG) in the same buffer: 15% of ILP input was eluted with 30% EG (v/v), whereas the majority (62%) was eluted with 50% EG. Table 2 summarizes the various steps of purification, resulting in a 6300-fold purification, with a 24% of recovery, and an estimate specific activity of $10^{5.8}$ units/mg of protein.

TABLE 2. ILP purification steps.

	Units	Sp.activity	Purification	Recovery (%)
Crude	108,000	$10^{2.0}$	1	100
50-90% $(NH_4)_2SO_4$	53,500	$10^{2.7}$	5	49
SEPHADEX G-75	42,000	$10^{3.5}$	31	39
BLUE-SEPHAROSE				
30% EG	6,500	$10^{5.0}$	1,000	6
50% EG	26,000	$10^{5.8}$	6,300	24

3.3. ILP effects on cellular functions.

IFNs, in addition to the antiviral activity, are biological compounds that influence several cell functions, among which particularly important are those related to the immune system (8-9). To investigate whether also ILP possesses cell modulatory functions, preliminary experiments were carried out to search for some of these effects. On Table 3 are reported data on the capability by purified ILP of modulating HLA Class II antigen expression by human M14 melanoma cells in comparison to the other IFN types. Class II values were obtained as indicated in Materials and Methods by radioimmunoassay. As already reported (5, 10, 11), IFN gamma is the most effective IFN in enhancing Class II expression, whereas IFN alfa is not active on this cell line (5). Also ILP is able to increase Class II expression by M14 cells, and the effect is higher after a 98 h-treatment than after 48 h, as it occurs with IFN gamma (5, 10). At variance from IFN gamma, however, ILP was not able to de novo induce class II expression in Class II-negative normal diploid fibroblasts (not shown) as it occurs with IFN beta.

TABLE 3. Increase of HLA Class II antigen expression by M14 melanoma cells treated with ILP and IFNs.

Treatment	Class II $IU_{50}/10^6$ cells	
	48h	96h
None	5.0	5.4
IFN gamma 500 IU/ml	17.6	79.4
IFN beta 5000 IU/ml	11.8	7.2
IFN alpha 5000 IU/ml	5.2	5.3
ILP 5 IU/ml	13.4	23.4

Cells were processed as described in Methods and in (5, 10). The concentrations of IFN beta and gamma were chosen in order to obtain maximum effects (plateau levels of increase).

IFNs, particularly IFN gamma, are known to enhance phagocytosis by macrophages (8, 9). We treated with ILP and IFN alpha the human monocyte cell line U-937 for 48h, then the phagocytic activity of the cell cultures against E.coli bacteria was tested as indicated in Methods according to (6). As shown on Table 4 U-937 cells treated with ILP cause a remarkable reduction of E.coli in the culture fluids, up to 2000-fold reduction in the absence of bacterial growth, with 1 U/ml of ILP).

TABLE 4. Increase of phagocytosis by U-937 monocyte cells treated with ILP and IFN alpha for 48h.

Treatment	E.coli fold reduction	
	E.coli untreated	E.coli opsonized
E.coli alone	0	0
+ U-937 untreated	3	3
+ U-937+ IFN alpha$_{500}$ IU/ml	3	9
+ U-937+ ILP$_1$ IU/ml	200	2000
+ U-937+ ILP$_5$ IU/ml	9	51

At the end of IFN treatment, cell were washed, and incubated with bacteria for 2 h, as described (6), with/without a non-agglutinating concentration of human anti-E.coli antiserum. The target to effector ratio was 100:1. During the incubation time bacteria doubled only once, if any.

4. DISCUSSION.

ILP possesses the main properties of the IFN system (2), but differs from the "classical" types of IFN for molecular weight and antigenic properties (2-3). It is not an inducer of a known type of IFN in human cells, since it induces a durable antiviral state even in the presence of antibody to IFN alpha, beta and gamma, applied individually or combined (2).

In this report further informations are provided on ILP production by normal or neoplastic cells of hemopoietic origin. Both freshly isolated or continuously growing in culture, cells of B, T and monocyte lineages are capable of producing significant levels of ILP when properly induced with inhibitors of transcription. In PBMC most of ILP seems to be produced by monocytes, but the results obtained with cell lines show that also other types of white cells, once stabilized, may be equally effective. This finding may suggest some type of cell interaction which may regulate ILP production under natural conditions. Obviously this hypothesis, which may give a clue to the physiologic role of ILP, requires to be investigated. The ILP produced by different cell subpopulations seems to exert similar characteristics, as for molecular weight, binding capacity to ionic and affinity columns; and for antiviral and cell-modulating activities (not shown).

Besides the antiviral activities of ILP, previously described (2-3), ILP is able to modulate at least some of the cellular functions shown to be affected by the "classical" IFN types, i.e. enhancement of expression of Class II molecules in human melanoma cells (see Table 3), and enhancement of E.coli phagocytosis by the human monocytic line U-937. No informations are available so far on the mechanism of this action. However, the fact that ILP effect is more prononunced when the bacteria are opsonized with a specific antiserum suggests that ILP could affect the process of phacocytosis via an increase of expression of the Fc receptor by the monocyte cells, as it was shown for the other IFNs (12-13). On the other hand preliminary experiments (not reported in this paper) suggest that ILP has little effect, if any, on NK cell activity. This property is not conflicting with the IFN-like characteristics of ILP, since also some subtypes of alpha IFN are uncapable to activate NK cytotoxicity (14).

5. ACKNOWLEDGEMENTS.

This work was supported by grants from CNR, Progetto Finalizzato Controllo Malattie da Infezione, Contratti 86, and Progetto Finalizzato Oncologia, Contratto 86, and by a grant from MPI, Project 40%. The authors wish to thank Dr. C. Juliano, F. Ameglio and C. Ausiello for their generous help given for phagocytosis, HLA and NK assays.

REFERENCES

1. Dianzani F, Monahan TM, Jordan CJ, Langford MP: Inhibition ofRNA synthesis in human lymphoid cellsinducesinterferon
2. Dianzani F, Dolei A, Di Marco P: A new type of human interferonproducedbyperipheral blood mononuclear cells treated with inhibitors transcription. J Interferon Res 6, 43-50, 1986.
3. Dianzani F, Dolei A, Di Marco P: An interferon-like activity released by human peripheral blood mononuclear cells treated with inhibitors of transcription: a new type of interferon? In The Biology of the Interferon system, 1985, Schellekens P Stewart WE (Eds), Amsterdam, Elsevier, pp. 45-54, 1986.
4. Lockart RZ: Criteria for acceptance of a viral inhibitor as an interferon and a general description of the biological properties of known interferons. In Interferon and Interferon inducers, Finter NB (Ed). Amsterdam: Elsevier, pp. 11-27, 1973.
5. Dolei A, Capobianchi MR, Ameglio F: Human interferon gamma enhances the expression and shedding of Class I and Class II MHC products in neoplastic cells better than interferon alpha and interferon beta. Infect Immun 40, 172-176, 1983.
6. Van Furth R, Van Zwet TL, Leith PCJ: In vitro determination of phagocytosis and intracellular killing by polymorphonuclear and mononuclear phagocytes. In Handbook of Experimental Immunology. Cellular Immunology, Weir DM(Ed). Oxford: Blackwell Scientific Publications, pp 32/1-32/19, 1978.
7. Drocourt JL, Thang DC, Thang MN: Blue-dextran Sepharose affinity chromatography: recognition of a polynucleotide binding site of a protein. Eur J Biochemistry 82, 355-359, 1978.
8. Dianzani F, Rossi GB (Eds): The Interferon System. New York:

Serono Symposia Publications from Raven Press, vol.**24**, 1985.
9. Hokland ME: Immunomodulatory effects of interferons on human mononuclear cells with special reference to the expression of cell surface antigens. Acta Pathol Microbiol Immunol Scand, Section C, Suppl **286**, 93-121, 1985.
10. Capobianchi MR, Ameglio F, Tosi R and Dolei A: Differences in the expression and release of DR, BR and DQ molecules in human cells treated with recombinant interferon gamma. Comparison to other interferons. Human Immunol **13**, 1-11, 1985.
11. Dolei A, Ameglio F: Effects of interferon on phenotypic expression by cells. Texas Rep Biol Med, in the press.
12. Guyre PM, Morganelli PM, Miller R: Recombinant immune interferon increases immunoglobulin G Fc receptors on cultured human mononuclear phagocytes. J Clin Invest **72**, 393-399, 1983.
13. Naray-Feyes-Toth A, Guyre PM: Recombinant immune interferon induces increased IgE receptor expression on the human monocyte cell line U-937. J Immunol **133**, 1914-1919, 1984.
14. Pestka S, Evinger M, Maeda S, Rehberg E, Familletti PC, Kelder B: Biological properties of natural and recombinant interferons. Texas Reps Biol Med **41**, 31-36, 1982.

APPARENT AUTO-INHIBITION OF MURINE INTERFERON-γ DUE TO A SERUM COMPONENT

G. FROYEN, H. HEREMANS, R. DIJKMANS and A. BILLIAU
Rega Institute, University of Leuven, B-3000 Leuven, Belgium

1. ABSTRACT

Recombinant mouse interferons (IFNs) α, β and γ were tested for their inhibitory activity on the replication of vesicular stomatitis virus (VSV) and mengovirus in JLSV5-cells. As expected, treatment of the cells with increasing doses of interferons α and β resulted in progressively decreasing yields of virus. The same dose response relationship occurred with relatively low dose ranges of MuIFN-γ. However, with high doses of this IFN, virus inhibition was partially undone. This apparent auto-inhibition effect of high doses of MuIFN-γ occurred with natural as well as with eukaryotic cell- and E.coli-derived recombinant MuIFN-γ; it was seen in JLSV5, but not in L929-cells. One possible source of inhibitor which may at least partially explain the apparent auto-inhibition was identified as the mouse serum that was added as a stabilizer to the purified IFN preparations.

2. INTRODUCTION

Preparations of IFN-γ have been described to contain an inhibitor of the action of IFN-γ (1,2). The presence of this inhibitor is apparent from a deviation of the titration curves for antiviral as well as anticellular effects. Attempts to dissociate the inhibitor from IFN-γ have not been successful, leading to the hypothesis that the IFN-γ molecule itself or an aberrant form of it may be auto-inhibitory (2). In the studies described here, we have confirmed the basic observation of "auto-inhibition" by murine IFN-γ preparations and we present evidence that one possible source of the inhibitor is serum naturally present in or added to the interferon preparations.

3. MATERIALS AND METHODS
3.1. Interferons

Recombinant murine IFN-α₁ (rMuIFN-α₁) and -β (rMuIFN-β) were derived, respectively, from CHO-pSV10EF-3 (3) and CHO-βAA₃ (4) cells. They were kindly provided as crude cell culture supernatants by Dr. J. Trapman (Erasmus University, Rotterdam, The Netherlands). Both interferons were purified by affinity chromatography on a monoclonal antibody against MuIFN-α₁ or MuIFN-β. Recombinant murine interferon-γ (rMuIFN-γ) was obtained from MICK-3 cells, a CHO-derived cell line developed in our laboratory (5). E.coli-derived MuIFN-γ was kindly provided by Dr. S. Pestka (Roche Institute Molecular Biology, Nutley, N.J.). Natural mouse splenocyte-derived MuIFN-γ was produced in our laboratory as described earlier (6). The methods for purification of IFN-γ by adsorption/desorption to silicic acid and by gel filtration were also described earlier (6). Some preparations, as indicated in the text, underwent further purification by affinity chromatography to a monoclonal antibody (F₃) obtained from a rat x mouse hybridoma isolated in our laboratory. Finally, we used a recombinant rat interferon-γ (rRaIFN-γ) kindly provided by Dr. H. Schellekens (Primate Center TNO, Rijswijk, The Netherlands). All purified IFNs were stabilized by addition of 1 % normal mouse serum, taken from NMRI-mice. The potency of the preparations was determined by CPE-inhibition assays on L929-cells challenged with mengovirus. The assays were standardized by inclusion of laboratory standard preparations consisting of the natural mixture of Mu-IFN-α/β or splenocyte-derived MuIFN-γ. Both laboratory standards had been assigned a unitage by calibration against NIH-standard preparation G002-904-511, using a CPE-inhibition assay on primary mouse embryo fibroblasts.

3.2. <u>Yield reduction assay for antiviral activity</u>

The cell lines used were mouse L929- and JLSV5-cells and the rat cell line Ratec. Confluent monolayers of cells in 24-well plates were incubated with serial dilutions of interferon and incubated for 24 h. Virus (VSV or mengovirus) was then added at m.o.i. > 1. Excess virus was removed after 1 h adsorption, and the cultures were refed and incubated at 37°C for 18 h. Virus yields were determined by plaque assay on L929-cells. Titration curves were obtained by plotting log_{10} virus yields against log_{10} interferon dose.

4. RESULTS AND DISCUSSION

The phenomenology of the auto-inhibition is illustrated in Fig. 1 which shows repeated titration curves on a preparation of murinic IFN-γ. To do these experiments serial dilutions were tested for inhibitory effect on single cycle virus growth of VSV in JLSV5-cells. It can be seen that, as the dose of interferon is increased, virus yields decrease up to a certain dose onwards from which the virus yields increased again.

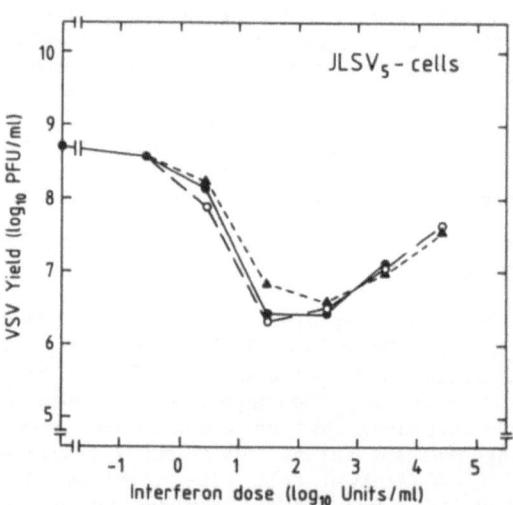

FIGURE 1. Virus yield reduction in JLSV5-cells treated with increasing doses of IFN-γ and challenged with VSV. The preparation of IFN-γ was derived from <u>E.coli</u> and purified by affinity chromatography to monoclonal antibodies. Curves represent results of three individual experiments done at different times.

The figure also shows good reproducibility of the titration curves. On the basis of the curves each IFN-γ preparation could be assigned a titer of antiviral activity, i.e. the dilution causing a 0.5 $\log_{10}$ yield reduction, and an auto-inhibitory titer, i.e. the dilution corresponding to the lowest point of the titration curve where the action of the inhibitor becomes apparent.

Fig. 2 shows similar titration curves illustrating that the auto-inhibitory effect was not seen with murine IFN-α or -β. Also, as shown in Fig. 3 (a and b), addition of a high, auto-inhibitory dose of IFN-γ to the dilution series of IFN-α or -β did not inhibit the antiviral activity of these interferons, but rather acted additively with them. It was concluded therefore that the inhibitor acted specifically on IFN-γ.

The action of the inhibitor was also specific for certain cells, in particular JLSV5-cells. Titrations of IFN-γ on L929-cells or on a rat cell line (RATEC) did not yield curves with an auto-inhibition zone effect, as seen in Fig. 4 (a and b). This suggested that the inhibitory effect was not due to a direct blocking of the IFN-γ receptor.

Different preparations of interferon-γ differed in the arbitrary units needed to cause auto-inhibition. Remarkably, however, all preparations that had undergone purification had the same end-point dilution of auto-inhibition, i.e. the auto-inhibitory titer (Table 1). The only exception to this was a preparation that had been concentrated 13-fold after purification and, accordingly, had a 10-fold higher auto-inhibitory titer. On the other hand, the crude IFN-γ preparation had an exceptionally low inhibitory titer. From these observations it was presumed that the inhibitor had its origin in a component added to the preparations during or after purification, in amounts that were fixed and independent of the antiviral titer. An obvious candidate was the 1 % normal mouse serum added as a stabilizer to all purified interferon preparations. The results of an

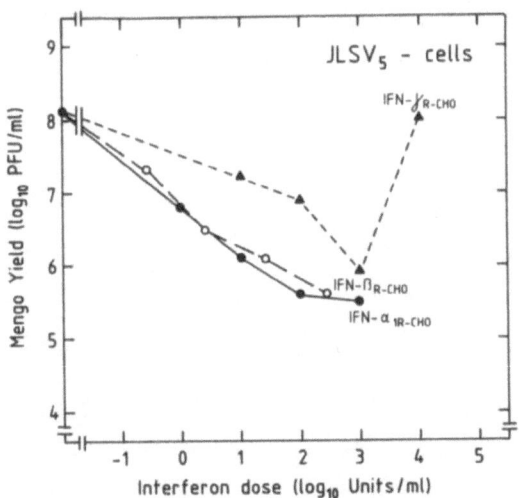

FIGURE 2. Virus yield reduction in JLSV5-cells treated with increasing doses of IFN-α, -β or -γ. All interferons were derived from CHO-cells expressing the appropriate genes, and were purified by immunoaffinity chromatography.

48

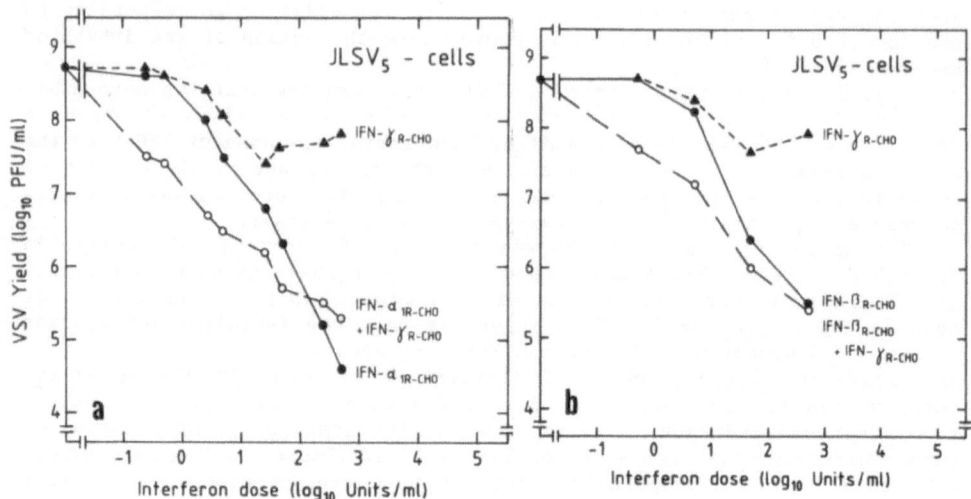

FIGURE 3. Effect of addition of a high dose of IFN-γ to the antiviral effect of increasing doses of IFN-α (a) or IFN-β (b). ●—●, IFN-α or -β only; ▲—▲, IFN-γ only; O—O, IFN-α or -β mixed with 10^3 units/ml of IFN-γ.

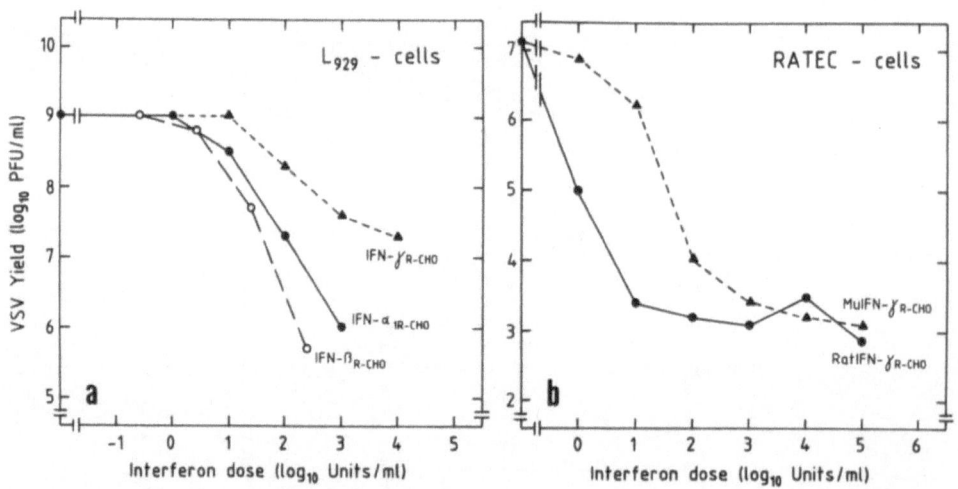

FIGURE 4. Virus yield reduction in mouse (a) and rat (b) cells treated with increasing doses of mouse or rat interferon.
(a) : Mouse L929-cells treated with MuIFN-α, -β or -γ.
(b) : Rat RATEC-cells treated with MuIFN-γ or rat IFN-γ.
All IFNs were of CHO-origin and purified by immunoaffinity chromatography.

TABLE 1. Ratio's between antiviral and auto-inhibitory activities of different MuIFN-γ preparations

Cellular origin of IFN preparation	Manipulations			Biological activity $(\log_{10})$	
	Purification	Addition of stabilizer (1 % mouse serum	Concentration	End-point dilution antiviral effect[a]	End-point dilution auto-inhibition[b]
Spleen cells	silica + aff. chrom. Mab	+	1 x	− 4.5	− 3.0
E.coli	aff. chrom. Mab	+	1 x	− 5.0	− 3.1
CHO-cells	crude supernatant	−	1 x	− 4.0	− 1.1
CHO-cells	silica + gel filtr. (45K)	+	1 x	− 4.5	− 3.0
CHO-cells	aff. chrom. Mab	+	1 x	− 4.5	− 3.0
CHO-cells	aff. chrom. Mab	+	13 x	− 6.0	− 4.0

[a] Dilution corresponding to 0.5 $\log_{10}$ virus yield reduction.
[b] Dilution corresponding to lowest point of titration curve where "auto-inhibition" becomes apparent.

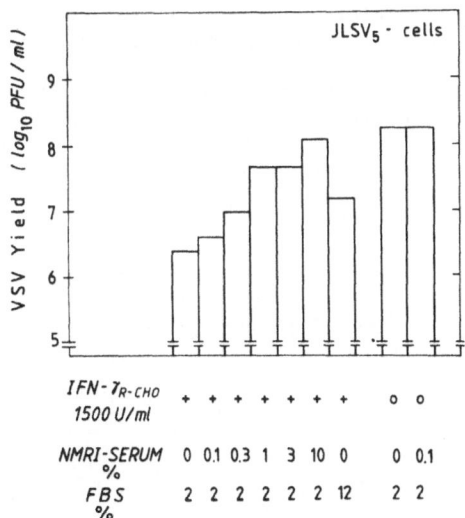

FIGURE 5. Inhibitory effect of normal mouse serum (NMRI) and fetal bovine serum (FBS) on the antiviral effect of MuIFN-γ. Mixtures of MuIFN-γ (1500 units/ml final concentration) and serum (final concentrations as indicated) were applied to JLSV5-cells and a regular yield reduction assay was done with VSV as a challenge.

this possiblity are shown in Fig. 5. Increasing concentrations of normal mouse serum were added to a fixed amount of IFN-γ (1500 units/ml final concentration), and the mixtures were tested for antiviral activity on JLSV5-cells. As can be seen in Fig. 5, mouse serum inhibited the antiviral effect of IFN-γ in a dose-dependent fashion. A concentration of 0.3 % mouse serum was still able to cause a significant reduction in the antiviral effect of 1500 units/ml of IFN-γ.

The fact that media used for applying IFN-γ to cell cultures contained fetal bovine serum (FBS), raised the problem of the presence of inhibitor(s) in this serum. As can be seen in Fig. 5, FBS did contain inhibitory activity for IFN-γ, albeit at a much lower concentration than mouse serum (end-point at about 12 %).

We conclude that serum added as a stabilizer of purified IFN-γ or used as a component of cell culture medium contains (an) inhibitor(s) for IFN-γ, and may thereby disturb antiviral assays. While the presence of the inhibitor(s) must certainly be taken into account when interpreting the results of in vitro experiments, a possible in vivo role of the inhibitors remains to be investigated.

ACKNOWLEDGEMENT

The present study was supported by grants from the Cancer Research Foundation of the Belgian ASLK/CGER and from the Ministry of Science Administration (Concerted Research Action). Editorial help from Christiane Callebaut is gratefully acknowledged.

REFERENCES

1. Fleischmann WR, Georgiades JA, Osborne LC, Dianzani F, Johnson HM: Infect. Immun. 26: 949-955, 1979.
2. Lefkowitz EJ, Fleischmann WR: J. Interferon Res. 5: 101-110, 1985.
3. Zwarthoff EC, Bosveld IJ, Vonk WP, Trapman J: J. Gen. Virol. 66: 685-691, 1985.
4. Zwarthoff EC, Van Heuvel M, Mooren ATA, Van Der Korput GA, Bosveld IJ, Trapman J: The Interferon System. Cerona Symposia Publications. New York : Raven Press, 24: 9-14, 1985.
5. Dijkmans R, Volckaert G, Van Damme J, De Ley M, Billiau A, De Somer P: J. Interferon Res. 5: 511-520, 1985.
6. Heremans H, De Ley M, Billiau A, De Somer P: Cell. Immunol. 71: 353-364, 1982.

MODES OF ACTION

NUCLEAR LOCALIZATION OF INTERNALIZED INTERFERONS-β AND -γ: A ROLE FOR NUCLEAR RECEPTORS

Sidney E. Grossberg, Vladimir M. Kushnaryov, Hector S. MacDonald, and J. James Sedmak

Department of Microbiology, Medical College of Wisconsin, Milwaukee, Wisconsin 53226 USA

INTRODUCTION

The mechanism by which IFN treatment alters gene expression remains unknown, although IFN binding to plasma membrane receptors appears to be a crucial step in processing (1). This binding has been thought to be sufficient to trigger intracellular activities by still unidentified transmembrane signals or messengers. Recent reports indicate that newly induced RNA transcripts can appear within several minutes after the cells are treated with IFN (2 - 5). Since different lines of evidence indicate that IFN molecules are transported into the nucleus within a few minutes by facilitated receptor-mediated endocytosis (6,7), IFNs may act as their own messengers, alone or in combination with their receptors.

The uptake of ^{125}I-MuIFN-β and -γ into mass populations of mouse L_{929} cells reveals rapid internalization of the ^{125}I label (6,8). Previously, measurement of IFN biological activity after fractionation of IFN-treated cells showed an increase in activity associated with the nuclear fraction ten minutes after IFN treatment (9). Such methods, however, cannot clearly demonstrate the fate of the internalized IFN molecules. The use of a new electron microscopy technique employing post-embedding labeling demonstrates unequivocally that natural, unmodified IFN molecules enter and are processed within the cell to enter the dense chromatin. The procedure consists essentially of treating samples of cells with interferon at 4°C, after which they are washed and moved to 37°C; at brief intervals, samples are rapidly fixed and embedded in Lowicryl K4M (6,7,10). Ultrathin sections are then exposed to polyclonal or monoclonal anti-IFN antibody, after which colloidal gold coated with protein A (to bind to the Fc portions of the immunoglobulin) is added.

RESULTS AND DISCUSSION

Figure 1 shows electron micrographs of ultrathin sections of L_{929} cells incubated at 37°C for different periods with natural MuIFN-β. Within one minute, IFN molecules are detected on coated areas of the plasma membrane and in coated pits, which then invaginate to form coated vesicles and receptosomes. This sequence of events, which is characteristic of receptor-mediated endocytosis, takes place within 1-2 minutes of incubation at 37°C. The colloidal gold label is located very close to the membrane of these organelles, implying that interferon is not detached from its receptor. Receptosomes carrying interferon molecules can be seen emptying into the perinuclear space. Within 2-3 minutes of incubation, the majority of IFN molecules are detected on and in the cell nucleus, in or near nuclear pores, and in the dense chromatin.

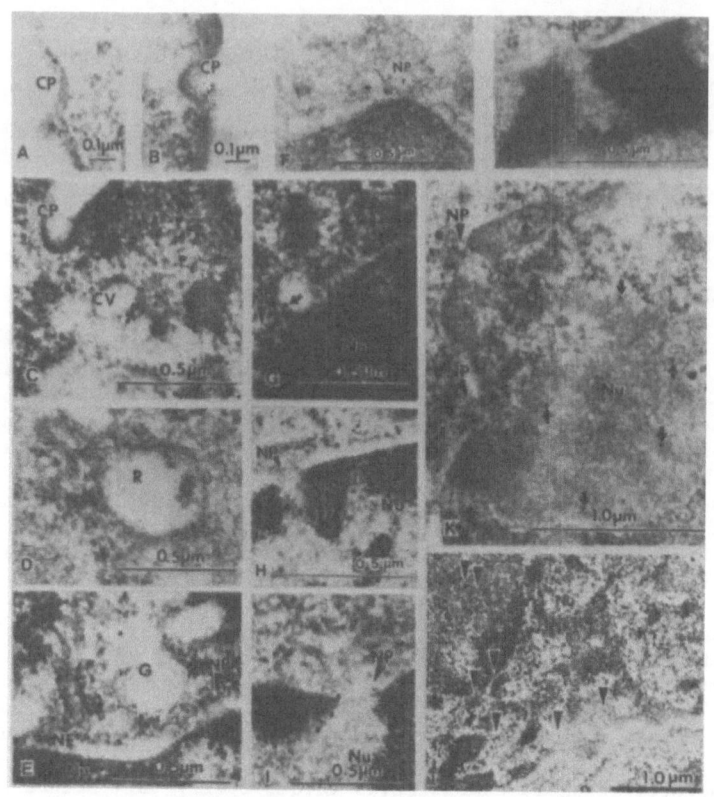

FIGURE 1. Electron photomicrographs showing rapid receptor-mediated endocytosis of MuIFN-β through the cytoplasm and into the nucleus of L cells. After treatment at 4°C with MuIFN-β, cells were incubated at 37°C for different periods, fixed, and embedded in Lowicryl K4M. Ultrathin sections were treated with anti-MuIFN-β monoclonal antibody and then colloidal gold-protein A. A and B: IFN, identified by the colloidal gold, located in coated pits (CP) at 4°C; C through E: 1-2 min at 37°C, IFN in coated vesicles (CV), receptosomes (R), Golgi cisterns (G); F and G: IFN at nuclear pores (NR) of the nuclear envelope (NE); H and I: IFN entering nucleus (Nu) through the NP; K and L: intranuclear IFN.

TABLE 1. Number and affinity of MuIFN-β receptors in L cell membranes

	Receptor Number	Dissociation Constant (kD)
Plasma membrane	5,000	9.8×10^{-10}
Nuclear membrane	14,000	1.4×10^{-10}

Treatment of purified, isolated nuclei with purified IFNs -β and -γ has demonstrated the presence of specific, high-affinity receptors on the nuclear envelope (7,8). Table 1 provides comparative estimates of MuIFN-β receptor number and affinity of receptors in the plasma and nuclear membranes. Scatchard analysis of MuIFN-γ binding provides an estimate of 24,000 high-affinity binding sites per nucleus, having a K_d of 2×10^{-10} M, with some low-affinity sites as well. Brief treatment of isolated nuclei with trypsin reduced binding of ^{125}I-MuIFN-β and -γ, suggesting that these nuclear receptors contain protein. The lack of competition for nuclear binding by ^{125}I-MuIFN-β by excess unlabeled MuIFN-γ, and of ^{125}I-MuIFN-γ by excess unlabeled MuIFN-β, indicates that β- and γ-IFNs have separate nuclear receptors (7,8), a distinction noted for plasma membrane receptors of these IFNs (1).

To determine the possible association between interferon molecules and putative membrane receptors intracellularly, L cells were treated with a lipophilic, photoactivable, cross-linking agent, dithiobisphenylazide, and ^{125}I-MuIFN-β. After further incubation, the cells were exposed to a flash of light to cross-link the ^{125}I-MuIFN-β with adjacent molecules less than 11 Å distant, and the cells fractionated into nuclei and cytoplasmic portions. These fractions were then treated with micrococcal nuclease, ribonuclease, and 1% SDS. The resultant fractions were chromatographed on Sephacryl S-500 and their radioactivity measured. The cytoplasmic fraction of cells held either at 37°C or at 4°C contained a ^{125}I-MuIFN protein complex of 106 kDa as a single peak. The nuclear fraction of cells warmed to 37°C prior to cross-linking contained approximately 50% of the cell-associated radiolabel (whereas only 5.8% was found in the nuclei of the cells held at 4°C, probably as contaminating ^{125}I-MuIFN-β); the nuclear fraction further contained two complexes of 80 kDa and 235 kDa. These results suggest that interferon molecules alone, or in combination with plasma membrane receptors, may bind to nuclear membrane receptors. These chromatographic results were confirmed by SDS-PAGE.

The physiological significance of the very rapid transport of immunologically detectable IFN-β and -γ into the nucleus, which may be facilitated by nuclear membrane receptors, must still be determined. We have recently demonstrated that brief treatment of isolated nuclei of L cells with MuIFN-β rapidly decreases the energy-dependent RNA efflux from nuclei in a dose-related manner (11). In contrast, the treatment of isolated nuclei with insulin increases nuclear mRNA efflux, nucleotide triphosphatase activity, and protein phosphorylation (12). Human IFN-α and -β have been shown to induce new synthesis of several mRNAs within 5-10 minutes, and recombinant HuIFN-γ within 30 minutes, following IFN treatment of the cells (2-5).

The very rapid uptake and vesicular transport of immunologically recognizable molecules of IFN-β and -γ through the cytoplasm and into the dense chromatin areas of the nucleus suggest that IFN molecules may act as their own messengers, either alone or bound to receptor, where they may be directly involved in genome regulation.

ACKNOWLEDGMENTS

We thank Janet DeBruin for technical assistance. This work was supported in part by grants from the Walter Schroeder Foundation, Stackner Family Foundation, and the NIH.

REFERENCES

1. Zoon, KC, and Arnheiter, H: Studies of the interferon receptors. Pharmacol. Ther. 24:259-278. 1984.

2. Friedman, RL, Manly, SP, McMahon, M, Kerr, IM, and Stark, GR: Transcriptional and posttranscriptional regulation of interferon-induced gene expression in human cells. Cell 38:745-755. 1984.

3. Friedman, RL, and Stark, GR: α-Interferon-induced transcription of HLA and metallothionein genes containing homologous upstream sequences. Nature 314:637-639. 1985.

4. Larner, AC, Jonak, Q, Cheng, Y-S, Korant, B, Knight, E, and Darnell, JE, Jr: Transcriptional induction of two genes in human cells by β-interferon. Proc. Natl. Acad. Sci. USA 81:6733-6737. 1984.

5. Luster, AD, Unkeless, JC, and Ravetch, JV: γ-Interferon transcriptionally regulates an early-response gene containing homology to platelet proteins. Nature 315:672-676. 1985.

6. Kushnaryov, VM, MacDonald, HS, DeBruin, J, LeMense, GP, Sedmak, JJ, and Grossberg, SE: Internalization and transport of mouse β-interferon into the cell nucleus. J. Interferon Res. 6: 241-245. 1986.

7. MacDonald, HS, Kushnaryov, VM, Sedmak, JJ, and Grossberg, SE: Transport of γ-interferon into the cell nucleus may be mediated by nuclear membrane receptors. Biochem. Biophys. Res. Commun. 138:254-260. 1986.

8. Kushnaryov, VM, MacDonald, HS, Sedmak, JJ, and Grossberg, SE: Murine interferon-β receptor-mediated endocytosis and nuclear binding. Proc. Natl. Acad. Sci. USA 82:3281-3285. 1985.

9. Stewart, WE: Uptake of interferon: The fate of cell-bound interferon during induction of antiviral and non-antiviral activities. In: Effects of Interferon on Cells, Viruses and the Immune System. Geraldes, A(ed): New York: Academic Press, pp. 75-98. 1975.

10. Altman, LG, Schneider, BG, and Papermaster, D: Rapid embedding of tissues in Lowicryl K4M for immunoelectron microscopy. J. Histochem. Cytochem. 32:1217-1223. 1984.

11. MacDonald, HS, Kushnaryov, VM, Hevner, R, Sedmak, JJ, and Grossberg, SE: Mouse β-interferon reduces RNA efflux from isolated nuclei. J. Interferon Res. 6:247-250. 1986.

12. Goldfine, ID, Kriz, BM, and Wong, KY: Entry of insulin into target cells in vitro and in vivo. In: Receptor-Mediated Binding and Internalization of Toxins and Hormones. Middlebrook, JL and Kohn, LD(eds): New York: Academic Press, pp. 233-250. 1981.

THE 5' END OF THE HUMAN (2'-5') OLIGO A SYNTHETASE GENE CONTAINS SEQUENCES RESPONSIBLE FOR TRANSCRIPTIONAL ACTIVATION BY INTERFERON

J. CHEBATH, P. BENECH, M. VIGNERON AND M. REVEL
Department of Virology, The Weizmann Institute of Science, Rehovot, Israel.

1. INTRODUCTION

The (2'-5') oligo A synthetase (OASE) is a multienzyme system formed of at least four polypeptides (40,46,67 and 100 Kd)(1) and multiple mRNAs (2, 3) which can be induced over a hundred fold in cells exposed to IFNs. We have shown that two mRNAs of 1.6 and 1.8 kb encode the two different small enzyme forms of 40 and 46 Kd, and are produced by differential splicing/ processing of transcripts from the same gene (4). To explore the molecular basis for the IFN-mediated regulation of OASE gene expression, we have now examined the functional promoter region in the 5' flanking sequence of this gene. We find that the bulk of the IFN-inducible RNA transcription in this OASE gene, is initiated at multiple sites distributed over a 50 bp in a region devoid of the classical TATA and CAAT consensus signals of RNA poly- merase II promoters (5). In the genomic sequence, these multiple RNA start sites directly precede by 96-46 bp the initiator methionine of the enzyme.

To determine whether the OASE genomic regions preceding the RNA starts contain functional promoter sequences which can be activated in response to IFN, we have examined chimeric gene constructs between the OASE 5' gene region and the bacterial CAT gene (6). In transiently transfected human cells, these constructs were IFN-dependent for the production of CAT activ- ity. Induction factors of 20 to 50-fold were observed. The RNA starts in these OASE-CAT constructs were the same as in the intact OASE gene, and these specific transcripts were nearly completely dependent on IFN treat- ment of the transfected cells. These experiments allow to define the 5' sequences of the OASE gene which function as IFN-activated promoter.

2. RESULTS

2.1. Analysis of the 5' end of the human OASE gene

The 5' end of cDNA clone 9-21 (4) extends 37 b upstream of the initiator methionine codon, second in the sequence ATGATG (4); in this work numbering is from this ATG. To localize the physical 5' end of the IFN-induced RNAs, primer extension analysis was carried out with two antisense DNA fragments, FI spanning nucleotides +76 to +4, and FII from +188 to +76 (Fig. 1). These 5'-labelled primers were hybridized to poly A^+ RNA from SV80 cells treated with 200 U/ml IFN-β for 12 hours. The size of the CDNA obtained by reverse transcription was determined by denaturing gel electrophoresis and auto- radiography (Fig. 1 A-B). With both primers, multiple bands were observed which ended between -27 to -98 in one case or -37 to -102 in the other (Fig. 1 A lane 1 and B lane 4). RNA with such 5' ends were not seen without IFN-treatment (Fig. 1 B lane 2).

S1 nuclease analysis was carried out on a 5'-labelled antisense DNA fragment FIII (+4 to -160) hybridized to poly A^+ RNA from IFN-treated SV80 cells. Denaturing gel electrophoresis revealed (Fig. 1C, lanes 4-6) that at least 10 fragments (a-j) were protected. These multiple bands are not an

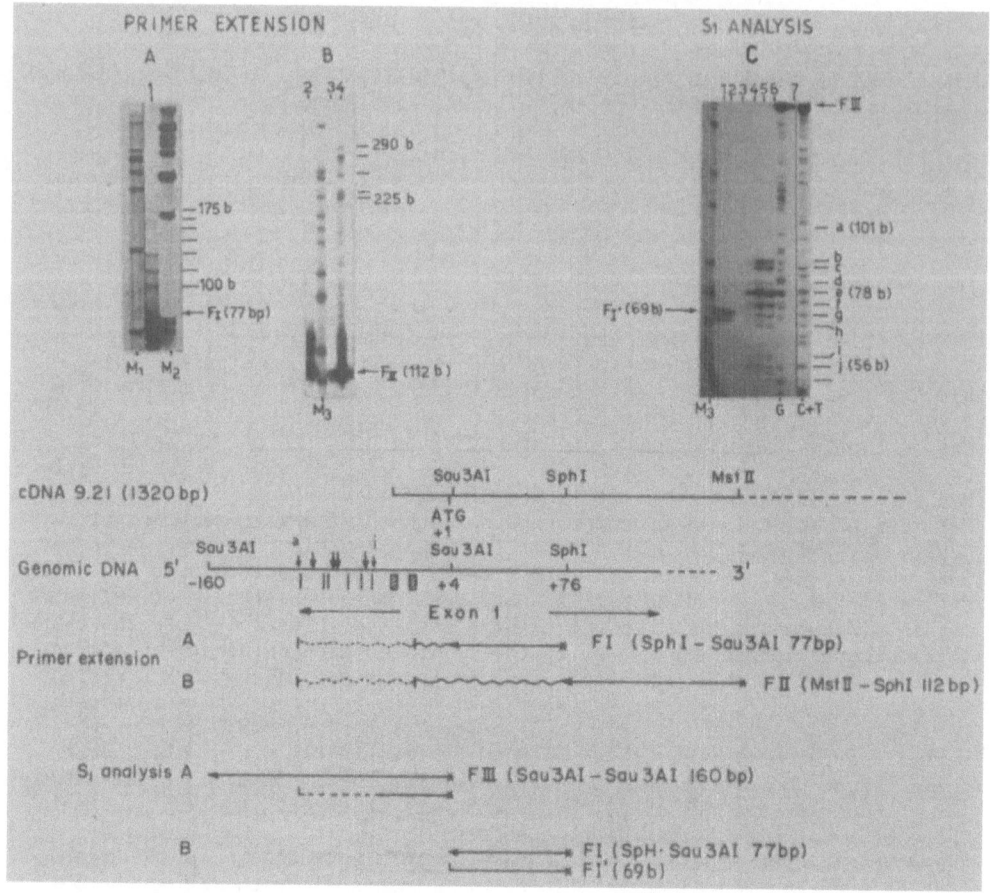

FIGURE 1. Mapping of the 5' end of OASE mRNAs on the gene.
Primer extension: (A) A 5' labelled (15) ds genomic fragment FI (1.1 pmol)
was hybridized overnight at 42°C (2) with 60 μg of poly A⁺ RNA from IFN-β
treated Wish cells (3), subjected to primer extension (15) and electrophoresis on a 8% polyacrylamide-urea gel (lane 1). Markers: ØX DNA HaeIII-
cut (M1) and pBR322 DNA HinfI-cut (M2). Position of FI and its extension
products are shown in the lower panel. (B) Primer extension with ss fragment FII and 4 μg poly A⁺ RNA from IFN-β treated SV80 cells (3), analyzed
on a polyacrylamide-urea gradient gel (lane 4). Lane 2: tRNA instead of
mRNA; lane 3, no Reverse Transcriptase. Marker was pBR322 DNA, MspI-cut
(M3). S1 analysis: (C) Ss probes FI' and FIII hybridized with 2 μg mRNA,
S1 nuclease digested (3) and run on gradient gels. Lanes 4,5,6: FIII
protected fragments with 135, 270, 540 units S1 nuclease respectively.
Lane 7: same no nuclease. Lane 3: mRNA replaced by tRNA. Lanes 1,2:
reactions with FI' fragment. FIII sequence of G and C+T is shown. Lower
panel: The 5' end of OASE cDNA 9-21 and gene (4) are shown. The RNA start
sites defined by the illustrated reactions are indicated on the genomic
DNA by arrows (S1 analyses); the bars show the limits of primer extension.

```
  -200              -190              -180              -170              -160              -150
   *                 *                 *                 *                 *    Sau3a        *
GTA AGT TTC AGG AAG CAG AAG AGT GAG CAG ATG AAA TTC AGC ACT GGG ATC AGG GGA GTG


  -140              -130              -120              -110              -100        ↓      -90    ↓
   *                 *                 *                 *                 *         a        *      b
TCT GAT TTG CAA AAG GAA AGT GCA AAG ACA GCT CCT CCC TTC TGA GGA AAC GAA ACC AAC
                                                                                                    +

                                ↓
 ↓  -80    ↓       -70    ↓   ↓   ↓-60            ↓ -50              -40                    -30
    c        d       e     f   g   h               i       j          *      lcDNA 9-21
AGC AGT CCA AGC TCA GTC AGC AGA AGA GAT AAA AGC AAA CAG GTC TGG GAG GCA GTT CTG
 +               ++      +


  -20              -10                1                10                20                30
   *                *                 *    Sau3a        *                 *                 *
TTG CCA CTC TCT CTC CTG TCA ATG ATG GAT CTC AGA AAT ACC CCA GCC AAA TCT CTG GAC
                              Met Asp Leu Arg Asn Thr Pro Ala Lys Ser Leu Asp


   40               50                60                70                80
    *                *                 *                 *          Sph1
AAG TTC ATT GAA GAC TAT CTC TTG CCA GAC ACG TGT TTC CGC ATG C
Lys Phe Ile Glu Asp Tyr Leu Leu Pro Asp Thr Cys Phe Arg Met
```

FIGURE 2. Human (2'-5') oligo synthetase gene: RNA starts and promoter
region: IFN activated RNA start sites defined by S1 analysis are shown by
letters a to j (arrows) The beginning of cDNA clone 9-21 and of the coding
region (4) is shown.

artefact of the method as shown by a control with fragment FI' entirely
contained in the 9-21 cDNA sequence (Fig. 1C lane 2). No protection was
seen without mRNA (lanes 3,7). By comparison with the sequence lanes in
the same gel, the 5' ends of protected fragments a-j can be mapped on the
gene at -96 to -46 (Fig. 2). Since these ends are in good agreement with
the primer extension data (Fig. 1, lower part), it is very likely that they
correspond to multiple RNA start sites. Site e at -71 gave consistently
higher intensity and may be preferentially used by RNA polymerase over some
of the upstream and downstream sites.

2.2 IFN-mediated activation of the OASE promoter fused to a CAT gene

The SphI-SphI fragment of 829 bp containing the 5' end of the OASE gene
(4), and extending from -753 to +76, was cloned in the polylinker of the
pGEM-3 vector (Promega) just 5' of a HindIII fragment from pSVOCAT (6) containing the bacterial chloramphenicol acetyltransferase coding sequences and
SV40 polyadenylation region. The resulting pGEM-CAT-E and control constructs are schematically shown in Fig. 3. CAT translation could start at
the OASE ATG in pGEM-CAT-E. Supercoiled plasmid DNAs were transfected into
human HeLa or hepatoma SK cells by CaPO4 precipitation (7), the cells were
treated by IFN-β 200 U/ml after 24 hours (or later as indicated), extracted
at 48 hours and the chloramphenicol acetylation measured in 2 hours incubations. CAT activity was very low in HeLa cells transfected with pGEM-CAT-E
but not treated by IFN, while it increased progressively 10-50 fold from
6 to 24 hours after IFN treatment (Fig. 3). Controls with pGEM-CAT-O lacking the OASE gene fragment, or with the fragment in inverse orientation
(pGEM-CAT-1/E where the multiple OASE RNA start sites point away from the
CAT gene), produced no activity. CAT expression from the SV40 promoter of
pSV2CAT (6) was active without and with IFN.

60

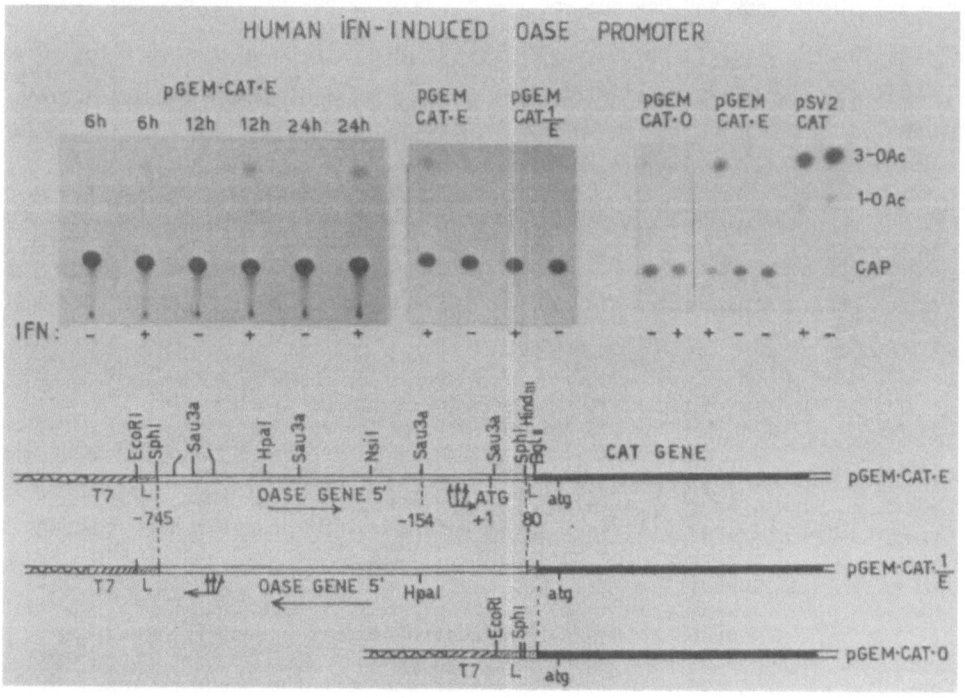

FIGURE 3. IFN-dependent expression of the CAT gene linked to OASE promoter. pGEM-3 derived constructs shown in the lower panel (dots indicate the poly-linker; T7 is the phage T7 promoter; triple arrow, the OASE RNA starts) were transfected in equal amounts into subconfluent HeLa cells by CaPO4 precipitation with glycerol shock. Cell extracts were prepared after 48 hours, equal amounts of proteins were incubated with ^{14}C-chloramphenicol and acetyl-CaO (6), and ethyl acetate-extracted products analyzed by thin layer chromatography and autoradiography. Treatment with 200 U/ml IFN-β is indicated by + and was for 6, 12 or 24 hours (when not indicated) before cell extractions.

Analysis of the RNA transcripts was made in plasmid-transfected hepatoma SK cells, which appear more efficient than HeLa for OASE-CAT expression. The S1 nuclease protection experiment of Fig. 4 shows transcripts starting on pGEM-CAT-E from the same b-j sites as in the resident human OASE gene (compare lane 6 to 3), only in SK cells treated by IFN. Thus, the multiple RNA start sites at -90 to -60 are specifically activated in IFN-treated cells and further experiments showed that less than 100 bp of upstream sequences are needed for IFN response (not shown). An additional upstream RNA start site (z on Fig. 4) was also observed in pGEM-CAT-E transfected SK cells, but was much less IFN-dependent than the proximal sites. Construct with the upstream region of the SphI fragment but lacking the proximal sites gave no CAT activity (not shown).

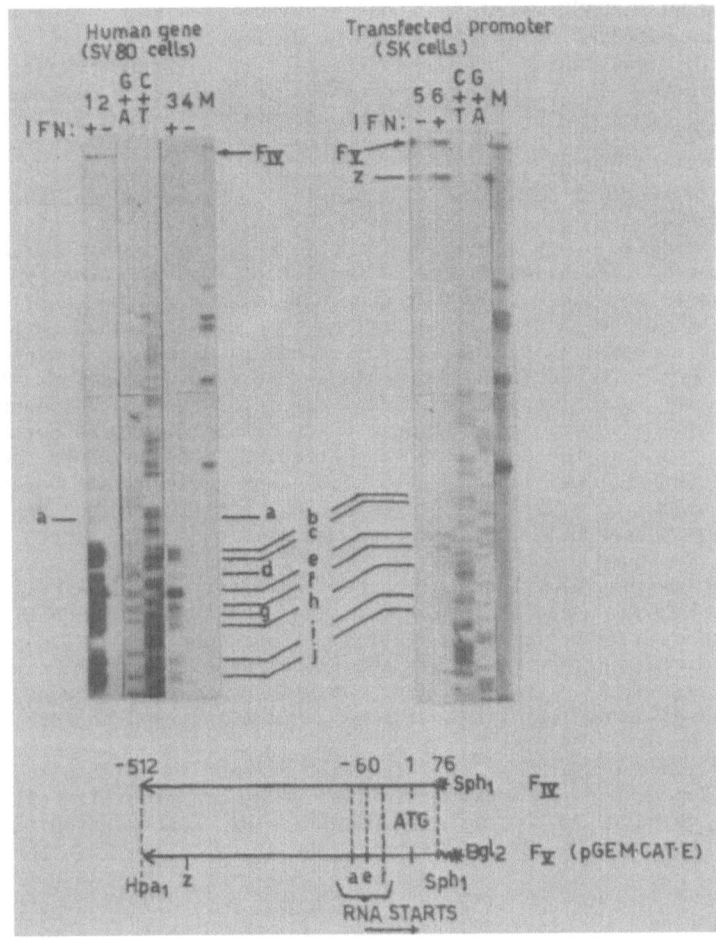

FIGURE 4. 5' ends of transcripts from resident OASE gene and transfected promoter. <u>Left</u>: Poly A$^+$ RNA (2 μg) from SV80 cells treated with IFN-β (lanes 1,3) or untreated (lanes 2,4), was hybridized to probe FIV 5'-labelled at the SphI site in exon 1 of the OASE gene. Reactions digested by S1 nuclease (as in Fig. 1) were run on 8% polyacrylamide-urea gels and autoradiographed for 16 hours (lanes 1,2) or 5 hours (lanes 3,4). Sequencing reactions of FIV are in central lanes; M is ØX DNA, HaeIII-cut. <u>Right</u>: Total RNA (20 μg) from SK cells transfected by pGEM-CAT-E plasmid and treated for 11 hours by IFN-β (lanes 6) or untreated (lane 5) was hybridized to probe FV 5'-labelled at the BglII site of pGEM-CAT-E DNA (see Fig. 3). Sequence of FV is shown on the gel electrophoresis. The lower panel shows the position of OASE RNA 5' ends which were the same on the endogenous and transfected genes.

3. DISCUSSION

The present study defines the IFN-dependent transcriptional initiation sites of the human OASE gene. We found that the OASE RNAs start at 10 or more sites spread over 50 bp from 48 to 96 nucleotides upstream of the initiator ATG codon (which itself was verified by in vitro translation and expression of the active 40 and 46 Kd OASE in E. coli)(1,4,8). No TATA box is present in front of the multiple OASE RNA start sites, a situation similar to the SV40 late (9) and HMG-CoA reductase genes (10). Interestingly, the mouse IFN-induced gene encoding the 56 Kd protein was also reported to have multiple start sites and no TATA box (11). Since we published the 9-21 cDNA sequence (4), we have isolated several full-length cDNAs for the 1.8 kb RNA, and two other sequences for the 1.6 kb RNA were published (12, 13). All these cDNAs ended in the a-j region of Fig. 2, strongly suggesting that these are the major starts for the OASE 1.8 and 1.6 kb RNAs. It is interesting that 5 of these starts fall on CAG sequences, but the role of this triplet in positioning the RNA polymerase is unknown. The nearest TATA-like signal precedes by 200 bp this a-j region. An additional RNA start may exist further upstream (band z of Fig. 3 seen in uninduced transfected SK cells and similar additional bands seen in long exposure of S1 analysis of non-transfected cells) which lead us to assume that the OASE RNA started further away from the ATG codon near a clear TATA box (4). The primer extension, S1 analysis and promoter-CAT constructs show now clearly that the IFN-induced OASE RNAs start in the same exon which contains the ATG codon. Nuclease protection experiments between mRNA and this DNA region generate several fragments because the first exon has variable lengths due to the spreading of the main RNA start sites. Such different fragments seen in Sp6 uniformly labelled, riboprobe/RNase analysis were probably wrongly interpreted as two additional exons in the OASE RNA untranslated region (4). The OASE gene appears now to have 6 exons, 5 forming the 1.6 kb RNA and the 6th additional exon being used to form the 1.8 kb RNA as previously reported (3,4).

The SphI genomic fragment including part of exon 1, is able to confer a strong IFN-dependence on the bacterial CAT gene, and probably contains DNA sequences necessary for the correct initiation of transcription and its regulation by some IFN-generated signal. No sequence similar to the 21 bp found conserved in a number of other IFN-stimulated genes, as HLA genes (14), is found in the SphI fragment of the OASE gene. This sequence could be in another part of the gene, but does not seem essential for the very strong activation by IFN of the OASE gene promoter described here.

REFERENCES

1. Chebath J, Benech P, Hovanessian A, Galabru J, Revel M: J Biol Chem in press. 1986.
2. Merlin G, Chebath J, Benech P, Metz R, Revel M: Proc Natl Acad Sci USA 80:4904-4908, 1983.
3. Benech P, Merlin G, Revel M, Chebath J: Nucleic Acids Res 13:1267-1281, 1985.
4. Benech P, Mory Y, Revel M, Chebath J: EMBO J 4:2249-2256, 1985.
5. Breatnach R, Chambon P: Ann Rev Biochem 50:349-383, 1981.
6. Gorman C, Moffat L, Howard B: Mol Cell Biol 2:1044-1051, 1982.
7. Van der Eb A, Graham F: Meth Enzymol 65:826-839, 1980.
8. Chebath J, Benech P, Mory Y et al: In: Williams BRG, Silverman RH(eds): The 2-5A System: Molecular and Clinical Aspects of the Interferon-Regulated Pathway. Alan R. Liss, New York, pp. 149-161, 1985.

9. Brady J, Bolen JB, Radonovich M, Salzman N, Khoury G: Proc Natl Acad Sci USA 81:2040-2044, 1984.
10. Reynolds G, Basu S, Osborne T, Chin D, Gil G, Brown M, Goldstein J, Luskey K: Cell 38:275-285, 1984.
11. Lengyel P, Tominaga S, Samanta H, Engel D, Chao H, Garciablanco M, Thakur A: J Cell Biochem. Supplement 10C Abstract M8 p 219, 1986.
12. Wathelet M, Moutschen S, Cravador L, Dewitt P, Defilippi D, Huez G, Content J: FEBS Lett 196:113-120, 1986.
13. Shiojiri S, Fukumaga R, Ichii Y, Sokawa Y: J Biochem 99:1455-1464, 1986.
14. Friedman R, Stark G: Nature 214:637-639, 1985.
15. Maniatis T, Fritsch EF, Sambrook J: Molecular cloning: a laboratory manual, Cold Spring Harbor Laboratory, Cold Spring Harbor, New York 1982.
16. Vigneron M, Barrera-Saldana HA, Baty D, Everett RE, Chambon P: EMBO J 3:2373-2382, 1984.

EXPRESSION AND PRELIMINARY DELETION ANALYSIS OF THE 42 kDa 2-5A SYNTHETASE INDUCED BY HUMAN INTERFERON

J. CONTENT, L. DE WIT, M. WATHELET, S. MOUTSCHEN and G. HUEZ.
Institut Pasteur du Brabant (Virology) Rue Engeland 642, 1180
Bruxelles ; Univ. Libre de Bruxelles (Biol. Mol.), 1640 Rhode-
St-Genèse (Belgium).

1. INTRODUCTION

Among the multiple proteins induced by interferon, the 2-5A synthetase has been shown to be involved in some of its antiviral actions. In the presence of dsRNA, this enzyme catalyzes the synthesis of 2'-5' linked oligomers of adenosine, $ppp(A'2p)_nA$. Viral growth is inhibited through mRNA degradation mediated by a latent endoribonuclease which is activated by these 2-5A oligomers. This effect is transient. The 2-5A oligomers are rapidily degraded by a 2'-phosphodiesterase (1). In addition to this antiviral role, the 2-5A system seems to be involved in the regulation of cell growth and differentiation (2,3). 2-5A activities associated with proteins of different sizes were detected in the cytoplasm and nucleoplasm (4, 5). Two functional mRNAs (1.6 and 1.8 kb in human cells (6, 7) and 1.8 and 3.6kb in mouse cells (5) were shown to encode 2-5A synthetase activity. The relationship between the different mRNAs, which seem to derive from the same gene by tissue specific differential splicing (6, 7) and the different proteins, as well as their cellular localization is still not clear. In human cells, a cytoplasmic IFN-inducible 2-5A synthetase activity mediated by a 80-105 kDa polypeptide, as detected in SDS-polyacrylamide gels, has been described (8, 9). This observation fits well with the detection of a 100 K component by Western blot analysis ; however, the latter technique reveals also two additional proteins of 67 and 69 K (Chebath et al., 1986, personal communication). Considering those different forms of 2-5A synthetase it would be of interest to sort them out according to their regulation, structural characteristics, cellular localization and specific function.

We have isolated and sequenced a full-length cDNA clone corresponding to one form of human 2-5A synthetase from a cDNA library prepared with mRNA from Hu-IFN-α induced amniotic cells (the UAC cells). An expression construction was obtained by inserting the entire 42 kDa 2-5A synthetase coding region downstream the SP6 RNA polymerase promotor of a SP64 vector to synthetise 5ᴸcapped, non polyadenylated 2-5A synthetase mRNA. This mRNA was translated in rabbit reticulocyte lysate into a single 42 kDA protein with 2-5A synthetase activity (10). Translation of this synthetic mRNA in Xenopus laevis oocytes allowed us to study the intracellular distribution of the 42 kDa enzymatic activity. Furthermore, sizing of the oocyte translated enzymatic activity in non denaturing glycerol gradient demonstrated that

```
     -94
     A ACT GAA ACC AAC AGC AGT CCA AGC TCA GTC AGC AGA AGA GAT AAA AGC AAA CAG GTC
     -36                                  1                                      (8)
     TGG GAG GCA GTT CTG TTG CCA CTC TCT CTC CTG TCA ATG ATG GAT CTC AGA AAT ACC CCA
                                           Met Met Asp Leu Arg Asn Thr Pro
     25                                                                       (20)
     GCC AAA TCT CTG GAC AAG TTC ATT GAA GAC TAT CTC TTG CCA GAC ACG TGT TTC CGC ATG
     Ala Lys Ser Leu Asp Lys Phe Ile Glu Asp Tyr Leu Leu Pro Asp Thr Cys Phe Arg Met
     85                                                                       (40)
     CAA ATC AAC CAT GCC ATT GAC ATC ATC TGT GGG TTC CTG AAG GAA AGG TGC TTC CGA GGT
     Gln Ile Asn His Ala Ile Asp Ile Ile Cys Gly Phe Leu Lys Glu Arg Cys Phe Arg Gly
     145                                                                      (68)
     AGC TCC TAC CCT GTG TGT GTG TCC AAG GTG GTA AAG GGT GGC TCC TCA GGC AAG GGC ACC
     Ser Ser Tyr Pro Val Cys Val Ser Lys Val Val Lys Gly Gly Ser Ser Gly Lys Gly Thr
     205                                                                      (88)
     ACC CTC AGA GGC CGA TCT GAC GCT GAC CTG GTT GTC TTC CTC AGT CCT CTC ACC ACT TTT
     Thr Leu Arg Gly Arg Ser Asp Ala Asp Leu Val Val Phe Leu Ser Pro Leu Thr Thr Phe
     265                                                                      (108)
     CAG GAT CAG TTA AAT CGC CGG GGA GAG TTC ATC CAG GAA ATT AGG AGA CAG CTG GAA GCC
     Gln Asp Gln Leu Asn Arg Arg Gly Glu Phe Ile Gln Glu Ile Arg Arg Gln Leu Glu Ala
     325                                                                      (128)
     TGT CAA AGA GAG AGA GCA TTT TCC GTG AAG TTT GAG GTC CAG GCT CCA CGC TGG GGC AAC
     Cys Gln Arg Glu Arg Ala Phe Ser Val Lys Phe Glu Val Gln Ala Pro Arg Trp Gly Asn
     385                                                                      (148)
     CCC CGT GCG CTC AGC TTC GTA CTG AGT TCG CTC AGC CTC GGG GAG GGG GTG GAG TTC GAT
     Pro Arg Ala Leu Ser Phe Val Leu Ser Ser Leu Gln Leu Gly Glu Gly Val Glu Phe Asp
     445                                                                      (168)
     GTG CTG CCT GCC TTT GAT GCC CTG GGT CAG TTG ACT GGC AGC TAT AAA CCT AAC CCC CAA
     Val Leu Pro Ala Phe Asp Ala Leu Gly Gln Leu Thr Gly Ser Tyr Lys Pro Asn Pro Gln
     505                                                                      (188)
     ATC TAT GTC AAG CTC ATC GAG GAG TGC ACC GAC CTG CAG AAA GAG GGC GAG TTC TCC ACC
     Ile Tyr Val Lys Leu Ile Glu Glu Cys Thr Asp Leu Gln Lys Glu Gly Glu Phe Ser Thr
     565                                                                      (208)
     TGC TTC ACA GAA CTA CAG AGA GAC TTC CTG AAG CAG CGC CCC ACC AAG CTC AAG AGC CTC
     Cys Phe Thr Glu Leu Gln Arg Asp Phe Leu Lys Gln Arg Pro Thr Lys Leu Lys Ser Leu
     625                                                                      (228)
     ATC CGC CTA GTC AAG CAC TGG TAC CAA AAT TGT AAG AAG AAG CTT GGG AAG CTG CCA CCT
     Ile Arg Leu Val Lys His Trp Tyr Gln Asn Cys Lys Lys Lys Leu Gly Lys Leu Pro Pro
     685                                                                      (248)
                 Sac I
     CAG TAT GCC CTG [GAG CTC] CTG ACG GTC TAT GCT TGG GAG CGA GGG AGC ATG AAA ACA CAT
     Gln Tyr Ala Leu Glu Leu Leu Thr Val Tyr Ala Trp Glu Arg Gly Ser Met Lys Thr His
     745                                                                      (268)
     TTC AAC ACA GCC CAG GGA TTT CGG ACG GTC TTG GAA TTA GTC ATA AAC TAC CAG CAA CTC
     Phe Asn Thr Ala Gln Gly Phe Arg Thr Val Leu Glu Leu Val Ile Asn Tyr Gln Gln Leu
     805                                                                      (288)
                                         Aha II
     TGC ATC TAC TGG ACA AAG TAT TAT GAC [TTT|AAA] AAC CCC ATT ATT GAA AAG TAC CTG AGA
     Cys Ile Tyr Trp Thr Lys Tyr Tyr Asp Phe Lys Asn Pro Ile Ile Glu Lys Tyr Leu Arg
     865                                    <5>                                (308)
     AGG CAG CTC ACG AAA CCC ACG CCT GTG ATC CTG GAC CCG GCG GAC CCT ACA GGA AAC TTG
     Arg Gln Leu Thr Lys Pro Thr Pro Val Ile Leu Asp Pro Ala Asp Pro Thr Gly Asn Leu
     925                                              <10>                    (328)
     GGT GGT GGA GAC CCA AAG CGT TGG AGG CAG CTG GCA CAA GAG GCT GAG GCC TGG CTG AAT
     Gly Gly Gly Asp Pro Lys Arg Trp Arg Gln Leu Ala Gln Glu Ala Glu Ala Trp Leu Asn
     985                                                         Div (348)
                                            Sac I
     TAC CCA TGC TTT AAG AAT TGG GAT GGG TCC CCA [GTG AGC TCC] TGG ATT CTG CTG GTG AGA
     Tyr Pro Cys Phe Lys Asn Trp Asp Gly Ser Pro Val Ser Ser Trp Ile Leu Leu Val Arg
     1045                                  <5>   (364)
     CCT CCT GCT TCC TCC CTG CCA TTC ATC CCT GCC CCT CTC CAT GAA GCT TGA GAC ATA TAG
     Pro Pro Ala Ser Ser Leu Pro Phe Ile Pro Ala Pro Leu His Glu Ala
     1105
     CTG GAG ACC ATT CTT TCC AAA GAA CTT ACC TCT TGC CAA AGG CCA TTT ATA TTC ATA TAG
     1165
     TGA CAG GCT GTG CTC CAT ATT TTA CAG TCA TTT TGG TCA CAA TCG AGG GTT TCT GGA ATT
     1225                       Eco RI
     TTC ACA TCC CTT GTC CA[G AAT TC]A TTC CCC TAA GAG TAA TAA TAA ATA ATC TCT AAC ACC
```

FIGURE 1. Nucleotide sequence of human 2-5A synthetase cDNA (adapted from
10)

The number on the left refer to the nucleotide number starting with the adenosine of the initiating ATG. The numbers in brackets on the right refer to
the amino acid number starting with the initiating methionine. It is not yet
known which one of the first two AUG is used *in vivo* (15,11). Rectangular
boxes indicate the restriction sites used to linearize the SP64 2-5A, expression plasmid (10) before transcription with SP6 RNA polymerase. Lozenge
shaped boxes indicate the approximate location of deletion obtained after
treating the EcoRI digested SP64 2-5A for 5,10 and 15 min. with nuclease
Bal 31.

the 42 kDa protein is fully active as a monomer (10).

Here we show that, from the construction described, it is possible to produce truncated 2-5A synthetase molecules in order to study the structural requirements for their double stranded RNA binding and enzymatic activity.

RESULTS AND DISCUSSION

In order to evaluate the effect of deleting some regions of the 42 kDa 2-5A synthetase, we first translated the corresponding synthetic, capped non polyadenylated mRNA (transcribed *in vitro* from the pSP64 2-5A DNA (10) in a reticulocyte lysate cell-free system. In the presence of ^{35}S-methionine the amount of 42 kDa protein could be estimated. Since the enzymatic activity could be also easily detected we could calculate the specific activity of the *in vitro* made 42 kDa 2-5A synthetase. This point is important since we observed that several deletions of the 3' end of the cDNA reduced considerably the efficiency of translation of the corresponding synthetic mRNAs, probably as a result of the complete deletion of the 3'-untranslated region.

Figure 1 indicates the location of the few convenient 3' restriction sites used to linearize the pSP64 2-5A DNA. In order to obtain more precise deletions, we also trimmed the same DNA from the EcoRI site, with the nuclease Bal 31 for different periods of time. The location of the resulting cleavages were calculated by comparison with those obtained with restriction enzymes (Fig. 2) and are indicated on the figure (Fig. 1). The effect of various deletions is demonstrated at the DNA and RNA level in figure 2.

Figure 3 shows the size of the proteins translated from the various truncated 42 kDa 2-5A synthetase mRNA's. In each case the size of the *in vitro* translated protein has been reduced as predicted from the location of the deletion sites indicated on Fig. 1, yielding a family of proteins of 29 kDa (Sac I), 33 kDa (Aha III) and 33 kDa, 36 kDa, 39 kDa for the Bal 31 digested DNA template. In addition, all the truncated mRNA's yield additional proteins of higher molecular weight. This phenomenon is presently unexplained. However, after binding of those ^{35}S-labeled translation products to poly(IC)-agarose, only one peptide corresponding to the size indicated above is retained in each lane (Fig. 3B) and we could evaluate, by densitometric scanning, that despite their reduced translation, the truncated proteins still bind poly(IC)-agarose with the same efficiency, excepted for the Sac I deletion where dsRNA binding is decreased to about 50 %. Figure 3C shows the 2-5A synthetase activity of the various truncated proteins. Deleting 30 amino acids, will reduce the specific activity of the enzyme to 50 % and any deletion, from 60 to 130 amino acids, will abolish completely the 2-5A synthetase activity. Figure 4 summarizes the data obtained so far.

It is interesting to note the strong dissociation between requirement for dsRNA binding and for 2-5A synthetase activity. Further studies and more precise deletions may help to define more accurately the limits of proteins domain(s) involved in dsRNA binding and 2-5A synthetase activity. It is also interesting that the "divergence site" between the 42 kDa and 46 kDa

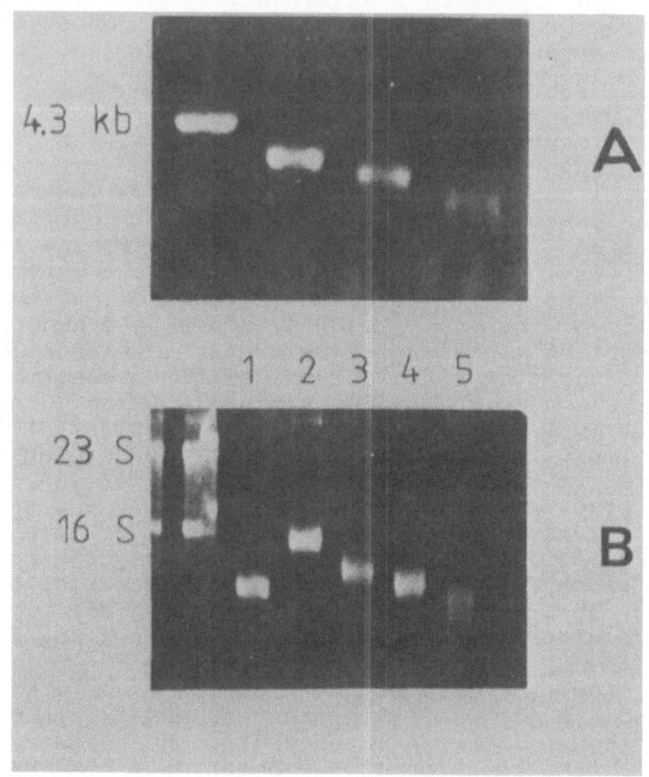

FIGURE 2. A. Agarose gel electrophoresis of SP64 2-5A DNA and Bal 31 deleted derivatives
Lane 1 : EcoRI linearized SP64 2-5A; lane 2, 3, 4: 12 μg of SP64 2-5A DNA were treated for 5, 10 and 15 min. respectively with 6 units of Bal31 nuclease (Boehringer) in 480 μl at 30°. 300 ng aliquots were analyzed by agarose gel electrophoresis and the molecular weight of the Bal31 digested DNA's was evaluated by comparison with molecular weight markers.
        B. Agarose gel electrophoresis of RNA transcripts obtained from SP64 2-5A DNA and its BAL31-deleted derivatives.
SP64 2-5A DNA was digested with EcoRI (lanes 2-5) or with AhaIII (lane 1) and then treated for 5 min (lane 3), 10 min (lane 4) or 15 min (lane 5) with nuclease Bal31 (see panel A). After phenol and ether extractions and ethanol precipitation, the various templates were transcribed with SP6 RNA polymerase in the presence of 5' cap precursor as described (10). 10% aliquots were analyzed by electrophoresis on 1% agarose gel and their molecular weight evaluated by comparison with that of standard rRNA markers (23S = 3.56 kb ; 16S = 1.77 kb).

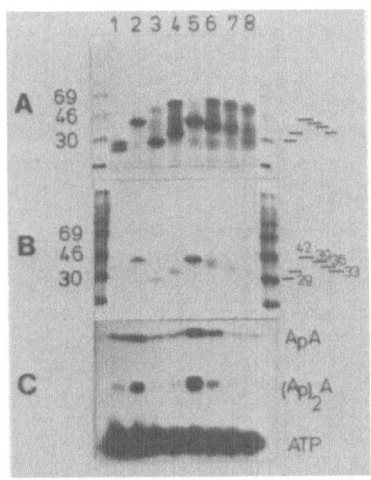

FIGURE 3. Cell-free translation of truncated 42kDa 2-5A synthetase(A), analysis of their binding to double stranded RNA(B) and 2-5A synthetase activity (C). A. *Polyacrylamide gel electrophoresis of the* [35]S *methionine labeled cell-free translation products of various synthetic 2-5A synthetase mRNA's.* Preparation of the DNA templates was as described in legend to Fig. 2. Cell-free translation, polyacrylamide gel electrophoresis and fluorography were as described in (10). Lane 1: endogenous activity of the reticulocyte lysate (no mRNA addded); lane 2: synthetic, full-size 42kDa mRNA; lane 3: synthetic 2-5A synthetase mRNA from SacI linearized SP64 2-5A DNA; lane 4: synthetic 2-5A synthetase mRNA from AhaIII linearized SP64 2-5A DNA; lane 5: non deleted synthetic 2-5A synthetase transcribed from EcoRI linearized SP64 2-5A DNA (see Fig. 2A, lane 1 & Fig. 2B, lane 2); lanes 6, 7, 8: EcoRI linearized SP64 2-5A DNA has been degraded for 5 (lane 6), 10 (lane 7) and 15 min (lane 8) with Bal31, and synthetic truncated 2-5A synthetase mRNA's prepared as described in Figs. 1 and 2. Number on the left side indicate the molecular weight in kDa, of the protein markers used to calibrate the gel.
B. *Poly(IC) agarose binding of the various forms of truncated 2-5A synthetases*. 50 µl of the reticulocyte lysate were bound to 200 µl of 50% poly(IC) agarose beads suspension as described in (10). 40% aliquots of the resins were boiled in electrophoresis sample buffer and analysed by polyacrylamide gel electrophoresis as in panel A, above. The sketch on the right side of panel B (and panel A) indicates the estimated molecular weight in kDa of the intact and truncated 2-5A synthetases observed in lanes 3 to 8.
C. *Enzymatic activity of various forms of truncated 2-5A synthetases*. 2-5A synthetase activity was measured on 50% aliquots of the Poly(IC) agarose resin samples (see panel B) by incubation in the presence of $(\alpha-^{32}P)ATP$, followed by calf intestine alkaline phosphatase treatment, polyethylene imine cellulose thin layer chromatography in 1M acetic acid, and autoradiography as described (10). The 8 lanes of the chromatogram have been aligned to those of the polyacrylamide gel electrophoresis fluorogram presented in panels A and B and correspond to the samples described under panel A.
ApA = dimers ; $(Ap)_2A$ = trimers.

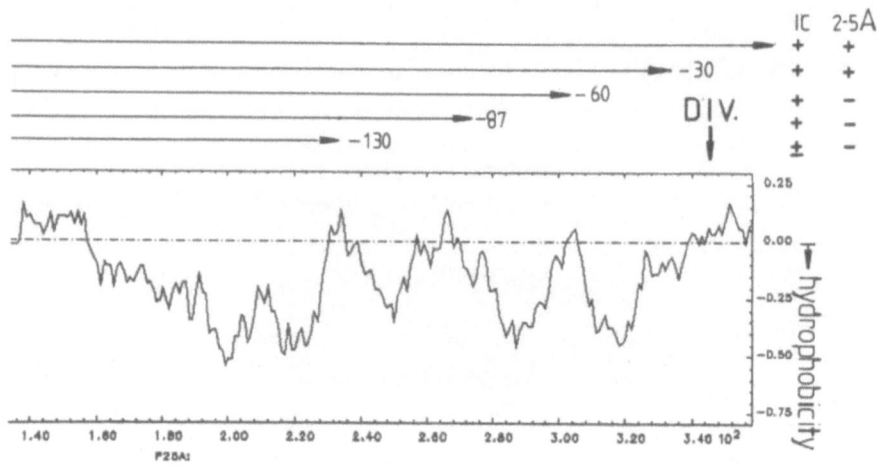

FIGURE 4. Summary of the results obtained on truncated 42kDa derived 2-5A synthetases.

The graph represents a computer-derived hydrophobicity plot for 2/3 of the 42kDa 2-5A synthetase amino acid sequence. The program used,progressively evaluates hydrophobicity of a span of 14 amino acids according to Eisenberg et al. (13). Abcissa values indicate the amino acid residue numbers. "DIV" indicates the location of leucine 346 : downstream this residue divergence occurs in the 46kDa 2-5A synthetase codes by the 1.8 kb mRNA (11, 14) (see also Fig. 1). "IC" symbolizes the dsRNA binding of the various enzymes. "2-5A" symbolizes the 2-5A synthetase activity of the various enzymes.

synthetase (6, 7, 11) lies within an hydrophylic region that may not be required for expressing the enzymatic activity. It has been suggested that carboxy-terminal variations of the protein in this region may affect the cellular distribution of the enzyme (11). Convertedly, our preliminary data suggest that at least a region located between residues 300 and 334 of the 42kDa 2-5A synthetase is essential for the enzymatic activity. This region corresponding to the 5' end of exon 7 (11) is also overlapping with the target of an antibody prepared against a synthetic peptide (284-303) which recognizes at least four different 2-5A synthetases in Western blot analysis (Chebath et al. 1986, personal communication) suggesting again that those four enzymes may share a common carboxy-terminal domain. Furthermore, variations in more proximal domains of the protein upstream residue 274 may alter the dsRNA binding properties of the various 2-5A synthetases. This is perhaps compatible with the recent observation by Ilson et al. (12) that a 33 kDa human 2-5A synthetase requires 1000 fold more dsRNA for its activation then a 110 kDal human 2-5A synthetase purified from the same HeLa cells.

ACKNOWLEDGEMENTS

This work was supported by the Fund for Joint Basic Research (Belgium) n° 2.9006.79 to J.C., by a grant from the Fund for Medical Scientific Research (Belgium) n° 34.560.85 to J.C. and G.H. and by ULB-Actions de Recherches Concertées of the Belgian Government to G.H. G.H. is Research Associate and M.W. is Research Assistant of the National Fund for Scientific Research (Belgium)

REFERENCES

1. Johnston MI and Torrence PF : in Interferon 3, Mechanisms of production and action. Friedman, RM (Ed) pp. 189-298, Elsevier North Holland, Amsterdam, New York, 1984.
2. Silverman RH : in Interferon 3, Mechanisms of production and action. Friedman, RM (Ed). pp. 177-188, Elsevier North Holland, Amsterdam, New York, 1984.
3. Zullo JN, Cochran BH, Huang AS and Stiles CD : Cell, 43, pp. 793-800, 1985
4. Nilsen TW, Wood DL and Baglioni C. J. Biol. Chem., 257, pp. 1602-1605, 1982.
5. St-Laurent G, Yoshie O, Floyd-Smith G, Samanta H, Sehgal PB and Lengyel P. Cell, 33, pp. 95-102, 1983.
6. Benech P, Merlin G, Revel M and Chebath,J. J. Nucleic Ac. Res., 13, pp 1267-1281 (1985).
7. Saunders ME, Gewert DR, Tugwell ME, Mc Mahon M and Williams BRG. EMBO J.,4, pp. 1761-1768, 1985.
8. Yang K,Samanta H, Dougherty J, Javaram B, Broeze R and Lengyel P. J. Biol. Chem., 256, pp. 9324-9328, 1981.

9. Wells JA, Swyryd E and Stark GB.   J. Biol. Chem., 259, pp. 1363-1370, 1984.
10. Wathelet M, Moutschen S, Cravador A, DeWit L, Defilippi P, Huez G. and Content J.   FEBS Lett.,196, pp. 113-120, 1986.
11. Benech P, Mory Y, Revel M and Chebath J.   EMBO J., 4, p. 2249, 1985.
12. Ilson D, Torrence PF and Vilcek J.   J. Interferon Res., 6, pp. 5-12, 1986
13. Eisenberg D, Weis RM, Tenmiller TCS and Wilcox W.   Faraday Symp. Chem. Soc., 17, pp. 109-120, 1982.
14. Saunders M, Gewert D, Castelino M, Rutherford M, Flenniken A, Willard H, Williams BRG. In :"The 2-5A System", Vol. 202 Williams BRC and Silverman RH, eds. Alan R. Liss, New York, 163-174 (1985).
15. Shiojiri S, Fukunaga R, Ichii Y and Sokawa Y.   J. Biochem., 99, pp. 1455-1464, 1986.

THE DOUBLE-STRANDED RNA DEPENDENT PROTEIN KINASE FROM HUMAN CELLS

A.G. HOVANESSIAN and J. GALABRU

Unité d'Oncologie Virale, Institut Pasteur
25 rue du Dr. Roux
75724 Paris Cedex 15, France

INTRODUCTION

The protein kinase activity dependent on double-stranded (ds)RNA is enhanced in mouse and human cells following treatment with interferon. This kinase activity is manifested by the phosphorylation of (a) a 65,000-67,000-Mr protein (p65) in mouse cells or a 68,000-72,000-Mr protein (p68) in human cells; (b) a 35,000-Mr protein which is the α-subunit of protein synthesis initiation factor eIF2; and (c) added calf thymus histones (HIIA). This kinase activity is independent of cyclic AMP or cyclic GMP and is markedly stimulated by Mn^{2+}. It phosphorylates p65 and p68 by their serine and threonine residues. The role of such protein kinase activity is considered to be the phosphorylation of the α-subunit of eIF2, thus mediating inhibition of the initiation of protein synthesis in cell-free system (for references see 1). The phosphorylation of p65 and p68 has been shown during virus infection in mouse and human cells, respectively (2-4). The enhanced phosphorylation of eIF2 has also been reported in interferon-treated cells infected with reovirus or encephalomyocarditis virus (4,5). Recent data in cells infected with adenovirus suggests that the dsRNA-dependent protein kinase plays a decisive role in the control of cellular and viral protein synthesis (6).

Previously, several attempts for purification of the dsRNA-dependent protein kinase and its precise characterization have been handicaped by the poor recovery of the kinase from interferon-treated cells. This has been due in part to degradation of the kinase and/or its inactivation. With the use of a specific monoclonal antibody against the dsRNA-dependent protein kinase (7), here we discuss the properties of this kinase. The dsRNA-dependent protein kinase from human cells will be referred to as p68 kinase.

RESULTS AND DISCUSSION

1. Monoclonal antibody specific for p68 kinase

Preparation of the monoclonal antibody specific for p68 kinase was described before (7). This antibody can identify p68 kinase by electrophoretic transfer blot analysis. The monoclonal antibody coupled to Sepharose (Mab-Sepharose) was used as an immunoadsorbant to purify p68 kinase and characterize its activity (8). The p68 kinase bound to Mab-Sepharose is active, i.e., incubation with γ-^{32}P-ATP and

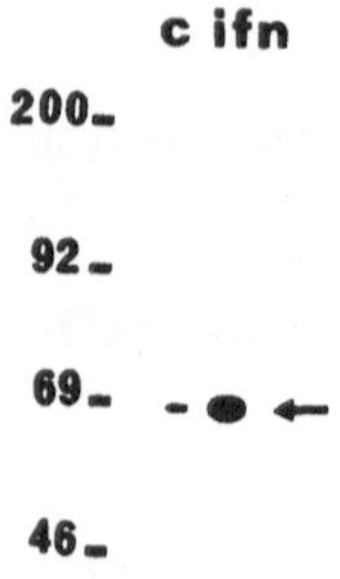

c ifn

200_

92_

69_ - ● ←

46_

Figure-1. The p68 kinase from control and interferon-treated Daudi cells. (^{35}S)methionine labeled extracts were purified on the monoclonal antibody column and analyzed by polyacrylamide gel electrophoresis. An autoradiograph is shown.

dsRNA results in the phosphorylation of p68 kinase as well as phosphorylation of exogenous substrates eIF2 or histone. It should be noted that here we are dealing with 2 step-phosphorylation reaction: the first step is required for the phosphorylation of p68 kinase whereas the second step is involved in the phosphorylation of exogenous substrates. In these experiments we used calf thymus histone as an exogenous substrate.

2. One single protein is purified on the immuno-affinity column

Previously, we have sugested that the dsRNA-dependent protein kinase is a complex of two interferon-induced proteins: a Mr 48,000 protein (p48) which appeared to be responsible for the phosphorylation of a Mr 68,000 protein (p68) in the presence of dsRNA. However, it has now become evident that our previous results were misinterpreted due to partial degradation of p68 kinase during preparation of cell extracts and in the course of its purification. As a matter of fact during isolation of the protein kinase on the immuno-affinity column, we came across to a powerful proteolytic activity which gets co-purified with the protein kinase and degrades it when this latter is activated by cations and polyanions. This proteolytic activity is inhibited by PMSF. For these reasons, all buffers used for extraction of the kinase and its purification were supplemented with EDTA and PMSF. In addition to these, the electrophoresis buffer was supplemented with PMSF since degradation of P68 might also occur during polyacrylamide gel electrophoresis. Under these latter experimental conditions, one single protein is found to be associated with Mab-Sepharose. Figure 1 shows the purified (^{35}S)methionine labeled p68 kinase from control and interferon-treated Daudi cells.

The p68 kinase is already present in control cells and it is enhanced in cells treated with interferon.The fact that p68 is the only protein detectable in the purified protein kinase preparation, then its phosphorylation should be the product of an autophosphorylation reaction.

3. Activation step required for autophosphorylation of the p68 kinase

Heparin at 5-10 units/ml can replace the requirement of dsRNA in the activation step necessary for autophosphorylation of p68 kinase. The common characteristic between these 2 activators is their polyanionic nature. Accordingly other polyanions such as dextran sulfate and poly-L(Glutamine) also activate the kinase. Heparin is a convenient activator since it could be inactivated by antithrombin III, a protein with high affinity for heparin. We used this property in order to find out whether heparin causes a stable change or a reversible change in the activity state of the protein kinase. The results show that phosphorylation of p68 kinase depends on the presence of the activator throughout the reaction, i.e., once the activator is blocked then p68 kinase can no longer be auto-phosphorylated. Thus, activation of p68 kinase is reversible.

4. Correlation between the degree of autophosphorylation of p68 kinase and kinase activity on exogenous substrate, histone

The p68 kinase has at least 10 phosphate binding sites. It is possible therefore to obtain p68 kinase preparations which have been saturated at different degrees with phosphate. A convenient procedure for the preparation of such samples is obtained by incubation of Mab-Sepharose bound p68 kinase with dsRNA at different concentrations of ATP (8). These samples could then be washed and assayed for phosphorylation of histone. The results indicated that pmol of PO_4 incorporated per mg of histone was highly correlated with pmol of PO_4 per p68 kinase sample (8). Therefore, histone kinase activity was increased with higher amounts of phosphate incorporated into p68 kinase. Further evidence that phosphorylated p68 kinase has protein kinase activity on an exogenous substrate was provided by experiments in which phosphate saturated p68 kinase was dephosphorylated by bacterial alkaline phosphatase (8). Dephosphorylation of p68 kinase resulted in a decreased level of protein kinase activity capable of phosphorylating histone.

5. Characterization of p68 kinase

The p68 kinase is characterized by two distinct protein kinase activities which involves at first an autophosphorylation reaction and secondly a protein kinase activity on exogenous substrates, such as eIF2 and histone (see Figure 2). Autophosphorylation of p68 kinase can occur when activated by polyanions (dsRNA or heparin) and cations, Mn^{2+} and Mg^{2+}. Mn^{2+} is more efficient than Mg^{2+} for the activity of the purified protein kinase. Phosphorylated p68 kinase is capable of catalyzing phosphorylation of the exogenous substrate, a reaction

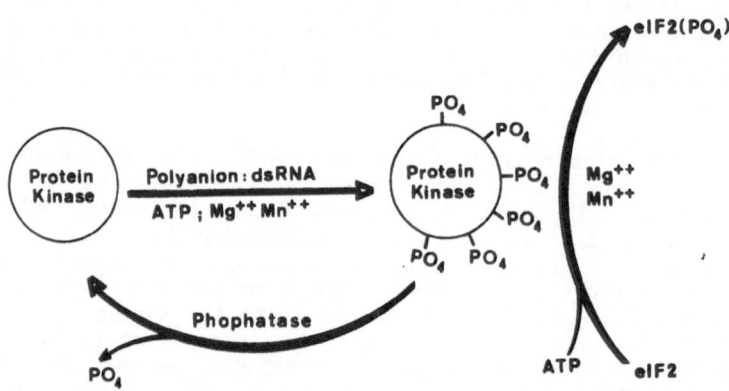

Figure-2.   The dsRNA-dependent protein kinase is defined by 2
distinct protein kinase activities: (a) the p68 kinase should be
activated in order to be autophosphorylated; (b) phosphorylated p68
kinase then catalyzes the phosphorylation of exogenous substrate, eIF2
and histone.

which is dependent on the phosphate content of p68 kinase.
Accordingly, dephosphorylation of p68 kinase results in a decreased
level of kinase activity on exogenous substrates, eIF2 and histone.

The two protein kinase activities of p68 can be distinguished by
the following criteria:

1) Phosphorylation of p68 is dependent on dsRNA or heparin whereas
phosphorylation of histone is independent of these activators;
2) Both protein kinase activities are stimulated by Mn^{2+},
autophosphorylation reaction however requires 10 fold more Mn^{2+} (1-2
mM) than phosphorylation of the exogenous substrate (0.1-0.2 mM);
3) Binding of ATP to p68 is dependent on dsRNA in contrast to
phosphorylated p68. Binding of ATP to phosphorylated p68 is dependent
on the phosphate-content of p68;
4) Incubation of the native (not-phosphorylated) and phosphorylated
p68 preparation in buffer containing 25% ethylene glycol results in a
complete inhibition of the autophosphorylation process whereas
phosphorylation of histone is not affected (unpublished data).
The last observation suggests that there might be two catalytic
sites in p68 kinase, one for autophosphorylation and the other for
phosphorylation of the exogenous substrate. The first catalytic unit
becomes inactivated by ethylene glycol while the second catalytic unit
is not affected. This second catalytic unit becomes activated when p68
is saturated with phosphate.

6. Assay of p68 kinase in peripheral blood mononuclear cells

The monoclonal antibody specific of p68 kinase is of great use for the precise characterization of the dsRNA-dependent protein kinase. Besides its use for the purification of p68 kinase to homogeneity, the monoclonal antibody can be used to assay the protein kinase activity since the immuno-affinity bound protein kinase is active. This technique is extremely efficient for the detection of low amounts of protein kinase activity. Its sensitivity is several fold superior compared with the assay of the protein kinase after partial purification on poly(I).poly(C)-Sepharose. For example, it has been impossible to detect the protein kinase activity in human peripheral blood mononuclear (PBM) cells by the poly(I).poly(C)-Sepharose method. On the other hand, recently we were able to show presence of p68 kinase in PBM cells after purification of cell extracts on Mab-Sepharose. In healthy individuals, the activity of the protein kinase remains constant whereas it is enhanced significantly in patients with viral infections (9). Thus the activity of p68 kinase in PBM cells can be used as a convenient marker to investigate the production of interferon during viral infections and monitor the response to interferon therapy.

REFERENCES

1. Galabru J, Hovanessian AG (1984). Interferon: Action antivirale et effets biologiques. Bull. Inst. Pasteur 82: 283.
2. Gupta SL, Holmes SL, Mehra LL (1982). Interferon action against reovirus activation of interferon induced protein kinase in mouse L-cells upon reovirus infection. Virology 120: 495.
3. Nilsen TW, Maroney PA, Baglioni C (1982). Inhibition of protein synthesis in reovirus-infected HeLa cells with elevated levels of interferon induced protein kinase activity. J. Biol. Chem. 257: 14593.
4. Samuel CE, Duncan R, Knutson GS, Hershey JWB (1984). Mechanism of interferon action. Increased phosphorylation of protein synthesis initiation factor eIF2 alpha in interferon-treated, reovirus-infected mouse L929 fibroblasts in vitro and in vivo. J. Biol. Chem. 259: 13451.
5. Rice AP, Duncan R, Hershey JWB, Kerr IM (1985). The dsRNA dependent protein kinase and 2-5A system are both activated in interferon-treated encephalomyocarditis virus-infected HeLa cells. J. Virol. 54: 894.
6. Reichel PA, Merrick WC, Siekierka J., Mathews MB (1985). Regulation of a protein synthesis initiation factor by adenovirus virus-associated RNAI. Nature 313: 196.
7. Laurent AG, Krust B, Galabru J, Svab J, Hovanessian AG (1985). Monoclonal antibodies to interferon induced 68,000-Mr protein and their use for the detection of dsRNA dependent protein kinase in human cells. Proc. Natl. Acad. Sci. 82: 4341.
8. Galabru J, Hovanessian AG (1985). Two interferon-induced proteins are involved in the protein kinase complex dependent on double-stranded RNA. Cell 43: 685.
9. Buffet-Janvresse C, Vannier JP, Laurent AG, Robert N, Hovanessian AG (1986). Enhanced level of double-stranded RNA-dependent protein kinase in peripheral blood mononuclear cells of patients with viral infections. J. Interferon Res. 6: 85.

HOMOLOGS TO MOUSE Mx PROTEIN INDUCED BY INTERFERON IN VARIOUS SPECIES

CH. MORTIER & O. HALLER

Institute for Immunology and Virology, University of Zürich, 8028 Zürich, Switzerland

SUMMARY

Polyclonal and monoclonal antibodies with specificity for the interferon-alpha/beta induced nuclear Mx protein of the mouse are known to crossreact with similarly induced proteins of man and rat. We demonstrate here crossreactivity to interferon-induced hamster and goat proteins and show that these proteins accumulate in the cytoplasm. No reactivity was observed with proteins from cat, chicken, guinea-pig, monkey, pig, and rabbit indicating either lack of Mx homologs or lack of crossreactive antigenic determinants in these species.

INTRODUCTION

Interferon (IFN) induces in mouse cells the synthesis of a number of proteins, some of which are thought to contribute to the antiviral state. One of these is the Mx protein, a 75'000 dalton protein induced by IFN-alpha/beta in cells from influenza virus resistant mice known to carry the Mx^+ allele at the Mx locus on chromosome 16 (1,2,3). The Mx protein accumulates in the cell nucleus (4) and is involved in inhibition of influenza virus replication presumably by affecting influenza viral mRNA synthesis (5) and/or translation (6). Many inbred strains of mice carry Mx^- alleles derived from the wild type Mx^+ allele by deletions (7). Cells from Mx^- mice are unable to synthesize the Mx protein and do not develop an adequate antiviral state towards influenza virus in response to IFN (8). However, when Mx^- cells are transfected with MxcDNA, constitutive expression of Mx protein (under control of the SV40 early promoter) correlates with resistance of individual cells to influenza virus even in the absence of IFN treatment (7). This indicates that the Mx protein alone, without help from other IFN-induced proteins, is inhibitory to influenza virus.

Antibodies to the Mx protein have been produced in congenic mice: when immunized with extracts from IFN-treated Mx^+ cells (containing the Mx protein), congenic Mx^- animals make polyclonal antibodies directed exclusively at the Mx protein. Using this protocol, monoclonal antibodies have also been produced (9).

Immunofluorescence staining with these antibodies allows the detection of the Mx protein in IFN-treated cells (4) as well as in organ sections of IFN-treated or virus-infected Mx^+ mice (10 and unpublished observations). We therefore used this technique to look for crossreactive proteins in cells from other species and subsequently found IFN-induced homologs to mouse Mx protein in man (11) and rat (12). We have now extended this study to include cells from 8 additional species.

MATERIALS AND METHODS

Cells: The following laboratory cell lines were used: BHK-21 (Syrian hamster kidney cells), CHO-K1 (Chinese hamster ovary cells), HKCC-1 (hamster kidney cells (13)), MDBK (bovine kidney cells), PK(15) (pig (sus scrofa) kidney cells), MPK (minipig kidney cells) obtained from the American Type 'Culture Collection, CV-1 (African green monkey kidney cells), BGM (African green monkey kidney cells) and COS-1 (SV40 transformed monkey cells). Human fetal lung cells were a gift from Dr. Peter Groscurth, Anatomisches Institut der Universität Zürich. Feline embryonic lung cells were purchased from Flow Laboratories. Primary cultures were established in our laboratory from the following organs as previously described (2): calf kidney, chicken embryo, goat kidney, guinea-pig kidney, minipig kidney, mouse embryo, rabbit kidney and rat embryo.

Interferons: Hu IFN-alpha 2 produced in E.coli and partially purified (10^5/U/mg protein) was a gift from Dr. Ch. Weissmann, Zürich. This preparation was active on bovine, cat, goat, guinea-pig, monkey, pig and rabbit cells (14). Its antiviral activity on these heterologous cells was assessed in a standard CPE inhibition assay using VSV as challenge virus. Chick IFN (15) was a gift from Dr. C. Jungwirth, Würzburg and rat IFN (16) was kindly provided by Dr. H. Schellekens, Rijswijk. Hamster IFN was induced by i.v. inoculation of Syrian hamsters with 10^8 EID50 of Newcastle disease virus (NDV). Sera collected 18 h later were pooled, treated for 72 h at ph2 and titrated on HKCC - 1 cells for antiviral activity against VSV. Sera taken from the same animals but before NDV treatment served as control serum devoid of IFN-activity.

Anti-Mx antibodies: Polyclonal antibodies and monoclonals 2C12, 5D11, and 6D4 were produced by immunizing BALB/c (Mx^-) mice with cell extracts from IFN-alpha/beta treated congenic BALB.A26-Mx (Mx^+) cells (9).

Immunofluorescence and immunoprecipitations were performed as previously described (4,9).

RESULTS

Anti-Mx antibodies detect IFN-induced proteins in hamster and goat cells. Immunofluorescence staining with monoclonal antibody 2C12 revealed the accumulation of an IFN-induced

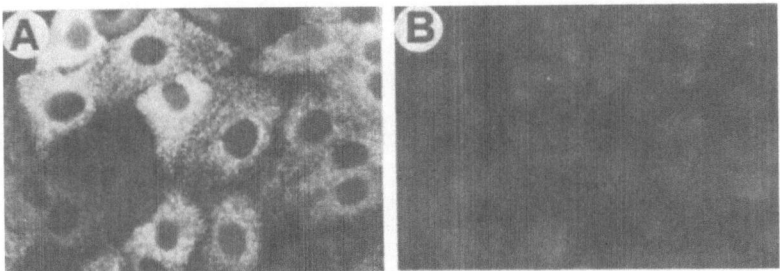

Fig. 1: Mx homologous protein in IFN-treated hamster cells. HKCC-1 cells were treated for 18 h with
40 U/ml of homologous IFN (A) or were left untreated (B). The cells were then fixed and stained
with monoclonal antibody 2C12 by indirect immunofluorescence.

protein in cells of the hamster kidney cell line HKCC-1.
Staining was most brilliant in the cytoplasm whereas the
nucleus was spared (Fig. 1A). No specific staining was observed in control cells not exposed to IFN (Fig. 1B). Polyclonal antibodies gave the same picture. In contrast, monoclonal
antibodies 5D11 and 6D4 showed no crossreactivity (not shown).
The same results were obtained with BHK-21 and CHO-K1 cells
(not shown).
   Next, primary cultures of goat kidney cells were analyzed.
Again, monoclonal antibody 2C12 and a polyclonal antiserum
stained IFN-induced protein(s) in the cytoplasm of goat cells
(Fig. 2), whereas monoclonal antibodies 5D11 and GD4 were
negative (not shown).

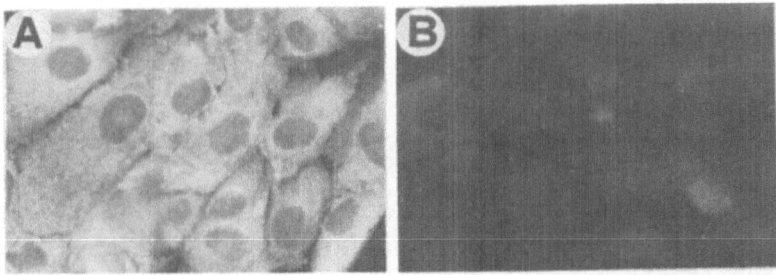

Fig. 2: Mx homologous protein in IFN-treated goat
cells. Primary goat kidney cell cultures were established. The cells were then treated for 18 h
with 1000 U/ml of human IFN-alpha 2 (A) or were
left untreated (B). Fixed cells were stained with a
mouse polyclonal anti-Mx antibody by indirect immunofluorescence.

Immunoprecipitation of IFN-induced proteins. Figure 3 shows that polyclonal or monoclonal antibodies directed against the murine Mx protein were capable of immunoprecipitating IFN-induced proteins of goat and hamster cells. These proteins were not detectable in extracts of untreated cells (lines 1 and 4). Their synthesis was clearly induced by treatment of cells with type I IFN (lines 2, 3 and 5) and their size was in the range of 70-75 kilodalton as expected for Mx homologous proteins (2, 11, 12).

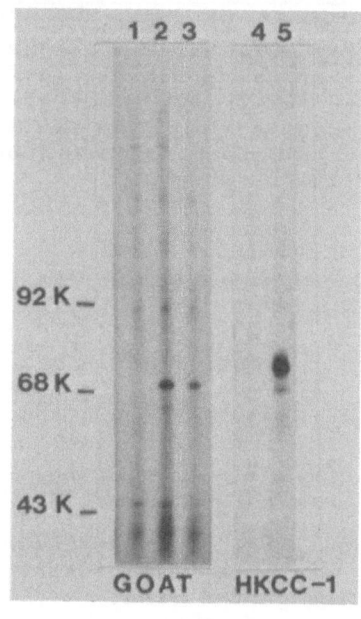

Fig. 3: Immunoprecipitation of IFN-inducible proteins by anti-Mx antibodies. Goat kidney cells (left panel) or hamster HKCC-1 cells (right panel) were treated for 5 h with either control culture medium (1,4), 1000 U/ml of human IFN-alpha 2 (2,3) or 40 U/ml of hamster IFN-alpha/beta (5). Cells were labeled with ^{35}S-methionine, cell extracts were prepared and aliquots were assayed for immunoprecipitable proteins with polyclonal antibodies (1,2) or monoclonal antibody 2C12 (3,4,5).

Screening cells from various species for Mx homologous proteins. Polyclonal and monoclonal antibodies were used to screen cells from various species for the presence of IFN-induced crossreactive proteins in immunofluorescence assays (4). Cells showing positive immunofluorescence are listed in Table 1. Southern blots of genomic DNA prepared from these species gave clear hybridization signals with mouse Mx cDNA probes (Grob, Meier and Haller, unpublished data) indicating the presence of corresponding nucleic acid sequences. Likewise, positive Southern blot hybridizations with Mx-specific probes were obtained with DNAs from bovine, chicken, and pig cells (not shown). However, the present polyclonal and monoclonal antibodies did not detect crossreactive IFN-induced proteins in MDBK cells, calf kidney cells, chicken embryo cells, pig PK(15) cells, minipig kidney cells or minipig MPK cells, nor in BGM cells, COS-1 cells, CV-1 cells, feline embryonic lung cells, guinea-pig kidney cells and rabbit kidney cells. This lack of reactivity most likely reflects the restricted specificity of the antibodies used rather than an absence of Mx homologs in these species.

TABLE 1

Species positive for IFN-induced proteins crossreactive with monoclonal antibody 2C12

Species	Cells analyzed	Localization of protein
goat	kidney cells	cytoplasm
hamster	HKCC-1	cytoplasm
	CHO-K1	cytoplasm
	BHK-21	cytoplasm
human	fetal lung cells	cytoplasm
mouse	embryo cells	nucleus
rat	embryo cells	cytoplasm and nucleus

DISCUSSION

The present results indicate that proteins with homology to the murine Mx protein are common in several species. In all instances analyzed, synthesis of these proteins is stringently regulated by type I IFN. The degree of sequence homology between these related proteins is presently unknown. However, the 2C12 crossreactive proteins described in Table 1 have at least one antigenic determinant in common. The specificity of the anti-Mx antibodies used here is by necessity restricted to those antigenic determinants of the mouse protein which are immunogenic in mice. Their crossreactivity to homologous proteins of other species may therefore be unduly limited. Using mouse Mx cDNA probes in Southern blot analyses of genomic DNA we have recently found Mx related sequences in a wide variety of species (Grob, Meier and Haller, unpublished). The possibility exists that immunizations across the species barrier or with synthetic peptides would yield antibodies of broader specificity capable of detecting Mx homologs in additional species.

Despite common structural features and apparently similar control of synthesis, the crossreactive proteins described here seem to accumulate in different cellular compartments. The significance of this finding is not clear at present. The functions of the various crossreactive proteins remain to be established. In any case, the Mx system would seem to be present in many species and may have a broader significance than hitherto suspected.

ACKNOWLEDGEMENT

We thank Drs C. Jungwirth, H. Schellekens and Ch. Weissmann for gifts of interferons , P. Groscurth for providing cells, and P. Staeheli for helpful suggestions. This work was supported by Grant 3.507-083 of the Swiss National Science Foundation.

REFERENCES

1. Haller O (1981) Curr Topics Microbiol Immunol 92:25-52
2. Horisberger MA, Staeheli P Haller O (1983) Proc Natl Acad Sci USA 80:1910-1914
3. Staeheli P, Pravtcheva D, Lundin L-G, Acklin M, Ruddle F, Lindenmann J, Haller O (1986) J Virol 58:967-969
4. Dreiding P, Staeheli P, Haller O (1985) Virology 140: 192-196
5. Krug RM, Shaw M, Broni B, Shapiro G, Haller O (1985) J Virol 56:201-206
6. Meyer T, Horisberger MA (1984) J Virol 49:709-716
7. Staeheli P, Haller O, Boll W, Lindenmann J, Weissmann C (1986) Cell 44:147-158
8. Haller O, Arnheiter H, Lindenmann J, Gresser I (1980) Nature 283:660-662
9. Staeheli P, Dreiding P, Haller O, Lindenmann J (1985) J Biol Chem 260:1821-1825
10. Wüthrich, R, Staeheli P, Haller O (1985) The Biology of the Interferon System, pp 317-323
11. Staeheli P, Haller O (1985) Mol Cell Biol 5:2150-2153
12. Staeheli P, Faeh J, Mortier C, Haller O (1986) The Biology of the Interferon System 1985, pp 115-118
13. Shaw MW, Haller O, Lamb RA, Choppin PW (1984) Negative Strand Viruses, pp 131-138
14. Stewart II WE (1979) The Interferon System, Springer Verlag, Wien, New York
15. Jungwirth C (1985) J Interferon Res 5:209-214
16. Schellekens H, Wilde G A, Weimar W (1980) J Gen Virol 46: 243-247

COMPARATIVE STUDIES OF THE GENOMIC STRUCTURE OF TWO HUMAN INTERFERON INDUCED GENES.

M. WATHELET*,L. DE VRIES*, P. DEFILIPPI°, C. NOLS*, P. VANDENBUSSCHE°, V. POHL*, G. HUEZ* and J. CONTENT°.
* Université Libre de Bruxelles, 67 rue des Chevaux, Rhode-St-Genèse 1640, BELGIUM and ° Institut Pasteur du Brabant, 642 rue Engeland, Bruxelles 1180, BELGIUM.

1. INTRODUCTION

The various biological responses observed in interferon(IFN)-treated cells are probably mediated by some of the proteins they induce. Several cDNA corresponding to these proteins have been cloned, allowing transcriptionnal and functionnal studies. All the genes studied so far seem to behave differently upon IFN induction: the accumulation of mRNA, as well as the time needed to reach maximal level and the extend of maximal induction, are different. Morover, this response is different for most genes depending on the type of IFN or on cell used (1-3). Finally, it has been shown that among these genes, some of them are still induced in IFN resistant lymphoblastoid (Daudi) cells, demonstrating the coexistence of at least two different pathways of induction (4).

However, the existence of a common inducer should be reflected in the promoter region sequences of some of these genes, and indeed, a 28 nucleotides long consensus sequence has emerged from the comparison of four human induced genes for which such data were available (5). Since then, the sequence was found in mouse class I transplantation antigens encoded in the major histocompability complex (H-2K, H-2L,...)(6).

The H-2K promoter region was studied more closely, and the Friedman& Stark consensus sequence (or IRS, for Interferon Responsive Sequence), was shown to be necessary for IFN induction to occur, but only when associated with a functional enhancer sequence (7).

We have previously described the cloning and sequencing of cDNA corresponding to two induced proteins, the human 42kDa 2-5A synthetase and a 56kDa protein(56K) of unknown function (8,9). Since the time course of accumulation of both mRNA appears to be similar after IFN treatement, we aimed to compare the promoter region of these genes.

2. RESULTS AND DISCUSSION.

2.1. Isolation of genomic clones.

A human cosmid genomic library (in pHC79-2cos/tk, kindly provided by W. Lindenmaier), was screened with radioactively labeled cRNA corresponding to our cDNA clones of 2-5A synthetase and 56K. While all six genomic 2-5A synthetase clones contained the complete cDNA sequence, the three 56K genomic clones did not bear the upstream part of the cDNA sequence.

Two human phage genomic libraries (in lambda gtWES/lambdaC and in lambdaCharon4A, a gift of J.M. Jeltsch) were screened with a radioactively labeled oligodeoxyribonucleotide complementary to nucleotide -48 to -9 of the 56K cDNA (9). One clone was isolated, which corresponds to about

50% of the cDNA (5' side) and which contains a segment of more than 2kb upstream of the cDNA 5'end.

The restriction map, as well as the intron-exon pattern of our genomic clones of 2-5A synthetase, determined by electron microscopy (10), appears to be very similar to those published previously (11, 12).

Fig. 1A

GENOMIC ORGANIZATION OF 2.5 A SYNTHETASE

	EXONS (e)	INTRONS (i)
5'	e_1 : 248 $\pm$ 29 BP (9)	i_1 : 1.400 $\pm$ 135 BP (17)
	e_2 : 340 $\pm$ 42 BP (14)	i_2 : 2.405 $\pm$ 266 BP (21)
	e_3 : 228 $\pm$ 34 BP (16)	i_3 : 5.446 $\pm$ 401 BP (18)
	e_4 : 240 $\pm$ 24 BP (17)	i_4 : 864 $\pm$ 99 BP (13)
3'	e_5 : 455 $\pm$ 119 BP (12)	

( ) : NUMBER OF SAMPLES

Fig. 1B

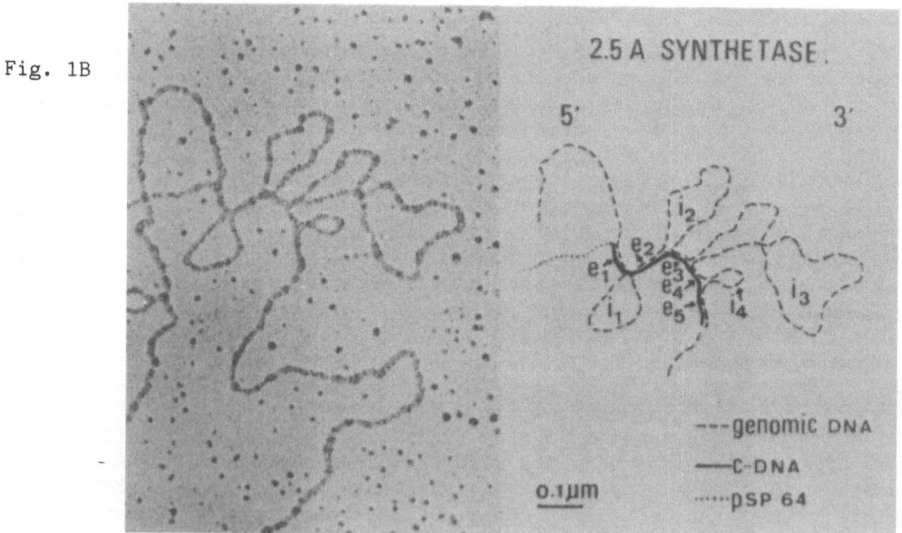

Fig. 1B: electron microscopy of a cDNA/genomicDNA hybrid for the 2-5A synthetase gene (10).

2.2 Promoter region structure.

The putative promoter region of both genes were subcloned into pSP64 vector for restriction analysis and overlapping fragments were sequenced according to the dideoxy termination method (13)(to be published elsewhere). For the 2-5A synthetase gene, we subcloned the 4.3kb EcoRI fragment containing a 0.85kb SphI fragment, where the 5' end of the cDNA was located. This fragment, as well as the 5' 0.65kb EcoRI-SphI fragment were sequenced. Surprisingly, the 5' end of our cDNA extends in what is described by Benech **et al** (11), as the second intron of the gene. This discrepancy could be explained by differential splicing or differential initiation of transcription in different cells. This last hypothesis is supported by the existence of another TATA box (T3), which is also compatible with primer extension and nuclease S1 mapping experiments (B.R.G. Williams, pers. commun.)(see Fig. 2A).

FIG. 2A

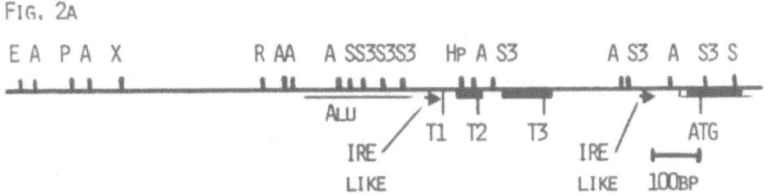

E=EcoRI, A=ALUI, P=PstI, X=XbaI, R=RsaI, S=SphI, S3=Sau3AI, Hp=HpaI.

Fig. 2A: T1,T2,T3 have TATA-box like sequences; black boxes correspond to exons described by Benech et al (11), and the white box corresponds to our cDNA 5' extension in the second intron described by the same authors (11); IRE stands for Interferon gene Regulatory Element (14), see chapter 2.3.

For the 56K gene, the situation is clearer. A 2.1kb HindIII and an overlapping 1.5kb EcoRI fragments, corresponding to the 5' end of the cDNA, were subcloned into pSP64 vector. 1.6kb encompassing the promoter region was sequenced and the exact location of the transcriptionnal start was mapped by nuclease S1 analysis 24 and 25 nucleotides upstream the 5' end of our cDNA clone. In the sequenced region, only one typical TATATAA sequence was found, which was located 29 and 30 nucleotides upstream of the determined cap site.

On the restriction map of the 56K promoter region (see next page), the white box corresponds to the first exon, IRS stands for Interferon Responsive Sequence (or Friedman&Stark consensus sequence) (5, 7) and IRE stands for Interferon gene Regulatory Element (14).

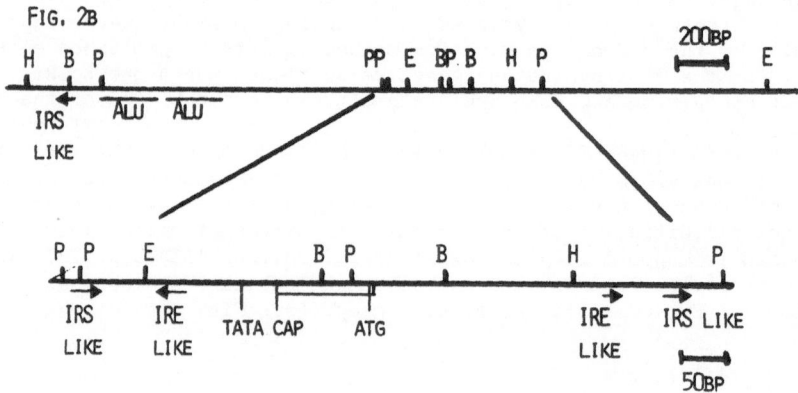

FIG. 2B

H=HINDIII, B=BGLII, P=PSTI, E=ECORI.

Fig. 2B: Restriction map of the 56K promoter region.

2.3. Sequence homologies.

Sequences similar to the 28 nucleotides consensus sequence of Friedman&Stark (IRS)(5,7) were found only in the 56K gene, but at three different locations: -1.7kb on the sense strand, -207bp on the antisense strand and +396bp on the antisense strand, in the first intron (see Fig. 2B for location and Fig. 3 for sequence comparison).

IRS-LIKE SEQUENCE IN THE 56K GENE.

CONSENSUS	TTCNᴳ𝒸NACCTCNGCAGTTTCTCᶜ𝑇TCT-CT	
MT-II	cTCTGGACCT--GCAGTTTCTCCTCT-CT	86%
HLA-DR	TGTTGAACCTCAG-AGTTTCTCCTCTcAT	82%
56K (-1.7KB)	TGCCAGATCTAAACAGTTTCTGCTCAACT	72%
56K (-207BP)	TTG-C-AGGTCTGCAGTTTATCTGTT-TT	65%
56K (+396BP)	TTC--TTTCcCATCAGATTCACTTCT-GT	65%

Fig. 3: three typical IRS modules were found in the 56K gene at various distances from the RNA cap site as indicated in parenthesis. Two other examples of such IRS are presented for comparison as well as the consensus sequence described by Friedman&Stark (5).

The promoter region of both genes were compared to all vertebrate and viral DNA sequences in the Gen Bank (release 40,0) using the algorithm of Goad&Kanehisa (15). Both the 2-5A synthetase and the 56K genes, but especially the latter, contain segments which are homologous to the Hu-β-IFN gene promoter. Functionnal studies using deleted promoter of the Hu-β-IFN gene identified the minimal region needed for inducibility by dsRNA or viruses(14). This region was termed IRE for Interferon gene Regulatory Element and was further dissected in a constitutive transcription element and a negative regulatory sequence that prevents enhancer activity prior to induction (14,16,17).

The 2-5A synthetase gene contains two DNA segments IRE-like on the antisense strand at position +815 and +1266 taking the first nucleotide of the sequence as +1(see Fig.2A for location and Fig.4A for comparison).

LOCALLY HOMOLOGOUS REGIONS BETWEEN 2-5A SYNTHETASE GENE AND β-IFN GENE

AATTGAAAGGGAAAA   +815 TO +829

** ********* ** 1.02×10^{-6}

AACTGAAAGGGAGAA   -87 TO -73 (IRE REGION)

AAAAGGA AAGTGCAAAG   +1266 TO +1282

*** *** ***** **** 3.01×10^{-6}

AAAGGGAGAAGTG AAAG   -82 TO -66 (IRE REGION)

Fig. 4A: number to the right of the sequence indicate the coordinates of the segment taking the first nucleotide for 2-5A synthetase and the RNA cap site for β-IFN gene as +1; underlined values represent the probability of finding such homology at random (15).

The 56K gene contains two longer DNA segments homologous to the IRE, on the sense strand at position -116 and on the antisense strand at position +340 in the first intron (see Fig. 2B for location and Fig. 4B for comparison).

LOCALLY HOMOLOGOUS REGIONS BETWEEN 56K GENE AND β-IFN GENE

AGAAGGGAAA TGGGAAA   +340 TO +356 (1ST INTRON)

***** **** ******* 7.64×10^{-8}

AGAAGTGAAAGTGGGAAA   -76 TO -59 (IRE REGION)

ACCTAGGGAAACCGAAAGGGGAAAGTGAAA -116 TO -87 SENSE STRAND,

** **** **** ******* ******** 6.32×10^{-11}

ACATAGGAAAACTGAAAGGGAGAAGTGAAA   -96 TO -67 (IRE REGION)

Fig. 4B: numbers to the right of the sequence indicate the coordinates of the segment taking the cap site as +1; underlined values represent the probability of finding such homology at random (15).

2.4. Inducers of the 2-5A synthetase and the 56K genes.

Two possible functions could be attributed to these IRE-like structures in the 2-5A synthetase and 56K genes.
a: They could act as enhancer sequences under the control of an element like the IRS, as in the H-2K gene (7).
b: They could constitute the target of polyIC or viruses induction. In such a case, both genes should be inducible by polyIC or viruses. To test this possibility, we treated FS-4 and Daudi cells with either alpha-IFN, polyIC or polyIC+cycloheximide, and we analysed the 2-5A synthetase and 56K mRNA accumulation by dot blot on cytoplasmic extracts.(18).

In a preliminary experiment, we observed a clear induction of 56K mRNA with all three treatements versus control cells in both cell lines. For the 2-5A synthetase mRNA, we observed a signal with polyIC and polyIC+ cycloheximide only in Daudi cells (see Fig. 5).

INDUCTION BY DSRNA WITH OR WITHOUT PROTEIN SYNTHESIS INHIBITOR.

	DAUDI/56K	FS-4/56K	DAUDI/2-5ASYNTHETASE
Co	1.0	1.0	1.0
PIC	5.6	3.0	5.8
ICY	6.4	10.0	5.9
IFN	6.8	25.0	6.2

Fig. 5: Co=control cells; pIC= polyIC 50 mg/1 during 4h; ICy= polyIC 100 mg/1 + cycloheximide 50 mg/1 during 1h and cycloheximide alone for four more hours; serial dilutions of cytoplasmic extracts were blotted on nylon membranes and hybridised with radioactively labeled cRNA corresponding to each cDNA; autoradiograms were subjected to densitometric scanning; values represent the extent of induction after treatment versus control cells; IFN= IFN-alpha2-Arg 500 u/ml for 6h. Both treatments with polyIC were performed in the presence of 1000 units/ml of polyclonal antibody raised against   -interferon.

3. CONCLUSIONS

Comparative studies of the promoter region of these IFN-induced genes, revealed the existence of both IRS-like and IRE-like sequences in the 56K gene and of only IRE-like sequences in the 2-5A synthetase gene.
The presence of IRE-like sequence is correlated, at least for the 56K gene, with inducibility by dsRNA as determined by cytoplasmic dot blot analysis.
More refined experiments are under way to determine the exact contribution of the various elements found in these genes in the inducibility by IFN, polyIC or other known inducers.

4. ACKNOWLEDGEMENTS

This work was supported by the fund for Joint Basic Research (Belgium) n°2.9006.79 to J.C., by a grant from the Fund for Medical Scientific Research (Belgium) n°34.560.85 to J.C. and G.H. and by ULB-Actions de Recherches Concertées of the Belgian Government to G.H.. G.H. is Research Associate and M.W. is Research Assistant of the National Fund for Scientific Research (Belgium).

5. REFERENCES

(1) Content, J. (1986) in: Molecular basis of viral replication (R. Perez-Bercoff Ed.) Plenum Press, New York.

(2) Kelly, J.M., Gilbert, C.S., Stark, G.R. and Kerr, I.M. (1985) Eur.J.Biochem. **153**, 367-371.

(3) Revel, M., and Chebath, J. (1986) TIBS, **11**, 166-170.

(4) McMahon, M., Stark, G.R., and Kerr, I.M. (1986) J.Virol., **57**, 362-366.

(5) Friedman, R.L., and Stark, G.R. (1985) Nature, **314**, 637-639.

(6) Kimura, A., Israël, A., Le Bail, O., and Kourilsky, P. (1986) Cell, **44**, 261-272.

(7) Israël, A., Kimura, A., Fournier, A., Fellous, M., and Kourilsky, P. (1986) Nature, **322**, 743-746.

(8) Wathelet, M., Moutschen, S., Cravador, A., DeWit, L., Defilippi, P., Huez, G., and Content, J. (1986) FEBS Lett. **196**, 113-120.

(9) Wathelet, M., Moutschen, S., Defilippi, P., Cravador, A., Colet, M., Huez, G., and Content, J. (1986) Eur.J.Biochem. **155**, 11-17.

(10) Davis, R.W., Simon, M., Davidson, N. (1971) Meth. in Enz., **21**, 413-428.

(11) Benech, P., Mory, Y., Revel, M., and Chebath, J. (1985) EMBO J., **4**, 2249-2256.

(12) Saunders, M.E., Gewert, D.R., Tugwell, M.E., McMahon, M., and Williams, B.R.G. (1985) EMBO J., **4**, 1761-1768.

(13) Sanger, F., and Coulson, A.R. (1978) FEBS Lett. **87**, 107-110.

(14) Goodbourn, S., Zinn, K., and Maniatis, T. (1985) Cell, **41**, 509-520.

(15) Goad, W.H., and Kanehisa, M. (1982) N.A.R., **10**, 183-196.

(16) Goodbourn, S., Burstein, H., and Maniatis, T. (1986) Cell **45**, 601-610.

(17) Zinn, K., and Maniatis, T. (1986) Cell **45**, 611-618.

(18) White, B.A., Bancroft, F.C. (1982) J.Biol.Chem., **257**, 8569-8572.

CHARACTERIZATION OF TWO NEW IFN-INDUCED PROTEINS

E. KNIGHT, JR., D. FAHEY, R.KUTNY, AND D.C. BLOMSTROM

INTRODUCTION
    The interferons (IFNs) cause a number of different biological
responses in cells. The two responses that have been studied the most
are the antiviral state and the inhibition of cell growth, but their
molecular mechanisms have not yet been fully elucidated. The IFNs
induce a number of new proteins in cells, however the biological roles
of most of these proteins remain unknown. A number of laboratories
are characterizing these IFN-induced proteins either by purification
of the protein itself or by recombinant DNA cloning of the mRNA and
subsequent deduction of the amino acid sequence from the cloned DNA
sequence. In this report we described the cloning and amino acid
sequence characterization of an IFN-induced 15 kDa cytoplasmic
protein. We also present the cellular location of an IFN-induced 20
kDa membrane protein whose induction correlates with the inhibition of
cell growth.

EXPERIMENTAL PROCEDURES
    N-terminal amino acid sequence was obtained from the purified
IFN-induced 15 kDa protein (1). Construction and screening of a cDNA
library in plasmid pBR322 was performed as described previously (2,3).
Characterization of a full length cDNA for the 15 kDa protein has been
described (1). Membranes were prepared from Daudi cells by standard
techniques (4) and cell surface proteins were labeled with ^{125}Iodine
catalyzed by lactoperoxidase.

RESULTS
    Our approach to isolating a cDNA clone for the 15 kDa IFN-induced
protein was to purify the protein, obtain N-terminal amino acid
sequence, synthesize DNA probes based on this amino acid sequence and
use the synthetic probes to identify a cDNA clone. A total of 384
20-mers were synthesized from a seven amino acid region of the
N-terminal sequence of the 15 kDa protein (Figure 1).
The 384 20-mers were divided into four pools at synthesis and used as
probes to detect mRNA from IFN-treated Daudi cells. The data in
Figure 1 show that the 20-mers in pool 4 hybridize to a mRNA from
IFN-treated cells but not to a mRNA from untreated cells. The mRNA of
approximately 700 bases (Figure 1) is large enough to contain the
complete coding sequence of the 15 kDa protein. The 20-mers in pool 4
were used as a radioactive probe (^{32}P) to identify a 15 kDa protein
cDNA clone in a clone bank, pBR322 in E. coli MM 294, prepared from
mRNA from IFN-β-treated Daudi cells (1). Approximately 15,000
colonies were screened and of 38 positive clones one strongly
hybridizing clone (clone 31) was purified, characterized and the
plasmid DNA sequenced. Clone 31 when used as a probe on Northern
blots hybridized to a mRNA of about 600-700 bases from IFN-β-treated
Daudi cells (Figure 2).

94

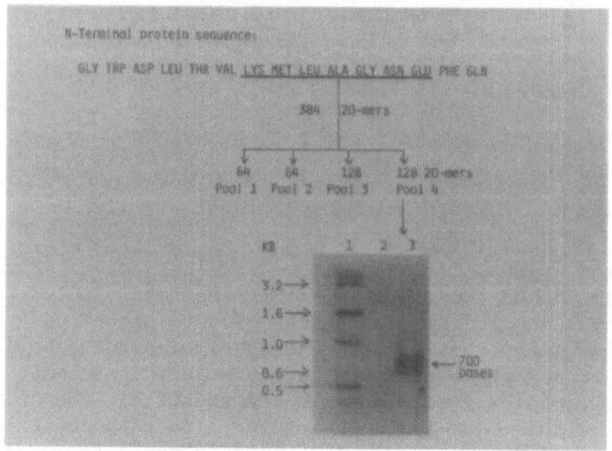

FIGURE 1. N-terminal amino acid sequence of 15 kDa protein and selection of oligonucleotide probes. The N-terminal sequence of the 15 kDa protein was determined, and the underlined sequence used to design all possible (384) 20-mers. The 20-mers in 4 pools were labeled with ^{32}P and used to hybridize with mRNA on Northern blots from untreated and IFN-β-treated Daudi cells. RNA was analyzed on 1.4% agarose gels. Hybridization was for 20 h at 45°C. Blots were washed extensively at 25°C, and then for 1 min at 60°C and exposed to film at −70°C. Lane 1, molecular weight markers; lane 2, 10 μg of mRNA from untreated Daudi cells; lane 3, 10 μg of mRNA from IFN-β-treated Daudi cells. KB, kilobases.

Figure 3 shows the amino acid sequence of the 15 kDa protein deduced from the DNA sequence of clone 31. The cDNA contains 635 base pairs containing an open reading frame for a protein of 145 amino acids. The 635 bases may contain almost all the base sequence in the 15 kDa protein mRNA (Figures 1 and 2). The sequence 3' to the coding sequence contains a polyadenylation signal site. An interesting three fold repeated sequence occurs 5' to the start ATG (Figure 3).

We have also obtained amino acid sequence on approximately 85% of the 15 kDa protein (1). Trypsin and chymotrypsin generated peptides have been sequenced in addition to the N-terminus. The underlined sequences in Figure 3 were obtained from the peptides and their sequences agreed at all positions with the amino acid sequence deduced from the cDNA.

In order to determine the genomic gene copy number for the 15 kDa protein, genomic DNA from Daudi and human diploid fibroblast cells was probed with the 635 base pair DNA from clone 31. Digestion of both Daudi and fibroblast DNA with EcoRI and BamHI restriction enzymes gave a predominant band at 4.3 kilobases (Figure 4) indicating a single copy gene for the 15 kDa protein.

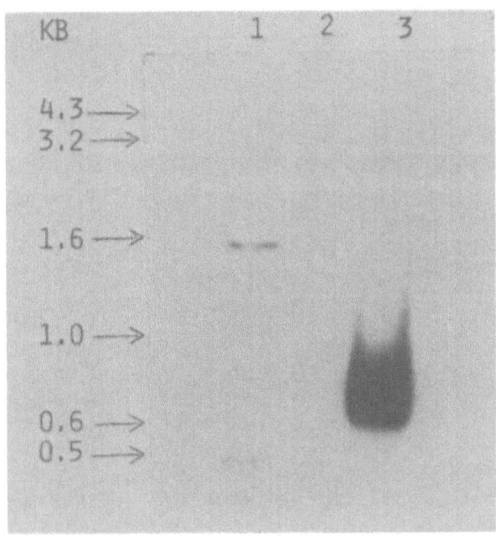

FIGURE 2. Analysis of mRNA from untreated and IFN-β-treated Daudi cells in a Northern blot using clone 31 as a probe. RNA was analyzed on a 1.4% agarose gel. Hybridization was for 20 h at 42°C. Blots were washed extensively at 25°C, 5 min at 50°C, and then exposed to film at -70°C. Lane 1, molecular weight markers in kilobases; lane 2, mRNA from untreated Daudi cells; lane 3, mRNA from IFN-β-treated Daudi cells.

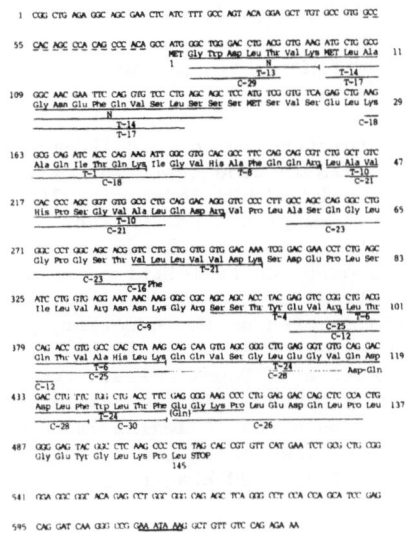

FIGURE 3. Nucleotide sequence and deduced amino acid sequence of clone 31 cDNA and amino acid sequence of 15 kDa protein. Numbers on left refer to nucleotide position and on the right to amino acid position. N-terminal (N), tryptic (T) and chymotryptic (C) sequences are underlined. The polyadenylation signal sequence AATAAA is designated by a bar. The 5' repeated sequence GCCCACA is designated by a dashed line.

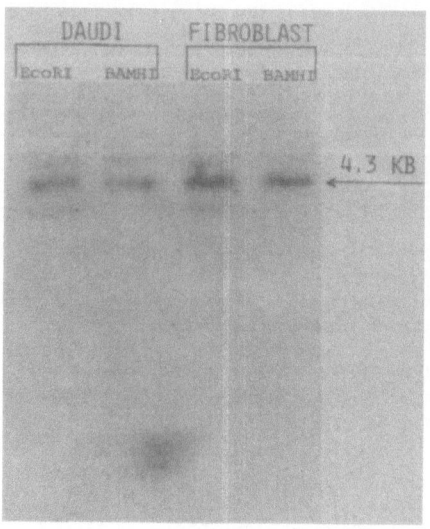

FIGURE 4. Analysis of genomic DNA for the gene for the 15 kDa
protein. Genomic DNA from Daudi and human diploid fibroblast cells
was digested with the restriction enzymes BamHI and EcorI. After
electrophoresis and transfer the genomic DNA was analyzed with
^{32}P-labeled clone 31 DNA.

We have been interested in another IFN-induced protein which
correlates with the IFN-induced cessation of cell growth (4). This
protein of approximately 20 kDa induced by IFN is located in the
membranes of cells whose growth can be inhibited by IFN but is not
induced in those cells whose growth can not be inhibited by IFN (4).
Daudi cells whose growth can be inhibited by IFN-α and β were labeled
with ^{125}Iodine catalyzed by lactoperoxidase. This procedure labels
only those proteins located on the exterior of the cell. Figure 5
shows the ^{125}Iodine-labeled proteins in the membranes isolated from
IFN-β-treated and untreated Daudi cells. Membrane proteins were
separated on an SDS-gel and ^{125}Iodine-labeled proteins were detected
by autoradiography. When the ^{125}Iodine-labeled proteins from
untreated and IFN-β-treated cells are compared (Figure 5) it can be
seen that a protein of 18 - 20 kDa is in the membranes from

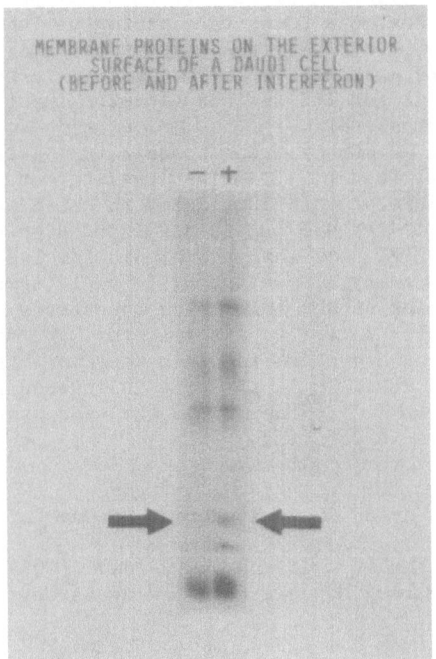

FIGURE 5. Labeling of surface proteins with ^{125}Iodine of
IFN-β-treated and untreated Daudi cells. Daudi cells were treated
for 16 hr with 50 U/ml of IFN-β. The IFN-β-treated and untreated
Daudi cells were labeled with ^{125}Iodine catalyzed by lactoperoxidase.
Membrane fractions were prepared, membrane proteins separated by
SDS-PAGE, and ^{125}Iodine labeled proteins identified by autoradiog-
raphy.

IFN-β-treated Daudi cells but is not in the membranes of untreated
cells. This result indicates that the IFN-induced 20 kDa membrane
protein is located on the exterior of the cell. Also, the data in
Figure 5 show that a 14 - 16 kDa protein induced by IFN-β is located
on the cell surface.

DISCUSSION
    We have isolated and characterized a cDNA clone which contains the
complete coding sequence for the 15 kDa IFN-induced protein. We
believe that this clone of 635 bases contains most of the 15 kDa mRNA
sequence since the 15 kDa mRNA is approximately 600 - 700 bases in
length as determined by Northern blot analysis. The amino acid
sequence deduced from the cDNA corresponds at every position to the
amino acid sequence obtained from the protein. The mRNA for the 15
kDa protein contains an interesting three fold repeated sequence
immediately 5' to the start ATG. This 5' repeated sequence is not
found 5' to the start ATG in any other reported mRNA (6). We
speculate that this repeated sequence may be involved in translational
control of the 15 kDa mRNA.

The 15 kDa protein is a very hydrophobic protein containing no cysteines or N-glycosylation sites. It is a new protein for no significant homologies were found in a search of the Dayhoff protein sequence data bank. Its biological role in the cell is at present unknown. We plan to express the 15 kDa protein in heterologous systems and to use it and its antibody in experiments directed toward defining its biological role.

The IFN-induced 20 kDa membrane protein is located on the exterior of the cell as determined by ^{125}Iodine labeling catalyzed by lactoperoxidase. This protein is induced in cells whose growth is subsequently inhibited by IFN but is not induced in cells whose growth is not inhibited by IFN. If this protein is involved in the cessation of cell growth induced by IFN we speculate that it may be a protein placed on the exterior of the cell which interferes with the binding of an essential growth factor to its receptor. We are currently purifying this membrane protein in order to study its role, if any, in growth regulation. The 12 - 14 kDa protein induced by IFN-β in Daudi cells and labeled with ^{125}Iodine may be the same protein reported by Burrone and Milstein in 1982 (5).

REFERENCES
1. Blomstrom DC, Fahey D, Kutny R, Korant BD and Knight E,Jr: (1986) J. Biol. Chem. 261, 8811-8816.
2. Maniatis T, Fritsch EF and Sambrook J(eds): (1982) Molecular Cloning: A Laboratory Manual, Cold Spring Harbor Laboratory, Cold Spring Harbor, N.Y.
3. Gubler U and Hoffman BJ: (1983) Gene 25, 263-269.
4. Knight E,Jr, Fahey D and Blomstrom DC: (1985) J. IFN Res. 5, 305-313.
5. Burrone OR and Milstein C: (1982) EMBO J. 1, 345-349.
6. Kozak M: (1984) Nucleic Acids Res. 12, 857-870.

EARLY EVENTS IN THE INTERACTION OF IFN-α WITH IFN-SENSITIVE AND IFN-RESISTANT DAUDI CELLS

LAWRENCE M. PFEFFER[1], DAVID B. DONNER[2], TSUTOMU ARAKAWA[3], NOWELL STEBBING[3], AND IGOR TAMM[1]

The Rockefeller University[1], and Memorial-Sloan Kettering Cancer Center[2], New York, NY and AMGEN[3], Thousand Oaks, CA (U.S.A.)

1. ABSTRACT

Daudi cells, and IFN-sensitive and -resistant Daudi subclones contain 2,300 (K_d=20 pM), 3,700 (K_d=52 pM), and 3,000 (K_d=45 pM) homogeneous IFN-α binding sites per cell, respectively. Thus, IFN-sensitive and -resistant Daudi cells have similar numbers of high affinity IFN-α receptors. The dissociation of IFN from its receptor on each Daudi cell line was resolved into fast and slow components. IFN-α dissociated from IFN-sensitive Daudi cells more slowly than from resistant cells. Also, a greater amount of the IFN-receptor complex becomes associated with the Triton-insoluble cytoskeletal matrix in the sensitive than in the resistant cells. Lectins, which increase the formation of Triton-insoluble IFN-receptor complexes in sensitive and resistant cells, decrease the dissociation of IFN-α from its receptor. Finally, IFN-α inhibits the anti-Ig-induced redistribution of surface Ig in IFN-sensitive but not in IFN-resistant cells. Our results suggest that association of IFN-receptor complexes with the Triton-insoluble matrix may be of functional significance. Such interaction may slow the dissociation of IFN-α from its receptor, impair mobility of cell surface receptors, and perhaps play an important role in transmembrane signalling through the IFN receptor.

2. INTRODUCTION

Interferons (IFNs) inhibit the proliferation of normal and transformed cells in culture, inhibit viral replication, and modulate cell structure and function. In the past few years the availability of pure IFN produced by recombinant DNA technology has permitted the extensive study of the structure, regulation and function of IFN receptors. It is now known that IFNs bind to high affinity receptors on the surface of target cells (1,2), however, the events that occur subsequently are poorly understood. Our own investigation has shown that treatment of human fibroblasts and tumor cells with human IFN elicits a coordinated cellular response involving the plasma membrane-cytoskeletal complex (3-6). In extending this work, we have undertaken studies aimed at defining, at the molecular level, how binding of IFN-α to its receptor elicits the pleiotropic action of this cytokine in target cells.

The Daudi line of human lymphoblastoid cells is highly sensitive to the antiproliferative action of human leukocyte IFN (IFN-α) (7) and, in general, appears to bind more IFN-α than other human cells (2). Our previous studies have lead us to speculate that the interaction between IFN-α receptors and the plasma membrane-cytoskeletal complex may play an important role in the inhibitory action of IFN-α on Daudi cell proliferation. We have now characterized IFN-α binding, dissociation, the formation of Triton-insoluble IFN-receptor complexes, and the ability of IFN-α to inhibit the lateral mobility of surface immunoglobulins on IFN-sensitive and IFN-resistant subclones of Daudi cells, as well as on the parental cell line.

3. MATERIALS AND METHODS

3.1. <u>Cell cultures</u>. The Daudi cell line and the IFN-sensitive (IFN-sen) and the IFN-resistant (IFN-res) subclones were grown as static suspension cultures at concentrations between 2×10^5 and 1.5×10^6 cells/ml in RPMI-1640 medium supplemented with 10% fetal bovine serum.

3.2. <u>IFN-α binding to intact Daudi cells and isolated plasma membranes</u>. [^{125}I] was coupled to IFN-α by using limiting amounts of chloramine T. Iodinated IFN-α (~130 Ci/g), separated from free [^{125}I] on a column of Sephadex G-25, retained >90% of its antiproliferative activity on Daudi cells. To isolate plasma membrane fractions, Daudi cells (5×10^7 cells/ml) were disrupted by hypotonic lysis. A crude plasma membrane fraction was then separated from other cellular organelles by centrifugation on a discontinuous sucrose gradient. For analysis of [^{125}I]IFN-α binding, cells (2×10^6 cells/ml) or plasma membranes (~20 μg/ml) were suspended in tissue culture medium and incubated for 100 min with [^{125}I]IFN-α (1.2×10^{-12} to 1.5×10^{-10} M) in the presence or absence of 2 nM unlabeled IFN-α at 15° and 37°. Triplicate 300 μl aliquots were centrifuged through a mixture of 2 parts dinonyl phthalate and 1 part dibutyl phthalate oil in a 500 μl microfuge tube at 4° at 10,000 x g for 1 min. The supernatant medium was aspirated, 300 μl distilled water was added to the tube, and then the water and oil were aspirated. The tip of the microfuge tube that contained [^{125}I]IFN bound to cells or plasma membranes was cut off and placed in a tube for assay of ligand binding. Specific binding of IFN-α to cells or membranes is equivalent to the difference between binding in the presence of [^{125}I]IFN-α alone and binding in the presence of 2 nM unlabeled IFN-α. The data from triplicate experiments were plotted according to the procedure of Scatchard (8) and curves were fitted by linear regression analysis.

Cell-bound [^{125}I]IFN-α resistant to extraction from the cell surface by dilute acid is considered internalized (9). To quantitate the amount of acid-resistant radioactivity, triplicate 500 μl aliquots of the cell suspension were pelleted (10,000 x g, 15 sec), incubated with 300 μl of 0.2 M acetic acid and 0.5 M NaCl for 6 min at 4° and recentrifuged (10,000 x g, 1 min). In other experiments, triplicate 500 μl aliquots of the cell suspension were pelleted (10,000 x g, 15 sec), suspended in 300 μl of 0.3 M sucrose, 50 mM NaCl, 1 mM MgCl$_2$, 10 mM HEPES, pH 7.4 and 0.5% Triton X-100 for 2 min at 4° and recentrifuged (10,000 x g, 3 min) (10). The radioactivity that remained in the pelleted Triton-insoluble material was considered a measure of the association of IFN-receptor complexes with the cytoskeletal matrix.

3.3. <u>Redistribution of surface immunoglobulins</u>. Aliquots of Daudi cells (2×10^6 cells) were pelleted (1,000 x g, 5 min) and resuspended in 1 ml of RPMI-1640 medium with 20 mM HEPES (pH 7.4). After 10 min incubation at 4°, fluorescein-conjugated goat antibodies to human immunoglobulins (anti-Ig), were added to the cell suspension to a final concentration of 20 μg/ml and the cells were maintained for 30 min at 4°. The cells were then warmed to 37° for 40 min and fixed with 3.7% formaldehyde in phosphate-buffered saline for 30 min. Cells were examined for the fluorescence pattern of anti-Ig and scored for the presence of intense fluorescence at one pole of the cell (cap formation). At least 500 cells were scored for each variable examined.

3.4. <u>Interferon</u>. Recombinant DNA-derived IFN-Con$_1$ (1 x 10^9 units/mg protein, AMGEN, Inc.), designed as a consensus of the known IFN-α subtypes, was purified to homogeneity and used as the source of IFN-α in these studies. IFN concentrations are expressed in terms of international reference units (u) per ml as assayed by protection against the cytopathic effects of vesicular stomatitis virus in human diploid fibroblasts.

4. RESULTS

4.1. <u>IFN-α binding to Daudi cells.</u> The specific binding of [^{125}I]IFN-α to Daudi cells was 40% greater at 37° than at 15° after 100 min of incubation. At both temperatures, the IFN-α binding capacity of Daudi cells was nearly saturated by 0.1 nM IFN-α.

To determine whether the greater binding at 37° than at 15° was due to internalization of IFN-α at the higher temperature, dilute acid extraction (0.2 M acetic acid, 0.5 M NaCl, 6 min) was used to quantitate the amount of IFN-α internalized during incubations (see Materials and Methods). After incubation of Daudi cells with 0.01 nM of [^{125}I]IFN-α for 60 min at 15°, <2% of the cell-associated radioactivity was internalized. In contrast, after incubation of Daudi cells for 60 min at 37°, ~50% of the cell-associated radioactivity was internalized. The rate of internalization of IFN-α at 37° was linear for 90 min. Thus, increased binding at 37° relative to 15° resulted from internalization of surface bound ligand. Scatchard analysis (8) of the IFN-α binding data obtained from experiments conducted at 15° yielded a linear plot, and suggested the presence of a single class of noncooperative sites with a binding constant of 20 pM and 2,300 IFN-α receptors/cell. Correction of the data from assays conducted at 37° to account for internalization of IFN indicated that the surface binding of [^{125}I]IFN-α was comparable to that at 15°.

4.2. <u>The dissociation of [^{125}I]IFN-α from Daudi cells.</u> Daudi cells (2 x 10^6 cells/ml) were incubated with 0.01 nM [^{125}I] for 100 min at 15° and pelleted (1,000 x g, 5 min). Dissociation of bound ligand was then initiated by resuspending the cells in medium that contained a 200-fold excess (2 nM) of unlabeled IFN-α and was equilibriated to 15°. The amount of [^{125}I]IFN-α remaining specifically bound was then measured as a function of time. The dissociation reaction was stopped by centrifugation (10,000 x g, 1 min) of the cells through a mixture of phthalate oils. Dissociation of [^{125}I]IFN-α from Daudi cells was resolved into fast and slow components. While ~20% of the bound IFN-α dissociates within 5 min, 35% remains bound after 120 min. The multicomponent course of dissociation contrasts with the monocomponent Scatchard plot derived from equilibrium binding data at 15°.

4.3. <u>Triton-insolubility of the IFN-receptor complex on Daudi cells.</u>
Experiments were conducted to determine whether the IFN-receptor complex becomes associated with the cytoskeletal matrix, as would be indicated by

the accumulation of Triton-insoluble radioactivity during incubation. Cells were incubated with 0.01 nM [^{125}I]IFN-α in the presence or absence of 2 nM unlabeled IFN-α at 15° and at various times the amounts of specific, cell-associated radioactivity and specific, Triton-insobuble radioactivity were quantitated. The binding of [^{125}I]IFN-α to Daudi cells and the association of IFN-receptor complexes with the Triton-insoluble matrix increased with time. Triton-insoluble complexes formed shortly after the initial binding of IFN-α to the cell surface. Both specific IFN-α binding and the amount of IFN-α in specific cytoskeleton-associated complexes reached steady state levels after 100 min at 15°. At this time, ~25% of the bound IFN-α was Triton-insoluble. Extraction with Triton X-100 prior to the addition of [^{125}I]IFN-α removed >95% of the IFN-α binding capacity of Daudi cells. These results suggest that upon IFN-α binding, a significant fraction of the IFN-receptor complexes become associated with the cytoskeletal matrix.

Lectins, such as concanavalin A (con A) and wheat germ agglutinin (WGA), that crosslink cell surface components increase the association of IFN-α with the Triton-insoluble matrix. Daudi cells were incubated with 0.1 nM IFN-α at 15° for 100 min, and then with 50 μg/ml of either lectin at 4° for 30 min. After such treatment, 21%, 61% and 71% cell-bound IFN was Triton-insoluble in control, con A-treated and WGA-treated cells, respectively. The increased formation of Triton-insoluble IFN-receptor complexes correlated with decreased dissociation of IFN from Daudi cells. Thus, in control and WGA-treated cells, 20% and 60% of the IFN that was originally bound remained so after 2 hr of dissociation, respectively.

4.4. **Characterization of the IFN-α receptor in isolated plasma membrane fractions.** The specific binding of [^{125}I]IFN-α to Daudi cell plasma membrane fractions, isolated after centrifugation on discontinuous sucrose density gradients, was measured at 15°. Scatchard analysis yielded a linear plot, indicating a single class of IFN-α binding sites (K_d=110 pM) on Daudi cell membranes. Furthermore, the dissociation kinetics of IFN-α and the cytoskeletal association of IFN-receptor complexes were similar on intact cells and plasma membrane fractions.

4.5. **IFN-α receptor on IFN-sensitive and -resistant Daudi subclones.** Scatchard analysis of IFN-α binding at 15° yielded linear plots indicating the presence of 3,000 receptors/cell (K_d=45 pM) on a Daudi subclone that is resistant to the antiproliferative action of IFN-α (IFN-res), and 3,700 receptors/cell (K_d=52 pM) on a subclone that has the IFN-sensitivity of the Daudi cell line (IFN-sen). The time course of dissociation of [^{125}I]IFN-α from IFN-sen cells was similar to that determined with Daudi cells, and was resolved into fast and slow components. In contrast, the dissociation from IFN-res cells was much more rapid than that from IFN-sensitive cells. Only 15% of the IFN remained bound on IFN-res cells after 120 min of dissociation as compared with 35% remaining bound on IFN-sen or on Daudi cells. Since experiments with Daudi cells showed that cytoskeletal association of IFN-receptor complexes correlated with decreased dissociation of IFN, we assayed for the formation of Triton-insoluble IFN-receptor complexes on IFN-res and IFN-sen cells. In IFN-sen cells, 21% of cell-bound IFN-α was Triton-insoluble, similar to the amount observed in Daudi cells, while in IFN-res cells a very low level (~5%) of the surface-bound IFN was Triton-insoluble. Failure of association of IFN-receptor complexes with the cytoskeletal matrix in IFN-resistant cells may be a significant factor in the decreased IFN sensitivity of such cells.

4.6. IFN-α effects on the redistribution of surface immunoglobulins on Daudi cells.

Daudi lymphoblastoid cells display immunoglobulins (Ig) on their cell surface. Addition of anti-Ig (5 μg/ml) to Daudi, IFN-res and IFN-sen cells induces the formation of caps at one pole of the cell in >85% of the cells. IFN-α inhibits the anti-Ig-induced cap formation by ~60% in IFN-sensitive but not in IFN-resistant cells. Inhibition of cap formation was obtained with IFN-α concentrations as low as 2.5×10^{-13} M. The inhibitory effect decreased with length of IFN pretreatment; in cells pretreated with IFN for 5 hr no inhibition of cap formation was observed.

5. DISCUSSION

To understand the mechanisms by which IFN-α inhibits Daudi cell proliferation, we have characterized IFN-α binding, dissociation, the formation of Triton-insoluble IFN-receptor complexes, and the ability of IFN-α to inhibit the lateral mobility of surface immunoglobulins on IFN-sensitive and -resistant Daudi cells. The Daudi cells, and IFN-res and IFN-sen cells contain 2,300 (K_d=20 pM), 3,000 (K_d=45 pM), and 3,700 (K_d=52 pM) IFN-α binding sites per cell at 15°, respectively. Thus, IFN-sensitive and -resistant Daudi cells have similar numbers of high affinity receptors. These observations show that while binding to specific high affinity cell surface receptors is an essential first step in IFN-α action, the presence of such sites is insufficient to insure cellular responsiveness. Furthermore, binding to only a small fraction of available IFN-α receptors is sufficient to elicit a maximal cellular response. This is demonstrated by the finding that while the proliferation of Daudi cells is maximally inhibited by 1 pM IFN-α, IFN binding sites are saturated by 100 pM IFN-α.

The observations summarized above focused our attention on the events that occur shortly after IFN binding and that might be important in IFN action. A role for the cytoskeleton in IFN action has been suggested in studies which show that drugs that impair cytoskeletal organization block the establishment or induce the decay of IFN-induced antiviral protection (11), while drugs that enhance cytoskeletal organization potentiate the antiviral and antiproliferative activities of IFN (12). IFN-induces a rapid (within 15 min), transient increase in the rigidity of the plasma membrane lipid bilayer of human tumor cells, an effect believed to reflect IFN interaction with cell surface receptors and the associated cytoskeletal matrix (4).

In the present study, the cytoskeleton is operationally defined as the complex of cellular elements, including the plasma membrane, the nucleus, fibrous structural elements, and associated proteins that are insoluble in a nonionic detergent (Triton X-100) (10). During incubation of [^{125}I]IFN-α with Daudi cells or isolated plasma membranes, a fraction of the bound IFN becomes Triton-insoluble. The formation of Triton-insoluble receptor complexes apparently is not a reflection of the process of internalization, since complexes form at 15°, a temperature at which negligible internalization of IFN occurs, and since the formation of complexes and IFN internalization are temporally separated. Futhermore, IFN internalization may not be required to elicit some, and possibly not any, of the IFN actions in Daudi cells (13). In the present study, a relationship has been demonstrated between cytoskeletal coupling of IFN-receptor complexes and the antiproliferative effect of IFN-α.

The dissociation of IFN-α from receptors is biphasic and was resolved into fast and slow components. The rate of dissociation from the slowly dissociating complex appears to partly reflect the extent of cytoskeletal association of IFN-receptor complexes. In apparent contrast to the kinetics of IFN-α release from Daudi cells or plasma membranes, Scatchard analysis of saturation binding data indicate the presence of a single high affinity IFN-α binding sites in equilibrium with the medium. These observations are consistent with a model in which, after IFN-α binding, a subpopulation of the IFN-receptor complexes becomes coupled to the cytoskeletal matrix, and is no longer in direct equilibrium with the medium.

Receptors for epidermal growth factor (14), nerve growth factor (15), membrane immunoglobulin (16), and chemotactic peptide (17) are associated with the cytoskeletal matrix prior to or become associated upon binding of the ligand. Such an association may account for the obervation that some receptors become progressively immobilized within the plane of the plasma membrane upon ligand binding. Crosslinking of ligand-receptor complexes may be a first step that leads to attachment to the cytoskeleton. In the present study, we report that IFN inhibits the lateral mobility of cell surface immunoglobulins on Daudi cells, an effect that could be mediated through IFN-receptor interaction with the cytoskeletal matrix. This effect is rapid and dissipates with length of pretreatment. Furthermore, crosslinking of NGF (18) or, in the present study, IFN-α with wheat germ agglutinin increase the fraction of receptors coupled to the cytoskeleton and decreases the rate of ligand dissociation from the ligand-receptor complex.

The present study establishes a relationship between interaction of receptor complexes with the cytoskeletal matrix and the ability of IFN-α to inhibit Daudi cell proliferation. This relationship was demonstrated by using IFN-sensitive and -resistant Daudi subclones. In IFN-sen cells the formation of Triton-insoluble receptor complexes was markedly increased in comparison to the formation in IFN-res cells. Thus, a possible role is established for the coupling of IFN-receptor complexes to the cytoskeletal matrix or a cytoskeleton-associated protein in the antiproliferative action of IFN-α.

6. ACKNOWLEDGEMENTS

We would especially like to thank Dr. A. Hovanessian of Institut Pasteur for generously providing us with IFN-sen and IFN-res Daudi subclones. This work was supported by research grants GM 36716, AM 22121 and AM 30788 from the National Institutes of Health. L.M. Pfeffer is a Leukemia of America Scholar and D.B. Donner is the recipient of a Research Career Development Award AM 01145 from the National Institutes of Health.

REFERENCES

1. Aguet, M. (1980) Nature 284, 459-461.

2. Branca, A.A. & Baglioni, C. (1981) Nature 294, 768-770.
3. Pfeffer, L.M., Wang, E. & Tamm, I. (1980) J. Cell Biol. 85, 9-17.
4. Pfeffer, L.M., Landsberger, F.R. & Tamm, I. (1981) J. Interferon Res. 1, 613-620.
5. Wang, E., Pfeffer, L.M. & Tamm, I. (1981) Proc. Natl. Acad. Sci. USA 78, 6281-6285.
6. Pfeffer, L.M. & Tamm, I. (1982) J. Interferon Res. 2, 431-440.
7. Adams, A., Strander, H. & Cantell, K. (1975) J. Gen. Virol. 28, 207-214.
8. Scatchard, G. (1949) Ann. N.Y. Acad. Sci. 51, 660-672.
9. Haigler, H.T., Maxfield, F.R., Willingham, M.C. & Pastan, I. (1980) J. Biol. Chem. 255, 1239-1241.
10. Ben-Ze'ev, A., Duerr, A., Solomon, F. & Penman, S. (1979) Cell 17, 859-865.
11. Bourgeade, M.F. & Chany, C. (1976) Proc. Soc. Exp. Biol. Med. 153, 501-504.
12. Bourgeade, M.F. & Chany, C. (1979) Int. J. Cancer 24, 314-318.
13. Branca, A.A., Faltynek, C.R., D'Alessandro, S.B. & Baglioni, C. (1982) J. Biol. Chem. 257, 13291-13296.
14. Landreth, G.E., Williams, L.K. & Rieser, G.D. (1985) J. Cell Biol. 101, 1341-1350.
15. Schechter, A. & Bothwell, M. (1981) Cell 24, 867-874.
16. Braun, J., Hochman, P. & Unanue, E. (1982) J. Immunol. 128, 1198-1204.
17. Jesaitis, A.J., Naemura, J., Sklar, L., Cochrane, C. & Painter, R. (1981) J. Cell Biol. 98, 1378-1387.
18. Vale, R.D. & Shooter, E.M. (1982) J. Cell Biol. 94, 710-717.

RECEPTORS

SOME FRESHLY EBV-TRANSFORMED B-CELL LINES LACK FUNCTIONAL IFN-RECEPTORS

ERIK LUNDGREN[1], MONA LUNDSTRÖM[1] and ALEX VON GABAIN[2]
[1]Unit of Applied Cell and Molecular Biology, University of Umeå, S-901 87 Umeå, Sweden
[2]Dept. of Bacteriology, Karolinska Inst., S-104 01 Stockholm, Sweden

INTRODUCTION

Various cell lines carrying the Epstein-Barr virus (EBV) genome differ in their responsiveness towards interferons (IFN). Among Burkitt's lymphoma derived cell lines some are growth inhibited by IFN-α at very low concentrations (e.g. Daudi, P3HR-1), while others are resistant even at high concentrations (1, 2). Primary B-cells triggered to growth by EBV react in a complex pattern to IFN:s. Thus, the initial growth response, the increased antibody secretion and the immortalization event are all inhibited by IFN (3). However, once the cells have been immortalized, they are no longer affected by cell growth inhibition (3).

In an attempt to identify on the molecular level the reason for abrogation of the IFN response we decided to study the IFN-binding sites of such EBV-immortalized B-cell lines, using radioiodinated IFN-α. We found that in contrast to some Burkitt's lymphoma lines, such cell lines displayed a relatively high number of low affinity binding sites, and in some cases also a low or a non-detectable number of receptors.

MATERIALS AND METHODS

The cells were grown and cell growth rate was estimated according to previously described methods (2).

A variant of IFN-α2 (4), was produced in E. coli, purified by immune affinity chromatography and subsequent FPLC to homogenity (specific activity 3×10^8 U/mg protein). The material was radioiodinated using the Bolton and Hunter reagent and a specific activity of 7×10^9 cpm/mg protein was obtained, without any significant loss of activity.

For binding assays, the labelled IFN was serially diluted in 0.1 ml culture medium on top of a gradient consisting of 0.1 ml 15 % silicon oil and 85 % mineral oil overlayered on 0.1 ml FCS supplemented with 10 % sucrose. 10^6 cells in 0.1 culture medium were added to the dilutions and the tubes were incubated at 25°C. After 2 hours with intermittent rotation, the tubes were centrifuged, the tube tips were cut off and counted separately from the supernatant fraction. The experimental values were corrected by subtraction of unspecific binding, determined in the presence of 100-fold molar excess of unlabelled IFN. The calculated values of the number of binding sites per cell were estimated by Scatchard analysis of equilibrium binding data. The calculations for the Scatchard plots were done using a curve fitting program. The unspecific binding, which amounted to 5 to 15 % is not indicated in the figures.

RESULTS

The previously reported growth inhibition of Daudi cells by IFN was reproduced in the experiment illustrated in Fig. 1. Half maximal inhibition is achieved at an IFN concentration of about 0.1 pM, corresponding to roughly 0.4 U/ml. The same dose dependence was found when cell counts were estimated (data not shown). On the other hand, no inhibition was seen on all the tested EBV-immortalized cell lines, here exemplified by the lack of any growth inhibition of EBV-8 cells in the presence of more than 10 000-fold higher IFN concentration.

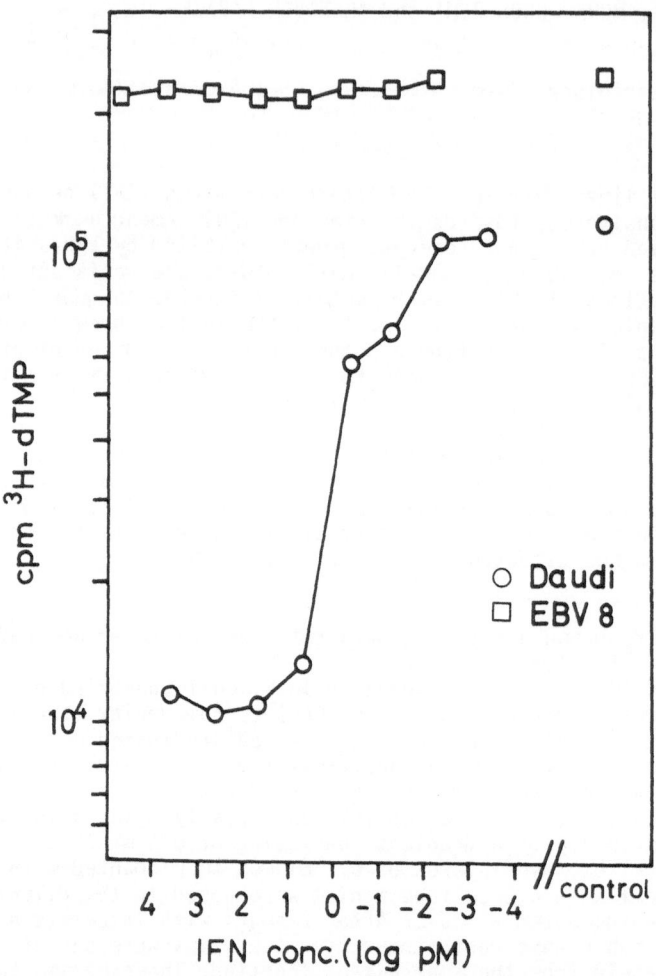

FIGURE 1. Effect of IFN on thymidine uptake in Daudi cells (○) and the EBV-immortalized line EBV-8 (□). Graded doses of IFN was added to exponentially growing cells in a concentration of 200 000 cells/ml and thymidine uptake was measured after 72 hours.

The same lack of response was observed for induction of the IFN-inducible enzyme 2'-5'oligo(A) synthetase. This was readily achieved in Daudi cells, but not in any of 5 tested EBV-immortalized lines (data not shown).

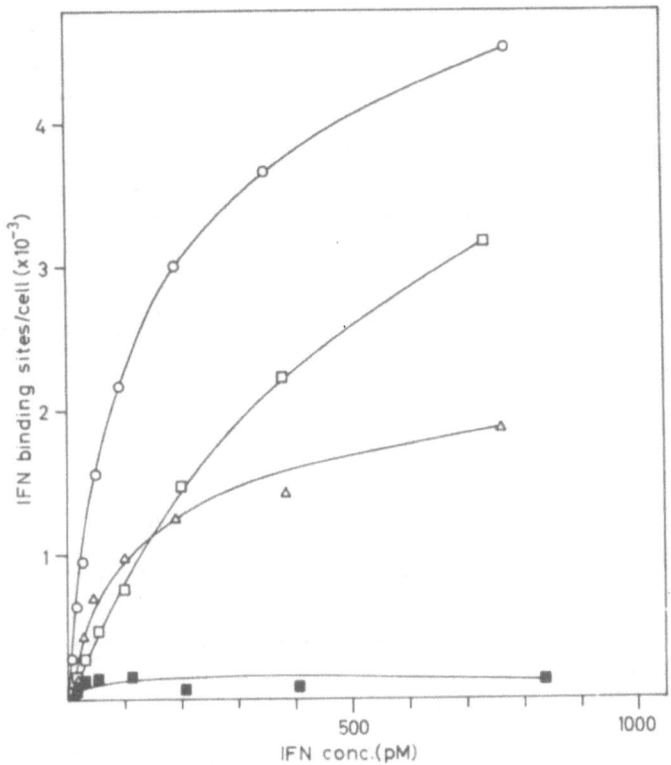

FIGURE 2. Steady state binding of radioiodinated IFN-α. Binding was performed at 25°C according to Materials and Methods. (○) Daudi, (△) Namalwa, (■) EBV-7 and (□) 158 B4.

Receptor binding of radioiodinated IFN is shown in Fig. 2. The profiles of the curves indicate that most of the cells were nearly saturated at the maximal ligand concentrations used, however, with some marked differences between cell lines. Data obtained from transformation of the binding curves according to Scatchard are presented in Table 1. The tested Burkitt's lymphoma lines disclosed binding to IFN independent of whether they carried EBV-genome (Daudi, Namalwa) or not (Ramos), the Kd values were in the range of 30-80 pM. As reported by others, Daudi cells also display a low affinity receptor at certain conditions (5). Also the cell line U-266, with plasma cell characteristics, revealed presentation of receptors similar to those found on the Burkitt's lymphoma lines.

TABLE 1. Binding characteristics of radioiodinated IFN-α to some B lymphocyte lines. Figures within parenthesis indicate biphasic binding and the
value for low affinity receptor is given.

Cell lines	Kd(pM)	Binding sites/cell
Burkitt's lymphoma		
Daudi	55 (1 410)	1 800 (4 560)
Namalwa	78	1 860
Ramos	32	1 330
Plasma cell		
U-266	98	5 776
EBV-immortalized lines		
EBV-2	ND[1]	ND
EBV-7	1 305	430
EBV-8	1 635	595
158 B1	550	3 100
158 B4	715	6 470
158 A4	280	2 800
216-1	415	900
120-1	420	15 600

[1]ND, nondetectable

The EBV-immortalized lines showed a different pattern. The lines EBV-2,
7 and 8 had very low or nondetectable number of receptors, and only in the
presence of high amounts of IFN some binding could be observed. All other
lines were found to carry receptors, although the apparent Kd differed up
to one order of magnitude from that of the Burkitt's lymphoma lines.
The dynamics of ligand receptor turnover also exhibited differences
between the EBV lines and Daudi. Fig. 3 shows that degradation of ligands
measured as appearance of TCA-soluble radioactivity increased in
supernatants from Daudi cells, while no increase was detectable in
EBV-transformed cells, here exemplified for the line 464:1.

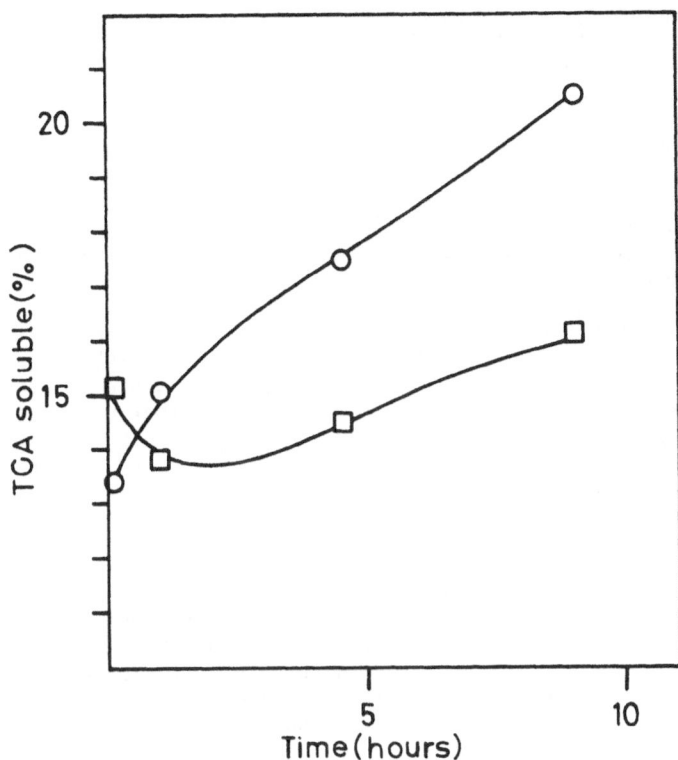

FIGURE 3. Degradation of radioiodinated IFN at 37°C. Cells were exposed to the ligand for indicated time periods and the fraction of TCA-soluble radioactivity in the supernatant was estimated. (○) Daudi, (□) 464:1.

DISCUSSION
  The major conclusion of this report is that the minimal growth inhibition by IFN-α of EBV-immortalized B-lymphocytes is associated with a receptor presentation, which differed from that of Burkitt's lymphoma derived cell lines and a plasma cell line. Thus, in one case, binding sites were nondetectable (EBV-2) and in some other cases the receptor number was low with only low affinity binding (EBV-7 and 8). In cases where the number of receptors was comparable with that of the tumor lines the apparent binding affinities were lower and indicate conditions of receptor occupancy, which falls outside what has been considered to be physiological concentrations of IFN.
  The receptors on the EBV-immortalized lines also showed a qualitative difference, in that the ligands were not detectably degraded, which is considered to be a consequence of receptor internalization. It is not known what role internalization plays for signal transduction, however,

our finding points to a change in receptor physiology. The fact that the cells did not respond to IFN in the two biological assays used suggest, that this is a consequence of downregulated receptors. Direct proof for this possibility can only be obtained once the signal transduction pathway for the IFN receptor has been elucidated.

Several possibilites could explain the observed differences in receptor presentation. Functions encoded by EBV that antagonizes receptor function is one way of understanding the results. However, as Burkitt's lymphoma lines contained the virus genome, at least no simple model could assemble the findings.

Another way of interpretation is based on the observation that IFN-α is produced by B lymphocytes during the initial phase of EBV infection and the frequency of immortalization is low. Therefore, selection effects could be imagined. In this study no levels of IFN-α or β was detectable in the culture media, nor any activity, which competed with the binding of the radioiodinated material.

It is established that cells constituting Burkitt's lymphoma and EBV-immortalized lines represent distinct differentiation stages, the Burkitt's lymphoma cells being more immature than the EBV-immortalized cells (6). It is not known whether IFN receptor presentation varies with differentiation or growth stage. However, we note that U-266 cells, a line with mature plasma cell phenotype, show receptor binding of the same type as the Burkitt's lymphoma cells. The role of growth and differentiation stage will be further explored.

The biological significance of the presence of a large number of IFN-α genes in mammalians is not yet established. Many of the variants can be assumed to represent alleles, although direct evidence is lacking so far. Genetic polymorphism on the receptor level has not been possible to detect. Such studies are easily done on EBV-immortalized lines derived from family material and the obvious, although unlikely, possibility that our finding just represent differential bindings of cloned IFN-α:s on receptors with allelic variation will be the object for further studies. It is striking that the lines EBV-2, 7 and 8 all are derived from the same family.

ACKNOWLEDGEMENTS
   This study was supported by a grant from KabiVitrum AB, Stockholm. The skilful technical assistance of Elisabeth Ivarsson, Chrissie Roth, Christina Strömberg and Jenny Larsson is acknowledged.

REFERENCES

1. Adams A, Strander H and Cantell K: Sensitivity of the Epstein-Barr virus transformed human lymphoid cell lines to interferon. J. Gen. Virol., 28: 207-217, 1975.
2. Leandersson T and Lundgren E: Cell growth regulation during density inhibition and interferon treatment. Exp. Cell Res., 138: 167-174, 1982.
3. Lotz M, Tsoukas CD, Fong S, Carson DA and Vaughan JH: Regulation of Epstein-Barr virus infection by recombinant interferons. Selected sensitivity to interferon-γ. Eur. J. Immunol., 15: 520-525, 1985.
4. Lund B, von Gabain A, Edlund T, Ny T and Lundgren E: Differential expression of interferon genes in a substrain of Namalwa cells. J. Interferon Res., 5: 229-238, 1985.

5. Hannigan GE, Gewert DR and Williams BRG: Characterization and regulation of α-interferon receptor expression in interferon-sensitive and -resistant human lymphoblastoid cells. J. Biol. Chem., 259: 9456-9460, 1984.
6. Nilsson K: The nature of lymphoid cells and their relationship to the virus. Epstein MA and Achong BG (eds). In: The Epstein-Barr virus, pp 227-281, Springer Verlag Berlin, 1979.

PRODUCTION OF ANTIBODIES AGAINST THE MURINE IFN-γ RECEPTOR.

S. Landolfo, F. Cofano, A. Fassio, L. Fava and G. Cavallo.

Institute of Microbiology, University of Torino, 10126 – Torino, .ce
Italy.

INTRODUCTION

Stimulation of T-lymphocytes with antigens or mitogens triggers the
release of several lymphokines, that regulate the immune response (1).
Of these, IFN-γ has been shown to control the activity of several
cell populations, namely T- and B-lymphocytes, natural killer cells
and macrophages (2). The initial IFN-target cell interaction occurs at
specific membrane receptors (3). Previous data have demonstrated that
the number of binding sites varies from 2,000 to 25,000, with
dissociation constants (Kd) between 10^{-9} and 10^{-11} (3). One of the
major obstacles, however, to characterization and purification of the
receptor molecule is that it is expressed in a number of binding sites
per single cell insufficient for its purification and biochemical
characterization.

This paper describes some preliminary studies performed in our
laboratory in order to characterize and purify the membrane receptor
for murine IFN-γ from cell lines expressing relatively high amounts
of IFN-γ binding sites.

RESULTS AND DISCUSSION

The binding of ^{125}I-MuIFN-γ to various cell lines was measured at
23° C to determine the binding characteristics of IFN-γ and cell
surface receptors, as previously described (4). As shown in Table 1,
the IFN-γ binding varied considerably among the cell lines tested.
EL-4 and L1210-C7 cells displayed the highest number of binding sites
per cell (90,000 and 50,000 respectively), whereas P388-D1 and the
L-929 displayed very few (450 and 1,000). A Scatchard plot analysis
gave a Kd between 2×10^{-9} for P388-D1 and 8×10^{-10} for EL-4 cells.
Being linear, moreover, this analysis showed a single class of
high-affinity receptors (data not shown). These results substantiate
previous reports that high-affinity receptors specific for IFN-γ

exist on the cell surface (3).

Table 1. Binding properties of ^{125}I-rMuIFN-γ to different murine cell lines.

Cell line	Origin	No of receptors	Kd (M)
L929	Fibroblast	2,000	3×10^{-9}
L1210-C7	B lymphoc.	50,000	2×10^{-10}
EL-4	T lymphoc.	90,000	8×10^{-10}
CTLL	T lymphoc.	20,000	6×10^{-9}
P388-D1	Macrophage	500	2×10^{-9}
TS/A	Fibroblast	1,500	5×10^{-9}
B16	Melanocyte	8,000	1×10^{-10}

Our next step was to see whether EL-4 cells with their many IFN-γ binding sites, could be employed as immunogen to produce antibodies recognizing the receptor, that would greatly facilitate the biochemical characterization and purification of this molecule. Four rabbits and 16 rats were extensively immunized with EL-4 cells.

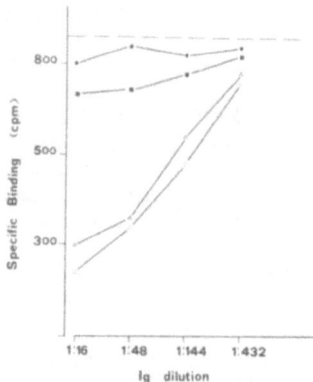

Fig. 1. Inhibition of ^{125}I-MuIFN-γ binding to L1210-C7 cells by Ig from rabbits immunized with whole L1210-C7 cells. One million of cells were mixed with serial dilutions of Ig from immunized ($\triangle$, $\bigcirc$) or normal rabbits ($\blacksquare$) or from immunized rats ($\bullet$). After 30 min at 23° C, ^{125}I-MuIFN-γ (30 x 10^3 cpm) was added for 2 hours. Specific binding was calculated by adding a 100-fold excess of cold MuIFN-γ.

Immunoglobulins (Ig) purified from their sera by ammonium sulfate precipitation were tested for their ability to inhibit ^{125}I-IFN-γ binding to EL-4 cells. As shown in Fig. 1, Ig from two rabbits inhibited (60% specific inhibition) IFN-binding to the target cells, whereas neither the Ig from immune rats nor these from normal rabbits displayed any inhibitory activity. The rabbit Ig recognizing the IFN-γ receptor were then used in immunoprecipitation studies with target

cells biosynthetically-labelled with ^{35}S-methionine to see whether the
receptor molecule could be purified. As expected anti-L1210 cell
antibodies immunoprecipitated from EL-4 cells about a dozen of surface
proteins . Only one of them, however, in the region of 90,000 could be
clearly inhibited by addition of cold IFN-ɣ in large excess ,
suggesting that this surface protein can bind IFN-ɣ and may thus be
the cell receptor for MuIFN-ɣ (5,6).

In their attempt to purify the cell receptor for interleukin-2 ,
Leonard et al. (7) covalently cross-linked the ligand to the cell
receptor and then immunoprecipitated the complex by adding
anti-interleukin-2 antibodies. We decided to adopt a similar strategy.
EL-4 cells, cross-linked by Disuccinimidyl suberate (DSS) (Pierce)
with ^{125}I-IFN-ɣ , were incubated separately with two different
anti-MuIFN-ɣ monoclonal antibodies: a) AN-18, produced in our
laboratory (8) and b) H5.102.1, kindly provided by S. Russell,
Gainesville, USA, in the attempt to immunoprecipitate the entire
complex. As expected, both MoAb precipitated free ^{125}I-IFN-ɣ not
cross-linked on the cell membrane, both the monomeric (Mr=15,000) and
the dimeric (Mr=30,000) form , but one MoAb, namely AN-18, also
identified a third band (Mr 105,000) (Fig.2). This band migrates in
the same region of the gel as the receptor cross-linked to ^{125}IIFN-ɣ
suggesting that the two bands are very similar if not identical. The
marked density of the bands after MoAb treatment is simply due to the
fact that immunoprecipitation is followed by enrichment of ^{125}I-IFN-ɣ
in both the cross-linked and free forms.

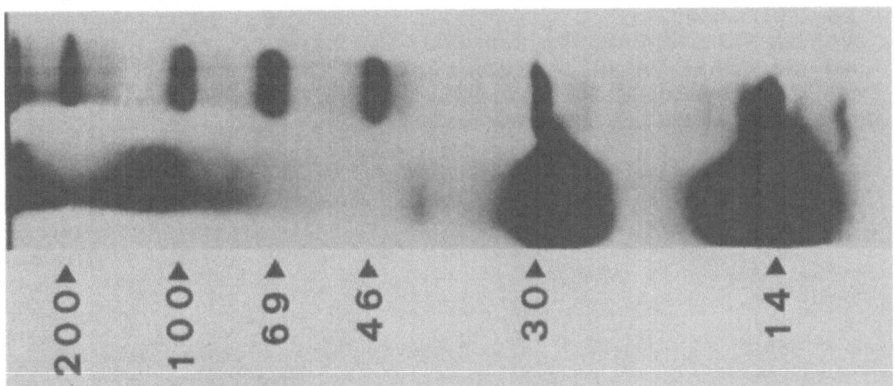

Fig. 2. Gel electrophoresis of ^{125}I-MuIFN- covalently linked to
receptors of EL-4 cells and immunoprecipitated with rat anti-MuIFN-ɣ
monoclonal antibodies. EL-4 cells were cross-linked to ^{125}I-MuIFN-ɣ
with 1mM DSS and the IFN-receptor complex was immunoprecipitated with
anti-MuIFN-ɣ monoclonal antibodies (Left line). Molecolar weights are
shown in the right line.

In conclusion, in this study we describe two cell lines, namely

EL-4 and L1210-C7, expressing relatively high levels of IFN-γ receptors. L1210-C7 cells, used as immunogen, induced in rabbits the production of Ig that prevented IFN-γ from binding to its receptor and immunoprecipitated from metabolically-labelled EL-4 cells a surface protein identical to the IFN-γ receptor. As an alternative approach, the receptor was cross-linked to ^{125}I-IFN-γ and immunoprecipitated with an anti-IFN- MoAb capable of recognizing the ligand bound to its own receptor. These two systems thus represent useful models for further purification and biochemical characterization of the IFN-γ cell receptor.

This work was supported by grants from the National Research Council, PF 'CMI' and 'Oncologia, and from the Ministry of Public Education (40%).

REFERENCES

1. Paul W. (ed): Fundamental Immunology. New York:Raven Press, 1984.
2. Vilcek J., Gray P.W., Rinderknecht E., and Savastopoulos C.G. Lymphokines 11: 1, 1985.
3. Zoon K.C. and Arnheiter H.: Pharmac. Ther. 24: 259, 1984.
4. Cofano F., Fassio A., Cavallo G. and Landolfo S.: J. Gen. Virol. 67: 1205, 1986.
5. Littman S.J., Faltynek C.R., and Baglioni C.: J. Biol. Chem. 260: 1191, 1985.
6. Wietzerbin J., Gaudelet C., Aguet M., and Falcoff E.: J. Immunol. 136: 2451, 1986.
7. Leonard W.J., Depper J.M., Robb R.J., Waldmann T.A. and Green W.C.: Proc. Natl. Acad. Sci. USA 80: 6957, 1983.
8. Prat M., Gribaudo G., Comoglio P.M., Cavallo G. and Landolfo S.: Proc. Natl. Acad. Sci. USA 81: 4515, 1984.

PURIFICATION AND CHARACTERIZATION OF THE RECEPTOR FOR HUMAN INTERFERON-γ

D. NOVICK, P. ORCHANSKY, M. REVEL AND M. RUBINSTEIN
Department of Virology, The Weizmann Institute of Science, Rehovot, Israel.

1. SUMMARY

The receptor for human interferon-γ (IFN-γ) was purified from foreskin fibroblasts. Triton X-100 extracts obtained from either intact cells or membrane preparations were passed through an immobilized interferon-column. Sodium dodecyl sulfate-polyacrylamide gel electrophoresis (SDS-PAGE) of eluted fractions revealed a major band of $Mr=95,000$ and minor bands of $Mr=80,000$ and $60,000$. Further purification was obtained by lectin chromatography. The purified receptor retained the ability to bind ^{125}I-IFN-γ with a Kd of $2.2 \times 10^{-10}M$, a value close to that obtained with intact fibroblasts $(5 \times 10^{-10}M)$. A complex of $Mr=105,000-125,00$ was visualized by immunoprecipitation of ^{125}I-IFN-γ cross-linked to the purified receptor followed by SDS-PAGE and autoradiography. A similar complex was obtained when ^{125}I-IFN-γ was cross-linked to intact cells.

2. INTRODUCTION

Interferon-gamma (IFN-γ) is a product of activated T-lymphocytes and it exerts antiviral activity, growth inhibitory effect and several immuno-regulatory activities on a variety of cell types (1). As all other poly-peptide hormones and lymphokines it acts on various cells via a specific cell surface receptor. The existence of a specific receptor for IFN-γ on various cells was demonstrated in several studies (2-12). In all cases, specific receptors were demonstrated by high affinity binding of ^{125}I-labelled IFN-γ (Kd $<10^{-8}M$) and by specific competition with unlabelled IFN-γ. No competition was noticed with the other types of human inter-ferons, namely IFN-α or IFN-β (4,6,7,10). Cross-linking experiments of labelled IFN-γ to intact cells yielded a complex with an apparent molecular weight of $90,000-125,000$ (6,7,9,12). This complex was related to the membranal receptor of IFN-γ. We have recently presented evidence that human cells of hematopoietic origin may have an IFN-γ receptor that is structurally and functionally different from the receptors in cells of non-hematopoietic origin (11). Isolation followed by structural analysis of these receptors would help to understand their mode of action, to establish their diversity and to identify the respective genes.

In the present study we describe methods for purification of IFN-γ receptor from human fibroblasts. Initial characterization indicates that the purified receptor exhibits the same affinity for IFN-γ and the same molecular weight as those of the cell bound receptor.

3. PROCEDURE
3.1. Materials

IFN-γ was produced in recombinant Chinese hamster ovary cells (13) and purified to homogeneity on a monoclonal antibody column (14). Pure IFN-γ $(5 \times 10^{7}$ units/mg) was labelled with the Bolton and Hunter reagent (4,19) to a specific activity of 300-1000 cpm/unit ($3 \times 10^5 - 10^6$ Ci/mol). Anti IFN-γ serum was prepared by immunizing rabbits with pure IFN-γ and exhibited a neutralizing titer of 300,000 IFN-γ units/ml. Affigel-10 was purchased from BioRad.

3.2 Methods

Preparation of cell membranes - All procedures were performed at 0-4°C. Fibroblasts (3×10^{10} cells) were washed with Dulbecco balanced salt solution and suspended in a hypotonic buffer (final concentration: 10 mM Tris-HCl pH 7.5, 1 mM $MgCl_2$, 1 mM $CaCl_2$, 1 mM PMSF, 20 µl/ml aprotinin). The suspension was vigorously mixed on a Vortex shaker and allowed to stand for 5 min before being spun at three separate stages: 700xg for 5 min, 3500xg for 20 min and 40,000xg for 1 h. The 40,000xg precipitate was collected.

Solubilization of the receptor and affinity chromatography - Intact cells or their membranes were suspended in a solubilization buffer (final concentration: 1% Triton X-100, 10 mM Hepes, 150 mM NaCl, 1 mM PMSF, 20 µl/ml aprotinin, pH 7.5) and left at 4°C for 1 h with occassional shaking. The mixture was then spun (10,000xg, 15 min, 4°C) and the supernatant was collected and spun in an ultracentrifuge (100,000xg, 60 min, 4°C). The supernant was applied to an immobilized IFN-γ column (7 mg IFN-γ coupled to 1 ml of Affigel-10 according to a procedure provided by the manufacturer) at a flow rate of 15 ml/hr at 4°C. The column was washed with 30 ml phosphate buffered saline (PBS) containing 0.1% Triton X-100 and then the receptor eluted with a buffer consisting of 50 mM Na_2CO_3, 500 mM NaCl, 0.05% Triton X-100, pH 11. Ten fractions of 1 ml were collected and immediately neutralized to pH 7.5 with 3M acetic acid.

Lectin affinity chromatography - Peak fractions eluted from the IFN-γ column were applied to an RCA-120 agarose column (0.5 ml) at a flow rate of 0.2 ml/min at 4°C. The column was washed with 20 ml wash buffer (20 mM Hepes, 150 mM NaCl, 10 mM $MgSO_4$, 0.1% Triton X-100, pH 7.5). The receptor was then eluted with a solution of 0.3M β-methyl galactopyranoside in the wash buffer and fractions of 0.5 ml were collected.

Binding of radiolabelled IFN-γ to intact cells - HeLa cells in 35 mm wells (2.4×10^6 cells/well) were incubated for 2.5 hrs at 4°C with increasing concentrations (1-1000 pM) of ^{125}I-IFN-γ in Dulbecco balanced salt solution. The cells were then washed three times with cold Dulbecco balanced salt solution, detached with trypsin and the cell associated radioactivity was counted. The dissociation constant was calculated by a Scatchard analysis using the LIGAND program (15).

Binding of radiolabelled IFN-γ to a soluble receptor - Aliquots (20-30 µl) of the solubilized receptor from various purification steps were mixed with ^{125}I-IFN-γ (250 units) either with or without unlabelled IFN-γ (100,000 units) in 20 mM Hepes buffer pH 7.5 containing 0.1% BSA (200 µl). The mixture was incubated for 2 hrs at 4°C, rabbit IgG (0.1% in 0.1M phosphate buffer pH 7.5, 0.5 ml) was then added, followed by PEG-8000 (22% in 0.1M phosphate buffer pH 7.5, 0.5 ml). The mixture was left for 10 min at 4°C and then passed through a 0.45µ filter (25 mm HAWP, Millipore). The filters were washed with cold PEG-8000 solution (5% in 0.1M phosphate buffer, pH 7.5), and counted. Background counts were determined in the presence of excess unlabelled IFN-γ and were subtracted. The affinity of the solubilized receptor was determined in a similar manner but with a constant amount (0.2 µg) of receptor and increasing amounts (0-6000 pM) of ^{125}I-IFN-γ. The dissociation constant was calculated with the aid of a LIGAND program (15).

Cross-linking of radiolabelled IFN-γ to intact cells - Either HeLa or WISH cells (1×10^8 cells obtained from EDTA-treated monolayers) were suspended in 1 ml of cold Dulbecco balanced salt solution. ^{125}I-IFN-γ (10,000 units,

2×10^6 cpm) was added with our without a 100-fold excess of unlabelled IFN-γ and the suspension was left for 90 min at 4°C. DSS dissolved in dimethylsulfoxide (DMSO) was then added to a final concentration of 1 mM and the suspension was left for 20 min at 4°C. The cells were then washed with Dulbecco balanced salt solution and then solubilized in 20 mM Tris buffer containing 1.5% Triton X-100, 1 mM PMSF and 20 μl/ml aprotinin (0.5 ml, pH 6.5). The suspension was left for 1 h at 4°C, and then spun (27,000xg, 20 min, 4°C). The supernatant was immunoprecipitated by incubation with rabbit anti IFN-γ serum (1:50) for 2 hrs at room temperature followed by addition of Protein-A Sepharose. The sepharose beads were washed three times with 10 mM phosphate buffer pH 7.5 containing 1% BSA, 1% NP-40 and 2M KCl and then twice with PBS. The beads were suspended in a sample buffer containing 2% β-mercaptoethanol and the supernatant analysed by SDS-PAGE followed by autoradiography.

Cross-linking of radiolabelled IFN-γ to the soluble receptor – Preparations of the soluble receptor (0.5-2.5 μg were mixed with ^{125}I-IFN-γ (250 units, 2×10^5 cpm) and left for 1 h at room temperature. DSS (dissolved in DMSO) was added to a final concentration of 0.3 mM. The cross-linking was stopped after 20 min at 4°C by the addition of Tris-HCl buffer pH 7.5 to a final concentration of 20 mM. Aliquots were either analysed directly by SDS-PAGE and autoradiography or immunoprecipitated with rabbit anti IFN-γ serum as described for cross-linking with intact cells.

Protein determination – Protein was determined by the fluorescamine method (16). Crystalline bovine serum albumin was used as a standard.

Gel electrophoresis and autoradiography – Polyacrylamide gel electrophoresis was carried out according to the method of Laemmli (17) using 7.5% acrylamide (final concentration). Either radiolabelled protein standards (Amersham) or low molecular weight protein standards (Pharmacia) were used. Gels were either stained by a silver stain method (18) or dried and autoradiographed.

4. RESULTS

Purification of the IFN-γ receptor – Affinity chromatography on an immobilized IFN-γ column was used as the main step of IFN-γ receptor purification. A purification profile of the receptor from a Triton X-100 extract of whole cells is shown in Fig. 1. This figure demonstrates the ratio of specific activities in the eluted fractions vs. the load fraction indicating a high degree of purification. However, the degree of purification could not be accurately determined since in most of the cases no specific binding could be demonstrated in the load fraction.

SDS-PAGE of aliquots from the various fractions is shown in Fig. 2. Under reducing conditions a major band of apparent Mr=95,000 was visualized by silver staining of the gel. Additional bands of Mr=80,000 and 60,000 were seen. Under non reducing conditions the major protein band was shifted from Mr=95,000 to Mr >200,000.

Similar patterns of binding activity and protein bands were obtained when a Triton X-100 extract of isolated membranes, rather than the whole cells was used as a starting material for the purification of the receptor. Additionally, binding activity was demonstrated in the crude membrane extracts in some preparations. A purification factor of 300-600 was obtained in these cases. However, since the yields of purified receptor were lower than those obtained with whole cell extracts, the step of membrane isolation was omitted. Additional three-fold purification of the receptor was obtained by lectin chromatography (Fig. 3). Representative purification experiments are summarized in Table 1.

TABLE 1. Purification of human IFN-γ receptor

Exp.	Step	Protein (mg)	IFN-γ binding (pmole)	Specific activity (pmol/mg)	Purifi- cation (-fold)	Recovery %
1.	Solubilized cells[a]	445	40	0.09	1	100
	IFN-γ agarose eluate	0.025	6	240	2670	15
2.	Solubilized cells[a]	1224	0	–	–	–
	IFN-γ agarose eluate	0.08	13	148	–	–
3.	Membrane fraction	156	48	0.3	1	100
	Solubilized membranes[a]	48	16	0.3	1	33
	IFN-γ agarose eluate	0.08	8	100	325	17
4.	IFN-γ agarose eluate	0.036	6	167	1	100
	RCA agarose eluate	0.014	6	438	2.6	100

a. 100,000xg supernatant.

Fig. 1

Fig. 2

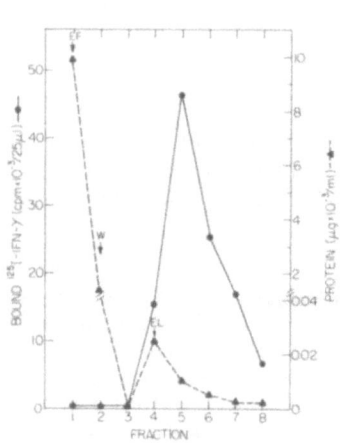

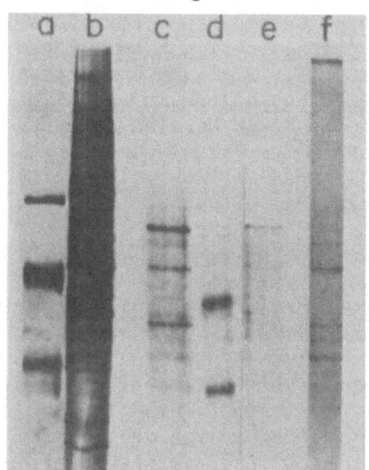

FIGURE 1 (left). Ligand affinity chromatography of the IFN-γ receptor.
Triton X-100 extract (100 ml) of whole cells (3×10^{10}) was applied to an IFN-
γ agarose column (1 ml). The column was then washed and eluted as described
under Methods. Aliquots (25 µl) of individual fractions (effluent EF, 100
ml); wash (W, 2x15 ml) and elutions (EL, 5x1 ml) were assayed for ^{125}I-IFN-γ
binding (●——●) and for protein by the fluorescamine method (▲---▲). Background counts (8500 cpm) were subtracted from all readings.
FIGURE 2 (right). SDS-PAGE of the purified IFN-γ receptor. Lanes: a & d,
molecular weight markers (from top: phosphorylase 94,000; bovine serum
albumin 67,000, ovalbumin 43,000. In lane d, phosphorylase is barely seen);
Triton X-100 cell extract (100,000xg supernatant); c, purified receptor from
the IFN-γ agarose column (concentrated by ultrafiltration); e, purified receptor as in c but prior to concentration; f, purified receptor as in lane c
but under non reducing conditions. Proteins were visualized by silver
staining.

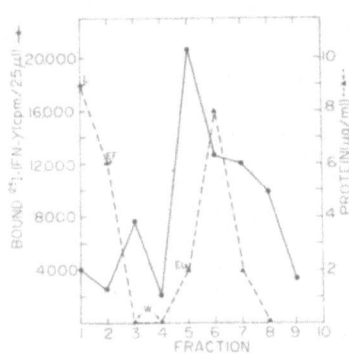

FIGURE 3. Chromatography of the receptor on RCA-120 agarose. Purified receptor from IFN-γ agarose chromatography (36 μg, 4 ml) was loaded on a RCA-120 agarose column (0.5 ml). The column was washed and the receptor eluted as described under Methods. ^{125}I-IFN-γ binding activity (●——●) was determined in 25 μl aliquots of the various fractions (L: load, EF: effluent, W: wash, EL: eluate). Background counts (7000 cpm) were subtracted from all readings. Protein (▲----▲) was determined by the fluorescamine method.

Molecular weight determination of the IFN-γ receptor complex - Cross-linking experiments of ^{125}I-IFN-γ to intact cells and to the purified receptor preparations were performed. The specificity of the interaction was demonstrated by competition with an excess of unlabelled IFN-γ. SDS-PAGE of the cross-linking reaction products from intact cells followed by autoradiography revealed a broad band of Mr=105,000-125,000. A similar analysis performed on the cross-linking products of ^{125}I-IFN-γ and the purified receptor from a membrane preparation gave two bands of Mr=105,000 and 125,000 (Fig. 4). An additional broad band of Mr=65,000-85,000 was seen, however, this band was not immunoprecipitated with the anti-IFN-γ serum.

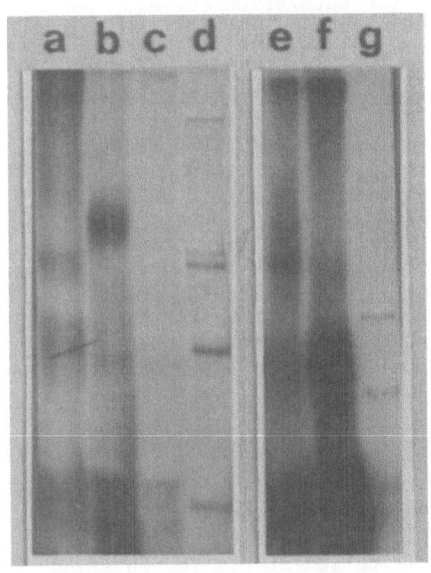

FIGURE 4. SDS-PAGE of ^{125}I-IFN-γ cross-linked to the receptor. ^{125}I-IFN-γ was cross-linked to itself (lane a), to WISH cells in the absence (lane b) or presence of unlabelled IFN-γ (lane c). ^{125}I-IFN-γ was also cross-linked to the purified receptor (lanes e and f). All samples, except for the one in lane f) were immunoprecipitated prior to electrophoresis. ^{14}C-methylated molecular weight markers (lanes d and g) were (from top): myosin 200,000; phosphorylase B 100,000 and 92,000; bovine serum albumin 69,000 and ovalbumin 46,000.

Determination of the affinity of ^{125}I-IFN-γ to intact cells and to the solubilized receptor - Binding experiments with ^{125}I-IFN-γ were performed on a variety of intact cells including foreskin fibroblasts, transformed fibroblasts (HeLa) and epithelial cells. A dissociation constant (Kd) of $3-5\times10^{-10}$M and receptor density of 10,000-20,000 molecules per cell were calculated in all cases with the aid of the LIGAND program. A typical saturation curve and Scatchard analysis are shown in Fig. 5. The linear Scatchard plot indicates a single type of high affinity binding sites.

Experiments done in order to determine the affinity of ^{125}I-IFN-γ to the purified receptor yielded a typical saturation curve (Fig. 6, inset). Scatchard analysis performed by the LIGAND program revealed a second order reaction, corresponding to the single class of the high affinity binding sites found on intact cells (Fig. 6). A Kd value of 2.2×10^{-10}M was calculated from the binding data. The active receptor represented 2.5% (by weight) of the total protein in the purified receptor preparation.

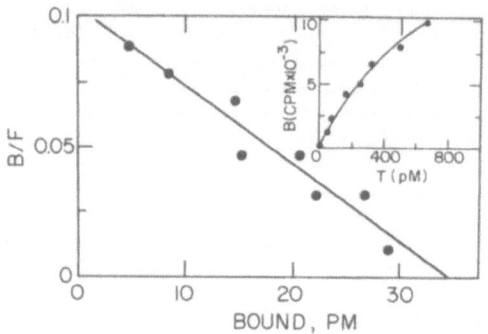

FIGURE 5. Equilibrium binding of ^{125}I-IFN-γ to HeLa cells. HeLa cells (2.4×10^6) were incubated with various concentrations of ^{125}I-IFN-γ (12.2 µCi/ug) for 2.5 hr at 4°C and assayed for binding as described under Methods. The binding data were analysed by LIGAND program. Non specific binding was 0.02±13% and the Kd was $3.3\cdot10^{-10}$M±18%. Inset: Plot of the saturation curve.

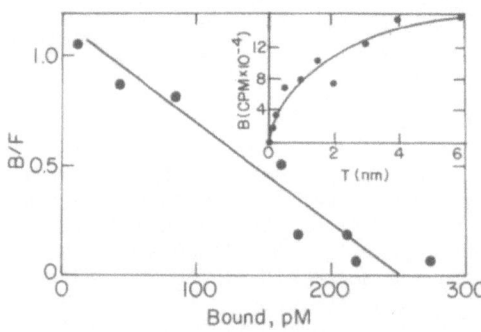

FIGURE 6. Equilibrium binding of ^{125}I-IFN-γ to the purified receptor. Aliquots of the receptor from the IFN-γ agarose column (0.2 µg) were incubated together with increasing concentrations of ^{125}I-IFN-γ for 2 hr at 4°C as described under Methods. Binding data were analyzed by the LIGAND program. Non specific binding was 0.33±22% and the Kd was 2.2×10^{-10}±16%. Inset: Plot of the binding curve.

5. DISCUSSION

In the present study, solubilization and purification of the IFN-γ receptor from human fibroblasts are described. Affinity chromatography on an immobilized IFN-γ column is the main purification step. The purified receptor exhibits a major band of Mr=95,000 and retains the same affinity to IFN-γ as the receptor on intact cells.

Based on the calculated number of receptors on intact cells and an assumed molecular weight of the receptor (95,000) a batch of 2×10^{10} cells should contain at least 50-100 µg of receptor and therefore a 10,000-fold purification (starting with cell extract) is required in order to obtain a pure receptor. Our affinity chromatography yielded 25-90 μg of a partially purified receptor. Specific binding of ^{125}I-IFN-γ to the solubilized cell extract (100,000xg supernatant) could not be demonstrated in most cases and a purification factor could not be calculated. However, in a few cases where specific binding was obtained, a purification factor of 2500-3000 was achieved (Table 1, Exp. 1). Specific binding was obtained and a purification factor of 300-600 was calculated when membrane fraction rather than whole cells was prepared and used for extraction of the receptor (Table 1, Exp. 3). However, the intermediate step of membrane preparation did not provide any advantage in terms of the overall yield or purity and therefore was omitted. The lectin chromatography on ricin-agarose which was used as a second purification step indicated that the receptor is a glycoprotein but gave only a marginal purification (2-6 fold). Based on these results, receptor preparations obtained from whole cell extracts and purified by a single step of chromatography on an IFN-γ agarose column were used for characterization.

Cross-linking experiments of ^{125}I-IFN-γ to intact cells yielded a complex of molecular weight 105,000-125,000 (refs. 6,7,9,12 and Fig. 4). Similarly, bands of Mr=125,000-105,000 were visualized by autoradiography when the purified receptor was cross-linked to ^{125}I-IFN-γ and the product was analysed by SDS-PAGE. The additional broad band of Mr=65,000-85,000 could either stem from degradation of the receptor or represent a subunit of the receptor molecule. Assuming a 1:1 molar ratio of ^{125}I-IFN-γ and its receptor, the net molecular weight of the receptor is 80,000-95,000. Indeed, these calculated values are in agreement with the bands observed in silver stained polyacrylamide gels of the purified receptor, namely, a major band of Mr=95,000 and a band of 80,000 (Fig. 3). The Mr=60,000 band may correspond to the broad band of Mr=65,000-85,000 which was observed by autoradiography. The purified receptor exhibited the same affinity for ^{125}I-IFN-γ as the receptor on intact cells. Based on the Scatchard analysis only 2.5% of the protein in the purified receptor preparation represented active receptor molecules. The other proteins in the preparation probably include inactive receptor molecules which migrate as a band of Mr=95,000 on SDS-PAGE under reducing conditions. The additional bands are probably related to degradation products as well as to non-receptor proteins which were bound to the IFN-γ agarose column.

In conclusion, purification of active human IFN-γ receptor was demonstrated and the purified receptor was characterized and used for production of antibodies. This study will enable further structural and functional analysis of the human IFN-γ receptor.

6. ACKNOWLEDGEMENTS
    The authors acknowledge InterYeda Ltd,, Ness-Ziona, Israel for the generous supply of fibroblasts. The assistance of Mr. Z. Palash and Mr. Y. Papeer is acknowledged. M.R. has the Edna and Maurice Weiss Chair in Interferon Research.

7. REFERENCES

1. Trinchieri G and Perussia B: Immunol Today 6, 131-136, 1985.
2. Anderson J, Yip YK, and Vilcek J: J Biol Chem 257, 11031-11304, 1982.
3. Anderson J, Yip YK, and Vilcek J: J Biol Chem 259, 6497-6502, 1983.
4. Orchansky P, Novick D, Fischer DG, and Rubinstein M: J.INterferon Res. 4, 275-282,
5. Celada, A, Gray PW, Rinderknecht E, and Schreiber RD: J.Exp. Med.

160, 55-74, 1984.
6. Sarkar, FH and Gupta SL: Proc Natl Acad Sci USA 81, 5160-5164, 1984.
7. Littman SJ, Faltyneck CR, and Baglioni C: J Biol Chem 260, 1191-1195, 1985.
8. Finbloom DS, Hoover DL, and Wahl LM: J Immunol 135, 300-305, 1985.
9. Celada A, Allen, R, Esparza I, Gray PW, and Schreiber RD: J Clin Invest 76, 2196-2205 (1985).
1 . Rashidbaigi A, Kung H, and Pestka S: J Biol Chem 260, 8514-8519, 1985.
11. Orchansky P, Rubinstein M, and Fischer DG: J Immunol. 136, 169-172, 1986.
12. Rashidbaigi, A, Langer JA, Jung V, Jones C, Morse HG, Tishfield JA, Trill JJ, Kung H, and Pestka S: Proc Natl Acad Sci USA 83,383-388, 1986.
13. Mory, Y, Ben-Barak J, Segev D, Cohen B, Novick D, Fischer, DG, Rubinstein M, and Revel M: In: The Biology of the Interferon System (Kirchner H and Schellekens H. eds), pp. 33-41, Elsevier, Amsterdam Science Publ., B.B. 1985.
14. Novick D, Eshhar Z, Fischer DG, Friedlander J, and Rubinstein M. EMBO J, 2, 1527-1530, 1983.
15. Munson PJ and Rodbard D: Anal. Biochem 107, 220-239, 1980.
16. Stein S and Moschera J: In: Methods Enzymol (Petska, S, ed). Vol 79, pp. 7-16, Academic Press, New York, 1981
17. Laemmli UK. Nature 227, 680-685, 1970.
18. Oakley BR, Kirsch DR, and Moriss NR: Anal BIochem 105, 361-363, 1980.
19. Bolton AE and Hunter WM: Biochem J 133, 519-539, 1973.

INDUCTION

INTERFERON INDUCTION BY TRANSFECTION OF SENDAI VIRUS C GENE cDNA

H. Taira, T. Kanda, T. Omata, M. Kawakita[+], H. Shibuta[+],
K. Iwasaki.
The Tokyo Metropolitan Institute of Medical Science,
[+]University of Tokyo, Tokyo, Japan

1. INTRODUCTION

To elucidate the mechanism of interferon (IFN) induction
on virus infection, we constructed two types of plasmids by
inserting a part of the cDNA of the Sendai virus into a
SV40-derived expression vector (pSV2-0); one, pSV2-PC,
contains the P+C gene, which codes for the P and C proteins
in overlapping reading frames, and the other, pSV2-C, codes
for only the C gene. After transfection them into mammalian
cells, we determined the IFN activity in the culture medium.
We found that the level obtained with pSV2-PC was
significantly positive but very low, while that with pSV2-C
was as high as or even higher than that observed in the
culture medium after Sendai virus infection. By cleaving
pSV2-C between the SV40 promotor and the C gene or by
inserting a stop codon within the C gene, induction of IFN
was greatly diminished. These results indicated that both
transcription and translation of the C gene is necessary for
IFN induction after Sendai virus infection.

2. PROCEDURE

2.1. Materials and methods

2.1.1.Construction of P+C gene-containing expression
plasmids. pSV2-0: an expression vector, pSV2-0, was
constructed from the pSV2-RaβG plasmid by removing the
rabbit β-globin cDNA from the latter and adding the HindIII
linker to both its ends, followed by ligation. Thus, pSV2-0
includes the SV40 replication origin and the early promotor
(0.71-0.65 map units), the small tumor-antigen intron
(0.56-0.44 map units) and the SV40 poly (A) addition site
from the early gene (0.19-0.1 map units).

pSV2-PC: the NcoI/HhaI fragment (2 kb) of cDNA clone H33
containing the Sendai virus P+C gene, was isolated and
ligated to HindIII-cleaved pSV2-0 plasmid DNA using T4 DNA
ligase.

pSV2-PC(Rl): pSV2-PC was cleaved with PvuII and SalI at
the 5' side of the P+C gene, yielding a 5,700 bp fragment (A
fragment) and 806 bp fragment (B fragment). The B fragment
was further partially cleaved with DdeI, yielding a PvuII-
DdeI fragment of 285 bp (C fragment) and two DdeI-SalI
fragments of 307 bp (D fragment) and 336 bp fragment (E
fragment). pSV2-PC(Rl) was constructed by ligating the
isolated A, C and E fragments.

pSV2-C: pSV2-PC was cleaved with PvuI and SalI to yield

three fragments; a PvuI-SalI fragment of 314 bp (C'
fragment) containing the first ATG codons of the P+C gene, a
SalI-PvuI fragment of 4017 bp (A' fragment) containing the
rest of the P+C gene, the SV40 small tumor-antigen intron
and the poly(A) addition site, and a PvuI-PvuI fragment of
2161 bp (B' fragment) containing the SV40 early promotor,
the replicating origin and pBR322. The PvuI-SalI fragment
(C') was treated with Sau3A. The resulting Sau3A-SalI
fragment was ligated to the B' and A' fragments to construct
the pSV2-C plasmid. Thus, pSV2-C lacks the first ATG codon
of the P gene, but retains that of the C gene.

pSV2-C(HindIII/BAP): pSV2-C was treated with HindIII and
bacterial alkaline phosphatase (BAP) to cleave the linkage
between the SV40 promotor and the inserted fragment.
pSV2-C(stop): an ochre codon was inserted into pSV2-C at
about 230 nucleotides down stream from the first ATG codon.
This was prepared by introducing the HpaI linker into the C
gene at its SalI (=HincII) site. pSV2-C was partially
cleaved with HincII followed by complete digestion with
SalI. The HincII cleaved full length DNA was isolated and
its 5' phosphate was removed with calf intestine alkaline
phosphatase. The HpaI linker (5'pGTTAAC) was introduced at
the HincII site using T4 DNA ligase. After transforming
E. coli HB101 with this preparation, the plasmid containing the HpaI linker was selected. A part of the nucleotide
sequence of P+C gene is also shown in Fig. 1, in which the
ATG codons of either the P gene or the C gene are
underlined. R1 (AGGGTGAAAG) and R2 (TAAGAAAAA) are the
consensus sequences observed in the Sendai virus genome.

2.1.2.Expression of the P+C gene under the control of the
SV40 promotor in monkey COS-1 cells. COS-1 cells were grown
in a 35-mm tissue culture dish and then transfected with 10
µg of plasmids (pSV2-0, pSV2-PC, pSV2-C) by calcium
phosphate precipitation methods(1). After 4 hrs the cells
were washed with Dulbecco's modified Eagle's medium (DME)
treated with 15% glycerol-Hepes buffered saline for 90 sec,
washed with DME, and then fed with 1.5 ml of DME supplemented with 10% calf serum. After incubation at 37°C for 40
hrs, the cells were labeled with [^{35}S]methionine (30 µCi/
dish) for 3 hrs. The labeling was stopped by removal of the
medium and washing with ice-cold phosphate buffered saline.
The cells on each dish were lysed with 300 µl of extraction
buffer[50 mM Tris-HCl, pH7.5 150 mM NaCl, 1% deoxycholate,
1% Triton X-100, 0.1% SDS, 1.7 mM PMSF]. The lysates(100
µl) were immunoprecipitated with 10 µl of antiserum from
Sendai virus-infected guinea pig or 40 µl of C-peptide
antiserum and protein A Sepharose CL-4B.

2.1.3.Northern blot analyses of RNAs(2). Total cellular
RNA was prepared from transfected COS-1 cells. Poly(A) RNA
was recovered from total cellular RNA by oligo(dT) cellulose
column. Poly(A) RNAs (10 µg) were electrophoresed through
1.3% agarose gel containing formaldehyde and transferred to
nitrocellulose. The [^{32}P]-labeled HindIII cDNA insert of
pSV2-PC was used for hybridization.

3. RESULTS AND DISCUSSION

Since the discovery of IFN, much progress has been made in the elucidation of the mechanism of IFN induction on virus infection, and several hypotheses have been presented, such as that it is a result of the production of virus-induced double-stranded RNA, or of depletion of the repressor due to the virus infectin(3), but the nature of the inducer molecule remained unknown. Thus, we investigated the IFN induction on transfection of plasmids containing a part of the cDNA of the Sendai virus genome RNA. Firstly, we constructed an expression vector, pSV2-0, from pSV2-RaβG, which is under the control of a SV40 promotor, by removing rabbit β-globin cDNA sequences from it followed by the addition of the HindIII linker to both its ends. As a source of the P+C gene, we chose cDNA clone H33, which contains the entire P+C gene(4). The NcoI/HhaI fragment of H33 containing 2 kilobases (kb) was inserted into pSV2-0 to produce pSV2-PC (Fig. 1). pSV2-PC(Rl) was constructed from pSV2-PC by removing the remaining NP gene, the 3'-untranslated region and the 3'-consensus sequence, R2, for the NP gene. pSV2-C is a plasmid that was constructed from pSV2-PC by deleting a portion containing the first ATG codon, which should express only the C gene. The structures of these plasmids are schematically illustrated in Fig. 1.

One μg of each plasmid was transfected into mouse L cells or monkey COS-1 cells (10^6 cells) by the calcium phosphate precipitation method(1). After incubation at 37°C for 4 hrs, the cells were rinsed with culture medium minus serum, treated with 15% glycerol-Hepes buffered saline for 90 sec, washed with culture medium, fed with culture medium supplemented with 5% calf serum, and then incubated for 16 hrs. Human lymphocytes were purified on Ficoll-paque and grown in RPMI medium supplemented with 10% fetal bovine serum. For transfection, lymphocytes (10^6 cells/ml) were incubated with 2 μg of plasmid DNA and DEAE-dextran at a concentration of 100 μg/ml (5). After incubation at 37°C for 30 min, the cells were washed with serum-free RPMI, fed

TABLE 1.Induction of interferon on transfection with plasmid

	Mouse L$_{929}$	Monkey COS-1	Human Lymphocytes
No DNA	< 1 IU/ml	< 1 EU/ml+	6 IU/ml
pSV2-0	< 1	< 1	6
pSV2-PC	1 ~ 10	< 1	12
pSV2-PC(Rl)	100	1	6
pSV2-C	1,000	128	100
pSV2-C(HindIII/BAP)	1	< 1	6
pSV2-C(stop)	1	< 1	12
Sendai virus	100	32	2,000

+In the case of COS-1 cells, IFN levels were assayed with human FL cells and expressed in experimental units (EU).

134

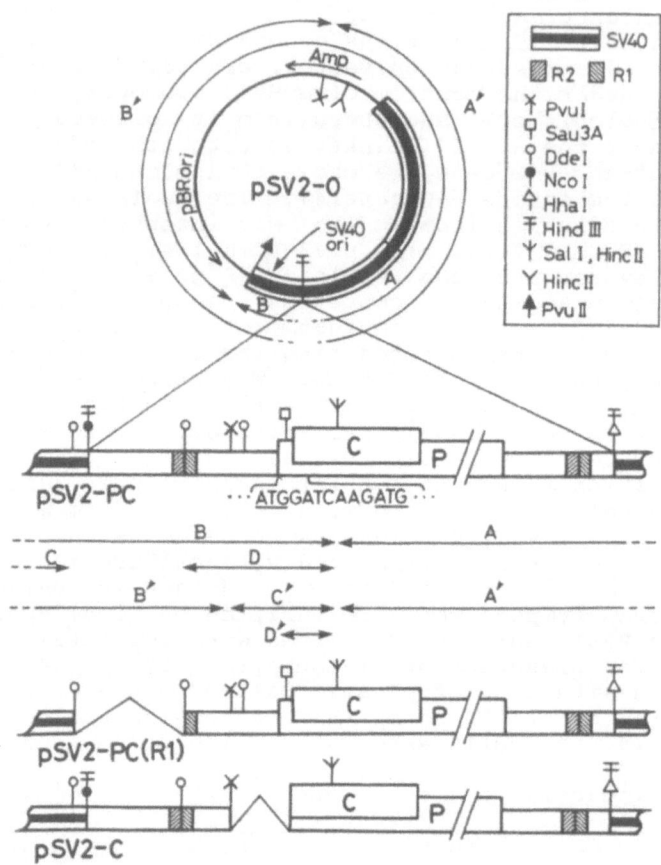

FIGURE 1. Construction of P+C gene-containing expression plasmids

with 1 ml of RPMI supplemented with 10% fetal bovine serumn, and then incubated for 20 hrs.

As shown in Table 1, the culture medium of L cells after 20 hrs transfection with pSV2-C showed very high antiviral activity, which was comparable to or even higher than that observed in the culture medium after Sendai virus infection. The activity, however, was low in the culture medium of cells transfected with pSV2-PC or pSV2-PC(Rl), being 1-0.1% or 10% of that with pSV2-C, respectively, and almost nil with pSV2-O. In the culture medium of monkey COS-1 cells transfected with pSV2-C, a significant increase in the activity was detected, which was expressed in the experimental unit since the international standard of the monkey IFN was unavailable. The similar observation as above was made with the culture medium of human lymphocytes, although the values were rather low as compared to that in the case of L cells, which was presumably due to the toxic effect of DEAE-dextran used for transfection, and to the low efficiency of transfection. The viability of lymphocytes determined by Trypan blue staining became about 10-20% after the DEAE-dextran treatment, whereas that of L cells was almost 100% after the calcium phosphate gel treatment. Furthermore, the uptake of $[^{32}P]$cDNA fragments of pSV2-PC into the viable lymphocytes measured according to Loyter et al. (6) was below 0.1% of total DNA added as compared to 2% in the case of L cells.

The antiviral activity produced on transfection of L cells with pSV2-C was examined. The activity was completely neutralized by anti-mouse IFN-α,β antibody, lost on trypsin treatment, and stable against acid treatment. Thus, we concluded that this antiviral activity is due to the presence of mouse IFN in the medium.

To elucidate the mechanism of the induction of IFN by the C gene, the following plasmids were constructed: pSV2-C (HindIII/BAP), a linear form of pSV2-C obtained by cleaving the latter at the junction between the SV40 promotor and the inserted C gene, and pSV2-C(stop), a derivative of pSV2-C, which has an insertion of the termination codon, TAA(ochre), in the frame of the C gene. The IFN activity in the culture medium of L cell transfected with either pSV2-C(HindIII/BAP) or pSV2-C(stop) showed a 1,000-fold reduction as compared to that with pSV2-C (Table 1). Similar results were obtained in COS-1 cells as well as lymphocytes. These results strongly indicated that both transcription and translation of the C gene are indispensable for the IFN induction. The possibility that the dsRNA structure is responsible, or acts as an inducer could be excluded, because pSV2-PC and pSV2-C have almost identical structures, except for small region within the untranslated sequence (PvuI-Sau3A, 87bp). In addition to these findings, L cells underwent strong cytopathic effects upon transfection with pSV2-C, and almost all the cells in the culture became rounded. Similar but much weaker effects was detected when the cells were transfected with pSV2-PC(Rl), whereas no effect was observed with the cells transfected with pSV2-0 or treated with purified IFN. From the survey of the results of the

repeated experiments, it was suggested that the extents of
the morphological changes induced in the cells seemed to
parallel the antiviral activity titers observed in the
medium. As shown in Fig. 2(A), the analysis of the mRNA by
the Northern blotting method revealed that mRNA hybridizing
with the cDNA probe specific for the P+C gene could be
identified in COS-1 cells transfected with pSV2-PC (lnae 2),
pSV2-C(lane 3) or pSV2-C(stop)(lane 4), but not pSV2-0 (lane
1). A transcript of 2.9kb was observed in the case of
transfection with pSV2-PC, pSV2-C and pSV2-C(stop), but not
in the case of with pSV2-0 or without plasmid, and its size
was consistent with the expected value of mRNA synthesized
from these plasmids. While a main transcript of 1.9kb was
observed in Sendai virus-infected cells. Synthesis of the P
protein as well as that of C protein in COS-1 cell
transfected with pSV2-PC or pSV2-C was examined using
antiserum from Sendai virus-infected guinea pig or antiserum
raised against the C terminal peptide of the C protein(7).
As shown in Fig. 2(B), the P protein and the C protein were
synthesized within Sendai virus-infected COS-1 cells as well
as COS-1 cells transfected with pSV2-PC, which were detected
by antiserum from Sendai virus-infected guinea pig (lanes 1
and 3), and also by anti-C serum (lanes 6 and 7).
Unexpectedly, synthesis of the C protein in COS-1 cells
transfected with pSV2-C was hardly detected with either
antiserum against Sendai virus (lane 4) or anti-C serum (not
shown). After long exposure, however, a faint band
corresponding to the C protein was visible, although not
very much convinsing. This may be to the improper
experimental conditions, since we could not exclude the
possibilities such as that the amount of the C protein
synthesized was very low, or that the synthesized C protein
had already been metabolized.
    From these observations, we concluded that the ability of
Sendai virus to induce IFN production resides within the C
gene, and both transcription and translation of the C gene
seemed to be indispensable for the IFN induction, although
there remains a discrepancy between the level of the
expression of the C protein and the titer of IFN induced by
transfection with pSV2-PC or pSV2-C, which is to be studied
further. It is interesting to note that recent studies on
the sequence of the genomes of some RNA viruses other than
Sendai virus, revealed the existence of an overlapping
reading frame similar to the P+C gene of Sendai virus(8)(9),
such as the P+C gene of the measles virus(10), the S1 gene
of the reovirus(11)(12) and the N+N$_s$ gene of the
bunyavirus(13)(14), and that the proteins encoded by these
second open reading frames are nonstructural proteins,
although little is known about their precise functions.
Our results offer the first insight into the biological
significance of the C gene in IFN induction. However, the
precise elucidation of its mechanism has to await further
studies with either purified C protein or its specific
antibody.

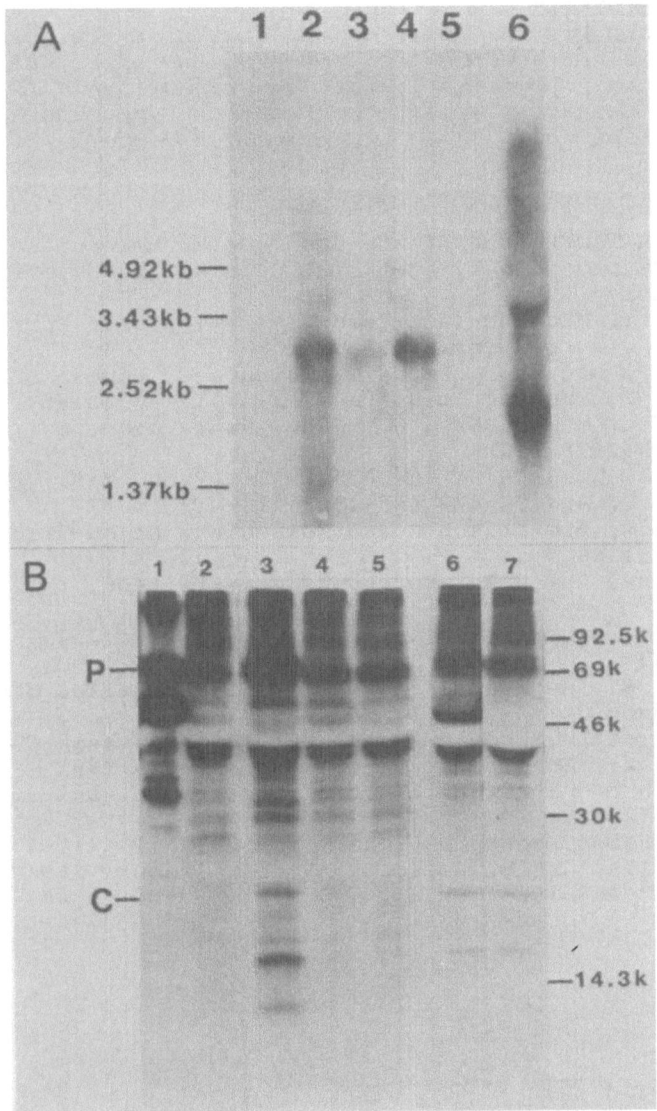

FIGURE 2. A)Northern blot analysis of RNAs: The poly(A) RNAs were prepared from COS-1 cells transfected with pSV2-0(lane 1), pSV2-PC(lane 2), pSV2-C(lane 3), pSV2-C(stop)(lane 4) and from untreated cells(lane 5). Total RNA from Sendai virus-infected cells served as a positive control(lane 6). B)Expression of the P or C protein: The cell extracts were prepared from Sendai virus-infected cells(lanes 1,6), from mock-infected cells(lane 2) and from cells transfected with pSV2-PC(lanes 3,7), pSV2-C(lane 4) and pSV2-0(lane 5). Viral proteins were recovered by immunoprecipitation with antiserum from Sendai virus-infected guinea pig (lanes 1~5) and with C-peptide antiserum (lanes 6,7).

4. ACKNOWLEDGEMENT

We thank Drs. J Curran and D. Kolakofsky for peptide antiserum raised against the C/C' proteins of Sendai virus. This study was supported in part by Grants-in-Aid for Scientific Research from the Ministry of Education, Science and Culture of Japan.

REFERENCES

1. Graham FL and van der Eb AJ: Virology **52**, 456-467 1973
2. Maniatis T, Fritsch EF and Sambrook J: Molecualar cloning a laboratory manual, New York: Cold Spring Harbor Laboratory, 1982
3. Stewart II WE: The Interferon System, p58-76 Springer-Verlag, Wien New York 1979
4. Shioda T, Hidaka Y, Kanda T, Shibuta H, Nomoto A and Iwasaki K: Nucleic Acids Res. 11, 7317-7330 1983
5. Sompayrac LM and Danna KJ: Proc. Natl. Acad. Sci. U. S. A **78**, 7575-7578 1981
6. Loyter A, Scangos GA and Ruddle FH: Proc. Natl. Acad. Sci. U. S. A. **79**, 422-426 1982
7. Curran JA, Richardson C and Kolakofsky D: J. Virol. **57**, 684-687 1986
8. Dethlefsen L and Kolakofsky D: J. Virol. **46**, 321-324 1983
9. Giorgi C, Blumberg BM and Kolakofsky D: Cell **35**, 829-836 1983
10. Bellini WJ, Englund G, Rozenblatt S, Arnheiter H and Richardson CD: J. Virol. 53, 908-919 1985
11. Cashdollar LW, Chmelo RA, Wiener JR and Joklik WK: Proc. Natl. Acad. Sci. U. S. A. **82**, 24-28 1983
12. Ernest H and Shatkin AJ: Proc. Natl. Acad. Sci. U. S. A. **82**, 48-52 1985
13. Akashi H and Bishop DHL: J. Virol. **45**, 1155-1158 1983
14. Cabradilla, Jr CD, Holloway BP and Obijeski JF: Virology **128**, 463-468 1983

STUDIES ON THE EXPRESSION OF SPONTANEOUS INTERFERON-LIKE ACTIVITY IN MOUSE PERITONEAL CELLS

F. BELARDELLI[1], S. GESSANI[1], P. BORGHI[1], E. PROIETTI[1], D. BORASCHI[2], Y. KAWADE[3] and I. GRESSER[4]
[1]Istituto Superiore di Sanità, Rome, Italy. [2]Sclavo Research Centre, Siena, Italy. [3]Institut for Virus Research, Kyoto, Japan. [4]Institut de Recherches Scientifiques sur le Cancer, Villejuif, France.

INTRODUCTION

Macrophages exert important functions in non specific natural resistance to virus infections (1-3). Several animal viruses do not multiply in macrophages when these cells are first placed in culture suggesting that in vivo macrophages may limit virus spread by inhibiting virus multiplication. The resistance of resting mouse peritoneal macrophages to vescicular stomatitis virus (VSV) and encephalomyocarditis virus (EMCV) can be abolished by injection of mice with sheep antibody to interferon α/β (4). An inverse correlation was observed between the intracellular levels of (2'-5')oligo-adenylate (2-5A) synthetase activity in peritoneal macrophages and the permissivity of these cells for VSV (5). Thus, the levels of 2-5A synthetase activity in peritoneal macrophages from mice injected with antibody to interferon were markedly lower with respect to macrophages freshly harvested from control mice (5). Moreover, although interferon could not be generally recovered in the peritoneal washings or tissue extracts of the majority of normal mice (6-8), we have recently demonstrated that peritoneal cells or macrophages freshly harvested from young mice were capable of transferring an antiviral state (to VSV and to EMCV) to syngeneic in vitro "aged" mouse macrophages as well as to different mouse cell monolayers permissive for virus replication (9). This transfer of the antiviral state was abolished by the addition of sheep antibodies to mouse α/β interferon (9). The ensemble of these data strongly suggested that interferon was produced constitutively in normal mice.

In this article we present further evidence indicating that α and β mouse interferons are spontaneously expressed in peritoneal cells from normal pathogen-free young mice and we show that this expression requires in some manner a normal responsiveness of peritoneal cells to lipopolysaccharide (LPS) from Gram negative bacteria.

MATERIALS AND METHODS

Mice.

Male and female DBA/2, Swiss, C57B1/6, C3H/HeN and BALB/c mice 5 to 8 weeks old were obtained from Charles River, Italia S.p.A. (Milan, Italy).

C3H/HeJ mice (Lps^d (10)) were obtained from Jackson Laboratories (Bar Harbor, Maine, U.S.A.).

C57B1/10ScCR (Lps^d (11)) were purchased from Bomholtgaard Limited (DK 8680 RY, Denmark).

Reagents.

RPMI 1640 medium (MA Bioproducts, Walkersville, Md.) was supplemented with penicillin (100 U/ml), streptomycin (100 µg/ml) glutamine (2 mM) and fetal calf serum (Gibco Laboratories, Grand Island, N.Y.) at a final concentration of 10% (vol/vol). All tissue culture reagents were purchased as "endotoxin-free" lots, as assessed by the Limulus amoebocyte assay.

Preparation of donor mouse cells.

The peritoneum was washed with 2.5 ml of RPMI 1640 nutrient medium containing 10% fetal calf serum (FCS). Cells were centrifuged and resuspended in RPMI medium containing 10% FCS before using them as donor cells.

Preparation of target cells.

For preparation of virus-permissive macrophage monolayers, peritoneal washings were prepared as described above and seeded in 24-well plastic plates (Nunc), each well containing approximately 10^6 cells per ml. Cells were allowed to attach to the plastic culture dish at 37°C for 3 h, and non-adherent cells were discarded. The criteria used to define these firmly attached cells as macrophages have been previously described (4). Macrophages were cultivated in vitro for 4 to 5 days, at which time they became permissive for VSV and EMCV and were used as target cell monolayers. Monolayer cultures of "aged" macrophages showed the same sensitivity as monolayer cultures of L cells to the antiviral action of mouse interferon α/β.

Antibodies.

All sera were decomplemented and extensively absorbed on murine cells. The immunoglobulin fractions were separated by ammonium sulfate precipitation (protein concentration varied between 10 and 33 mg/ml) and shown to be devoid af any toxicity for mouse peritoneal macrophage.

Sheep anti mouse α/β globulin (sheep n. R5/4 (9)) had a neutralizing titer of 1.0×10^{-5} against 4-8 units of α/β mouse interferon. Rat monoclonal antibody to mouse γ interferon (Kindly provided by Dr. Havell (Trudeau Institute Inc., New York) had a neutralizing titer of 5×10^{-4} against 4 units of mouse γ interferon.

Hybridomas producing rat monoclonal anti-mouse α (clone 4E-A1) and anti-mouse β (clone 7F-D3) antibodies were obtained by Dr. Y. Watanabe and Dr. Y. Kawade (Institute for Virus Re-

search, Kyoto University, Kyoto). Cells were passaged in a-
scitic form in Balb/c nude mice. Ascitic fluids were concentrated by ammonium sulfate precipitation. Monoclonal antibody
to mouse α interferon had a titer of 8 X 10^{-3} against 1 unit
of mouse α interferon. Monoclonal antibody to mouse β interferon had a titer of 1.0 X 10^{-6} against 1 unit of mouse β interferon.

Assay for induction af antiviral state in target cell monolayers. VSV yield assay.
    A 0.1-ml sample of medium containing freshly harvested
total peritoneal cells (usually 2.0x10^{6}) was first mixed with
0.2 ml of soft agar (Bacto-agar (Difco) 0.33% in 5% FCS-RPMI
1640) to prevent the donor peritoneal cells from adhering to
the underlying target cell monolayers. After solidification
for 5 min at 4°C, a second layer of soft agar was applied.
After incubation for 24 h at 37°C, agar and donor peritoneal
cells were discarded, and the remaining monolayer was washed
three times with medium (2% FCS-RPMI 1640). A 0.2-ml sample
of a viral dilution (multiplicity of infection af approximately 0.2) was added to each well. After 1 h of incubation at
37°C the cell sheet was washed thoroughly, and 1 ml of nutrient medium was added. After incubation for 18 h at 37°C in
a 5% CO_2-air incubator, the cell-free nutrient medium was assayed for virus in mouse L929 cells.

RESULTS

    Freshly harvested peritoneal cells from normal young
mice transfer an antiviral state to syngeneic "in vitro aged"
macrophages permissive for virus replication (9).
    Fig. 1 shows the effects of addition of different anti-
bodies to mouse interferons in the coculture medium on the
transfer of antiviral state (to VSV) from freshly explanted
mouse peritoneal cells to syngeneic in vitro "aged" target
macrophages. Cocultivation of donor peritoneal cells with
macrophage monolayers resulted in a 100-fold reduction of VSV
yield in target cells with respect to control in vitro "aged"
macrophages. Addition of sheep antibody to mouse α/β interferon in the coculture medium completely abolished the transfer
of the antiviral state. Monoclonal antibodies to mouse interferon α and β did markedly inhibit the transfer of the antiviral state, whereas antibody to mouse interferon γ was ineffective in abrogating the transfer of antiviral state.

142

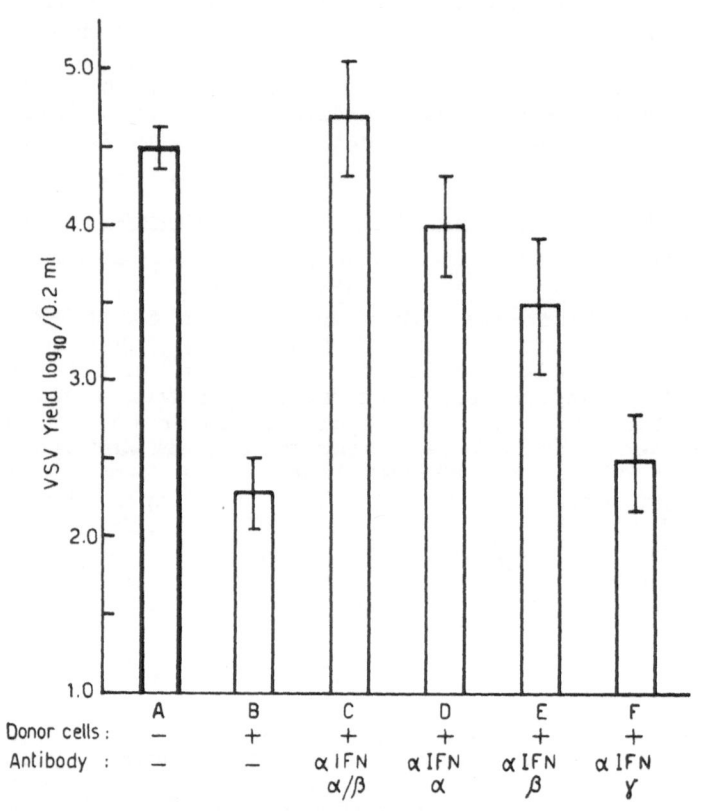

FIGURE 1. Effects of different antibodies to mouse interferons on the transfer of antiviral state from freshly harvested peritoneal cells to "in vitro" aged mouse macrophages.

Peritoneal macrophages explanted from 5 week-old male Swiss mice were cultivated in vitro to render them permissive for virus replication and used as target monolayer cultures. After 4 days, macrophages were co-cultivated with 2 X 10^6 peritoneal cells freshly explanted from 6 week-old male Swiss mice in the presence or absence of different antibody preparations. After 18 h of cocultivation at 37°C, peritoneal cells were removed and macrophages were infected with VSV. Virus yields were determined as described in Materials and Methods. There were three infected macrophage cultures for each experimental conditions.

A: control target macrophages;

B: + donor peritoneal cells;

C: + donor peritoneal cells in the presence of sheep antibody to mouse interferon α/β (final titer: 10^4 against 1 unit);

D: + donor peritoneal cells in the presence of rat monoclonal antibody to mouse interferon α ; final titer 5 X 10^3 against 1 unit of mouse α interferon;

E: + donor peritoneal cells in the presence of rat monoclonal antibody to mouse interferon β ; final titer: 10^4 against 1 unit;

F: + donor peritoneal cells in the presence of rat monoclonal antibody to mouse γ interferon (final titer: 2 X 10^4 against 1 unit).

The transfer of antiviral state from donor cells to mouse target cells occurred using donor peritoneal cells from the large majority of the strains of normal young mice tested (i.e. DBA/2, Swiss, C57B1/6, C57B1/beige, BALB/c nude and C3H/HeN mice). In the course of these experiments we observed that donor peritoneal cells from lipopolysaccharide (LPS)-hyporesponsive mice (i.e. C3H/HeJ and C57B1/10ScCR mice) were not capable of efficiently transferring an antiviral state to mouse target cells. As shown in Table 1 donor peritoneal cells from LPS-responsive C3H/He N mice (Lps^n) were capable of efficiently transferring an antiviral state to _in vitro_ aged target macrophages from C3H/HeN and C3H/HeJ mice. The transfer of antiviral state occurred also in the presence of Millipore filters between donor cells and target macrophages. In contrast in all instances peritoneal cells freshly explanted from LPS-hyporesponsive C3H/HeJ mice (Lps^d) were not capable of transferring an antiviral state to target macrophage monolayers (Table 1). Similar results were obtained using peritoneal cells from C57BL/6 (Lps^n) and C57BL/10ScCR (Lps^d) mice (data not shown) further indicating that a normal responsiveness to LPS was required for the transfer of antiviral state (data not shown).

TABLE 1. Transfer of antiviral state by peritoneal cells from LPS-responsi ve and LPS-hyporesponsive mice in presence or absence of millipore filters between donor cells and target cells.

Target macrophages* (strain of mice)	Control target cells	Millipore filter**	Donor cells*** (strain of mice)	
			C3H/HeN	C3H/HeJ
C3H/HeN	3.5 ± 0.3	–	0.7 ± 0.2	3.8 ± 0.2
C3H/HeN	3.5 ± 0.3	+	1.2 ± 0.3	4.4 ± 0.6
C3H/HeJ	6.0 ± 0.5	–	1.6 ± 0.6	5.6 ± 0.8
C3H/HeJ	6.0 ± 0.5	+	5.0 ± 0.3	6.0 ± 0.5

VSV Yield/0.2 ml ($\log_{10}$) $\pm$ SEM

* In vitro aged macrophages from 7 week-old male mice were used as tar get cell monolayer permissive for virus replication (after 4 days of in vitro culture).
** 0.45 μm Millipore filter.
*** 2×10^6 donor peritoneal cells explanted from 6 week-old male mice were used as described in Materials and Methods.

Injection of mice with antibody to mouse α/β interferon resulted in a marked increase of VSV or EMCV replication in peritoneal macrophages when these cells were infected with viruses a few hours after cell seeding (4). Injection of mice with monoclonal antibody to mouse α or β interferon also resulted in a marked increase in VSV multiplication in perito neal macrophages, provided that mice were given 2-3 injections of at least 10^3 - 10^4 neutralizing units of monoclonal antibody (data not shown).

VSV does not multiply in the majority of peritoneal macrophages freshly explanted from 4-to 8-week-old male or female mice. However, when peritoneal macrophages are cultivated in vitro for 3 to 5 days these cells become permissive for virus replication (9). In vitro addition of sheep antibody to mouse α/β interferon to freshly explanted mouse peritoneal macrophages markedly enhanced the decay of the antiviral state to VSV. This effect was also exerted to some extent by monoclonal antibodies to mouse α or β interferon. Fig. 2 shows the in vitro decay of antiviral state to VSV in peritoneal macrophages from C3H/HeN mice in the presence or absence of a sheep antibody to mouse α/β interferon. The increased permissivity to VSV in peritoneal macrophages treated with antibodies to interferon was not observed in macrophage monolayers treated with control sheep globulins (data not shown).

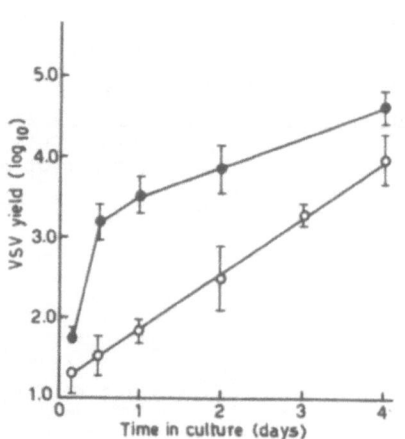

FIGURE 2. In vitro decay of antiviral state in mouse peritoneal macrophages in the absence or in the presence of antibody to mouse interferon α/β.

Peritoneal cells were harvested from the peritoneal cavity of 6 week-old C3H/HeN mice and seeded in 24-well plastic plates (10^6 cells per ml of medium). Cells were allowed to fix to the plastic culture dish at 37°C for 3 h and non-adherent cells were discarded. Peritoneal macrophages were then incubated at 37°C with 1 ml of nutrient medium with or without antibody to interferon. At different times of incubation, peritoneal macrophages were infected with VSV, and virus yields were determined. There were three macrophage cultures for each experimental sample.
Open symbols: in the absence of antibody to interferon.
Solid symbols: in the presence of antibody to interferon α/β (sheep n. R5/4; 1:20, final dilution).

DISCUSSION

In the present study we provide further experimental evidence indicating that α and β interferons are constitutively expressed in peritoneal cells freshly explanted from normal pathogen-free young mice. Polyclonal antibody to mouse α/β interferon and monoclonal antibodies to mouse α or β interferons markedly inhibited the transfer of antiviral state from donor peritoneal cells to in vitro "aged" macrophages permissive for virus replication (Fig. 1). Likewise injection of mice with these antibody preparations significantly increased virus multiplication in peritoneal macrophages infected with VSV a few hours after cell seeding. These effects were not exerted by a rat monoclonal antibody to mouse γ interferon.

Peritoneal cells or purified macrophages could transfer an antiviral state to target cells even when the two cell populations were separated by millipore filters, indicating that donor cells can release to some extent interferon-like molecules. It is of interest to underline that pretreatment of donor peritoneal cells with trypsin did not abolish the capacity of donor cells to transfer an antiviral state to target cells (data not shown), suggesting that interferon is constitutively produced by peritoneal cells and not simply bound to the cell membrane. Moreover the results reported in this article indicate that peritoneal cells freshly explanted from LPS-hyporesponsive mice were not capable of transferring an antiviral state to target cells (Table 1). It has been recently reported that macrophages bearing the Lps^d genotype (i.e. from LPS-hyporesponsive mice) exhibit a lower level of differentiation as assessed by a decreased Fc receptor capacity in vitro (12, 13); this defect in phagocytic capacity was completely normalized by treatment of cultures with interferons (14) as well as by co-cultivation with peritoneal macrophages bearing the LPS^n genotype (15), indicating that "endogenous interferon production by endotoxin-responsive macrophages provides an autostimulatory differentiation signal" (15). The ensemble of these results reported by Vogel and Fertsch (15), although obtained with completely different experimental approaches, appears to be in agreement with our data suggesting that LPS-responsive and LPS-hyporesponsive peritoneal cells differ in their capacities to produce spontaneous interferon under physiological conditions. It is possible to envisage that the expression of spontaneous interferons in mouse peritoneal cells may play some important roles in vivo, not only in the natural defense mechanisms against virus infections, but also in some other physiological responses.

ACKNOWLEDGMENTS

This work was supported in part by grants from Consiglio Nazionale delle Ricerche (Progetto Finalizzato Controllo delle Malattie da Infezione N. 85.02797.52). One of us (I.G.) is grateful to Richard Lounsbery Foundation for support.

REFERENCES

1. Mims C. Bacteriol. Rev. 28: 30-71, 1964.
2. Mogensen S.C. Microbiol. Rev. 43: 1-26, 1979.
3. Morahan P.S., Connor J.R. and Learly K.R. British Medical Bulletin 41: 15-21, 1985.
4. Belardelli F., Vignaux F., Proietti E. and Gresser I. Proc. Natl. Acad. Sci. USA 81: 602-606, 1984.
5. Gresser I., Vignaux F., Belardelli F., Tovey M.G. and Maunoury M.T. J. Virol. 53: 221-227, 1985.
6. Gresser I., Belardelli F., Maury C., Maunoury M.T. and Tovey M.G. J. Exp. Med. 158: 2095-2107, 1983.
7. Sen G.S. Prog. Nucleic Acid Res. Mol. Biol. 27: 105-156, 1982.
8. Galabru J., Robert N., Buffet-Janvresse C., Riviere Y. and Hovanessian A.G. J. Gen. Virol. 66: 711-718, 1985.
9. Proietti E., Gessani S., Belardelli F. and Gresser I. J. Virol. 57: 456-463, 1986.
10. Watson J., Kelly K., Largen M. and Taylor B.A. J. Immunol. 128: 380-387, 1978.
11. Coutinho A. and Meo T. Immunogenetics 7: 17-24, 1978.
12. Vogel S.N. and Rosenstreich D.L. J. Immunol. 123: 2842-2850, 1979.
13. Vogel S.N. and Rosenstreich D.L. Lymphokines 3: 149-180, 1981.
14. Vogel S.N., Weedon L.L., Moore R.N. and Rosenstreich D.L. J. Immunol. 128: 380-387, 1982.
15. Vogel S.N. and Fertsch D. Infect. Immun. 45: 417-423, 1984.

CELLULAR FACTORS MEDIATE INDUCTION AND SUPERINDUCTION OF THE HUMAN INTERFERON-ß GENE

Dirks, W., Dinter, H., Ausmeier, M., Collins, J. and Hauser, H.

Department of Genetics
GBF - Gesellschaft für Biotechnologische Forschung
Mascheroder Weg 1,
D - 3300 Braunschweig, F. R. G.

SUMMARY

Inducers like double stranded RNA and Virus mediate their action by transcriptional activation of the human Interferon-ß gene. The regulatory element of this gene is contained within 110 base pairs of 5' flanking DNA sequences. The same DNA fragment is a target for superinduction, i. e. the enhancement of Interferon expression by inhibitors of translation. These regulatory sequences interact with cellular factors, mediators in the induction process. As shown by promoter competition assays the interception of repressing factors leads to constitutive transcription of the human Interferon-ß gene. In addition, induced transcription is promoted by activating factors specifically interacting with the Interferon promoter DNA.

INTRODUCTION

The human Interferon-ß (IFN-ß) gene is predominantly expressed in fibroblast cells upon viral infection or treatment with double stranded RNA (dsRNA) (1). Specific mRNA is increased after a short time lag due to transcriptional activation and reaches a maximum several hours after induction (2). Following this period a rapid decrease of specific mRNA indicates its short half life time (t $_{1/2}$ = 30 min) (3). In human primary fibroblasts the induction with dsRNA is drastically enhanced by a simultaneous dose of cycloheximide (CHX) or other inhibitors of cellular translation (superinduction), whereas CHX alone does not induce the gene (4).

Gene transfer experiments with heterologous cells have shown that the DNA - sequences responsible for viral or dsRNA induction of the human IFN-ß gene are localized within the 5' flanking region (5,6,7). By analysis of promoter mutants we have shown that several regulating DNA sequence elements within 110 base pairs (bp) upstream of the transcriptional start site act together in a cooperative way to achieve high inducibility (8). These elements consist almost exclusively of purines (Figure 1). Duplication of element I leads to a strong enhancement of inducible expression (9). The region from -280 to -37 is able to confer inducibility to a heterologous promoter in 5' position, but does not serve as an enhancer in 3' position of this promoter (8).

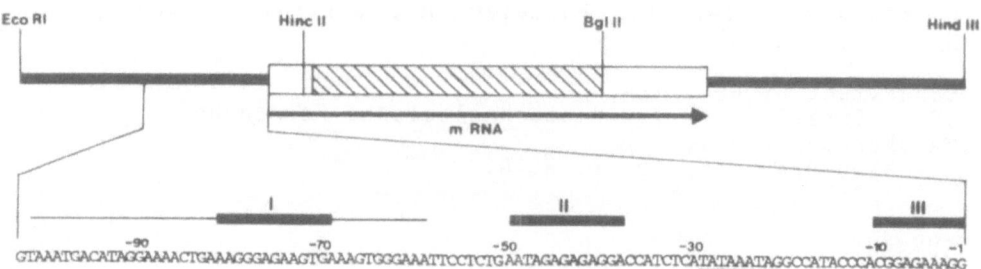

Figure 1: Schematic drawing of the human IFN-ß gene and flanking DNA sequences.
The transcribed part of the gene is indicated by an open box and an arrow. The translated part is depicted as a hatched region; flanking sequences are indicated by solid bars. The 5' flanking region (DNA sequence) contains the regulatory part of the promoter. The bars I, II and III indicate 3 purine-rich regions which act cooperatively to achieve full promoter activity in mouse L cells.

RESULTS

Initial studies focussed on the problem of superinduction regulation, specifically as to whether it was acting transcriptionally or post-transcriptionally. Formerly it was believed that superinduction is mainly due to a stabilization of IFN specific mRNA (1). A labile metabolic factor was postulated to be responsible for IFN-ß specific mRNA degradation. Prevention of translation would reduce the amount of this factor and IFN-ß mRNA could accumulate and become translated after removal of the translation-inhibitor.

We have studied hybrid gene expression in the mouse cell line Mo 57/2, in which endogeneous mouse IFN is superinducible (10). Transfer of the cloned human IFN-ß gene and defined gene hybrids into these cells allowed us to test the above mentioned hypothesis. As expected, the human wild type gene was superinducible in these cells (Figure 2). Because the same gene is not superinducible upon transfer in mouse Ltk⁻ cells (as for the endogeneous mouse L cell IFN), this result indicates that the superinduction phenomenon is governed by a cell-specific induction machinery. However, the expression

from a gene hybrid containing a constitutive promoter (ADIF)
was not affected by superinduction. On the other hand, a
hybrid composed of the human IFN-ß promoter and 5' flanking
sequences controlling the bacterial Chloramphenicol-acetyl-
transferase (CAT) gene was inducible by the superinduction
protocol. These results, indicating that the 5' flanking DNA
sequences are responsible for superinduction, could be reproduced in similar experiments by gene transfer into several
murine and primate cell lines including human FS-4 cells (data
not shown). A detailed biological response analysis of several

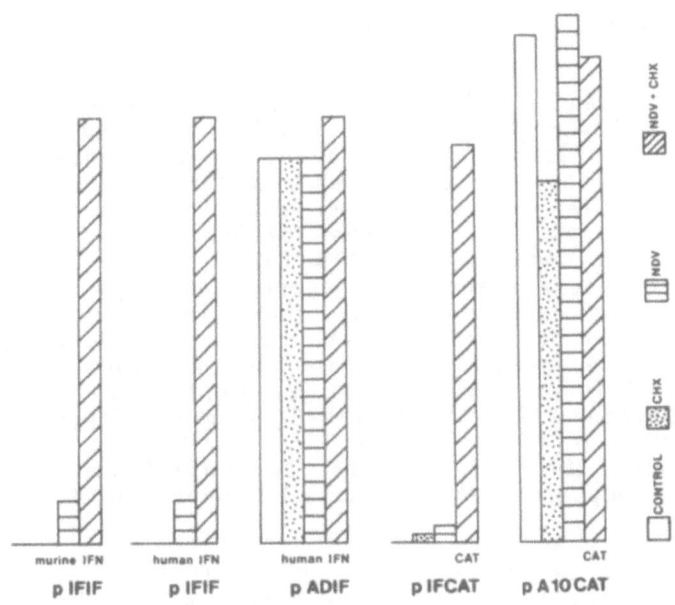

Figure 2: Superinduction of IFN and CAT from the human IFN-ß
promoter.
Mouse Mo 57/2 cells were stably transformed with the indicated
DNA constructs by transfection (15). Transformed cell clones
were pooled and treated with different induction protocolls.
20 hours after induction murine and human specific IFN was
tested in the cell supernatants on mouse L and monkey Vero
cells, respectively (16). Determination of CAT activity was
carried out 14 hours after induction in total cellular
extracts (13). pIFIF: Human IFN-ß gene controlled by its own
promoter; pADIF: Human IFN-ß structural gene under the
control of a constitutive Adenovirus promoter; pIFCAT: Human
IFN-ß promoter controlling the bacterial CAT gene; pA10CAT:
The CAT gene under the control of an enhancer-less SV 40 early
promoter (14). NDV: Newcastle Desease Virus. Relative IFN-
and CAT-values are given on a linear scale.

promoter mutants in mouse Mo 57/2 cells indicates that the sequences responsible for superinduction cannot be separated from the sequences responding to normal induction.

Recently, Goodbourn et al. could demonstrate that the regulatory element within the human IFN-ß 5' flanking sequences comprizes a constitutive and a negative regulatory element (11). The negative element prevents the action of the 5' flanking constitutive element. By DNAse I footprinting Zinn et al. have presented evidence for the binding of factors to this regulatory DNA region (12). Prior to induction factors binding to DNA are detected within the negatively regulating element. After induction these factors disassociate and other factors binds to the region responsible for constitutive expression.

We have tried to demonstrate the existence of cellular DNA-binding factors which are functionally involved in the transcription regulation of this gene. We took advantage of the fact that the endogeneous IFN of a monkey cell line, CV-1, is poorly inducible by induction or superinduction. In transient expression assays as established by Schöler and Gruss (13) we have tried to compete the wild type promoter with promoter mutants for activation and repression. The outlines and results of the experiments are shown in Figure 3. First, we tested that in our experimental system increasing DNA input (pSrM2CAT) (14) was accompanied by a linear increase in expression (Figure 3A). Second, we have shown that a hybrid gene containing the human IFN-ß promoter (pIFCAT) can be induced and superinduced in CV-1 cells. However, the nonlinear dose-response curves indicate the existence of limiting essential compounds which must be specific for this gene (Figure 3B).

Competition of the indicator DNA (pIFCAT) with the natural IFN gene (pIFIF) leads to a spontaneous activation of the promoter, which is of the same order of magnitude as found for normal induction (Figure 3C). This result can be explained by a titration of a repressing factor which is present in a limiting amount. Derepression by dilution of the factor due to increasing amounts of competing promoter DNA would uncover a constitutive promoter element. Heterologous eukaryotic promoter DNA was not able to activate the IFN promoter. Mapping of the DNA fragment which competes for the repressor-binding located the target site in the region between -65 and -37 (data not shown).

Competition experiments with NDV induced or superinduced CV-1 cells transformed with pIFCAT showed no effect on titration with pIFIF. However, upon virus induction the CAT expression by pIFCAT could be reduced by increasing amounts of cotransfected duplication mutant DNA (duIFIF) (Figure 3D). This duplication mutant (sequences from -90 to -51 are duplicated) shows a severalfold higher inducible expression in different cell types compared to the wild type promoter.

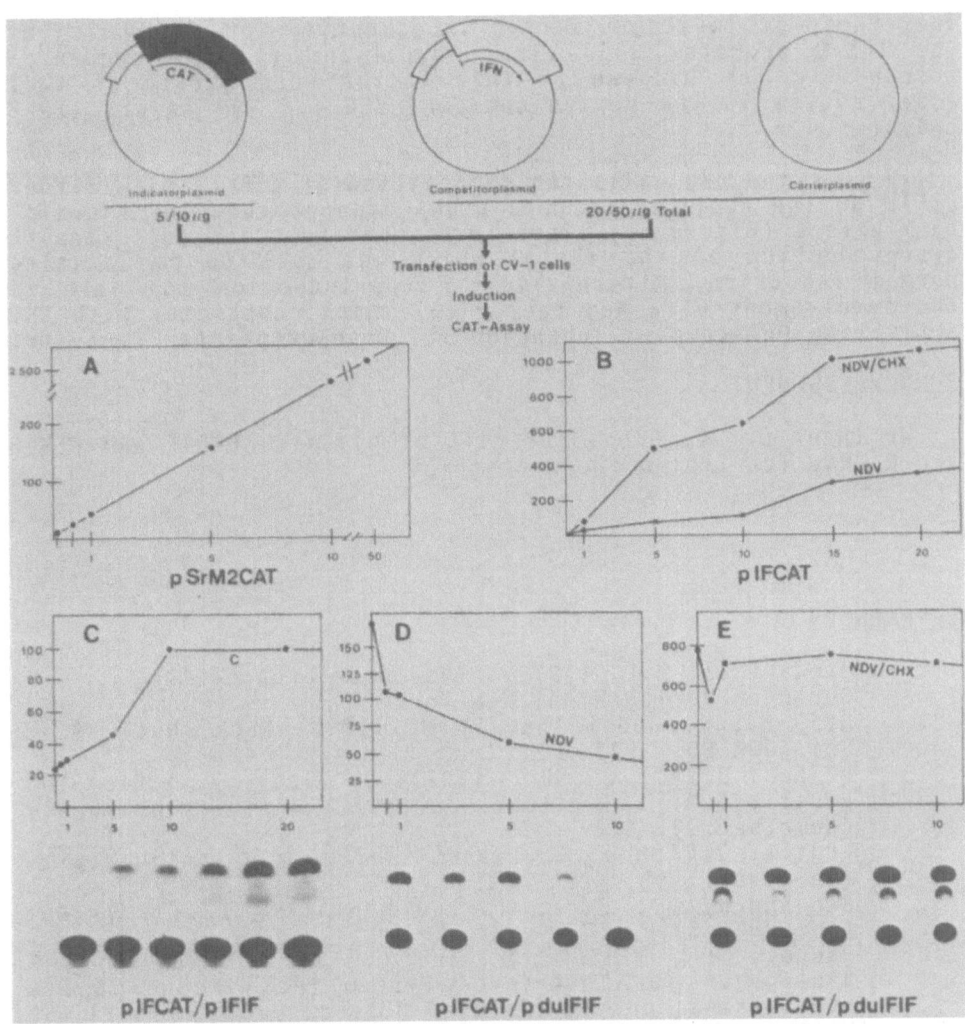

Figure 3: Competition for repressing and activating factors interacting with the human IFN-ß promoter DNA.
As indicated in the above scheme CAT-expression was determined in monkey CV-1 cells in transient expression experiments (13). Two days after transfection of the indicated DNA the cells were induced, superinduced or mock induced. Determination of CAT-activity was 24 hours after induction. Abscissa: CAT-expression [pMol acetylated CAM/hr x extract of 3×10^5 cells]. Ordinate: Transfected DNA [ug/3,5x10^5 cells]. A: Indicator: pSrM2CAT; the CAT gene is controlled by the constitutive promoter of Murine Sarcoma Virus. B: Indicator pIFCAT. C: Indicator 5ug pIFCAT; Competitor: pIFIF. D and E: Indicator: pIFCAT; Competitor: pduIFIF; the human IFN-ß gene under the control of its promoter containing a duplication as described in the text.

152

Therefore, at least one factor is essential for expression by the IFN-ß promoter upon viral induction. The DNA sequences responsible for binding of this factor were mapped in this assay system to the region between -280 and -37 of the IFN-ß promoter.

In superinduced cells the competition of pIFCAT with either pIFIF or the duplication mutant DNA did not show any significant effect (Figure 3E). Therefore superinduction must lead to overproduction of the factor which was seen to be limiting normal induction. Alternatively, superinduction may lead to the involvement of a new factor (5) which cooperates with the activating DNA-complex enhancing its transcriptional capacity.

ACKNOWLEDGEMENT

We thank Dr. G. Gross for gift of plasmids pADIF and pSVIF and R. Roy for typing the manuscript.

REFERENCES

1. Stewart II, W.E. (1979). The Interferon System, Springer Verlag Wien, New York
2. Raj, N.B.K. and Pitha, P. M. (1983) Proc. Natl. Acad. Sci. USA 80, 3923 - 3927
3. Vilcek, J. and Ng, M.H. (1971) J. Virol. 7, 588 - 594
4. Havell, E.A. and Vilcek, J. (1972) Antimicrob. Agents Chemother. 2, 476 - 481
5. Fujita, T., Ohno, S., Yasumitsu, H. and Taniguchi, T. (1985) Cell 41, 489 - 496
6. Goodbourn, S., Zinn, K. and Maniatis, T. (1985) Cell 41, 509 - 520
7. Hauser, H., Dinter, H. and Collins, J. (1985). The Biology of the Interferon System (H. Kirchner and H. Schellekens, eds.), Elsevier Science Publishers 21 - 32
8. Dinter, H. and Hauser, H. (1986) submitted for publication
9. Hauser, H., Dinter, H. and Collins, J. (1985). The Interferon System (G. Rossi, F. Dianzani, eds.) Raven Press 24, 29 - 34
10. Vilcek, J., Havell, E.A. (1973) Proc. Natl. Acad. Sci. USA 70, 3909 - 3913
11. Goodbourn, S., Burstein, H. and Maniatis, T. (1986) Cell 45, 601 - 610
12. Zinn, K. and Maniatis, T. (1986) Cell 45, 611 - 618
13. Schöler, H., Gruss, P. (1985) EMBO J. 4, 3013 - 3025
14. Laimins, L., Khoury, G., Gorman, C.M., Howard, B.H., Gruss, P. (1982) Proc. Natl. Acad. Sci. USA 79, 6453 - 6457
15. Graham, F. and van der Eb, K. (1975) Virol. 52, 445
16. Finter, N.B. (1969) J. Gen. Virol. 5, 419 - 425

INTERFERON BETA-2 IS CONSTITUTIVELY EXPRESSED IN HUMAN MONOCYTES

J. SANCEAU*, R. FALCOFF*, A. ZILBERSTEIN°, F. BERANGER*, M. REVEL° and C. VAQUERO*
* Unité INSERM 196, INSTITUT CURIE, Section Biologie, 26 rue d'Ulm, 75231 Paris cédex 05 (France) and ° WEIZMANN INSTITUTE OF SCIENCE, Department of Virology, Rehovot (Israel)

INTRODUCTION

In a previous report (1) we described the constitutive presence in human peripheral blood leukocytes (PBL), in absence of PHA activation, of a 15S IFN-β2 mRNA. This messenger appears to be slightly larger than the 14S IFN-β2 mRNA from fibroblasts (2). In contrast, IFN-γ mRNA could never be detected in absence of mitogen stimulation. Nevertheless, when PBL were PHA-activated, expression of IFN-γ gene was induced and that of IFN-β 2 was increased, although at different kinetics.

These results support the assumption that expression of the two IFN genes could take place in different subsets of cells. We, therefore, investigated this possibility studying expression of IFN genes in enriched T cell and monocyte populations isolated from human PBL. This IFN-β2 protein was further characterized and compared to the IFN-β 2 protein synthesized by superinduced fibroblasts (3).

MATERIALS AND METHODS

Antibodies were anti-Hu-IFN-γ, a monoclonal antibody from our laboratory (4) and anti-Hu-IFN-β2, a specific polyclonal antiserum raised against purified native IFN-β2 from superinduced fibroblasts and provided by A. Zilberstein.

Batches of human PBL (T lymphocytes 80 %, B lymphocytes 10 % and monocytes 10 % approximately) were obtained from healthy donors (G. Andreu, Hôtel-Dieu, Paris). Monocytes were isolated by adherence to plastic tissue culture plates, by incubation in RPMI 1640 medium containing 10 % FCS (20 X 10^6 cells X ml^{-1}) during 3 hours at 37°C. After three washes with cold medium, the monocyte enriched preparation (more than 95 % monocytes) was incubated at 1 X 10^6 cells X ml^{-1} in RPMI 1640 medium containing 5 % FCS with or without PHA as already described (5). The non adherent cells, mostly T lymphocytes with roughly 10 % of B lymphocytes, were resuspended in fresh medium (5 X 10^6 cells X ml^{-1}) with or without PHA.

Supernatants from adherent and non adherent cells were assayed for IFN activity (5). RNA isolation, Northern blot analysis, translation in reticulocyte lysates and immunoprecipitation were described previously (1,5).

RESULTS

IFN-β 2 mRNA accumulates in human monocytes

IFN-β2 mRNA was clearly detected in human monocytes prior to culture (Fig. 1). It accumulated early after the onset of the culture with a maximal level reached at 2.5 hours. PHA treatment increased the synthesis of the IFN-β2 mRNA without modification of the kinetics. Nevertheless, it could be argued that the 3 hour incubation needed to separate adherent cells from T lymphocytes would stimulate the IFN-β2 gene expression before culture. Since we have already shown (1) that IFN-β2 mRNA was present in PBL, even though at a low level and before any incubation, and since it is

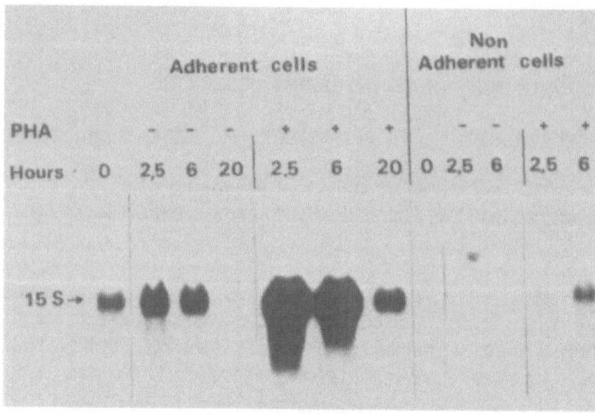

Figure 1. Kinetics of the IFN-β2 mRNA accumulation in adherent and non adherent cells. Total RNA (25 µg) extracted at the indicated times from control or PHA-activated cells, were run on agarose gel and, after transfer, probed with |^{32}P|-IFN-β 2 sequence (2).

not detected in non-activated non adherent cells (Fig. 1), IFN-β2 mRNA is probably present in monocytes prior any in vitro treatment of the cells. In contrast, some IFN-β2 mRNA appeared in non adherent cells only after PHA stimulation up to 6 hours.

Regarding IFN-ɣ, no messenger was detected in absence of PHA activation (Fig. 2)(1,5), while in the presence of PHA, the gene was expressed in both adherent and non adherent cells although with different kinetics. In non adherent cells, the IFN-ɣ mRNA peaked at 6 hours like in PHA-activated PBL (5). In adherent cells, this messenger increased up to 20 hours.

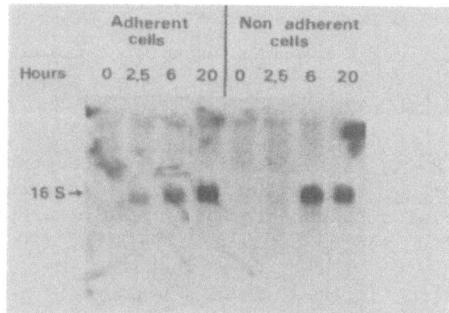

Figure 2. Kinetics of IFN-ɣ mRNA accumulation in adherent and non adherent cells. Total RNA (25 µg) were analysed by Northern blot hybridization with the labelled IFN-ɣ probe (5) as in Figure 1.

Human monocytes produce an antiviral protein neutralized by an anti-IFN-β2 antiserum

Adherent cells secreted constitutively low level of an antiviral protein. The maximal level of antiviral activity in the supernatant (128 U X ml^{-1}) was reached after 16 hours of culture. This activity was 10 fold higher when the cells were cultured in the presence of PHA.

After 20 hours, non adherent cells produced only 32 units per ml of antiviral activity while PHA treatment highly increased IFN level up to 1,000 fold (Fig. 3).

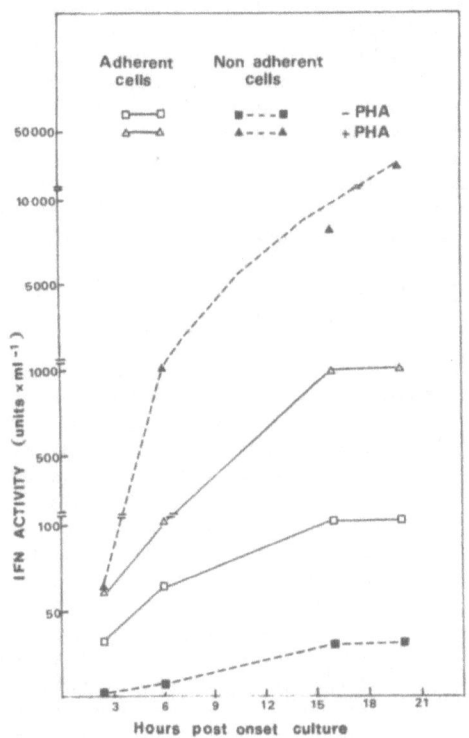

Figure 3. Antiviral activity secreted by adherent and non adherent cells treated or not with PHA. Cell supernatants were assayed for IFN activity on Wish cells with VSV as challenge (1).

Table 1. Antigenic characterization of IFN activity secreted by adherent and non adherent cells cultured without or with PHA.

		Residual IFN activity after antibody addition (U X ml^{-1})**					
		Adherent cells			Non adherent cells		
Hours	PHA	–	Anti–Hu–IFN–β2	Anti–Hu–IFN–γ	–	Anti–Hu IFN–β2	Anti–Hu–IFN–γ
2.5	–	32*	8	16	4	–	–
	+	32	8	16	16	8	8
6	–	32	4	16	4	–	–
	+	64	12	32	128	32	16
15	–	32	4	16	4	–	–
	+	64	16	32	128	48	4
20	+	64	32	8	128	128	4

Supernatant dilutions (*) were incubated 24 hours at 4°C with an excess of the different antibodies. The residual activity was then titrated (**) as in Figure 3.

156

Table 1 shows the neutralization analysis of the various IFN preparations with potent and highly specific antibodies namely a polyclonal anti-
IFN-β2 antiserum and a monoclonal anti-IFN-γ antiserum. Antibody directed
against IFN-α did not neutralize the antiviral activity (not shown).

When supernatants were from non-activated monocytes, the antiviral activity was mostly of the IFN-β2 type. When the monocytes were PHA-activated,
IFN-β2 was essentially secreted up to 15 hours. However, at 20 hour post-
activation, most of the antiviral activity appeared to be of the IFN-γ type. The level of secreted antiviral activities is in good agreement with
the kinetics of the accumulation of the two IFN messenger (see Fig. 1 and
Fig. 2). As for non adherent cells, the antiviral activity could only be
characterized in PHA-treated cultures and was, as expected, predominantly
IFN-γ.

The in vivo and in vitro translation products encoded by the monocyte
IFN-β2 mRNA are immunoprecipitated specifically by an anti-IFN-β2 antiserum

RNA extracted at various times after PHA activation from adherent and
non adherent cells were translated in reticulocyte lysate and the translation products were immunoprecipitated with an anti-IFN-β2 antiserum. The
anti- IFN-β1 antiserum was unable to immunoprecipitate any protein (data
not shown). Fig. 4 shows that
in adherent cells, IFN-β2 mRNA
obtained at 2.5 hours post activation translated efficiently
a 27 kD protein immunoprecipitated by the anti-IFN-β2 antiserum. This protein has the
same estimated M.W. than the
IFN-β2 protein immunoprecipitated under the same conditions,
from the translation products
obtained with poly(A)$^+$ RNA extracted from superinduced fibroblasts (3). In addition,
they were similarly processed
to a 21 kD protein in presence
of microsomes (data not presented).
RNA extracted from non adherent
cells at 6 hours post activation encoded also a slight level of this 27 kD protein.

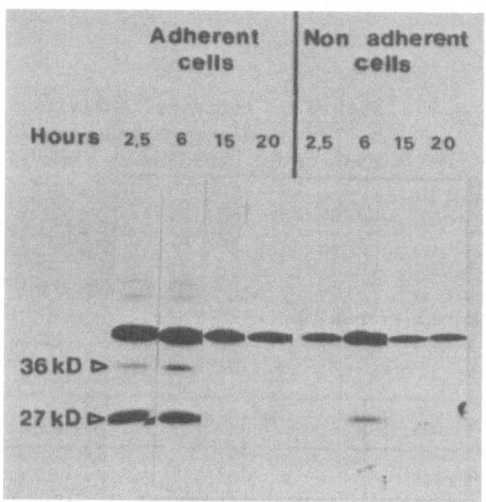

Figure 4. **Immunoprecipitation with the anti-IFN-β2 antiserum of in vitro
translation products of RNA extracted from adherent and non adherent cells
at various times of PHA activation.** Total RNA (200 µg/ml) were translated
in reticulocyte lysate in the presence of $|^{35}S|$ methionine. Translation
products (200,000 cpm) were thereafter immunoprecipitated with the anti-
IFN-β2 antiserum (1) and run in SDS-PAGE.

These results are in good correlation with the IFN-β2 mRNA accumulation
demonstrated by Northern blots (Fig. 1).
$|^{35}S|$ methionine was added at time 0 to cultures of monocytes and the
supernatants were immunoprecipitated at the indicated times with the anti-
IFN-β2 antiserum (Fig. 5a). IFN-β2 protein was first detected at 6 hours.
PHA treatment increased its level (specially at 15 hours).

Unlabelled supernatants of monocytes were prepared as mentioned above and used in a competition experiment performed with labelled IFN-β2 from superinduced fibroblasts (Fig. 5b). As expected from the first set of experiments, monocyte supernatants collected at 6 hours were able to compete efficiently with the labelled fibroblast IFN-β2.

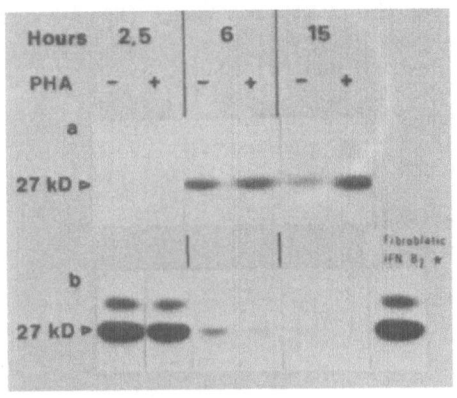

Figure 5. Time course of IFN-β2 protein accumulation secreted by adherent cells activated or not with PHA. a)|^{35}S| methionine labelled supernatants (200,000 cpm) were immunoprecipitated at the indicated times with anti-IFN-β2 antiserum. b) Cold supernatants obtained as in a) were used to compete with labelled fibroblast IFN-β2 (*) immunoprecipitated with the specific antiserum.

CONCLUSION

IFN-β2 mRNA is constitutively present in human monocytes and accumulates very rapidly simply upon culture. The maximal level is reached at 2.5 hours and is subsequently increased by PHA activation. Also, a low level of IFN-β2 mRNA synthesis takes place in T lymphocyte fraction and could be due to monocyte contamination. Nevertheless, in these T cells, expression of IFN-β2 gene occures only after PHA activation and reaches its maximal level later in activation. Taken together, these data favour a low expression of this gene in T lymphocytes.

In contrast, IFN-β mRNA is not present in adherent and non adherent cells. However PHA triggers IFN-β gene expression in both cells, although with different kinetics.

The protein coded by the 15S IFN-β2 mRNA is antigenically distinct from IFN-β1, similar to the IFN-β2 protein produced in fibroblasts (3) and is neutralized by the anti-IFN-β2 antiserum. Moreover, the in vitro and in vivo labelled translation products are immunoprecipitated specifically by the anti-IFN-β2 antiserum and share the M.W. characteristics displayed by IFN-β2 protein produced by fibroblasts. In addition, monocyte IFN-β2 protein is capable to completely displace the labelled fibroblast IFN-β2 from binding to the specific antibodies.

IFN-β2 mRNA is present in peripheral monocytes from healthy donors. The in vivo translation of this messenger would allow the synthesis of IFN-β2 under physiological conditions. The role of this protein in the hematopoietic system (differenciation, immune response...) remained to be determined. It was recently reported that IL1 as well as TNF were able to induce IFN-β2 in fibroblasts (6,7). However, it remains to be clarified whether these two cytokines produced predominantly by mononclear cells are the physiological inducers of IFN-β2 in monocytes.

REFERENCES

1. Vaquero C, Sanceau J, Weissenbach J, Beranger F and Falcoff R: J. Interferon Res. 6:161-170, 1986.
2. Weissenbach J, Chernajovsky Y, Zeevi M, Shulman L, Soreq H, Nir U, Wallach D, Perricaudet M, Tiollais P and Revel M: Proc. Natl. Acad. Sci. USA, 77:7152-7156, 1980.
3. Revel M, Ruggieri M and Zilberstein A: The Biology of the Interferon System, Serono Symposia, Eds Schellekens H and Stewart WE II (Elsevier, Amsterdam) pp. 207-216, 1985.
4. Stefanos S, Wietzerbin J, Catinot L, Devos R and Falcoff R: J. Interferon Res. 5:455-463, 1985.
5. Vaquero C, Sanceau J, Sondermeyer P and Falcoff R: Nucleic Acids Res. 12:2629-2640, 1984.
6. Kohase M, Henriksen-De Stefano D, May LT, Vilcek J and Sehgal PB: Cell, 45:659-666, 1986.
7. Zilberstein A, Ruggieri J, Korn J and Revel M: The 1986 TNO-ISIR Meeting on the Biology of the Interferon System.

REGULATION OF HUMAN FIBROBLAST GROWTH BY AUTOCRINE INTERFERON-BETA

M. Kohase[1], M. Tsujimoto[2], L.T. May[3],
P.B. Sehgal[3] and J. Vilček[1]

[1]New York University Medical Center, New York, NY 10016 (U.S.A.),
[2]Suntory Institute for Biomedical Research, Osaka 618 (Japan),
and [3]The Rockefeller University, New York, NY 10021 (U.S.A.)

1. INTRODUCTION

Several studies suggested a role for autocrine IFN-beta in the regulation of cell cycle progression and cellular differentiation. Spontaneous production of a beta-type IFN was shown to occur in growth-arrested leukemic cells during the process of terminal differentiation (1,2), in density-arrested, quiescent human melanoma cells (3), or during the progression of synchronized murine fibroblasts through the cell cycle (3). In addition, secretion of a beta-type IFN was demonstrated in cultures of murine bone marrow cells stimulated to differentiate by the addition of colony stimulating factor-1 (2,5). The role of endogenously produced beta-type IFN in these earlier studies was documented by the demonstration that the addition of neutralizing antibodies specific for IFN-beta either delayed growth arrest or inhibited terminal differentiation of the cells in culture.

In earlier studies, we demonstrated that tumor necrosis factor (TNF) had a marked mitogenic activity in human diploid fibroblasts (6). TNF also stimulated IFN-beta$_2$ production in human fibroblasts (7). Human IFN-beta$_2$ is the translation product of a 1.3 kb mRNA (8,9) derived from an intron-containing gene located on chromosone 7 (10). [Some investigators termed IFN-beta$_2$ "26-kDa protein" (11).] Although the IFN-beta$_1$ and -beta$_2$ genes do not cross-hybridize at the nucleic acid level, the two proteins show amino acid sequence homology and are cross-neutralized by polyclonal or monoclonal antibodies (12). Induction of IFN-beta$_2$ was also demonstrated with interleukin-1 (IL-1), a monocyte-derived cytokine sharing several biological activities with TNF (11-13). Transcription of the IFN-beta$_2$ gene can also be activated by platelet-derived growth factor (PDGF), IFN-beta$_1$ (13) or poly(I).poly(C) (7,11-13). Induction of IFN-beta$_2$ by these different agents may be related to their ability to affect cell growth and differentiation.

In this article we shall briefly review experimental studies in which we continued to examine the growth stimulating and antiviral activity of TNF and related activities of other cytokines and growth factors. Our results show that the biological effects resulting from the generation of autocrine IFN in cultures of human fibroblasts are influenced by a network of complex interactions.

2. COMPARISON OF THE MITOGENIC AND ANTIVIRAL ACTIVITIES OF TNF AND IL-1

Recent studies have indicated that TNF and IL-1, two distinct cytokines usually produced by monocytes/macrophages, show a number of similarities in their biological activities. One of the activities shared by TNF and IL-1 is the mitogenic effect in cultured fibroblasts (6,14). Both TNF (7) and IL-1 (11-13) also stimulate the induction of IFN-beta$_2$ in some cell lines. When recombinant human TNF, recombinant IL-1-alpha and recombinant IL-1-beta were compared in their mitogenic activity in two different lines of human foreskin fibroblasts, all three cytokines showed very similar characteristics (data not shown). However, in producing a mitogenic effect both forms of IL-1 showed a higher specific activity than TNF. Although at optimal concentrations all three cytokines produced a mitogenic effect with similar kinetics and also resulted in a similar magnitude of growth stimulation, the concentrations of IL-1-alpha or IL-1-beta required were approximately 10 to 100-fold lower than those of TNF.

Van Damme et al. (15) demonstrated that a 22K protein from activated human mononuclear cells, later identified as IL-1-beta (16), induced antiviral activity in cultures of the human MG-63 osteosarcoma cell line. We showed that TNF inhibited virus multiplication in human diploid fibroblasts (6). Antiviral activity of both TNF and IL-1 appears to be mediated by the induction of IFN-beta$_2$. To further explore the similarities and differences in the antiviral activities of TNF and IL-1, we compared the ability of the cytokines to induce antiviral activity in the MG-63 line of human osteosarcoma cells and in the FS-4 line of human diploid fibroblasts. Both IL-1-alpha and IL-1-beta readily induced antiviral activity against vesicular stomatitis virus (VSV) in MG-63 cells. TNF was also found to induce protection against infection with VSV virus in MG-63 cells. In contrast, only TNF induced protection against EMC virus in FS-4 cell cultures (Fig. 1). Both IL-1-alpha and IL-1-beta not only were ineffective in inducing an antiviral state in the diploid fibroblasts, but they actually inhibited the antiviral action of TNF when added to FS-4 cultures along with TNF (13).

IL-1-alpha and IL-1-beta failed to induce an antiviral state against EMC virus in FS-4 cells despite the fact they very effectively stimulated IFN-beta$_2$ mRNA synthesis (13). Furthermore, Zilberstein et al. (12) have recently shown that both TNF and IL-1 induce the secretion of the IFN-beta$_2$ protein in human diploid fibroblasts. Failure of IL-1 to induce an antiviral state and the ability of IL-1 to inhibit antiviral protection stimulated with TNF may be due to the fact that IL-1 can inhibit the antiviral action of exogenously added human IFN-beta (Fig. 1). This inhibition of the antiviral action of IFN is reminiscent of an earlier demonstrated inhibitory effect of several growth factors on IFN action (17).

3. COMPLEX INTERACTIONS INVOLVING GROWTH-STIMULATORY CYTOKINES AND THE IFN SYSTEM

The fact that TNF, but not IL-1 alpha or IL-1 beta, induced an antiviral state against EMC virus in FS-4 cells -- although all three cytokines induced IFN-beta$_2$ -- suggests that complex interactions take place among these cytokines at the level of their production and at the level of their actions. These complex interactions were further explored in experiments whose salient results are summarized in Fig. 1.

ACTIVITY	TNF	IL-1α	IL-1β	EGF	PDGF	Insulin	Dexamethasone
Antiviral Action in MG-63 Cells	+	+	+	−	−	−	ND
Antiviral Action in FS-4 Cells	+	−	−	−	−	−	−
Inhibition of IFN-β-Induced Antiviral Action*	−	+	+	+	+	ND	−
Inhibition of TNF-Induced Antiviral Action*	✕	+	+	+	+	−	+
Inhibition of Poly(I)·Poly(C)-Induced IFN Production*	−	+	+	+	ND	−	+

✳ In FS-4 Cells

FIGURE 1. Complex interactions of growth-stimulatory cytokines with the interferon system. Details of experiments that form the basis of the results summarized in the figure will be published separately (ref. 13; Kohase, Henriksen-DeStefano, Sehgal and Vilček, in preparation). To determine the induction of antiviral state EMC virus was used in FS-4 cells and VSV in MG-63 cells.

In the first series of experiments epidermal growth factor (EGF), platelet-derived growth factor (PDGF), insulin and dexamethasone were examined for their ability to induce antiviral protection against EMC virus in MG-63 cells and in FS-4 cells. EGF, PDGF, insulin and dexamethasone were chosen for these experiments because, like TNF and IL-1, all of these agents have been shown to be mitogenic in human diploid fibroblasts (18). Although no antiviral activity was seen with any of these agents, EGF and PDGF resembled IL-1 in their ability to inhibit the antiviral action induced by exogenously added IFN-beta or by TNF. Dexamethasone inhibited TNF-induced antiviral action, but not the antiviral action due to exogenously added IFN-beta. Inhibition of the TNF-induced antiviral action correlated with the ability of dexamethasone to inhibit TNF-induced IFN-beta$_2$ mRNA synthesis (data not shown). Dexamethasone was also shown to inhibit poly(I).poly(C)-induced IFN-beta$_1$ mRNA synthesis (Y. Zhang and J. Vilček, unpublished data). Finally, all of the agents were examined for their ability to affect poly(I).poly(C)-induced IFN production, determined on the basis of antiviral activity generated in FS-4 cells. While TNF and insulin showed no inhibitory effect, all the other agents examined suppressed IFN-beta production stimulated by poly(I).poly(C).

162

4. CONCLUSIONS

Our studies further illustrate the complex interrelationships involving the interferon system and cytokines/growth factors. It has been postulated on the basis of several studies that in various cell systems growth factors can stimulate not only gene functions required for cell cycle progression, but also may simultaneously activate the expression of IFN-beta genes (2,5,7,19). Such induction of IFN-beta was shown to serve as a negative feedback control mechanism leading to decreased cell proliferation (2,3,7,18). Although the mechanism of the antiproliferative action of IFNs has not been fully elucidated (20), some of the activity appears to correlate with an inhibitory effect of IFNs on the expression of proto-oncogenes, e.g., c-myc (21,22). Our results also illustrate the earlier recognized fact (17,20) that growth factors can act as inhibitors of interferon activity. In addition we show that under some conditions one growth factor can actually inhibit the generation of IFN stimulated by another growth factor. In view of the many interactions involved, the biological effects resulting from the generation of autocrine IFN, stimulated by various cytokines and growth factors, will depend on the cumulative outcome of these various interrelationships.

Acknowledgements. We thank Drs. Peter Lomedico, Charles Dinarello and their colleagues for the supply of IL-1 preparations, Drs. Teruhisa Noguchi and Shudo Yamazaki for valuable support and encouragement, Dorothy Henriksen-DeStefano and Edward Kenny for technical assistance and Cristy Minton for the typing of the manuscript. The present address of M. Kohase is: National Institute of Health, Nakato-Musashimurayama, Tokyo 190-12, Japan.

REFERENCES

1. Friedman-Einat M, Revel M, Kimchi A: Initial characterization of a spontaneous interferon secreted during growth and differentiation of Friend erythroleukemia cells. Mol. Cell. Biol. 2, 1472-1480, 1982.
2. Resnitzky D, Yarden A, Zipori D, Kimchi A: Autocrine β-related interferon controls c-myc suppression and growth arrest during haematopoietic cell differentiation. Cell 46, 31-40, 1986.
3. Creasey AA, Eppstein DA, Marsh YV, Khan Z, Merigan TC: Growth regulation of melanoma cells by interferon and (2'-5') oligo-adenylate synthetase. Mol. Cell. Biol. 3, 780-786, 1983.
4. Wells V, Malucci L: Expression of the 2-5A system during the cell cycle. Exp. Cell. Res. 159, 27-36, 1985.
5. Moore RN, Larsen HS, Horohov DW, and Rouse BT: Endogenous regulation of macrophage proliferative expansion by colony-stimulating factor-induced interferon. Science 223, 178-181, 1984.
6. Vilcek J, Palombella VJ, Henriksen-DeStefano D, Swenson C, Feinman R, Hirai M, Tsujimoto M: Fibroblast growth-enhancing activity of tumor necrosis factor and its relationship to other polypeptide growth factors. J. Exp. Med. 163, 632-643, 1986.
7. Kohase M, Henriksen-DeStefano D, May LT, Vilcek J, and Sehgal PB: Induction of β2-interferon by tumor necrosis factor: a homeostatic mechanism in the control of cell proliferation. Cell 45, 659-666.
8. Sehgal PB and Sagar AD: Heterogeneity of poly(I).poly(C)-induced human fibroblast interferon mRNA species. Nature 288, 95-97 (1980).

9. Weissenbach J, Chernajovsky Y, Zeevi M, Shulman L, Soreq H, Nir U, Wallach D, Perricaudet M, Tiollais P, and Revel M: Two interferon mRNAs in human fibroblasts: In vitro translation and Escherichia coli cloning studies. Proc. Natl. Acad. Sci. USA 77, 7152-7156, 1980.

10. Sehgal PB, Zilberstein A, May LT, Ferguson-Smith A, Slate DL, Revel M, Ruddle FH: Human chromosome 7 carries the 2-interferon gene. Proc. Natl. Acad. Sci. USA (in press).

11. Content J, DeWit L, Poupart P, Opdenakker G, Van Damme J, and Billau A: Induction of a 26-kDa-protein mRNA in human cells treated with an interleukin-1-related, leukocyte-derived factor. Eur. J. Biochem. 152, 253-257, 1985.

12. Zilberstein A, Ruggieri R, Korn JN, Revel M: Structure and expression of cDNA and genes for human interferon-beta₂, a distinct species inducible by growth-stimulatory cytokines. EMBO J. (in press).

13. Kohase M, May LT, Tamm I, Vilcek J, Sehgal PB: A cytokine network in human diploid fibroblasts: interactions of interferons, tumor necrosis factor, platelet-derived growth factor and interleukin-1. (Submitted for publication.)

14. Schmidt JA, Mizel SB, Cohen D, and Green I: Interleukin 1, a potential regulator of fibroblast proliferation. J. Immunol. 128, 2177-2182, 1982.

15. Van Damme J, Opdenakker G, Billiau A, De Somer P, DeWit L, Poupart P, and Content J: Stimulation of fibroblast interferon production by a 22 K protein from human leukocytes. J. Gen. Virol. 66, 693-700, 1985.

16. Van Damme J, DeLey M, Opdenakker G, Billiau A, DeSomer P: Homogeneous interferon-inducing 22 K factor is related to endogenous pyrogen and interleukin-1. Nature 314 266-268, 1985.

17. Oleszak E, Inglot AD: Platelet-derived growth factor inhibits antiviral and anticellular action of interferon in synchronized mouse or human cells. J. Interferon Res. 1, 37-48, 1980.

18. Vilcek J, Kohase M, and Henriksen-DeStefano D: Mitogenic effect of double-stranded RNA in human fibroblasts: role of autogenous interferon. J. Cell. Physiol. (in press).

19. Zullo JN, Cochran BH, Huang AS, Stiles CD: Platelet-derived growth factor and double-stranded ribonucleic acids stimulate expression of the same genes in 3T3 cells. Cell 43, 793-800, 1985.

20. Taylor-Papadimitriou J: Effects of interferons on cell growth and function. In: Interferon 1980, Volume 2 (I. Gresser, Editor), p. 13-46. Academic Press, London, 1980.

21. Jonak GJ, Knight E Jr.: Selective reduction of c-myc mRNA in Dauui cells by human βinterferon. Proc. Natl. Acad. Sci. USA 81, 1747-1750, 1984.

22. Einat M, Resnitsky D, Kimchi A: Close link between reduction of c-myc expression by interferon and G0/G1 arrest. Nature, 313, 597-600, 1985.

BIOLOGICAL ACTIVITIES OF HUMAN INTERFERON-β₂ PRODUCED BY cDNA EXPRESSION IN HAMSTER CELLS AND POSSIBLE AUTOCRINE FUNCTIONS OF THIS CYTOKINE-INDUCED IFN

A.ZILBERSTEIN, R.RUGGIERI, J.H.KORN, L.CHEN, Y.MORY, J.CHEBATH, L.SHULMAN, AND M.REVEL
Department of Virology, The Weizmann Institute of Science, Rehovot, Israel

1. INTRODUCTION

Human fibroblasts induced by poly (rI)(rC) and sequential cycloheximide/ Actinomycin D treatment produce, in addition to IFN-β_1 and its 0.9 kb mRNA, another mRNA of 1.3 kb encoding IFN activity neutralized by anti-IFN-β antibodies and hence designated IFN-β_2 (1,2). An IFN-β_2 cDNA was cloned (1) and used to screen a human genomic library from which two intron-containing genes IFA-2 (IFN-β2a) and IFA-11 (IFN-β2b) were identified (3). The IFN-β2a gene is 4.8 kb long, and was mapped on human chromosome 7 (4). Both genomic clones have been expressed in rodent cells and produced human IFN antiviral activities (5). The recombinant IFN-β_2 induces (2'-5') oligo A synthetase mRNA and HLA mRNAs in the presence of cycloheximide (3,6) and has antiviral activity on mouse-human hybrid cells containing human chromosome 21 (but not 9), excluding the possibility that IFN-β2 acts through IFN-β_1 induction (3,6). By immunocompetition with the in vitro translation product (23-26 Kd) of IFN-β2 mRNA, the native form of IFN-β_2 secreted by human cells, was identified as a 21-22 Kd glycoprotein (3). Native IFN-β_2 can be separated from IFN-β_1 because it is not retained on Blue-Sepharose and elutes from DEAE-cellulose pH 7.4 at lower salt (150 mM NaCl) than IFN-β_1. Antibodies raised against either IFN-β_2 or IFN-β_1 can immunoprecipitate specifically only their cognate species, but nevertheless cross-neutralization of the two antiviral activities is observed (3,7), suggesting conservation of a common active-site epitope. About half the aminoacid residues which are conserved in all human type I IFNs, are also conserved in IFN-β2a (5); the overall sequence homology with IFN-β_1 is around 15%. Similarity in the hydropathy profile of IFN-β_2 and IFN-β_1 were also noted (5). In this work, we report the constitutive expression of IFN-β2a cDNA, in hamster CHO cells, and the biological activities of the rIFN-β2a produced.

The most significant observation on the possible physiological function of IFN-β_2 is that it is induced under several conditions where IFN-β_1 is not found. Thus, cycloheximide treatment (1,8) or simply aging of the culture, induces IFN-β2. Moreover, TNF-α has been shown to induce IFN-β_2 mRNA, but not IFN-β_1 in human fibroblasts, and addition of anti-IFN-β to TNF-treated fibroblasts abolished the antiviral action of this cytokine and enhanced markedly the mitogenic action of TNF on these cells (9). IFN-appears, therefore, to be one of the autocrine IFN-β species (distinct from the virally induced IFN-β_1), which are made by cells during growth-transitions and differentiation (10,11) and cause their growth-arrest (12). Spontaneous and PHA-induced synthesis of IFN-β_2 mRNA, but not IFN-β_1, was also observed in human peripheral blood mononuclear cells (13) and IFN-β_2 mRNA was found to be made in much larger amounts than the IFN-β_1 mRNA in response to an IL-1-like inducer of IFN-β (14). Using a specific immunoassay we show here that both TNF-α and IL-1α indeed induce in human fibroblasts

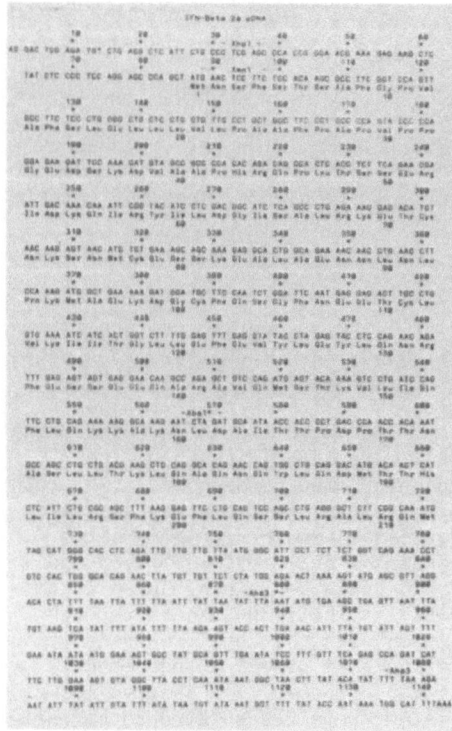

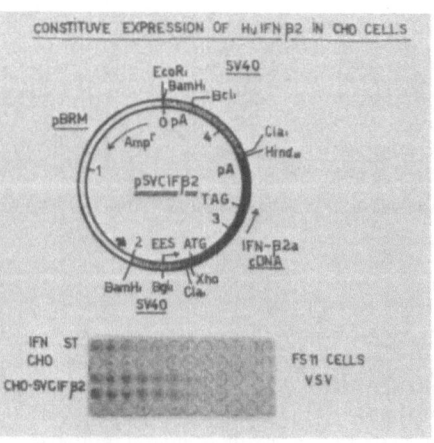

FIGURE 1. Human IFN-β_2a cDNA sequence. Expression vector used to transform hamster CHO cells. Inhibition of VSV cytopathic effect by recombinant rIFN-β_2 from clone B-131 (see text).

the secretion of the IFN-β_2 protein, which in turn activates the typical interferon-response genes.

2. RESULTS

2.1. Biological activities of recombinant human IFN-β_2

The sequence of the full-length IFN-β_2a cDNA is shown in Fig. 1. The open reading frame of 212 aminoacids (23.7 Kd) was verified by _in vitro_ transcription-translation using various cDNA deletions (7). There are two potential N-glycosylation sites in the sequence and since IFN-β_2 is secreted as a 21 Kd glycoprotein (3,7), it is likely that processing of the hydrophobic N-terminus occurs. The IFN-β_2a cDNA sequence was introduced in an expression vector pSVCIF-β_2 (Fig. 1) following the SV40 early gene promoter (EES) and 60 bases of the T-ag mRNA as described (7). A genomic segment was used to restore the IFN-β_2a polyadenylation site (pA), located just upstream of the HindIII site. The T-ag splicing region and its own pA site was placed following the IFN-β_2a cDNA. Hamster CHO DHFR⁻ cells were co-transfected with pSVCIFβ2 and a pSVDHFR plasmid (7). DHFR⁺ clones growing in nucleoside-free medium were further subjected to amplification-selection with methotrexate (MTX) and the medium of clone cultures was assayed for antiviral activity. Figure 1 shows results with clone B-131 selected by 250 nM MTX, which produces constitutively 300 U/ml/24 hrs of IFN activity demonstrable by inhibition of the cytopathic effect of VSV on human diploid FS11 fibroblasts. Further amplification of such clones by 500 nM MTX raised the IFN yield to 800 U/ml/24 hrs. The specific activity of IFN-β_2 was

TABLE 1. Herpes Simplex Virus Type 2 Plaque Reduction Assay on FS11 Cells

IFN U/ml	HSV-2 pfu x 10^{-6}/ml		IFN U/ml	HSV-2 pfu x 10^{-5}/ml III	IFN U/ml	HSV-2 pfu x 10^{-5}/ml III
	I	II				
No IFN	8.6	14	No IFN	3	No IFN	3
IFN-β_1 10	3.2	8	IFN-β_1 20	1.5	IFN-β_1 50 +IFN-β_2 10	0.4
IFN-β_1 100	1.4	1.4	IFN-β_1 60	0.9		
					IFN-β_1 50 + IFN-γ 10	0.9
IFN-β_2 10	1.6	5.4	IFN-β_2 20	0.9		
IFN-β_2 100	0.2	0.4	IFN-β_2 60	0.7	IFN-β_2 50 +IFN-γ 10	0.3
			IFN-γ 60	2.2		

I, II and III are different experiments and HSV-2 strains isolated from patients. Cells were infected at various virus dilutions and the number of plaques counted at 48 hrs.

estimated to be about 50–100 times lower than that of IFN-β_1(1,5). Metabolic labeling of such clones with ^{35}S-methionine resulted in the secretion of the 22 Kd protein immunoprecipitated by anti-IFN-β_2, in amounts comparable to those of human fibroblasts (7).

The IFN-β_{2}a antiviral activity is pH 2-stable and human cell-specific (relative titers, 0.1% on mouse and 10% on monkey cells). Polyclonal sera and a monoclonal antibody to human IFN-β_1 neutralized the rIFN-β_{2}a activity, which was not inhibited by monoclonal anti-IFN-α or anti-IFN-γ(7). Virus yield assays showed that rIFN-β_{2}a inhibits the replication of VSV (10 U/ml producing a 2 log virus reduction). IFN-β_2 was also active against Mengo virus and Herpes simplex virus. Table 1 shows that the number of plaques formed on FS11 cells by several HSV-2 strains was more strongly inhibited by IFN-β_2 than by IFN-β_1. In addition, a 2-3 fold synergy was seen when IFN-β_2 was given in combination with IFN-β_1 or IFN-γ. IFN-β_2 was also more efficient to inhibit HSV DNA synthesis in these cells (not shown).

In addition to the antiviral effect, IFN-β_2 appears to have the same biological activities as type I IFNs. A strong antimitogenic effect was observed by measuring the inhibition of thymidine incorporation in FS11 fibroblasts starved and refed with serum (not shown). Synthesis of prostaglandin E2 was stimulated in FS7 cells. One antiviral unit per ml of IFN-β_2, efficiently induces (2'-5') oligo A synthetase as does IFN-β_1 (5,7). Cell-surface HLA class I antigens, but not class II, were increased by rIFN-β_2 on human fibroblasts. HLA-A,B,C mRNA induction may be stronger with IFN-β_2 than with IFN-β_1(3). The induction of both synthetase and HLA mRNAs appears to be direct and does not require protein synthesis (3). Induction of the C56 mRNA was also observed (6). Table 2 shows that IFN-β_2 action is inhibited by polyclonal antibodies to the type I IFN receptor, produced in mouse by injections of mouse-human hybrid cells containing human chromosome 21 (15). However, monoclonal antibodies prepared from such mice and selected for inhibition of IFN-β_1 action (16) did not inhibit IFN-β_2 but also not IFN-αA and αD, suggesting that the various type I IFNs interact differently with the receptor or that chromosome 21 encodes more than one type of IFN receptor.

TABLE 2. Inhibition of IFN-β_2 Action by Polyclonal and Monoclonal Antibodies to Human Type I IFN Receptor

IFN added to FS11 fibroblasts:	Inhibition of (2'-5') oligo synthetase induction by	
	Polyclonal Anti-21 (% inhib.)	Monoclonal Anti-21 (% inhib.)
IFN-β1	87%	70%
IFN-β2	97%	0%
IFN-αA	86%	4%
IFN-αD	83%	3%

Antibodies (origin described in text) were added 1 hr before adding the various IFNs at 16 U/ml. NP-40 cell-extracts were prepared 24 hrs and assayed for (2'-5') oligo A synthetase activity. Without antibodies the various IFNs produced a 3-5 fold increase in enzyme over the untreated cell level (300 cpm) which was substracted.

TABLE 3. The Effect of TNF and IL-1 on (2'-5') Oligo A Synthetase Induction is Mediated by IFN-β

Human FS11 fibroblast treatment:	(2'-5') oligo A synthetase level		
	Enzymatic Activity[a] cpm	Immunoblot analysis[b] 40 Kda	67 Kda
None	150	0	0
TNF-α (400 U/ml, 24 h)	41,165	615	1,475
TNF-α + anti-IFN-β serum[c]	325	0	70
TNF-α + non-immune serum	29,860	485	1,125
None	85	0	0
IL-1α (3 U/ml, 18 h)	2,000	670	965
IL-1α + anti-IFN-β serum[c]	90	5	85
IL-1α + non-immune serum	710	610	825

a/ Enzyme activity in 2 μg protein of NP-40 S15 extracts.
b/ Western blots reacted with peptide antibodies to (2'-5') oligo A synthetase and ^{125}I-Protein A were scanned for areas of 40 Kda and 67 Kda bands, two main forms of the IFN-induced synthetase in FS11 cells (Chebath J, et al., in press).
c/ Rabbit antiserum B: 4 μl neutralizes 100 U/ml of IFN-β1. Amounts used: 25 μl/ml for TNF experiment, 2 μl/ml for IL-1 experiment.

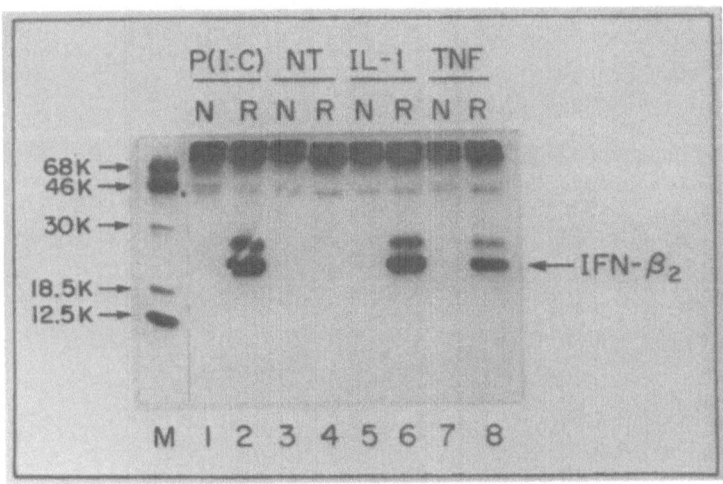

FIGURE 2.   Secretion of IFN-β$_2$ in response to various inducers.
Confluent FS11 fibroblasts were treated for 4 hrs with 50 µg/ml poly (rI)
(rC), or for 18 hrs with 4 U/ml IL-1α or 400 U/ml TNF-α, or left untreated
(NT). All cultures were labeled with ^{35}S-methionine (100 µCi/ml) from
4.5 to 18 hrs. 0.25 ml aliquots of medium were immunoprecipitated with
20 1 anti-IFN-β$_2$ serum R (7) or normal serum (N).
rIL-1α (3x10^7 U/mg) was a gift from Hoffmann-La Roche.
rTNF-α (10^7 U/mg) was a gift from Cetus Corporation.

2.2. Autocrine IFN-β2 action in response to cytokines

Both rTNF-α (300 U/ml) and rIL-1α (4 U/ml) induce the synthesis and
secretion of IFN-β$_2$ by FS11 fibroblasts as shown by an immunoassay based on
competition with the in vitro translation product of IFN-β$_2$ mRNA (7). The
amount of IFN-β$_2$ secreted was dose-dependent and optimal after 8 hours with
both cytokines. We calculated that IFN-β$_2$ accumulates to levels of 10-20
U/ml in the medium of TNF or IL-1 treated cells. Metabolic labeling and
immunoprecipitation showed that the same 21-22 Kd IFN-β$_2$ molecule is
secreted in response to the cytokines as with poly (rI)(rC)(Fig. 2).
A strong induction of (2'-5') oligo A synthetase was observed   FS11
cells, exposed to TNF or to IL-1 (Table 3). In both cases, the amount of
IFN-β$_2$ secreted (10-20 U/ml) would be enough to account for the synthetase
induction (7) by TNF and by IL-1 of the (2'-5') oligo A synthetase, prevent
ing the synthesis of the large 67 Kd and the small 40 Kd enzyme forms in
FS11 cells (Table 3). Kohase et al. (9) have reported that the TNF-induced
antiviral activity was also inhibited by anti-IFN-β antibodies but not by
anti-IFN-α or anti-IFN-γ. As found in other cells (9,14), TNF and IL-1
induce IFN-β$_2$ mRNA in FS11 fibroblasts (half maximum reached in 3-4 hrs),
but we did not detect IFN-β$_1$ mRNA during these experiments, even with IL-1
(Fig. 3) unlike previous reports (14).

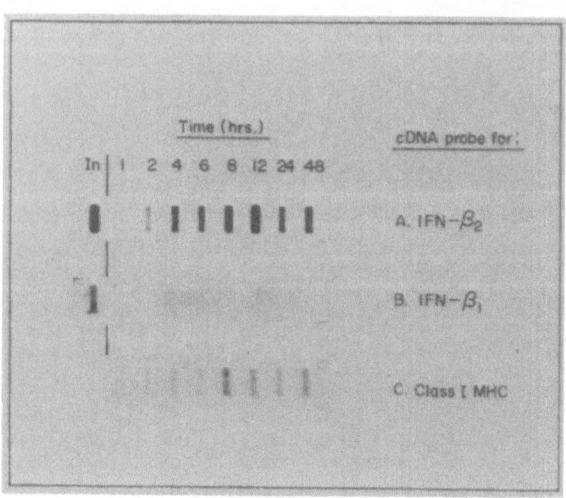

FIGURE 3. IL-1 induces IFN-β_2 mRNA but not IFN-β_1 mRNA.
Slot-blot hybridization of 14 µg total RNA from FS11 cells at indicated
time after 4 U/ml IL-1 . In = 0.3 µg of poly A$^+$. mRNA from poly (rI)(rC)-
cycloheximide-induced cells.

These results demonstrate that the (2'-5') oligo A synthetase induction by
IL-1 and TNF, as the induction by TNF of an antiviral stage (9) and also
probably the increase in class I HLA mRNA (Fig. 3) and antigens (17), are
mediated by the autocrine action of IFN-β_2 produced in response to the
cytokines. During growth transitions accompanying terminal differentiation
of cells, an induction of (2'-5') oligo A synthetase is observed which results from the autocrine action of an IFN-β species secreted by the differentiating cells (10,11). Addition of anti-IFN-β antibodies prevents the
growth-arrest of differentiating myeloleukemia cells (11). Similarly, the
anti-IFN-β antibodies prolong and enhance the mitogenic effect of TNF on
diploid fibroblasts (9). An important function of autocrine IFN-β may be
therefore, to insure an efficient growth control in response to growth stimuli. Since anti-IFN-β antibodies also prevented the switch-off of the myc
oncogene during growth arrest of differentiating cells (12), the inhibition
of cell proliferation by autocrine IFN-β may be mediated by the regulation
of oncogene activity. Inhibition of several oncogenes by IFN may be an
important property of cells showing normal growth-regulation, but may be
often lost in malignant cells (18).

The autocrine IFN observed in human and mouse cell system (10,11) has
been shown to be β-type IFN on the basis of neutralization by anti-IFN-β_1
antibodies. Characterization of these autocrine IFN-β showed, however, that
they were not the virally induced IFN-β_1 species (10,12). The demonstration
that IFN-β_2 is produced in response to growth-stimulatory cytokines such as
TNF-α and IL-1 in fibroblasts, identifies IFN-β_2 as a first cloned example
of this class of autocrine IFN-β like molecules.

ACKNOWLEDGEMENTS.
Work supported by Inter-Yeda, Israel.

REFERENCES

1. Weissenbach J, Chernajovsky Y, Zeevi M, Shulman L, Soreq H, Nir U, Wallach D, Perricaudet M, Tiollais P and Revel M: Proc Natl Acad Sci USA 77:7152-7156, 1980.
2. Sehgal PB and Sagar AD: Nature 287:95-97, 1980.
3. Zilberstein A, Ruggieri R and Revel M: In: Rossi GB and Dianzani E (eds.), The Interferon System, Serono Symposia,(Raven Press)NY, Vol. 24, pp. 73-83, 1985.
4. Sehgal PB, Zilberstein A, Ruggieri R, May LT, Ferguson-Smith A, Slate DL, Revel M and Ruddle FH: Proc Natl Acad Sci USA 83:5219-5222, 1986.
5. Revel M, Ruggieri R and Zilberstein A: In: The Biology of the Interferon System 1985,(Schellekens H and Stewart WE eds.), Elsevier Press, pp. 207-216, 1986.
6. Zilberstein A, Nissim A, Ruggieri R, Shulman L, Chen, L and Revel M: In: The Biology of the Interferon System,(Schellekens H and Stewart WE II eds.), Elsevier Press, pp. 119-124, 1986.
7. Zilberstein A, Ruggieri R, Korn, JH and Revel M:EMBO J, 5, in press,1986
8. Content J, De Wit L, Pierard D, Derynck, R, De Clerck E, and Fiers W Proc Natl Acad Sci USA 79:2768-2772, 1982.
9. Kohase M, Henriksen-De Stefano D, May LT, Vilcek J and Sehgal PB: Cell 45:659-666, 1986.
10. Friedman-Einat M, Revel M and Kimchi A: Mol. Cell. Biol. 2:1472-1480, 1982.
11. Yarden A, Shure-Gottlieb H, Chebath J, Revel M and Kimchi A: EMBO J 3:969-973, 1984.
12. Resnitzky D, Yarden A, Zipori D and Kimchi A: Cell 46:31-40, 1986.
13. Vaquero C, Sanceau J, Weissenbach J, Beranger F and Falcoff R: J Interferon Res 6: 161-170, 1986.
14. Content J, De Wit L, Poupart P, Opdenakker G, Van Damme J and Billiau A: Eur J Biochem 152:253-257, 1985,
15. Revel M, Bash D and Ruddle FH: Nature 260:139-141, 1976.
16. Shulman LM and Ruddle FH: In: The Biology of the Interferon System, (Kirchner H and Schellekens H, eds.), Elsevier Press, Amsterdam, pp. 333-337, 1984.
17. Collins T, Lapierre LA, Fiers W, Strominger JL, and Pober JS: Proc Natl Acad Sci USA 83:446-450, 1986.
18. Einat M, Resnitzky D and Kimchi A: Nature 313:597-600, 1985.

CELL BIOLOGY

OVERVIEW OF LONG-TERM IFN EFFECTS ON THE PHENOTYPE OF TRANSFORMED AND
TUMOR CELLS

D. BROUTY-BOYE
Institut de Recherches Scientifiques sur le Cancer, Villejuif, France

1. INTRODUCTION
   Prolonged in vitro treatment with interferon (IFN) has been shown to
suppress transformed and tumorigenic properties of certain cell types (1-
6). Using 12 different cell lines either transformed in vitro or derived
from human spontaneous solid tumors, we demonstrate that long-term IFN
treatment induces varied and even opposite cellular changes that do not
necessarily contribute to the phenotypic normalization of neoplastic cells.

2. MATERIALS AND METHODS
2.1. Cells and IFNs. The origins of the different cell lines and IFNs used
in their treatment are presented in Table 1.

TABLE 1.   Sources of cells and IFNs

Cells	Agent of transformation	Supplied by	IFNs
Murine C3H/10T1/2			
Clone 5T1	X-rays	Our laboratory (7)	C-243 cell α/β *
Clone 5T3	X-rays	Our laboratory (7)	C-243 cell α/β †
Murine C3H/10T1/2	MCA	ATCC	C-243 cell α/β *
Murine C3H/10T1/2	SV_{40}	Our laboratory (10)	C-243 cell α/β *
Murine NIH embryonic fibroblasts	Ki-MuSV	Dr. A. Morris (4)	C-243 cell α/β *
Human osteosarcoma			
OHA		Dr. H. Suarez (11)	Leucocyte-α °
G272		ATCC	Namalva-α °
Human gastric sarcoma			
SHAC		Dr. H. Suarez (15)	r-α_2 (16)
Human carcinoma			
Bladder HT 1376		ATCC	Leucocyte-α °
EJ/T24		ATCC	Leucocyte-α °
Colon CO 205		ATCC	Namalva-α °
Ovary OD 572		Dr.Magdelenat (17)	Namalva-α °

*, partially purified (8); †, highly purified (9); °, highly purified (12-
14).

2.2. *Long-term treatment of cells with IFNs.* Cells were seeded at densities of 2×10^5 cells per 25 cm^2 flask. IFNs were added 24 hr after seeding at the 1st passage and thereafter on the day of subculture. The cultures were passaged once every week with a medium change at either day 4 or 5. At each passage level, the cell concentration was adjusted to 2×10^5 cells/flask. Cells were treated with 640 units of murine IFN α/β or 5.000 units of human IFN α for periods $\geqslant$ 6 months.

2.3. *Cellular parameters related to in vitro and in vivo proliferation.* Parameters related to in vitro proliferation such as cell density, morphology and colony formation in soft agar were evaluated at different passage levels as previously described (2, 7). In vivo proliferation was assayed by tumor formation after inoculation of 2×10^6 cells into nude mice (2, 7).

2.4. *C-ras oncogene expression.* Persistence and expression of activated c-ras oncogenes present in human tumor cells was evaluated at different stages of IFN treatment by :
   2.4.1. Southern-blot experiments : High molecular weight (HMW) DNAs were extracted by the method of Gross-Bellard et al.(18), digested with restriction enzymes and resolved by electrophoresis in 1 % agarose. Nitrocellulose blotting was carried out by the method of Southern (19) and hybridization was performed with ^{32}P-labeled specific DNA probes (20).
   2.4.2. Dot and northern blot experiments :Poly(A)$^+$ cellular RNAs were prepared according to standard procedures (20). RNA blotting to nitrocellulose was performed as described by Thomas (21).
   2.4.3. Radioimmunoprecipitation experiments : Lysates of ^{35}S-methionine labeled cells were immunoprecipitated by monoclonal antibody Y13-259 specific for ras protein p21 (22). The immunoprecipitates were dissolved and electrophoresed in 12 % SDS-acrylamide gel.
   2.4.4. DNA transfection experiments : HMW DNAs were transfected into Swiss mouse 3T3 cells as described by Wigler et al. (23).

3. RESULTS
   The data summarized in Table 2 illustrate the inconsistent and opposite effects of IFN on transformed and tumor cell properties according to the cell type and the duration of treatment. It appears that the antiproliferative effects were most pronounced during the early times of IFN treatment. With prolonged treatment these effects decreased or did not persist. Several cell passages in the absence of IFN were usually sufficient for initial sensitivity to be reacquired. On the other hand, in one of the cell lines, the human bladder carcinoma EJ cells, stimulation of proliferation was seen during long-term IFN treatment.
   At the level of morphologic effects,IFN induced and maintained greater adherence and more normal appearance in murine transformed cells while there was generally no drastic change in human tumor cells. Two exceptions, however, were noted : murine SV$_{40}$ transformed cells and human osteosarcoma G272 cells both displayed opposite morphologic responses when compared to the other murine and human cells.
   In regard to tumorigenicity, only in murine X-ray transformed cells was long-term IFN treatment effective in suppressing tumor formation. With human tumor cells the best response was a reduced rate of tumor growth and a reduced number of animals bearing tumors, but not complete suppression. These effects were generally achieved in human tumor cells during short-term IFN treatment and were no longer present during long-term treatment.

TABLE 2. Comparison of short and long-term effects of IFN on different transformed and tumor cells

Cells	Extent of effects of IFN on									
	Proliferation				Cell piling or cell clumping		Morphology		Tumor formation	
	Liquid		Agar							
	ST*	LT†	ST	LT	ST	LT	ST	LT	ST	LT
Murine cells :										
X-ray/10T1/2 (clones 5T1 and 5T5)	+ °	±			+	+	+	+	±	+
MCA/10T1/2	+	-	+	+	+	+	+	+	±	±
SV40/10T1/2	+	±	+	+	-	-	-	-	-	↑
KiMuSV/MEF	+	±	+	+	+	+	+	+	↑	↑
Human cells :										
OHA	+	±	+	+	+	+	-	-	±	±
G 272	+				+		+		±	
SHAC	-	-			-	-	-		↑	↑
HT 1376	+				+		-		-	
EJ/T24	±	↑			±	↑	-	-	↑	↑
Co 205 (clone S/C13)	+	-			↑	-	-	-	±	-
OD 562	+				-		-		-	-

* ST = short-term IFN treatment (< 3 months)
† LT = long-term IFN treatment (> 6 months)
° +, pronounced effect ; ±, moderate effect ; -, no effect; ↑, stimulatory effect

178

One exception was the osteosarcoma OHA cells which still grew less in vivo after more than 1-year-treatment. Finally, in 4 out of the 12 cell lines studied, IFN-treatment increased the ability of the cells to form tumors in vivo. Of particular interest was that this increase in tumorigenicity in vivo occured despite reduced cell proliferation in vitro (in both liquid and agar) in two of the murine cell lines tested.

That the effects of IFN differed from one cell type to another was in itself not surprising in view of the diversity of the processes of transformation and tumorigenesis. To determine whether the action of IFN was related to effects on these processes, we analyzed in parallel experiments, human tumor cell phenotype and expression of oncogenes related to the human c-ras family. As indicated in Table 2, in OHA cells under prolonged IFN treatment, reduced proliferation with lack of cellular overlapping at confluency, reduced ability to form colonies in agar and tumors in animals, were maintained. In contrast, the same IFN treatment exerted stimulatory effects on proliferation in EJ cells but not in SHAC cells and yet the tumorigenicity of both cell types was increased. These IFN-induced phenotypic effects were, however, not associated with alterations in oncogene expression.

Southern blot analysis (Fig. 1) indicated that both OHA and EJ cells retained the c-Ki-ras and c-Ha-ras oncogene respectively, with no apparent structural alterations under long-term treatment with IFN. In addition

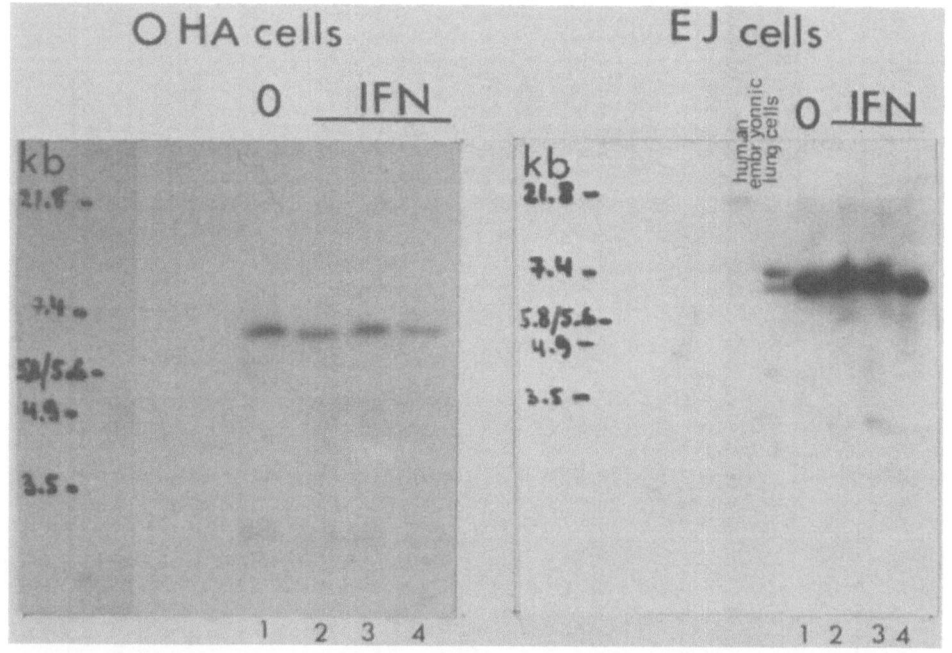

FIGURE 1. Southern blot analysis of 20 μg cellular HMW DNAs digested with EcoRI (OHA cells) or Bam HI (EJ cells) and hybridized with ^{32}P-labeled Ki ras PKBE 2 clone (24) and Ha-ras p EJ probe (25) DNAs. Lanes 1 = no treatment, lanes 2, 3, 4 = treatment with IFN for 7, 13 and 28 weeks.

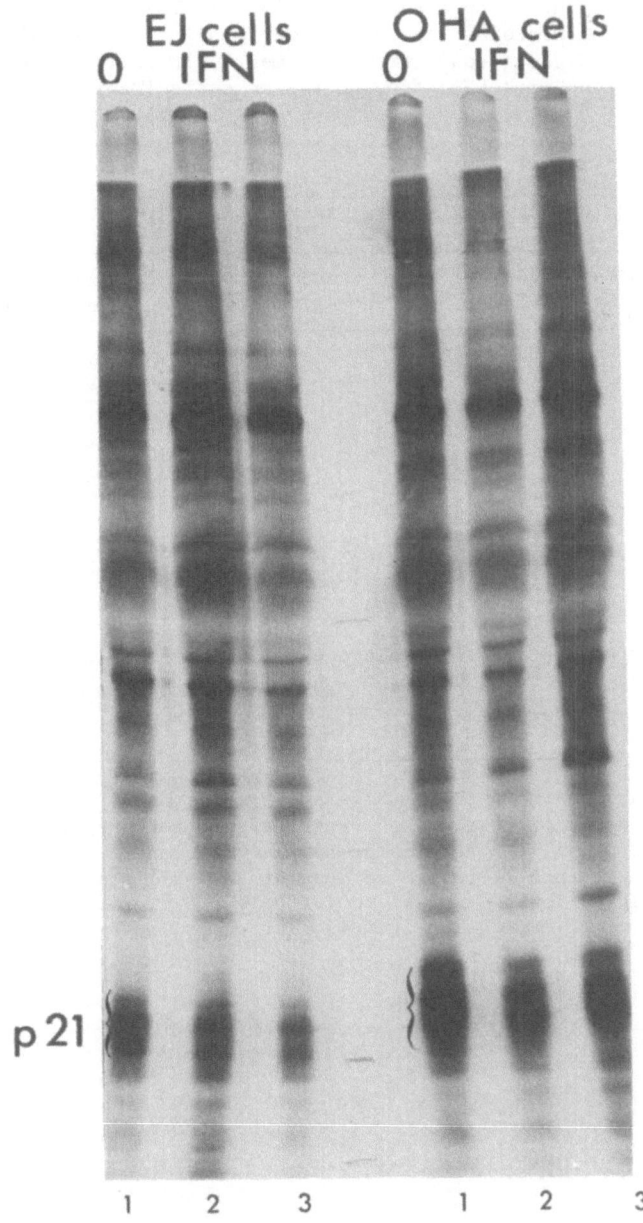

FIGURE 2. Immunoprecipitation of p21 with rat anti p21 antibody extracted from ^{35}S-methionine labeled cells and analyzed on 15 % SDS polyacrylamide gels. Lanes 1 = no treatment ; lanes 2, 3 = treatment with IFN for 1 and 40 weeks.

both IFN-treated cells appeared to contain the same elevated levels of
ras-protein, the p21, than untreated cells (Fig. 2). The c-ras oncogenes
and their transforming proteins were also functional since the DNAs from
IFN-treated OHA and EJ cells could transform mouse 3T3 cells at levels
comparable to that seen with DNAs from untreated cells (Table 3).

TABLE 3.   Transforming activity of IFN-treated OHA and EJ cell DNAs

Cells	Exp. n°	N° of passages with IFN-α	Total n° of foci[a]/ Total n° of dishes		Transforming efficiency [b]	
			0	+ IFN	0	+ IFN
OHA	I	6	5/10	3/10	0.025	0.015
	II	6	10/10	12/10	0.050	0.060
	III	12	7/10	10/10	0.035	0.050
	IV	12	3/10	9/10	0.015	0.045
EJ	I	5	2/10	3/10	0.010	0.015
	II	5	5/10	5/10	0.025	0.025
	III	15	18/15	22/15	0.060	0.073

a   Each dish of Swiss 3T3 cells was exposed to 20 µg of HMW DNA and
    transformed foci were scored after 6 weeks in culture.

b   Number of foci/µg of DNA

Finally, dot blot and northern blot analysis (Fig. 3) revealed no significant difference in the expression of N-ras oncogene between untreated
and IFN-treated SHAC cells. A surprising result with regard to c-myc
expression and the antiproliferative effects of IFN (27) was obtained
in SHAC cells. Despite lack of inhibition of cell proliferation, IFN
treatment reduced c-myc expression in these tumors cells (Fig. 3).

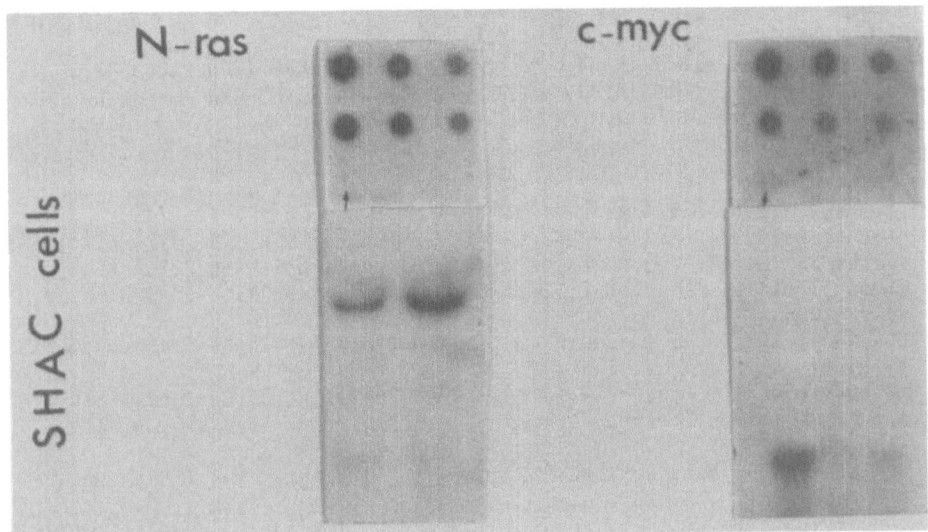

FIGURE 3. a) Dot blot analysis of poly A⁻ RNAs. The blots were hybridized with either ^{32}P-labeled human N-ras (p52 c⁻ clone (26)) or human c-myc (Eco RI-Cl AI fragment) probes. Lanes 1 = no treatment; lanes 2 = treatment with IFN for 10 weeks. Arrow indicates the spot for control human embryonic lung cells (HEL) which contains 2 µg of RNA. b) Northern blot analysis of poly A⁺ RNAs (3 µg). Lanes 1 = no treatment; lanes 2 = treatment with IFN for 10 weeks.

4. CONCLUSIONS

From this study, it would appear that among the different phenotypic properties which define neoplastic transformation those related to proliferation in vitro are more efficiently affected by IFN. In contrast, the characteristics related to proliferation in vivo are hardly inhibited.

As also described by others with cells transformed in vitro, phenotypic normalization can be achieved (1-6). In contrast, results obtained herein suggest that in cells derived from spontaneous solid tumors, IFN can only reduce transformed and tumorigenic properties.

The observation that IFN can increase the ability of 4 out of 12 cell lines to form tumors in vivo suggests that this process is not a rare exception and may have profound implications for the therapeutic use of IFN. The mechanisms underlying the opposite effects of IFN are at present unclear. However, we have shown that in conditions of either increased or decreased tumorigenicity, the product of 3 oncogenes which displays similar functions, was not altered by IFN. This result suggests that IFN is acting on tumor cell properties at a post-transcriptional level. How in one case, tumorigenicity is increased and in other decreased provides an area of investigation which should reveal important insights into the mechanisms of the process of transformation.

182

ACKNOWLEDGEMENTS
This work was aided in part by grants from INSERM (contracts 82 20 02
and 85 20 06) ; CNRS (ATP 959 050) and the Richard Lounsbery Foundation.

REFERENCES
1. Chany C, Vignal M (1970). J. Gen. Virol. 7, 203-210.
2. Brouty-Boyé D, Gresser I. (1981). Int. J. Cancer 28, 165-173.
3. Brouty-Boyé D, Chen YE, Chen LB (1981). Cancer Res. 41, 4171-4184.
4. Hicks NJ, Morris AG, Burke DC (1981). J. Cell Sci. 49, 225-236.
5. Brouty-Boyé D, Puvion-Dutilleul F, Gresser I (1982). Experientia 38,
   1292-1296.
6. Samid D, Chang EH, Friedman RM (1984). Biochem. Biophys. Res. Commun.
   119, 21-28.
7. Brouty-Boyé D, Gresser I, Baldwin C (1979). Int. J. Cancer 24, 261-265.
8. Tovey MG, Begon-Lours J, Gresser I (1974). Proc. Soc. Exp. Biol. Med.
   146, 809-815.
9. De Maeyer-Guignard J, Tovey MG, Gresser I, De Maeyer E (1978).Nature
   (London) 271, 622-625.
10.Brouty-Boyé D, Wybier-Franqui J, Calvo C, Feunteun J, Gresser I (1984).
   Int. J. Cancer 34, 107-112.
11.Suarez HG, Grosjean L, Nardeux P (1986). In : R. Gallo, D. Stehelin,
   O. Variner (eds.), Retroviruses and Human Pathology. Humana Press,
   Clifton, pp. 509-515.
12.Mogensen KE, Cantell K (1977). Pharmacol. Ther.(C) 1, 369-381.
13.Strander H, Mogensen KE, Cantell K (1975). J. Clin. Microbiol. 1, 116-
   117.
14.Mogensen KE, Bandu MT, Vignaux F, Aguet M, Gresser I (1981). Int. J.
   Cancer 28, 575-582.
15.Andeol Y, Nardeux P, Daya-Grosjean L, Landin RM, Suarez HG (1985).
   In : Proceedings of the 1st Annual Meeting in Oncogenes, Frederick,
   Maryland, July 10-14, pp. 127.
16.Weissmann C (1984). In Gresser I (ed.) Interferon 3, Academic Press,
   London, pp. 101-134.
17.Brouty-Boyé D, Mogensen KE, Gresser I (1985). Eur. J. Cancer Clin.
   Oncol. 21, 507-514.
18.Gross-Bellard M, Oudet P, Chambon P (1978). Eur. J. Biochem. 36, 32-38.
19.Southern EM (1975). J. Mol. Biol. 98, 503-517.
20.Maniatis T, Fritsch EF, Sambrook KJ (1982). Molecular Cloning. A
   Laboratory Manual, Cold Spring Harbor Laboratory, Cold Spring Harbor,
   New York.
21.Thomas PS (1980). Proc. Natl, Acad. Sci. USA 77, 5201-5205.
22.Furth ME, Davis LJ, Fleurdelys B, Scolnick EM (1982). J. Virol. 43, 294-
   304.
23.Wigler M, Pellicer A, Silverstein S, Axel R, Urlaub G, Chasin L (1979).
   Proc. Natl. Acad. Sci. USA 76, 1373-1376.
24.Ellis RW, DeFeo D, Shih TY, Gonda MA, Yound HA, Tsuchida N, Lowy DR,
   Scolnick EM (1981). Nature (London) 292, 506-511.
25.Tabin CJ, Bradley SM, Bargmann CI, Weinberg RA (1982). Nature (London)
   300, 143-149.
26.Murray MY, Cunnigham JM, Parada LF, Dautry F, Lebowitz P, Weinberg RA
   (1983). Cell 33, 749-757.
27.Jonack GJ, Knight E (1984). Proc. Natl. Acad. Sci. USA 81, 1747-1750.

MECHANISMS FOR ANTI-INVASIVE EFFECT OF INTERFERON

GEIR BUKHOLM, METTE BERGH and MIKLOS DEGRÉ
Kaptein W. Wilhelmsen og Frues Bakteriologiske Institutt, University of
Oslo, Rikshospitalet and Natl. Inst. Publ. Health., Oslo, Norway.

INTRODUCTION

The ability of penetrating the cell membrane and establish an intracellular state of infection is recognized as one of the important pathogenetic factors for bacteria belonging to the Salmonella (16, 21, 22) and Shigella (15, 17, 18, 23, 26, 27) genera and for enteroinvasive Escherichia coli (13, 14, 19). The penetrating potential of these bacteria can be tested in cell cultures such as HeLa cells, HEp2 cells, A-549 cells and the murine fibroblasts L-929 (9).

Invasion may be divided into three phases: reversible adhesion, irreversible adhesion an invasion (21); all of which are dependent of host cellular and agent factors. The invasion phase can be blocked by cytochalasin B (4) pointing to the active host cell role: endocytosis is the mechanism for uptake of bacteria. Also coxsackie B (6,10) and measles virus (11) infection of the cells enhanced the uptake of bacteria. This enhancement is evident before cytopathic effect can be recognized and was also achieved with UV inactivated virus particles. On the other hand: the presence of high molecular weight plasmids of the bacteria seems to be necessary for expression of invasiveness (19, 22, 26); and invasive properties can be transferred from one bacterium to another by transferring the invasion plasmid (27).

Interferons are known to have effect on numerous functions of normal mammalian cells. Such functions include cell mechanisms involved in host defence against different non-viral infectious agents i.e. bacteria, chlamydia, mycoplasma and protozoa (1). Pretreatment of cells with interferons reduces the ability of salmonella bacteria to penetrate the cell membrane (5, 7, 9). This effect is dose-dependent. Compared to the antiviral effect gamma interferons are more potent inhibitors of invasiveness than type I interferons (7). Like other non-antiviral effects of interferons the anti-invasive effect follows a biphasic curve. Interferon treatment of HEp2 cells, however, seemed only to affect salmonella bacteria and not shigella bacteria or enteroinvasive E. coli (7, 11).

Our further efforts in this project were to study the mechanisms behind the anti-invasive effect of interferon. Firstly, we wanted to investigate whether cell cultures different from the HEp2 would respond differently to interferon treatment with respect to resistance to Shigella flexneri infection. Invasiveness was therefore first studied in a modified HEp2 cell system: the HEp2 cells were measles virus infected. Secondly, A-549 cells were interferon treated and infected with Sh. flexneri; and thirdly, the effect on bacterial adhesiveness and the dependence of ribososmal activity were studied.

MATERIALS AND METHODS

Bacterial strains. One strain of Salmonella typhimurium (SIFF S4575/81, one strain of Sh. flexneri (SIFF R662/81) and one strain of Sh. sonnei (SIFF S3351/81) was used. The strains of S. typhimurium, Sh. flexneri and Sh. sonnei were originally isolated from patients with gastroenteritis and were invasive in Hep-2 cell cultures (8). The Sh. flexneri and the Sh. sonnei strains were able to produce kerato-conjunctivitis in guinea pigs (Sereny test) (28). The strains were maintained at -70°C.

Virus. Measles virus, Edmondson strain, was grown in HEp-2 cell monolayers. The concentration of virus was $10^{6.3}$ TCID50/ml according to the Reed and Muench equation.

Interferons and interferon globulins. Partially purified human leukocyte interferon was obtained from K. Cantell, Helsinki. Recombinant human alpha-2 interferon was obtained from S. Pestka, Nutley, N. J. The antiviral activity was tested by means of an infectivity inhibition microtest with human embryo fibroblast cells and vesicular stomatitis virus.

Toxins. Cycloheximide, Sigma, St Louis, M.O., USA was used. Purified shigella toxin and abrin was kindly provided by Dr. K. Sandvig, The Norwegian Radium Hospital, Oslo.

Cultivation of cell monolayers. Monolayers of the human epithelial cell lines HEp-2 and A-549 were grown on glass coverslips (14 mm in diameter) in 24 well tissue culture plates (Linbro, Flow Laboratories Inc., Inglewood, Cal.) as described in earlier publications (8, 12). The cell cultures were recognized as suitable for inoculation when they formed an almost continous monolayer with intermediate blank areas (leopard spots).

The cell cultures were treated with interferons for 24 h at various concentrations. After 24 h they were washed with PBS and incubated for one h in fresh medium without interferon.

Preparation of bacterial inocula. Salmonella and shigella strains were cultivated on chocolate agar for 24 h at 37°C under aerobic conditions. The bacteria were suspended in PBS to an optical density of 0.75 at 520 nm. A sterile filtrate of the Sh. flexneri was prepared as follows: Bacteria from overnight growth on chocolate agar were supended in PBS to OD_{520} of 0.75 and filtered through a sterile Millipore filter with pore size 320 nm.

Challenge of cell monolayers. The cell monolayers were challenged with three different types of inocula: (i) 200 ul of either S. typhimurium, S. flexneri or S. sonnei, (ii) 200 ul of S. typhimurium plus 100 ul of S. flexneri filtrate, (iii) 200 ul of S. typhimurium and 100 ul shigella toxin, abrin or cycloheximide. Control cultures not treated with interferon were included for each of the combinations i-iv. After inoculation the cells were incubated in a 5% CO_2 atmosphere at 37°C for 3 h.

Fixation and staining. The cells were fixed overnight in 2% glutaraldehyde in 0.1 M cacodylate buffer, pH 7.2 and stained for 4 min by acridine orange (50 mg/l) in cacodylate buffer as described earlier (8).

Microscopy. The preparations were examined at a final magnification of 1250 using a combination of two optical systems applied on the same microscope: (i) Nikon equipment for Nomarski differential interference contrast microscopy and (ii) for UV incident light microscopy. A detailed description of the methods employed was reported earlier (12). This approach enabled effective discrimination between extracellular (adhesive) and intracellular (invasive) bacteria. The accuracy and the sensitivity of this method has been established by comparison with scanning electron microscopy (12). About 200 cells in each of 3 parallel preparations were examined for intracellular bacteria.

Statistical analyses were based on the Pearson Chi-square test applied on the different frequency distributions of cells with intracellular bacteria. Level of significance was p=0.01.

RESULTS

Interferon effect on invasiveness of S. typhimurium in HEp-2 and A-549 cell cultures.

The S. typhimurium strain was initially tested for invasiveness in HEp-2 and A-549 cells treated with human leukocyte interferon or recombinant alpha-2 interferon. Invasiveness of S. typhimurium was significantly reduced in a dose dependent manner in both of the cell systems. The effect of the interferons seemed to be comparable in the two cell types. Compared to its anti-viral activity the recombinant interferon was somewhat less potent that the natural interferon. In both cell systems the maximum inhibitory dose of human leukocyte interferon was found at 100 U/ml and for recombinant alpha-2 interferon the corresponding value was 500 U/ml.

Since we could not observe any significant effect of interferon on invasiveness of Sh. flexneri in the HEp2 cells, it was logical to examine whether the effect of a sterile filtrate from our Sh. flexneri strain would have any modulating effect on invasiveness of S.typhimurium in HEp2 cells. When this filtrate was added to the HEp-2 cell cultures together with the interferon susceptible strain of S. typhimurium no anti-invasive interferon effect was expressed.

TABLE 1. EFFECT OF RECOMBINANT ALPHA-2 INTERFERON ON INVASIVENESS OF SALMONELLA TYPHIMURIUM IN HEp2 CELLS.

	% infected cells	Number of bacteria per infected cell	Number of bacteria per 100 cells
Control	60	5.7	347
10	53	5.9	317
100	34	4.1	137
500	17	4.3	72
1000	40	5.7	229

Interferon effect on invasiveness of Sh. flexneri and Sh. sonnei

Treatment of HEp2 cells with human leukocyte, recombinant alpha-2 or gamma interferon in concentrations from 1 to 1000 U/ml for 24 h did not influence the invasiveness of Sh. flexneri. When the cells were incubated with measles virus before bacterial inoculation, invasiveness was significantly enhanced as a function of virus dose and incubation time. The increase in bacterial invasiveness was evident before a cytopathic effect was evident. Treatment of cells with interferon 24 h before measles virus infection reduced invasiveness of Sh. flexneri and developement of the virus infection. However, whilst the interferon effect on viral multiplication followed a linear dose-response curve, the effect on bacterial invasiveness followed a biphasic curve with a maximum activity at 100 U/ml. At this concentration the interferon treatment almost completely eliminated the enhancement produced by the virus infection.

Invasiveness of Sh. flexneri was also inhibited in A-549 cells. Human leukocyte interferon in doses of 100 U/ml to 1000 U/ml with a maximum dose at 500 U/ml protected A-549 cells against invasion. Similiar to the results obtained with S. typhimurium the dose of recombinant alpha-2 interferon that gave the maximum inhibition of Sh. flexneri was higher (1000 U/ml) than the human leukocyte interferon dose necessary to give an equal effect (500 U/ml).

TABLE II. INTERFERON EFFECT ON INVASIVENESS OF SHIGELLA FLEXNERI IN A-549 CELLS

	% infected cells	Number of bacteria per infected cell	Number of bacteria per 100 cells
Control	43	5.6	240
100 leukocyte	27	3.8	103
500 "	9	5.8	53
1000 "	14	7.6	108

Interferon effect on bacterial adhesiveness

Treatment of HEp2 cells and measles virus-infected HEp2 cells with human leukocyte interferon or human gamma interferon in concentrations from 1 to 1000 IU/ml for 24 h did not influence the adhesive ability of Sh. flexneri.

TABLE III. EFFECT OF INTERFERON ON ADHESIVENESS OF SALMONELLA TYPHIMURIUM TO CYTOCHALASIN B-TREATED HEp2 CELLS.

IFN titer	Cells with adhesive bacteria	Adhesive bacteria per cell	Bacteria per 100 cells
0	59	2.9	183
10	58	2.7	166
100	52	2.5	139
1000	55	2.3	133
10000	57	2.7	166

Because interferon inhibited invasion of S. typhimurium into HEp2 cells, invasiveness had to be blocked to study any effect on adhesiveness of this bacterium. This was achieved by pre-treatement of the cells with dihydrocytochalasin (H$_2$CB) (5 ug/ml). After 3h of incubation a conciderable adherence of bacteria was observed. After interferon treatment the percentage of cells with adherent bacteria was reduced from 59 to 52 at an interferon concentration of 100 U/ml (Table III). The number of bacteria per cell was also reduced from 2.9 in the controls to 2.3 at an interferon concentration of 1000 IU/ml. The dose-response curve showed a biphasic shape with a maximum inhibitory concentraion at 1000 IU/ml. The anti-adhesive effect of interferon was reduced with anti–human leukocyte interferon globulin.

Effect of low temperature and ribosomal inhibitors on invasiveness of S. typhimurium in interferon–treated cells.

Cell cultures were treated with 100 and 500 IU of human leukocyte interferon and recombinant alpha-2 interferon for 24 h and were then inoculated with S. typhimurium and incubated at 4 °C. After 3 h of incubation many of the cells were infected with intracellular bacteria, although the number of bacteria was less than in cultures incubated at 37 °C. However, there was no significant difference between interferon–treated cultures and the control cultures. It was then likely to assume that the presence of metabolic activity of the host cells was necessary for the expression of the anti–invasive state of the cell.

It therefore became natural to investigate whether the effect was dependent on protein synthesis. Protein synthesis was blocked at the ribosomal level by treatment with cyloheximide in concentrations of 1, 10 and 100 mg/ml. Interferon–treated HEp2 cells were not able to protect themselves against salmonella invasion when bacteria were added to cultures together with cyloheximide. Similiar effects were achieved by adding two other protein synthesis inhibitors, abrin (1 or 10 ug/ml) or shigella toxin (100 pg/ml). In control cultures that were not interferon treated addition of cyloheximide or toxins did not exhibit any effect on invasiveness of S. typhimurium.

TABLE IV. EFFECT OF CYCLOHEXIMIDE, ABRIN AND SHIGELLA TOXIN ON INVASIVENESS OF SALMONELLA TYPHIMURIUM IN HEP-2 CELLS TREATED WITH INTERFERON

IFN	Toxin type	Toxin conc (ug/ml)	Percent inhibition of invasiveness
0	–	–	0
500	–	–	82
500	cycloheximide	1	0
500	shigella tox	0.0001	0
500	abrin	10	0

DISCUSSION

The importance of interferons as modulators of the host defense system against bacterial infections has been increasingly recognized. These activities include several immuno–specific and non–specific host defence factors such as enhanced phagocytosis and augmented cell mediated immunity The in vivo effect of interferons in infections with shigella and salmonella bacteria is uncertain. We have in a previous study shown that mouse fibroblast interferon can inhibit invasiveness of S. typhimurium in infant mice (9). Izadkah et al. also showed that gamma interferon can protect mice against infection with S. typhimurium (20). In several earlier studies we have shown that interferon inhibits the uptake of S. typhimurium in HEp-2 cells (5, 7). We have also shown that both S. typhimurium and Sh.

flexneri is probably endocytosed in these cells (4).

The mechanisms behind this interferon action is, however, not clear. Bacterial invasiveness can be divided into several stages where interferon may act: bacterial attachment, irreversible adhesiveness, endocytosis of the bacteria by the host cell and a later stage of intracellular multiplication.

Interferon treatment of HEp2 cells does not influence on the invasiveness of Sh. flexneri in our system. And filtrates from the shigella strain inhibit interferon to express an anti-invasive effect against S. typhimurium. Interferon treatment of measles virus infected cells reduced uptake of Sh. flexneri (11). This reduction of invasiveness could be due to interferon effect on virus multiplication; however, the anti-invasiveness dose-response curve was biphasic and not linear as was the anti-viral effect curve. Thus, interferon treated, measles virus infected HEp2 cells were resistant to shigella invasion in the same manner as interferon-treated HEp2 cells were to salmonella invasion. Measles virus infection had added new properties to the cells regarding their ability of expressing interferon induced anti-invasiveness. Our next question was then: would other cell cultures show different interferon susceptibilities when shigella bacteria or shigella filtrates was present? Because we suspected shigella toxins to be involved in the interferon interaction, we chose a cell type that was not susceptible to shigella toxin. When these cells, A-549 cells, were treated with interferon invasiveness of shigella bacteria was inhibited in a dose dependent manner.

What was the mechanisms behind these observations? From studies by Harris et al. (19) and Sansonetti et al. (27) we knew that plasmids of shigella or invasive E. coli correlated to virulence probably encode polypeptides that induce receptor mediated endocytosis by the host cell. From other studies we know that interferon may alter several functions of the cell membrane (2, 3), and interferons might possibly alter the affinity of the polypeptides to the specific receptors. On the other hand interferons might interfere at a later stage of the endocytosis process. It was therefore natural to study the interferon effect on adhesiveness. In HEp2 cells neither invasiveness nor adhesiveness was influenced by interferon treatment, an observation that was well correlated with the effect of shigella filtrates in these cell cultures. However, in measles virus infected HEp2 cells interferon had some effect on invasiveness, but not on adhesiveness. Therefore we proceeded to a system where we had demonstrated a clear interferon effect: S. typhimurium and HEp2 cells. To control the effect of invasiveness we blocked this activity by cytochalasin. This treatment might moderate interferon action. However, we observed a small, but significant interferon effect on adhesiveness. The effect was, however, too small to explain the much more prominent effect on bacterial adshesiveness after 3 h.

From earlier studies we knew that a reduction of temperature (4 °C) reduced the anti-invasive effect of interferon (7) and we knew that this effect could be inhibited in some cell cultures by Sh. flexneri or sterile filtrates from this bacterium. These results strongly pointed to an interferon effect on host cell metabolism as the pathway of its anti-invasive action. Because shigella bacteria secrete toxic substances that are able to block protein synthesis at the 60 S ribosomal subunit level the effect of purified shigella toxin was tested. Also the toxin abrin and cycloheximide completely blocked interferon action. These three drugs have different receptors on the 60S subunit, but they all block protein synthesis. It is also important that the shigella toxin did not exhibit any effect on interferon action in the A-549 cells which are not sensiible to shigella toxin.

Our cell cultures were treated with interferon for 24 h before bacteria were introduced to the cultures together with the toxins. This shows that a continuous protein synthesis is necessary for expression of interferon effect. However, the toxins did not alter invasiveness in non-interferon-treated cells. Therefore it is natural to assume that the anti-invasive state of the cell is dependent on interferon-induced proteins with a short half-life.

Niesel et al. reported in a recent study that interferon treatemnt of HEp2 cells inhibited invasiveness of Sh. flexneri (25). However, the effect was only significant when shigella inoculum was extremely small and incubation time was less than 60 min. This finding is consistent with our results: with a low number of bacteria the amount of toxins free in medium is below the level necessary to inhibit interferon action.

Thus, interferon inhibits invasiveness of salmonella, shigella and enteroinvasive E. coli in a dose dependent manner following a biphasic curve. The interferon effect on bacterial adhesiveness is relatively small, and the main action of interferon is likely to be directed on to the endocytosis part of the invasive process. The anti-invasive effect is mediated through the action of interferon induced proteins, because continuous ribosomal activity is required. However, which proteins that are essential in this process is not yet clear. Further experiment are in progress in our laboratory on this topic.

REFERENCES
1. Baron S, M Langford, EM MacDonald, GJ Stanton, J Ritmeyer and DA Weigert: Induction of interferon by bacteria, protozoa and viruses. Tex Rep Biol Med 41:150–157, 1982.
3. Brouty-Boye D and MG Tovey: Inhibition by interferon of thymidine uptake in chemostat cultures of L1210 cells. Intervirology 9:243–252, 1978.
4. Brouty-Boye D and RB Zetter: Inhibition of cell motility by interferon. Science 108:516–518, 1980.
5. Bukholm G: Effect of cytochalasin B and di-hydrocytochalasin B on invasiveness of entero-invasive bacteria in HEp-2 cell cultures. Acta Path Microbiol Scand Sect B 92:145–149, 1984.
6. Bukholm G and M Degre: Effect of human leukocyte interferon on invasiveness of Salmonella species in HEp-2 cell cultures. Infect Immun 42:1198–1202, 1983.
7. Bukholm G and M Degre: Invasiveness of Salmonella typhimurium in HEp2 cell cultures preinfected with coxsackie B1 virus. Acta Path Microbiol Immunol scand Sect. B 92:45–51, 1984.
8. Bukholm G and M Degre: Effect of human gamma interferon on invasiveness of Salmonella typhimurium in HEp-2 cell cultures. J Interferon Res 5:45–53, 1985.
9. Bukholm G and J Lassen: Bacterial adhesiveness and invasiveness in cell culture monolayer. 2. In vitro invasiveness of 45 strains belonging to the family Enterobacteriaceae. Acta Path Microbiol Immunol Sect B 90:409–413, 1982.
10. Bukholm G, BP Berdal, C Haug and M Degre: Mouse fibroblast interferon modifies Salmonella typhimurium infection in infant mice. Infect. Immun. 45:62–66, 1984.
11. Bukholm G, M Holberg-Petersen and M Degre: Invasiveness of Salmonella typhimurium in HEp2 cells pretreated with UV-inactivated coxsackie virus. Acta Path Microbiol Immunol Scand Sect B 93:61–65, 1985.
12. Bukholm G, K Modalsli and M Degre: Effect of measles-virus infection

and interferon treatment on invasiveness of Shigella flexneri in HEp2 cell cultures. J Med Microbiol 22: in press, 1986.

13. Bukholm G, BV Johansen, E Namork and J Lassen: Bacterial adhesiveness and invasiveness in cell culture monolayer. 1. A new light optical method evaluated by scanning electron microscopy. Acta Path Microbiol Immunol Scand Sect B 90:403-408, 1982.

14. DuPont HL, SB Formal, RB Hornick, MJ Snyder, JP Libonati, DG Sheahan, EH LaBrec and JP Kalas: Pathogenesis of E. coli diarrhea. N Engl J Med 285:1-9, 1971.

15. Formal SB, HL DuPont, R Hornick, MJ Snyder, J Libonati and E LaBrec: Experimental models in the investigation of virulence of dysentery bacilli and Escherichia coli. Ann NY Acad Sci 176:190-196, 1971.

16. Gerber DF and HM Watkins: Growth of Shigella in monolayer tissue cultures. J Bacteriol 82:815-822, 1961.

17. Gianella RA, O Washington, P Gemski and SB Formal: Invasion of HeLa cells by Salmonella typhimurum: A model for study of invasiveness of Salmonella. J Infect Dis 128:69-75, 1973.

18. Hale TL and PF Bonventre: Shigella infection of Henle intestinal epithelial cells: Role of the bacterium. Infect Immun 24:879-886, 1979.

19. Hale TL, RE Morris and PF Bonventre: Shigella infection of Henle intestinal epithelial cells: Role of the host cell. Infect Immun 24:887-894, 1979.

20. Harris JR, K Wachsmuth, BR Davis and ML Cohen: High-molecular-weight plasmid correlates with Escherichia coli enteroinvasiveness. Infect Immun 37:1295-1298, 1982.

21. Izadkhah Z, AD Mandel and G Sonnenfeld: Effects of treatment of mice with sera containing gamma interferon on the course of infection with Salmonella typhimurium. J Interferon Res 1:137-145, 1980.

22. Jones GW, LA Richardson and DB Uhlman: The invasion of HeLa cells by Salmonella typhimurium: reversible and irreversible bacterial attachment and the role of bacterial motility. J Gen Microbiol 127:351-360, 1981.

23. Jones GW, DK Rabert, DM Svinarich and HJ Whitfield: Association of adhesive, invasive and virulent phenotypes of Salmonella typhimurium with autonomous 60-megadalton plasmids. Infect Immun 38:476-486, 1982.

24. LaBrec EH, H Schneider, TJ Magnani and SB Formal: Epithelial cell penetration as an essential step in the pathogenesis of bacillary dysentery. J Bacteriol 88:1503-1518, 1964.

26. Mehlman IJ, EL Eide, AC Sanders, M Fishbein and CG Aulisio. Methodology for recognition of invasive potential of Escherichia coli. J Ass Off Anal Chem 60:546-562, 1977.

27. Niesel DW, CB Hess, YJ Cho, KD Klimpel and GR Klimpel: Natural and recombinant interferons inhibit epithelial cell invasion by Shigella spp. Infect Immun 52:828-833, 1986.

28. Sansonetti PJ, DJ Kopecko and SB Formal: Shigella sonnei plasmids: Evidence that a large plasmid is necessary for virulence. Infect Immun 34:75-83, 1981.

29. Sansonetti PJ, TL Hale, GJ Dammin, C Kapfer, HH Collins Jr and SB Formal: Alterations in the pathogenicity of Escherichia coli K-12 after transfer of plasmid and chromosomal genes from Shigella flexneri. Infect Immun 39:1392-1402, 1983.

30. Sereny B: Experimental keratoconjunctivitis shigellosa. Acta Microbiol Hung 4:367-376, 1957.

COMPARISON OF DIFFERENT MODES OF INTERFERON DELIVERY BY TUMOR SPECIFIC MONOCLONAL ANTIBODIES

H.K. Hochkeppel, H. Towbin and S.S. Alkan, Research Department, Pharmaceuticals Division, Ciba-Geigy Ltd., CH-4002 Basel, Switzerland

INTRODUCTION

Interferons (IFN) as well as other lymphokines (LK), such as interleukin 2 (IL2), have very short biological half lives, if administered systematically. Most of the LKs are pleiotropic and exert multiple activities. In addition, they are rather interlinked in a network of interactions with poorly understood feedback mechanisms or synergistic and antagonistic activities (1). In order to circumvent such antagonistic effects and also to reduce the unwanted side effects of IFN in the clinic it would be favourable if one could significantly lower the required dose of IFN (or other lymphokines) by specifically delivering them to the site of a desired organ or cells.

An approach to specifically deliver active human recombinant IFNα-D to EBV transformed tumor cells by an anti-EBV monoclo-antibody (mAb) has been reported previously (2). It was demonstrated in vitro that the IFN-mAb conjugates exerted greater antiviral and antiproliferative activities than uncoupled IFN on target cells. These studies have now been extended to different modes of delivering IFN to target cells. In the present study bispecific mAbs were prepared by crosslinking two mAbs of different specificities: one being specific for tumor and the other for IFN. These mAb_1-mAb_2 or $Fab'_1-Fab'_2$ conjugates were tested for their ability to transport IFN to tumor cells. Furthermore, in an attempt to elucidate the mechanism of IFN-receptor interaction the potency of antiproliferative activity of IFN targeted by bispecific mAbs, direct IFN-mAb conjugates and uncoupled IFN was compared.

MATERIAL AND METHODS

Crosslinking of proteins with N-succinimidyl-3(2-pyridyl (dithio) proprionate) (SPDP) was performed according to a previously described method (3) and Fab' preparation as well as crosslinking of $Fab'_1-Fab'_2$ was performed as described (4). Human recombinant hybrid IFNα-BDDD which is active on mouse cells (5) was kindly provided by Dr. M. Grütter, Ciba-Geigy Ltd. (Basel). Generation and characterization of anti-human IFNα mAb (144 BS) (6) and antiidiotypic mAb 12S84-3, have been described previously (7). The 144BS mAb is a non-neutralizing Ab. Target cell line was a mouse hybridoma (10K44), producing mAbs against a hapten (8). The 10K44 hybridoma cells possessed on their surface idiotype positive Igs which served as a specific tumor marker. The antiidiotypic mAb 12S84-3 was specific for this idiotype. The determination of antiviral activity as well as the inhibition of ^{3}H-Thymidin incorporation of target cells by IFNs has been described previously (2).

RESULTS

Bispecific mAbs were prepared by crosslinking two mAbs with SPDP, mAb$_1$ being specific for the idiotype expressed on the mouse hybridoma 10K44 and mAb$_2$ specific for human IFNα. Alternatively Fab' fractions of mAb$_1$ and mAb$_2$ were coupled by disulfide formation (Fig. 1).

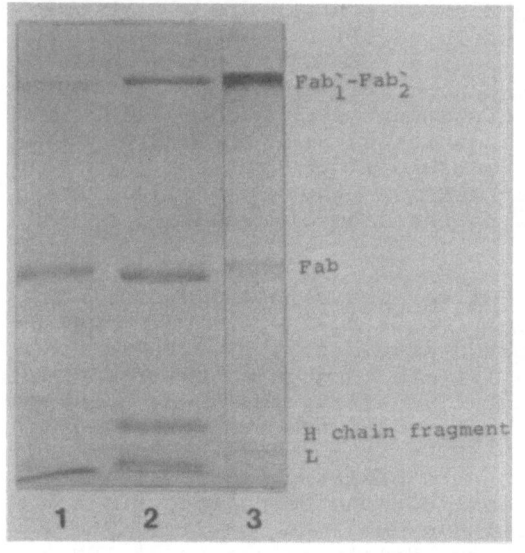

Fab'$_1$-Fab'$_2$

Fab

H chain fragment
L

1   2   3

Fig. 1 SDS polyacrylamide gel electrophoresis under non reducing conditions.
1. Fab'$_1$ preparation (antiidiotype) after treatment with Ellmann's reagent.
2. Fab'$_2$ preparation (anti-IFNα) after thiolation.
3. Crosslinked bispecific Fab'$_1$-Fab'$_2$ conjugate.

In order to test whether each mAb maintained its specificity after coupling, a solid phase RIA was performed. For this purpose the idiotype positive mAb (10K44, target antigen) was coated to the flexible microtiter plate and incubated with the antiidiotypic-anti-IFNα mAb conjugate as well as with the various control samples. After washing the wells were incubated with iodinated IFNα-BDDD and washed again. It was found that only the double mAb conjugate and not the individual antibodies or mixtures thereof bound significant amounts of iodinated IFNαBDDD (Table 1. Also, the mAb$_1$-mAb$_2$ conjugate bound only to the specific idiotype but not to other antigens. This experiment clearly demonstrated that the mAb_1-mAb_2 conjugate had kept its bispecificity after crosslinking.

In a pulse-targeting experiment it was then determined whether the direct mAb_1-IFN conjugate, the bispecific mAb_1-mAb_2 as well as Fab'_1-Fab'_2 conjugates would be able to transport radiolabelled IFN to idiotype positive 10K44 cells (Table 2). Cells were first incubated with Ab preparations and then pulsed with free or conjugated iodinated IFNα-BDDD for 15 min at 4°C, and cell-bound cpm determined following repeated washings. Table 2 illustrates that under these conditions all three conjugates are able to specifically target IFN to 10K44 mouse cells. Such pulse-targeted IFN also exerted a strong antiproliferative activity on the 10K44 cells (data not shown).

Having demonstrated that all three kinds of conjugates can deliver IFNα-BDDD to idiotype positive 10K44 cells the antiproliferative potency of IFNα-BDDD after targeting by different modes was tested (Table 3). 10K44 cells were incubated for 2.5 days at 37°C with IFN alone, direct IFN-mAb_1 conjugate, or bispecific mAb_1-mAb_2 or Fab'_1-Fab'_2 conjugates or a mixture of mAb_1/mAb_2s. IFN was then added at 0, 1, 4, 20 and 200 units IFN/ml, and the percent inhibition of H^3-TdR incorporation was measured at day 3. In comparison to free IFNα or IFNα plus mAb mixture, IFN targeted by all three modes, showed a clear enhanced antiproliferative activity at already 1 unit/ml. Nontargeted free IFN (1U/ml) exerted only marginal effects under these conditions.

TABLE 1

SPECIFIC BINDING OF DOUBLE MAB CONJUGATE TO THE TARGET ANTIGEN (IDIOTYPE) AND TO RADIOLABELLED IFNα BDDD SHOWN BY SOLID PHASE RADIOIMMUNOASSAY

SPECIFICITY OF SAMPLES PULSED WITH	NOMENCLATURE	CPM X 10^3			
		TARGET ANTIGEN: IDIOTYPE		TARGET ANTIGEN: ANTI-HAPTEN	
		RANGE	MEAN	RANGE	MEAN
ANTI IFN	144 BS	4,7-66	5.6	3,4-3.8	3.6
ANTI ID	12 S 84-3	0.35-0.45	0.4	0.6-0.8	0.7
ANTI IFN + ANTI ID MIXTURE	144 BS + 12 S 84-3	3,0-3.4	3.2	2,1-2.1	2.1
ANTI IFN - ANTI ID	144 BS - 12 S 84-3	31,3-33.9	32.6	6,2-7.4	6.8
ANTI-HAPTEN	3 A2-2	0,5-0.7	0.6	0,3-0.4	0.4

PLATES WERE COATED EITHER WITH IDIOTYPE POSITIVE MAB (10K44) OR WITH AN IDIOTYPE NEGATIVE MAB (3A2-2), BLOCKED, INCUBATED WITH VARIOUS CARRIER ABS, WASHED AND EQUAL AMOUNTS OF IFNα BDD-^{125}I (75.000 CPM) ADDED.

TABLE 2

SPECIFIC TARGETING OF RADIOLABELLED IFNα BDDD BY
ANTI-IDIOTYPIC MABs TO IDIOTYPE POSITIVE TARGET CELLS

	BINDING (CPM) TO 10^6 HYBRIDOMA CELLS	
MODE OF TARGETING	IDIOTYPE POSITIVE (10K44)	IDIOTYPE NEGATIVE (3A2-2)
IFN ALONE	120	NT
IFN-MAb$_1$ DIRECT CONJUGATE	1566	NT
IFN + MAb$_1$-MAb$_2$ CONJUGATE	2372	182
IFN + MAb$_1$ + MAb$_2$ MIXTURE	210	NT
IFN + FAb'$_1$-FAb'$_2$	1814	NT

MAb$_1$: ANTI IDIOTYPE (12 S 84-3)
MAb$_2$: ANTI IFNα (144BS)
IFN: α-BDDD-^{125}I (10^5 CPM ADDED TO 1-10^6 CELLS)
15 MIN. PULSE AT 4°; 3x WASH WITH HB101 MEDIUM 15% FCS.

TABLE 3

COMPARISON OF ANTIPROLIFERATIVE ACTIVITY OF IFNα TARGETED
BY MABS VIA DIFFERENT MODES

	CELLS INCUBATED WITH	IF α/DOSE ADDED (IU/ML)	H^3-TDR INCORP. CPM X 10^3	INHIBITION
1.	IFN ALONE	–	122 ± 5	0.0
		1	99 ± 5	19.6
		4	72 ± 1	41.5
		20	50 ± 1	58.8
		100	41 ± 1	66.3
2.	IFN-MAb$_1$ CONJUGATE	–	134 ± 3	0.0
		1	39 ± 3	71.0
		4	28 ± 3	78.5
		20	18 ± 2	86.6
		100	15 ± 1	89.0
3.	IFN+MAb$_1$-MAb$_2$ CONJUGATE	–	130 ± 2	0.0
		1	36 ± 3	72.5
		4	30 ± 2	77.2
		20	21 ± 2	84.0
		100	16 ± 4	87.6
4.	IFN+MAb$_1$+MAb$_2$ MIXTURE	–	137 ± 11	0.0
		1	100 ± 5	27.0
		4	75 ± 3	45.9
		20	62 ± 3	54.8
		100	52 ± 1	62.4
5.	IFN+FAb'$_1$-FAb'$_2$ CONJUGATE	–	129 ± 1	0.0
		1	61 ± 2	52.7
		4	54 ± 5	57.8
		20	35 ± 1	72.5
		100	25 ± 8	80.5

The specificity of this enhanced antiproliferative activity
of antibody-targeted IFN at low concentrations was demonstrated as follows: Idiotype positive 10K44 as well as idiotype
negative hybridoma cells, 3A2-2, and 1210 cells were incubated with first mAb conjugates and/or with uncoupled IFN (Fig.2)
and anti- proliferative activities measured. It was found
that the antiproliferative effect of targeted IFN was only
enhanced on the idiotype positive cells. Free IFN or IFN/mAb
mixture did not show this enhanced antiproliferative activity
on either cell lines.

FIGURE 2

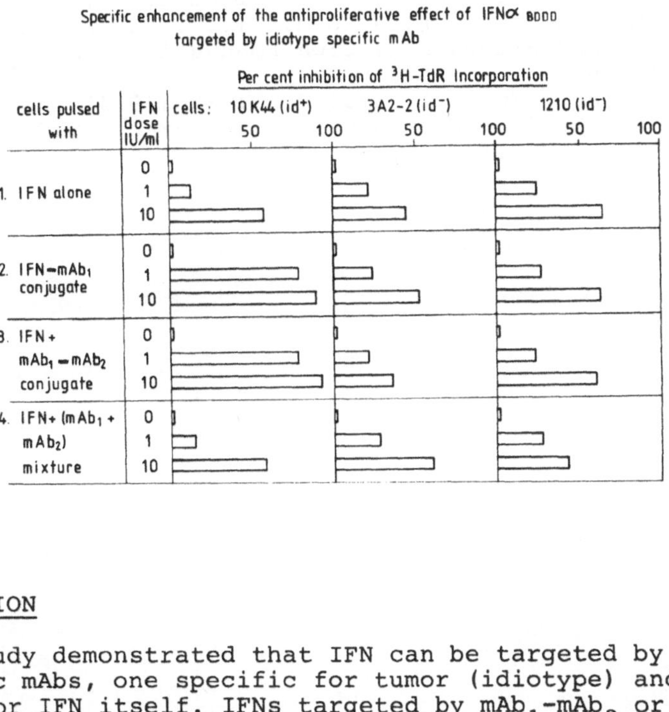

Specific enhancement of the antiproliferative effect of IFNα 8000
targeted by idiotype specific mAb

Per cent inhibition of ^{3}H-TdR Incorporation

cells pulsed with	IFN dose IU/ml	cells:	10 K44 (id$^+$) 50	100	3A2-2 (id$^-$) 50	100	1210 (id$^-$) 50	100
1. IFN alone	0							
	1							
	10							
2. IFN-mAb$_1$ conjugate	0							
	1							
	10							
3. IFN + mAb$_1$ - mAb$_2$ conjugate	0							
	1							
	10							
4. IFN+ (mAb$_1$ + mAb$_2$) mixture	0							
	1							
	10							

CONCLUSION

This study demonstrated that IFN can be targeted by using bispecific mAbs, one specific for tumor (idiotype) and the
other for IFN itself. IFNs targeted by mAb$_1$-mAb$_2$ or Fab'$_1$-Fab'$_2$
conjugates were shown to be biologically active. When the
antiproliferative activity of free versus targeted IFNs was
compared it was found that targeted IFNs were always superior
to free IFN. This enhanced antiproliferative effect was shown
to be target cell specific, because target cells lacking the
tumor (idiotype) marker were not influenced by mAb targeting.
The model described here which is an alternative to direct
targeting by mAbs could be useful for the specific targeting
of lymphokines which are too labile to be chemically modified.
Currently the above model is being used to investigate the
mechanism of enhanced antiproliferative action of mAb-target-
ed IFN.

ACKNOWLEDGEMENTS

We thank Remy Longato and Ralf Ilbertz for their excellent technical assistance and Yvonne Fischer for the preparation of this manuscript.

REFERENCES

1. Ling, P.D. , Warren, M.K. and Vogel., S.N. J. Immunol. 135, 1857-63 (1985)
2. Alkan, S.S., Miescher-Granger, S., Braun, D.G. and Hoch-keppel, H.K. J. Interferon Res. 4, 355-363 (1984)
3. Carlsson, J., Drevin, H. and Axen, R. Biochem. J. 173, 723-737 (1978)
4. Brennan, M., Davison P.F. and Paulus, H. Science 229, 81-83 (1985)
5. Meister, A., Uze,G., Mogensen, K., Gresser, I., Tovey, M.G., Grütter, M. and Meyer, F. J. Gen. Virol. (in press)
6. Alkan, S.S. and Braun, D.G. Ciba Foundation Symp. 119, 264-278 (1986)
7. Alkan, S.S. Ann. Immunol., 135 C, 31-38 (1984)
8. Alkan, S.S., Ball, R.K., Chang, J. and Braun, D.G. Molec. Immunol. 20, 203-211 (1983)

CHARACTERIZATION OF RAT MONOCLONAL ANTIBODIES TO MOUSE INTERFERON-α AND -β

Yoshimi KAWADE and Yoshihiko WATANABE
Institute for Virus Research, Kyoto University, Kyoto, Japan

1. INTRODUCTION

One of the features of the mouse interferon (IFN) system, in comparison with the human IFN system, is that "type I" IFNs, produced in various systems in vivo and in vitro in response to virus and other stimuli, contain both IFN-α and -β (1). Antiserum prepared using mouse type I IFN as immunogen therefore inevitably contain both antibodies to IFN-α and to IFN-β. This made it generally difficult to distinguish the α/β types of mouse IFN, and to selectively eliminate the activity of α or β IFN in various biological systems. Alleviating this situation, monoclonal antibodies to mouse IFN-β have recently been prepared (2,3), and were shown to be utilizable for one-step purification of IFN-β from crude materials containing both IFN-α and -β (4). However, monoclonal antibodies to mouse IFN-α have not been available. Here, we describe preparation of rat monoclonal antibodies to mouse IFN-α and -β, and initial characterization of their properties.

2. MATERIALS AND METHODS

2.1. Interferon
Mouse IFN-α,β was produced by Newcastle disease virus-induced L cells (5), and purified in two-step chromatography utilizing Controlled Pore Glass and polyclonal antibody (from a high-titered sheep antiserum against L cell IFN, kindly donated by Drs. B.J. Dalton and K. Paucker). It consisted of a mixture of electrophoretically pure IFN-α and -β with a specific activity of over 10^8 international units (IU)/mg protein (6).

IFN-α and -β separated from each other were prepared by SDS-polyacrylamide gel electrophoresis (PAGE) (7), or by means of affinity chromatography using the monoclonal antibodies described below. ^{125}I-labeled IFN-β was prepared by iodination of such IFN-β with Bolton-Hunter reagent.

Recombinant mouse IFN-α subtypes were generous gifts of Dr. H. Taira (α1 and α2), Drs. J. Trapman and E. Zwarthof (α4 and α6(T)), Dr. P. Pitha (α5(P)), and Dr. S. Pestka (αA). Nonglycosylated recombinant mouse IFN-β was a gift of Drs. H. Ozawa and S. Kobayashi (8).

IFN activity was assayed by measuring the inhibition of the cytopathic effect of vesicular stomatitis virus on L cells (9).

2.2. Rat-mouse hybridoma
Five week-old female ACI rats were immunized with purified L cell IFN, admixed with complete Freund's adjuvant. After ascertaining the emergence

of IFN-neutralizing activity in the serum, the rat spleen was removed and the cells were fused with mouse myeloma X63-Ag8.6.5.3 cells. Hybridomas were grown under standard conditions, and screened for antibody secretion by three methods: (1) immunoblotting assay, (2) immunoplate IFN-binding assay (see below), and (3) IFN neutralization assay. Positive cultures were cloned repeatedly by limiting dilution. For the screening by methods (1) and (2), L cell IFN containing both IFN-α and -β was used as antigen; to ascertain which antibody (anti-α or anti-β) was produced by the single-cell clones, the IFN activity recovered in the immunoplate IFN-binding assay was tested for neutralization by rabbit antisera specific to mouse IFN-α and -β (9).

2.3. Immunoplate IFN-binding assay

Wells of plastic microplates (96 wells) were coated with rabbit immunoglobulin G (IgG) against rat IgG, and blocked with bovine serum albumin (BSA). The hybridoma supernatants to be tested (or rat antibody solution) were added to each well and incubated at room temperature for 1 h. After washing, about 10^3 IU of crude L cell IFN (a mixture of IFN-α and -β) in 0.1 ml were added and incubated at room temperature for 1 h. After washing again, the bound IFN was extracted with 0.05 ml of 0.1 M acetic acid containing 0.1% BSA and antibiotics, and assayed for IFN.

3. RESULTS AND DISCUSSION

3.1. Specificity of antibodies

We could establish two and four clones of rat-mouse hybridomas that secrete antibodies to mouse IFN-α and -β, respectively. They exhibited no cross-reactions against each other, as tested with immunoplate IFN-binding assay. Cross-reactions against human (Hu)IFN-α and -β were also tested , using recombinant HuIFN-α2 (Essex Nippon), HuIFN-αA/D (Roche), natural HuIFN-α from Sendai virus-induced leukocytes (Kyoto Red Cross Blood Center), and natural HuIFN-β (Toray), but none was found.

3.2. Anti-α antibodies

The two anti-α antibodies, 4E-A1 and 9B-D12, were characterized as to their reaction with natural IFN-α from L cells and with several recombinant mouse IFN-α subtypes.

Since L cell IFN-α is a mixture of many different IFN-α subtypes (10),it is essential to test what proportion of total IFN-α reacts with the antibody. This was examined using each antibody conjugated to AffiGel 10. Most of the IFN-α activity (over 97%) was retained by 4E-A1, whereas only about a half was retained by 9B-D12. Thus, the former antibody seems to recognize an epitope common to most of the different mouse IFN-α subtypes (and yet is absent in HuIFN-α). Using the two antibody affinity columns sequentially, L cell IFN-α was fractionated into four fractions.

The L cell IFN-α activity adsorbed to the 4E-A1 antibody column (over 97%) could be quantitatively recovered by elution with acid. The eluted material appeared to be pure IFN-α protein, as analyzed by SDS-PAGE. Thus, for the purification of natural IFN-α, this antibody will serve most purposes, although some minor subtypes amounting to a few % may be lost in an unadsorbed fraction.

The antibody 4E-A1 was moreover found to neutralize well the major part of L cell IFN-α that was bound to its own affinity column. Hence, the antibody will be a useful reagent to substantially reduce the activity of

TABLE 1. Reaction of monoclonal antibodies to MuIFN-α with MuIFN-α subtypes.

| | | Reaction with antibody[a] | | |
Subtype	Producing cell	4E-A1	9B-D12	Ref. for IFN
MuIFN-α1	E. coli and CHO	+ (S)	+ (S)	11
MuIFN-α2	E. coli and CHO	+ (S)	+ (S)	11
MuIFN-α4	CHO	+ (S)	−	12
MuIFN-α5(P)	E. coli	+ (S)	+ (W)	13
MuIFN-α6(T)	CHO	+ (S)	+ (W)	12
MuIFN-αA	E. coli	+ (S)	−	14

a) + and − refer to results of immunoplate IFN-binding tests; (S) and (W) indicate strong and weak neutralization, respectively.

natural IFN-α, although polyclonal antisera are required for more complete elimination of the activity. The antibody 9B-D12, on the other hand, neutralized only weakly even the fraction of IFN-α that reacts with this antibody.

Several MuIFN-α subtypes were examined for reaction with the two antibodies (Table 1). 4E-A1 reacts with and neutralizes all the six subtypes examined; this is in harmony with the results on natural L cell IFN-α mentioned above. IFN-α1 and -α2 are glycoproteins (as are α4 and α5(P)) (11-13); their glycosylated and nonglycosylated forms produced by CHO and E. coli, respectively, showed only a minor difference in reaction with 4E-A1, indicating that the sugar moiety plays little role in the recognition of the antibody. The same was true with the other antibody, 9B-D12, which reacts with α1 and α2. IFN-α1 and -α2, being strongly neutralized by 9B-D12, probably do not represent the major component in L cell IFN-α, because the latter is not substantially neutralized by this antibody. This antibody reacts also with α5(P) and α6(T), but not with α4 and αA. Also, neutralization is weak against α5(P) and α6(T).

3.3. Anti-β antibodies
Mouse IFN-β is encoded by a single gene (15). L cell IFN-β protein consists of a main species with a molecular weight of 35,000, and a minor one with a molecular weight of 30,000 (7). This heterogeneity is attributable to a heterogeneity in glycosylation, there being three potential N-glycosylation sites in the polypeptide (15). The four kinds of anti-β antibodies we obtained reacted with both species of L cell IFN-β. Further, recombinant IFN-β produced by E. coli and lacking the sugar moiety also reacted with them. Thus, these antibodies apparently do not recognize the sugar moiety. Each of the four antibodies in immobilized form was found to allow us to purify IFN-β to electrophoretically pure form in one-step from crude L cell IFN (a mixture of IFN-α and- β).

The four antibodies neutralized L cell IFN-β, as well as E. coli-derived recombinant IFN-β, to extents widely different from one antibody to another.

We next examined how the antibody affects binding of IFN-β to cellular receptors. The antibodies with strong neutralizing capacity, designated 7F-D3, 7F-D9 and 4B-G12, completely inhibited the receptor binding of ^{125}I-labeled IFN-β, whereas the weakly neutralizing antibody (6D-G8) reduced the binding but did not completely inhibit it; about one-third of the control value remained bound in large excesses of antibody. These results are interpreted to mean that the main mechanism of neutralization is the inhibition of IFN-binding to cell receptors, the action of antibody being to reduce the affinity of IFN to its receptor; the larger the reduction, the greater is the extent of neutralization. More quantitative experiments are in progress for comparison with the theory (16).

REFERENCES

1. Kawade Y: Characterization of mouse interferon molecules. In: Humoral Factors in Host Defense, eds. Y Yamamura et al., Academic Press, 175-189, 1983.
2. Bosveld IJ, Vonk WP, Hekman RACP, Van Vliet PW, De Jonge P, Van Ewijk W and Trapman J: Preparation and properties of monoclonal antibodies against murine interferon-β. Virology 120, 235-239, 1982.
3. Murasko DM and Rusckowski M: Production, screening and characterization of a monoclonal antibody to murine interferon-β. J. gen. Virol. 64, 727-732, 1983.
4. Vonk WP and Trapman J: Large-scale, one-step purification of murine interferon-beta using a monoclonal antibody. J. Interferon Res. 3, 169-175, 1983.
5. Kawade Y and Yamamoto Y: Induction and production of L cell interferon. Methods in Enzymology 78, 139-143, 1981.
6. Watanabe Y and Kawade Y: Induction, production and purification of natural mouse IFN-α and -β. In: Interferon and Lymphokines-A Practical Approach, eds. MJ Clemens et al., IRL Press, in press, 1986.
7. Yamamoto Y and Kawade Y: Purification of two components of mouse L cell interferon: Electrophoretic demonstration of interferon proteins. J. gen. Virol. 33, 225-236, 1976.
8. Tanaka T, Matsuda S, Kawano G and Kobayashi S: Production, purification and crystallization of recombinant mouse interferon-β. In: The Biology of the Interferon System 1985, eds. WE Stewart II and H Schellekens: Elsevier, 65-80, 1986.
9. Yamamoto Y and Kawade Y: Antigenicity of mouse interferons: Distinct antigenicity of the two L cell Interferon species. Virology 103, 80-88, 1980.
10. Lemson PJ, Vonk WP, Van der Kopput JAGM, and Trapman J: Isolation and characterization of subspecies of murine interferon alpha. J. gen. Virol 65, 1365-1372, 1984.
11. Shaw GD, Taira H, Mantei N, Lengyel P and Weissman C: Structure and expression of cloned murine IFN-α genes. Nucleic Acids Res. 11, 555-573, 1983.
12. Zwarthoff E, Mooren ATA and Trapman J: Organization, structure and expression of murine interferon alpha genes. Nucleic Acids Res. 13, 791-804, 1985.
13. Kelley KA and Pitha PM: Characterization of a mouse interferon gene locus I. Isolation of a cluster of four α interferon genes. Nucleic Acids Res. 13, 805-823, 1985.
14. Daugherty B, Martin-Zanca D, Kelder B, Collier K, Seamans TC, Hotta K and Pestka S: Isolation and bacterial expression of a murine alpha leukocyte interferon gene. J. Interferon Res. 4, 635-643, 1984.
15. Higashi Y, Sokawa Y, Watanabe Y, Kawade Y, Ohno S, Takaoka C and Taniguchi T: Structure and expression of a cloned cDNA for mouse

interferon-β. J. Biol. Chem. <u>258</u>, 9522-9529, 1983.

16. Kawade Y: Neutralization of activity of effector protein by monoclonal antibody: formulation of antibody dose-dependence of neutralization for an equilibrium system of antibody, effector, and its cellular receptor. Immunology <u>56</u>, 497-504, 1985.

IMMUNOLOGY AND INTERACTION WITH OTHER LYMPHOKINES

MOLECULAR AND CELLULAR BIOLOGY OF TUMOR NECROSIS FACTOR, INTERFERON-γ, AND THEIR SYNERGISM

W. FIERS, P. BROUCKAERT, R. DEVOS*, L. FRANSEN*, G. HAEGEMAN, G. LEROUX-ROELS, A. MARMENOUT*, E. REMAUT, P. SUFFYS, J. TAVERNIER*, J. VAN DER HEYDEN* & F. VAN ROY
Laboratory of Molecular Biology, State University of Ghent, and *Biogent (A subsidiary of Biogen SA), Ghent, Belgium

1. INTRODUCTION

Tumor Necrosis Factor is produced upon stimulation, e.g. by lipopolysaccharide (LPS), of macrophages or monocytic cell lines. Human TNF cDNA has been cloned and sequenced by a number of groups (Pennica et al., 1984; Marmenout et al., 1985; Wang et al., 1985). The gene coding for the mature protein can readily be expressed in E. coli, and from this source virtually unlimited amounts of pure recombinant hTNF can be prepared. Recently, we have also been able to crystallize hTNF, so that now analysis by X-ray diffraction becomes a possibility. The TNF subunit molecular weight amounts to 17,356 dalton and the native structure is presumably a trimer. An internal disulfide bond is present in each subunit. The human genome contains a single gene for TNF and it is interrupted by three introns (Marmenout et al., 1985; Nedwin et al., 1985; Shirai et al., 1985); the TNF gene has been localized on chromosome 6 in the middle of the HLA region (Spies et al., 1986).

Also the mouse TNF cDNA gene has been isolated and expressed efficiently in E. coli (Fransen et al., 1985; Pennica et al., 1985). The mature polypeptides of human and mouse TNF show 79% homology. It should be noted that, although natural mouse TNF is a glycoprotein, nevertheless mTNF is synthesized in E. coli in a soluble form. Unlike interferons, TNF is very little species-specific, at least regarding its in vitro action (Fransen et al., 1986a); in vivo, however, species-specificity is more clearly in evidence (Brouckaert et al., 1986). Despite this high degree of homology in sequence and the lack of species-specificity, nevertheless antibodies raised in guinea pigs against human TNF or against mouse TNF virtually do not cross-react. It is readily possible to obtain mouse monoclonal antibodies directed against hTNF and these do not react with mTNF. An analysis of epitopes recognized by some of these monoclonal antibodies indicates that these epitopes are defined by the complex three-dimensional structure.

2. THE ACTION OF TNF ON MALIGNANT CELL LINES AND ON IN VIVO TUMORS

Using recombinant hTNF, earlier results on the selective action of this cytotoxic agent on malignant cells have been confirmed (Sugarman et al., 1985; Fransen et al., 1986b). Also the remarkable synergism with IFN-γ on many cell lines has been confirmed and further documented. As reported previously (Fiers et al., 1986a; Fiers et al., 1986b; Fransen et al., 1986b), we can summarize these results as follows: 1° Some transformed cells are very sensitive to TNF as such, other transformed cell lines become sensitive when also treated with IFN-γ, while primary, untransformed cells are not affected by hTNF even in the presence of IFN-γ. 2° Sensitivity to TNF is not correlated with a particular cell line or tumor type, e.g., some human breast carcinoma lines are sensitive while others are not, some colon carcinoma lines are sensitive but not others, etc. 3° Increased malignancy as determined by other parameters, e.g., invasiveness or tumorigenicity in immunodeficient animals, is sometimes correlated with increased sensitivity to TNF; the reverse, cell sublines selected for high malignancy which become less sensitive to TNF, has never been observed. 4° Very often inhibitors of RNA or protein synthesis such as actinomycin D or cycloheximide dramatically increase the sensitivity to the cytotoxic action of hTNF. This synergism with actinomycin D is independent of the synergism with IFN-ɣ, and vice versa. It has been reported by Hahn et al. (1985) that TNF may induce components responsible for protecting the cell against the toxic action of TNF and that many malignant cells or most other cells in the presence of actinomycin are unable to synthesize this protecting protein.

We have reported at the previous interferon meeting (Fiers et al., 1986a) that r-hTNF and r-mTNF are effective anticancer drugs, as complete regression of B16B16 tumors in syngeneic mice could be obtained. Also some long-term curing was observed, in which case mice remained tumor-free for periods of over three months (Brouckaert et al., 1986). Very promising results have also been obtained by Balkwill et al. (1986), who observed frequent regression or complete stabilization of human tumor xenografts in nude mice. Further results have shown a clear benefit of the combined treatment with IFN-ɣ, thus applying in vivo the remarkable synergism. As IFN-ɣ is highly species-specific, the latter results are clear proof that in these experiments the action of the effectors must have been direct. In other studies carried out in our laboratory we have observed, again using systems involving tumors derived from human cell lines in nude mice, that complete regression of a human ovarian carcinoma cell line could be obtained by treatment with r-hTNF in combination with actinomycin D. It should be noted that the actinomycin concentrations applied were very low and at a non-toxic level and also that actinomycin is being used in some chemotherapy protocols for particular types of human cancer.

Gresser et al. (1986) have studied tumors of Friend leukemia cells in syngeneic mice. Treatment with r-mTNF gave a very clear antitumor effect. This is remarkable because the malignant Friend cell line used is completely resistant to mTNF in tissue culture and further experiments have shown that in vivo there is no involvement of interferon. Hence it seems likely that in this case the antitumor effect of mTNF is (completely) due to indirect effects, e.g. on the vascular system supporting the tumor.

We conclude therefore that TNF is an effective and very promising antitumor drug. It is especially powerful when treatment is complemented, e.g., with IFN-ɣ or with a synergistic drug. However, especially when mTNF was used in mice, severe toxicity was often a problem and many animals died as a result of the treatment (of course, very high doses were used). The nature of this toxicity should therefore be investigated and one should try to dissociate the toxic effects from the therapeutic benefit.

3. TNF TOXICITY AND ITS ACTION ON NORMAL AND TRANSFORMED CELLS

Two components are present in the toxicity observed when tumor-bearing animals are treated with TNF. On the one hand there is a toxicity which is also seen by giving TNF to healthy experimental animals, and on the other hand, there is toxicity due to direct and indirect effects on the tumor and its ensuing necrosis. In a first approach one should try to understand the toxicity of the former type. Obviously this is due to the action of TNF on normal cells and tissues. Very soon after recombinant,

TABLE 1

BIOLOGICAL ACTIVITIES OF RECOMBINANT TUMOR NECROSIS FACTOR

CELL TYPE	BIOLOGICAL ACTION	REFERENCE
1. Fibroblast cells	growth factor	Fiers et al. (1986a) Vilcek et al. (1986) Sugarman et al. (1985)
2. Endothelial cells	induction of class I MHC Ag	Collins et al. (1986)
	induction of class I MHC Ag	Collins et al. (1986)
	procoagulant activity	Bevilacqua et al. (1986a)
	stimulation of adhesiveness	Gamble et al. (1985)
	induction of H4/18 Ag	Pober et al. (1986)
	reorganization	Stolpen et al. (1986)
	enhanced leukotriene production	Roubin et al. (1986)
3. Granulocytes		
4. Neutrophils	activation (phagocytosis and ADCC)	Shalaby et al. (1985) Gamble et al. (1985)
	stimulation of adherence	Bevilacqua et al. (1986b) Klebanoff et al. (1986)
5. Eosinophils	enhanced leukotriene (LTB4) synthesis	Roubin et al. (1986)
	enhanced toxicity	Silberstein et al. (1986)
	enhanced leukotriene (LTC4) synthesis	Roubin et al. (1986)
6. Macrophages	stimulation IL1 and PGE2 production	Bachwich et al. (1986)
7. Adipocytes (3T3-L1)	suppression of LPL activity	Beutler et al. (1985b)
	suppression of lipogenic enzymes	Torti et al. (1985)
8. Synovial cells	stimulation of collagenase and PGE2 production	Dayer et al. (1985)
9. Osteoclasts	stimulation of bone resorption	Bertolini et al. (1986)
10. Hepatocytes	depressed P450-system	Ghezzi et al. (1986)
11. Hybridoma mouse CTL x rat thymoma	induction of cytotoxicity	Plaetinck et al. (1986)

homogeneous and endotoxin-free TNF became available, studies were undertaken to investigate this action on various normal cells (Table 1). We observed that quite remarkably TNF has in fact a growth-promoting activity on normal fibroblasts; at confluency the cells reach at least a threefold higher cell density. TNF given to an experimental animal intravenously, first of all interacts with the endothelial system and a number of proteins and metabolic activities are induced in TNF-treated endothelial cells. Also, various types of leukocytes are affected by the TNF. Furthermore, various effects normally seen in a local inflammation are induced systemically by TNF treatment. One can also conclude that TNF exerts some effects without the need for RNA and protein synthesis (e.g. increase in the adhesiveness of polymorphonuclear cells; Bevilacqua et al. 1986b), while other effects are the result of the induction of new proteins in the treated cells. The latter effects can of course be blocked by actinomycin D or cycloheximide.

From an analysis of the various effects of TNF some hypotheses regarding the toxicity can already be formulated. A complex series of events takes place in the veins and capillaries as a result of the interplay between TNF action on white blood cells and on the endothelial cell layer (Fig. 1). The stimulated endothelial surface changes many of its properties, one of which being the increased adhesiveness for PMN. Furthermore, procoagulant activity appears, which leads to disseminated thrombosis, while at the same time mechanisms which are required for keeping the coagulation in balance are inhibited. Also various low-molecular weight intermediates are induced, which normally act locally, but which, as a result of TNF i.v. injection, now have a systemic effect.

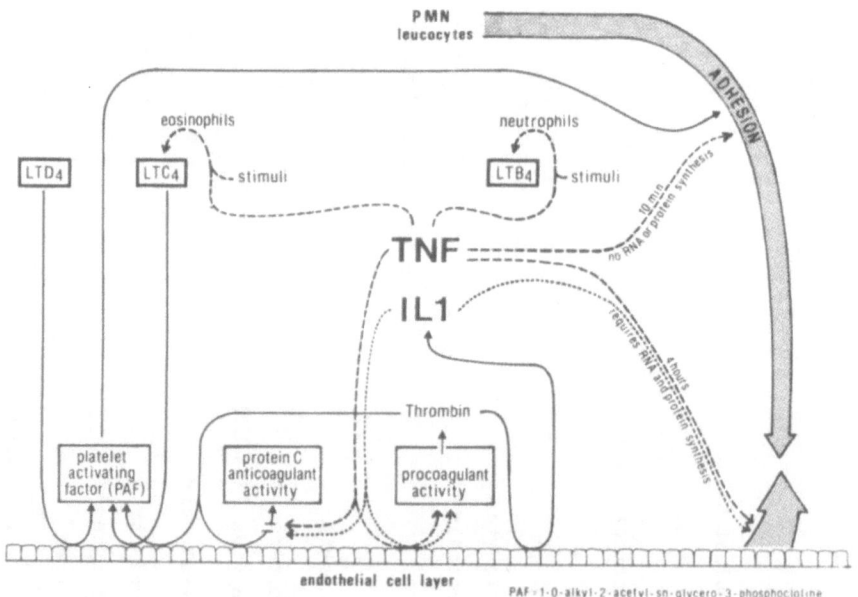

FIGURE 1

We know that the combination of IFN-γ with TNF, although much more effective on many types of malignant cells, is also much more toxic. Prolonged treatment of the endothelial cell layer with the combination TNF + IFN-γ leads to profound morphological changes (Stolpen et al., 1986) which may be accompanied by increased permeability. Other effects of TNF are exerted on the hypothalamus and perhaps the central nervous system and the liver.

These broad ranges of effects on various cell types are corroborated by studies aimed at the direct characterization of TNF receptors. The TNF receptors are found on a wide variety of malignant and non-malignant cells (Table 2). In general, there is a consensus that the binding constant is about $2-3 \times 10^{-10}$ M, and the number of receptors depends on cell type and is usually in the range of 1000-10,000 molecules per cell. Several groups have reported that the number of TNF receptors can be increased several times by IFN-γ action; however, this relative increase in TNF receptor number is in no way sufficient to explain the dramatic increase in susceptibility to TNF of many cells when also treated with IFN-γ (Fransen et al., 1986b).

It is interesting to note that many of the phenomena associated with TNF toxicity have previously been described and studied in cases of endotoxic shock and related affections. These effects can be alleviated by classical anti-inflammatory agents such as corticosteroids, and one of the main actions of the latter is precisely blocking the synthesis of endogenous TNF (Beutler et al., 1985a). Consequently, the many studies on the physio-pathology of endotoxin shock and septicemia and their treatment are relevant for the understanding of TNF toxicity; however, although TNF is the key mediator of LPS-induced toxicity, it should not be equated as LPS has many other effects as well.

4. TNF: MECHANISM OF ACTION

TNF interacts with its receptor and is subsequently internalized via the endosome pathway and becomes degraded (Baglioni et al., 1985). There is so far no compelling reason to believe that the various actions of TNF on different types of cells are due to more than one type of receptor. After activation of the receptor, pathways are triggered which on the one hand lead to metabolic events at the cytosol/cell membrane level, while another pathway of the primary trigger leads to the nucleus where new mRNAs are induced responsible for synthesis of new proteins. The former types of events cause cytotoxicity specifically in malignant cells as it is well known that these effects are in fact enhanced by blocking RNA or protein synthesis. But some effects in normal cells are protein synthesis-independent such as, e.g., the increased adhesiveness of neutrophils (Bevilacqua et al., 1986b; Klebanoff et al., 1986). In order to understand the key steps leading to the most severe aspects of the toxicity and to develop effective remedies, it is necessary to identify the primary biochemistry of the TNF action in various cells. Also, one would like to understand why some tumor cells are sensitive to TNF while others are not, and, moreover, what is the nature of the dramatic increase in sensitivity observed with many malignant cells when also treated with IFN-γ.

A proposed mechanism of action leading to cytotoxicity is shown in Fig. 2. The activated TNF receptor in some way may lead to the release of metabolizable, polyunsaturated fatty acids, such as arachidonic acid and perhaps others. It is known that the metabolism of these, at least in some cells, leads to reactive oxygen ($\cdot O_2^-$, $\cdot OH^-$, ...) (Rainsford, 1984). These reactive oxygen species may be responsible for the damage observed

TABLE 2.   **T N F - R E C E P T O R S**

	Cell line	n° of receptors	binding constant M	increase of n° by IFN-γ	reference
human rTNF	Hela S2	6000	2 x 10-10		Baglioni et al. (1985)
	Daudi, Raji	not detectable			
	Jurkat	1100	2.5 x 10-10		
	HT-29	800	2 x 10-10	2-3 x	Ruggiero et al. (1986)
	Hela D98/AH2	2400	2 x 10-10	2-3 x	
	SK-MEL-109	9000	2 x 10-10		
	L929	2200	6.1 x 10-10		Tsujimoto et al. (1985)
	FS-4	7500	3.2 x 10-10		
	Hela	15000	1.6 x 10-10		Tsujimoto et al. (1986)
	HT-29	4500	2.0 x 10-10		
	WISH	6800	2.4 x 10-10	1.5-2 x	
	A 673	5700	1.9 x 10-10		
	RPMI 7272	18000	2.3 x 10-10		
	WI-38	2000	2 x 10-10		Sugarman et al. (1985)
	T24	1500	2.5 x 10-10		
	ME 180	2000	2 x 10-10	2-3 x	Aggarwal et al. (1985)
hu LuKII	L(M)	200	1 x 10-10		Rubin et al. (1985)
	Hela	300	idem		
mu J774.1	L(M)	1100	3 x 10-12		Kull et al. (1985)
	L929	770	3 x 10-12		
	3T3A31	1100	2.5 x 10-12		
	J774.1	1100	3 x 10-12		
	CPAE (bovine)	1500	3 x 10-12		
mu RAW264.7	3T3 L1	10000	3 x 10-9		Beutler et al. (1985b)
	C2	10000	3 x 10-9		
	erythrocytes	not detectable			
	lymphocytes	not detectable			

Table 2: Summary of studies on TNF receptors. The type of TNF used in the various studies is indicated on the left: human rTNF, human LuKII, which is in fact lymphotoxin or TNF-   , while J774.1 and RAW264.7 refer to the cell lines from which natural murine TNF was isolated.

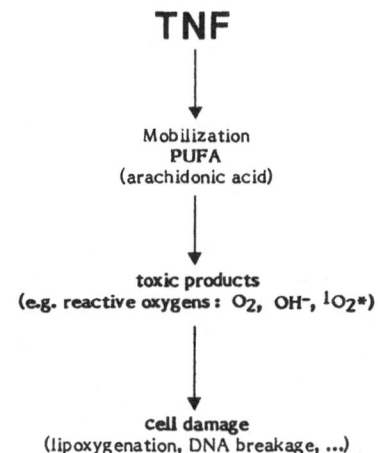

TNF

↓

Mobilization
PUFA
(arachidonic acid)

↓

toxic products
(e.g. reactive oxygens: O_2, OH^-, $^1O_2*$)

↓

cell damage
(lipoxygenation, DNA breakage, ...)

FIGURE 2.   Proposed mechanism of action   (PUFA = polyunsaturated fatty acids)

in TNF-sensitive cells: lipoxygenation leading to damage of the cell membrane, DNA fragmentation, etc.   Although this mechanism is certainly not proven, there are a number of observations supporting it.   That TNF causes metabolization of arachidonic acid is evident from a number of studies in various cell systems.   The cyclo-oxygenase pathway can be activated by TNF because synthesis of prostaglandin E2 has been observed in a number of studies (Table 1).   Blocking this pathway, e.g. by treatment with indomethacin, does not wipe out the cytotoxicity, but it does prevent the increased synthesis of prostaglandins.   The other known branch of arachidonic acid metabolization leading via lipo-oxygenase to leukotrienes is activated by TNF at least in neutrophils and eosinophils (Roubin et al., 1986).   Arachidonic acid and related polyunsaturated fatty acids can be released by various phospholipases and inhibitors of these such as dexamethasone (acting via lipomodulin?) and quinacrine do indeed interfere with TNF-induced cytotoxicity (P. Suffys, unpublished).   An important involvement of phospholipids and related metabolites can even be seen in vivo in tumors taken from animals treated with TNF for only 6 h (Gresser et al., 1986).   Undoubtedly, further dissection as to what metabolism occurs in what type of cells may shed light on the mechanism of the direct TNF-induced cytotoxicity.

On the other hand, TNF also acts on the genetic programming of the target cell and various proteins are induced in different cells.   We have mentioned before the proteins induced in endothelial cells and their possible role in the contribution of the endothelial system to the general toxicity.   Other protein(s) which could play a key role in TNF-induced effects are those which may be involved in conferring resistance or protection to the TNF action itself (Hahn et al., 1985).   A number of proteins that we have observed to be induced in TNF-treated cells are listed in Table 3.   c-fos is a nuclear protein which plays a key role in the cell cycle and is activated e.g. by serum stimulation.   As we see that it is also activated by TNF this could play a key role in the aforementioned mitogenic effect of TNF.   Another interesting protein, which is strongly stimulated by TNF, is the 26-K protein (also referred to by some other groups as IFN-β 2). This protein is strongly induced under the conditions normally applied for obtaining synthesis of IFN-β by fibroblast cell monolayers (Content et al., 1982). Hence, it is present as a contaminant in incompletely purified IFN-β preparations, and antisera raised with such preparations strongly react with the 26-K protein. In collaborative studies with J. Content and his colleagues we have further shown that 26-K protein can also be induced by cycloheximide treatment alone (Content et al, 1982). We now observe that 26-K is also strongly induced by TNF treatment as well as by IL1 (De Fillipi et al., this volume; Content et al., 1985). Recently, the complete nucleotide sequence of the gene corresponding to the 26-K protein was established (Haegeman et al., 1986). Using objective computer programs we could not establish any sequence homology to the known human interferons. Undoubtedly, 26-K protein, which is synthesized by fibroblasts, epithelial cells, lymphocytes and perhaps other types of cells as a result of various stimuli, must play a key regulatory role and its biological properties are very intriguing. Recently, Kohase et al. (1986) have also described the induction of 26-K and they proposed that the protein plays a homeostatic role in cell proliferation.

TABLE 3.   **PROTEINS (mRNAs) INDUCED BY TNF**

	Reference
26-K protein	Defilippi et al. (1986)
	Kohase et al. (1986)
2-5 A synthetase	Defilippi et al. (1986)
c-fos	Suffys et al. (unpublished)
H4/18-antigen	Pober et al. (1986)
HLA-A,B (class I)	Collins et al. (1986)

It is intriguing that many of the TNF effects on various cells can also be observed after treatment with IL1 (Table 4). Also, in vivo, a number of IL1 effects are equally seen upon TNF injection. However, at least in the case of endothelial cells, it has clearly been demonstrated that the receptor for TNF (identical to the receptor for lymphotoxin; Aggarwal et al., 1985) is clearly different from the IL1 receptor (Pober et al., 1986). The similarity in effects induced by TNF and by IL1 in many cells is quite impressive; possibly, the secondary messenger involved in the nuclear activation of a set of genes might be identical.

5. CONCLUSION

TNF, perhaps in combination with IFN-γ, is a very promising anticancer agent, it acts both on systemic neoplastic diseases and on solid tumors. These effects are due both to direct actions on the malignant cells and to indirect effects. The localized treatment with TNF either alone or in combination with IFN-γ or other drugs is at present a realistic therapy. Systemic treatment, however, with high doses of TNF for prolonged periods of time is at present limited by the toxicity. This toxicity is even more severe when combined with injections of IFN-γ.

An important component of the TNF-caused toxicity can be understood by studying the action of TNF on normal cells. Primary targets of TNF

TABLE 4. **IL1-LIKE ACTIVITIES OF TNF**

Endothelial cells

Procoagulant activity   Bevilacqua et al. (1984)
                         Bevilacqua et al. (1986a)
H4/18-antigen   Pober et al. (1986)
Leukocyte adhesiveness   Bevilacqua et al. (1986b)

Fibroblast cells

proliferation   Fiers et al. (1986a)
                         Vilcek et al. (1986)
26-K protein   Defilippi et al. (1986)

Osteoclasts

activation   Bertolini et al. (1986)

Adipocytes

inhibition of lipoprotein lipase   Beutler & Cerami (1985)

Melanoma cells

cytotoxicity   Onozaki et al. (1985)

Chondrocytes

stimulation of PGE_2 production   Dayer et al. (1985)
 and collagenase synthesis
cartilage resorption and inhibition Saklatvala (1986)
 of proteoglycan synthesis

T-Hybridoma cells

(mouse CTL x rat thymoma)   Plaetinck et al. (1986)
 induction of cytotoxicity

Brain (hypothalamus cells)

synthesis PGE_2 Dinarello et al., (1986)
 fever, anorexia

Neutrophils
activation   Oppenheim et al. (1986)
                         Shalaby et al. (1985)
                         Klebanoff et al. (1986)

---

action are the blood leukocytes and the endothelial walls of the vessels and capillaries. A number of biochemical events occurring in these cells have been identified. Various phenomena which could play a major role in TNF toxicity come more clearly into focus, such as adhesion of polymorphonuclear leukocytes to the endothelial cell layer, disturbances of the balance in coagulation-anticoagulation, etc. In order to understand these events, a better insight into the biochemical mechanism of TNF action is required. Many of the activities of TNF are also observed upon treatment with interleukin 1. Activation of the arachidonic acid pathway leading to prostaglandins and to leukotrienes is certainly one of the important mechanisms of TNF. A number of proteins are induced by TNF (and by IL1), and one of these, the 26-K protein, may play an interesting role in the regulation of cell proliferation. Further studies at the physiological level, the cell biological level, the biochemistry level and the molecular biology level will undoubtedly help in building an understanding of how TNF acts on normal cells and how it causes toxicity, and on the other hand how it selectively affects a number of malignant cells and how IFN-ɣ very often renders these cells susceptible to TNF cytotoxicity. On these bases effective and acceptable anticancer therapies may be developed.

ACKNOWLEDGMENTS

We thank our many colleagues who allowed us to refer to their results, either unpublished or in press. Research was supported by grants from the Fonds voor Geneeskundig Wetenschappelijk Onderzoek (F.G.W.O.), the Algemene Spaar- en Lijfrentekas (A.S.L.K.) and Biogen SA.

REFERENCES

1. Aggarwal BB, Eessalu TE & Hass PE: Nature 318:665, 1985.
2. Bachwich PR, Chensue SW, Larrick JW & Kunkel SL: Biochem Biophys Res Commun 136:94, 1986.
3. Baglioni C, McCandless S, Tavernier J & Fiers W: J Biol Chem 260:13395, 1985.
4. Balkwill FR, Lee A, Aldam G, Moodie E, Thomas JA, Tavernier J & Fiers W: Cancer Res 46:3990, 1986.
5. Bertolini DR, Nedwin GE, Bringman TS, Smith DD & Mundy GR: Nature 319:516, 1986.
6. Beutler B, Milsark IW & Cerami AC: Science 229:869, 1985a.
7. Beutler B, Mahoney J, Le Trang N, Pekala P & Cerami A: J Exp Med 161:984, 1985b.
8. Bevilacqua MP, Pober JS, Majeau GR, Cotran RS & Gimbrone MA Jr: J Exp Med 160:618, 1984.
9. Bevilacqua MP, Pober JS, Majeau, GR, Fiers W, Cotran RS & Gimbrone MA Jr: submitted, 1986a.
10. Bevilacqua MP, Wheeler ME, Pober JS, Fiers W, Mendrick DL, Cotran RS & Gimbrone MA Jr: submitted, 1986b.
11. Brouckaert PGG, Leroux-Roels G, Guisez Y, Tavernier J & Fiers W: Int J Cancer, in press, 1986.
12. Collins T, Lapierre LA, Fiers W, Strominger JL & Pober JS: Proc Natl Acad Sci USA 83:446, 1986.
13. Content J, De Wit L, Pierard D, Derynck R, De Clerq E & Fiers W: Proc Natl Acad Sci USA 79:2768, 1982.
14. Content J, De Wit L, Poupart P, Opdenakker G, Van Damme J & Billiau A: Eur J Biochem 152:253, 1985.
15. Dayer J-M, Beutler B & Cerami A: J Exp Med 162:2163, 1985.
16. Defilippi P, Poupart P, Haegeman G, Tavernier G, Fiers W & Content J: Proceedings of the 1986 ISIR-TNO meeting on the INTERFERON SYSTEM, Finland, this volume, 1986.
17. Dinarello CA, Cannon JG, Wolff SM, Bernheim HA, Beutler B, Cerami A, Figari IS, Palladino MA Jr & O'Connor JV: J Exp Med 163:1433, 1986.
18. Fiers W, Brouckaert P, Guisez Y, Remaut E, Van Roy F, Devos R, Fransen L, Leroux-Roels G, Marmenout A, Tavernier J & Van der Heyden J: In: The Biology of the Interferon System, Schellekens H & Stewart WE

(eds.), pp. 241-248, Amsterdam, Elsevier Science Publishers BV,
1986a.
19. Fiers W, Brouckaert P, Devos R, Fransen L, Leroux-Roels G, Remaut E,
    Suffys P, Tavernier J, Van der Heyden J & Van Roy F: In: Cold
    Spring Harbor Symposia on Quantitative Biology Vol 51: Molecular
    Biology of Homo sapiens, in press, 1986b.
20. Fransen L, Muller R, Marmenout A, Tavernier J, Van der Heyden J,
    Kawashima E, Chollet A, Tizard R, Van Heuverswyn H, Van Vliet A,
    Ruysschaert M-R & Fiers W: Nucl Acids Res 13:4417, 1985.
21. Fransen L, Ruysschaert MR, Van der Heyden J & Fiers W: Cell Immunol
    100:260, 1986a.
22. Fransen L, Van der Heyden J, Ruysschaert MR & Fiers W: Eur J Cancer
    Clin Oncol 22:419, 1986b.
23. Gamble JR, Harlan JM, Klebanoff SJ & Vadas MA: Proc Natl Acad Sci USA
    82:8667, 1985.
24. Ghezzi P, Saccardo B & Bianchi M: Biochem Biophys Res Commun 136:316,
    1986.
25. Gresser I, Belardelli J, Tavernier J, Fiers W, Podo F, Federico M,
    Carpinelli G, Duvillard P, Prade M, Maury C, Bandu M-T & Maunoury
    M-T: Int J Cancer, in press, 1986.
26. Haegeman G, Content J, Volckaert G, Derynck R, Tavernier J & Fiers W:
    Eur J Biochem: in press, 1986.
27. Hahn T, Toker L, Budilovsky S, Aderka D, Eshhar Z & Wallach D: Proc
    Natl Acad Sci USA 82:3814, 1985.
28. Klebanoff SJ, Vadas MA, Harlan JM, Sparks LH, Gamble JR, Agosti JM &
    Waltersdorph AM: J Immunol 136:4220, 1986.
29. Kohase M, Henriksen-Destefano D, May LT, Vilcek J & Sehgal PB: Cell
    45:659, 1986.
30. Kull FC Jr, Jacobs S & Cuatrecasas P: Proc Natl Acad Sci USA 82:5756,
    1985.
31. Marmenout A, Fransen L, Tavernier J, Van der Heyden J, Tizard R,
    Kawashima E, Shaw A, Johnson MJ, Semon D & Muller R: Eur J Biochem
    152:515, 1985.
32. Nedwin GE, Naylor SL, Sakaguchi AY, Smith D, Jarret-Nedwin J, Pennica
    D, Goeddel DV & Gray PW: Nucl Acids Res 13:6361, 1985.
33. Onozaki K, Matsusima K, Aggarwal BB & Oppenheim JJ: J Immunol
    135:3962, 1985.
34. Oppenheim JJ, Kovacs EJ, Matsushima K & Durum SK: Immunology Today
    7:45, 1986.
35. Pennica D, Nedwin GE, Hayflick JS, Seeburg PH, Derynck R, Palladino
    MA, Kohr WJ, Aggarwal BB & Goeddel DV: Nature 312:724, 1984.
36. Pennica D, Hayflick JS, Bringman TS, Palladino MA & Goeddel DV: Proc
    Natl Acad Sci USA 82:6060, 1985.
37. Plaetinck G, Declercq W, Tavernier J, Nabholz M & Fiers W: submitted,
    1986.
38. Pober JS, Bevilacqua MP, Mendrick DL, Lapierre LA, Fiers W & Gimbrone
    MA Jr: J Immunol 136:1680, 1986.
39. Rainsford KD: Aspirin and the Salicylates. London: Butterworth & Co.,
    1984.
40. Roubin R, Elsas P, Fiers W & Dessein A: submitted, 1986.
41. Rubin BY, Anderson SL, Sullivan SA, Williamson BD, Carswell EA & Old
    LJ: J Exp Med 162:1099, 1985.
42. Ruggiero V, Tavernier J, Fiers W & Baglioni C: J Immunol 136:2445,
    1986.
43. Saklatvala J: Nature 322:547, 1986.
44. Shalaby MR, Aggarwal BB, Rinderknecht E, Svedersky LP, Finkle BS &

Palladino MA Jr: J Immunol 135:2069, 1985.
45. Shirai T, Yamaguchi H, Ito H, Todd CW & Wallace RB: Nature 313:803, 1985.
46. Silberstein DS & David JR: Proc Natl Acad Sci USA 83:1055, 1986.
47. Spies T, Morton C, Nedospasov SA, Shakhov AN, Korobko VG, Fiers W, Pious D & Strominger JL: submitted, 1986.
48. Stolpen AH, Guinan EC, Fiers W & Pober JS: Am J Pathol 123:16, 1986.
49. Sugarman B, Aggarwal B, Hass P, Figari I, Palladino MA Jr & Shepard M: Science 230:943, 1985.
50. Torti FM, Dieckman B, Beutler B, Cerami A & Ringold GM: Science 229:867, 1985.
51. Tsujimoto M, Yip YK & Vilcek J: Proc Natl Acad Sci USA 82:7626, 1985.
52. Tsujimoto M, Yip YK & Vilcek J: J Immunol 136:2441, 1986.
53. Vilcek J, Palombella VJ, Henriksen-Destefano D, Swenson C, Feinman R, Hirai M & Tsujimoto M: J Exp Med 163:632, 1986.
54. Wang AM, Creasy AA, Ladner MB, Lin LS, Strickler J, Van Arsdell JN, Yamamoto R & Mark DF: Science 228:149, 1985.

INDUCTION OF 26kDa PROTEIN mRNA IN HUMAN CELLS TREATED WITH
RECOMBINANT HUMAN TUMOUR NECROSIS FACTOR (rTNF).

P. DEFILIPPI[1], P. POUPART[1], G. HAEGEMAN[2], J. TAVERNIER[3], W. FIERS[2]
and CONTENT,[1]J.

[1] Institut Pasteur du Brabant(Virology), 1180 Brussels,[2] Rijks-
universiteit Gent (Mol. Biol.), 9000 Ghent, [3]Biogent (subs of
Biogen S.A.) 9 000 Ghent (Belgium).

1. INTRODUCTION
    When human fibroblast cells are induced to produce IFN-β by
treatment with poly(I).poly(C) in the presence of cycloheximide,
synthesis of a protein of 26kDa (1),earlier described as IFN-β$_2$
by Weissenbach et al. (2)is stimulated. Peripheral blood lymphocytes (3) and epithelial cells (4) also express the 26kDa protein mRNA when appropriately induced. Recently, the nucleotide
sequence of the entire coding region of the 26kDa protein mRNA
has been described. Despite extensive searches, no statistically significant sequence homology with any known interferon species was found (5).
    Some of us also found that the 26K gene is also induced by a
highly purified Il-1 related protein obtained from Concanavalin
A stimulated human leukocytes (6). Since the 26kDal protein has
weak but consistent interferon related biological activities
(7), it is of interest to look for an alternative function of
the product of this strongly induced gene.
    rTNF is a cytokine exerting powerful cytostatic effect in
some tumour cells (13) and mitogenic effect in normal diploid
fibroblasts (9). Since in several instances TNF displays some
activities exerted mainly by IL-1 (8), it was of interest to
check whether TNF also induces the 26kDa protein in human fibroblasts.
Here we show that this is indeed the case. We also extend the
study to five other cell lines differing in their sensitivity
to the cytotoxic effect of TNF and describe a direct correlation
between the inducibility of the 26K gene by TNF and its mitogenic versus cytotoxic effect on these cells.

2.EXPERIMENTAL PROCEDURES
2.1 Materials
    Recombinant human TNF (h-TNF), expressed in E.coli, purified
to homogeneity, was at 1.57 mg/ml. Its final specific activity
was 2x10^7 U/mg.
    Recombinant human IL-1β (h-IL-1β) expressed in E.coli, 99%
pure as estimated by Coomassie blue staining, was at 5.69 mg/
ml. Recombinant IFN-α$_2$was a kind gift from Boehringer Ingelheim (5x10^7 U/ml).
2.2 Cell cultures
    The human transformed and non transformed cell lines (see
Table 1) were grown in DMEM + 10% heat inactivated newborn calf
serum. The breast carcinoma line MCF-7 was supplemented with
0.02% Na pyruvate.

TABLE 1.   26kDa protein mRNA induction in different transformed and non-transformed human cell lines

CELL LINE	Growth inhibition * assay TNF sensitivity (ng/ml) 50 %	26 K mRNA induction	
		TNF (30 ng/ml)	r IL-1β (100 ng/ml)
Cervix carcinoma			
HeLa H21	90	+	+++
HeLa-D98/AH2	3	−	−
Breast carcinoma			
BT-20	1.2	−	−
MCF-7	2.4	−	−
Osteosarcoma			
MG63	> 90	++	+++
Diploid fibroblast			
FS4	> 90	++	+++

*
This section of the table was modified from (13), Table 1 ; the sensitivity is now expressed in ng/ml TNF required to reduce cell survival to 50% ; > 90 = no 50% inhibition could be obtained at TNF concentration up to 90ng/ml.

+ : 3-5 fold induction   )
++ : 10 fold of induction } as determined by densitometric scan analysis
+++ : >10 fold of induction) of dot hybridization autoradiograms

All the treatments were performed for 18 hrs in parallel confluent cultures.

2.3 Cytoplasmic dot hybridization - Northern Blot

Dot hybridization analysis was slightly modified from White and Bancroft (10). For RNA binding, nylon membranes (Amersham, Hybond-N) were exposed for five minutes to UV light before hybridization (according to Amersham procedure). For Northern Blot analysis, total RNA preparations (5 μg) (extracted from whole cultures with the SDS proteinase K procedure, followed by two precipitations in 2MLiCl (11)) were glyoxalated, run on 1.1 % agarose gel and blotted as described before (4), but on nylon filter (Amersham, Hybond-N) as described above.

2.4 Hybridization

Hybridization was performed with a 26kDa protein cRNA probe prepared by transcribing either T_3 or SP6 plasmids constructions covering 800 bp from the beginning of cDNA 26_2 K-7 to the distal site Sau3A (5) in the presence of 80 μCI (α^3 P)UTP (800 Ci/mole, Amersham). The hybridization solution contained 1.0-1.5x10^7 cpm/ml of the cRNA probe and 10% (w/v) of PEG 6000(12). Otherwise the procedure was as described (6) except that hybridization was carried out for 4-6 hrs at 65°C.

3. RESULTS AND DISCUSSION

In initial experiments, the 26kDa protein mRNA level has
been studied by dot hybridization on cytoplasmic extracts of FS-4
foreskin fibroblasts with a high specific activity 26K cRNA
probe. A typical experiment is presented in Fig. 1A.

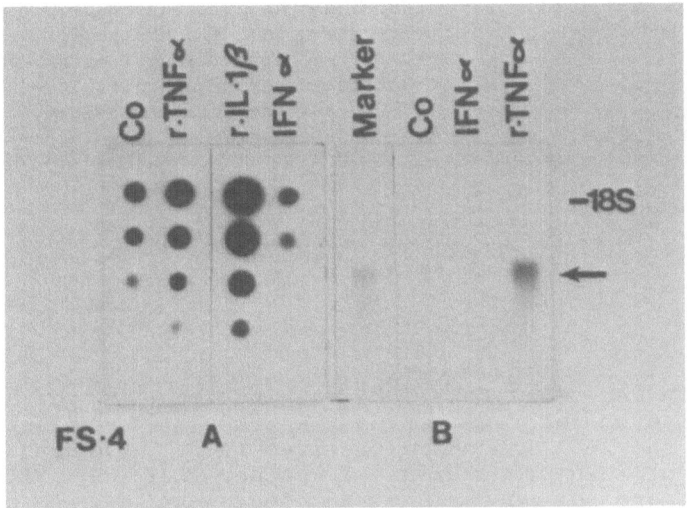

FIGURE 1 : rTNF and rIL-1β mediated induction of 26kDa protein
mRNA in FS4 cells :
A. Cytoplasmic dot hybridization of cellular extracts. The treatments were
performed for 6 hrs : rTNF = 30 ng/ml ; rIL-1β = 100 ng/ml ; α-IF : 10^3 U/ml.
B. Northern blot analysis : each lane corresponds to 5 μg of total RNA ;
the treatments were as in A., except that they were performed for 15 hrs.
Marker = 2 μg of poly A^+ mRNA extracted from MG63 cells superinduced for
IFN-β production.

The TNF treatment (30 ng/ml for 6 hrs) enhances the level of
the 26kDa protein mRNA more than 5 fold above the background level of untreated culture. A treatment with 100 ng/ml of r-IL1β
confirms unequivocally our earliest conclusion (6) that IL-1β is
a 26kDa protein mRNA inducer "per se" since indeed it enhances the
mRNA 5 to 10 fold more than TNF.
Conversely, α-IF (10^3 U/ml for 6 hrs) does not affect significantly the basal level of 26kDa protein mRNA. Northern blot analysis of total RNA extracted from untreated and TNF-treated cells
(30 ng/ml, 15 hrs of treatment) (Fig. 1B) has confirmed the high

level of induction of this mRNA, which migrates in agarose gel
as a single band of about 1.3 kb (4).
   In an other cellular system, human epithelial amniotic cells
UAC, we observed that TNF (30 ng/ml) induced 2-5 fold the mRNA
of 2-5A synthetase, after 1-4 hrs of treatment (Fig. 2).

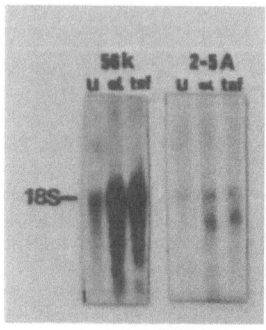

FIGURE 2 : rTNF-mediated induction of
2-5A synthetase mRNA in UAC cells.
Northern blot analysis : 5 µg of total RNA
of UAC cells treated for 15 hrs with 30ng/ml
rTNF. The hybridizations were performed respectively with : a 56kDa protein cDNA derived
cRNA probe, transcribed from a SP64-G construction as described in (16) ; a 2-5A synthetase cDNA derived cRNA probe, transcribed
from a SP64 construction as described in (17)

   Since in UAC cells this mRNA was induced even in the presence of antiserum reactive versus the 26kDa protein, our results
show that the 2-5A induction observed is not a consequence of
the 26kDa protein production (7).Whether this induction is at
transcriptional or post-transcriptional level has to be established. On the contrary, a 56kDa protein mRNA signal is also
enhanced by TNF in UAC cells, but is abolished when treating
the cells in the presence of the crude antiserum mentioned above, suggesting that this phenomenon is an indirect consequence
of TNF treament.
   Since the 26kDa protein gene appears as the first and best
TNF   responding gene known so far, we used a 26 K cRNA probe,
to compare TNF responses in different cell lines. Dot hybridization analysis of 26 kDA protein mRNA induction was extended
to a serie of five different human transformed cell lines, differing in their sensitivity to TNF cytolytic cytostatic effect
(13). The results are summarized in table 1. A clear TNF mediated 26kDa protein mRNA induction was observed in two transformed cell lines, characterized by a relatively low sensitivity
to treatment with TNF alone. The MG63 cells show a low sensitivity to TNF cytolytic effect, only slightly increased by IFN-γ
treatment, whereas the HeLa H21 cells are sensitive to treatment with TNF alone, but the cytolytic effect is greatly enhanced by concomitant treatment with 200U/ml of IFN-γ (13). The
extent of induction of 26kDa protein mRNA in HeLa H21 cells was
less pronounced than that observed in MG63 and FS4 cells (3-5
times fold-30 ng/ml TNF for 18 hrs as quantified by dot and

Northern blot analysis). rIL-1β treatment of HeLa H21 cells
(100 ng/ml for 18 hrs) confirms a greater extent of induction
than that obtained with TNF. Conversely, the cervix carcinoma
HeLa D°8/AH2 and the breast carcinoma lines BT20 and MCF-7,
for which 50% cytolysis occurs at a very low TNF concentration
do not show any expression of 26kDa protein mRNA after treatment with TNF. Even in the presence of rIL-1β (100 ng/ml) no
induction was observed (Table 1). These data indicate the occurrence of an inverse correlation linking the expression of 26kDa
protein mRNA in transformed cells with TNF cytotoxic effect.

This negative correlation could indicate either that :
1) depending on the cellular environment, the 26kDa protein expression may inhibit or counteract the cytolytic/cytotoxic
effect of TNF or even act as a mediator(s) of its mitogenic activity.
The mechanism(s) implied would be extremely interesting to clarify and may shed light on the biological activity of the 26kDa
protein.
2) alternatively or additionally, the 26kDa protein may constitute negative growth factors, dumping or limiting the mitogenic
effect of TNF when it occurs (as suggested for the 2-5A synthetase and IFN-β in PDGF induced mitogenesis (14)).

It is not possible to discriminate between these alternatives
at this stage. Further experiments may delineate the functional
role of the 26kDa protein.

When this manuscript was ready for publication, two articles
of different groups confirmed some of the data presented here :
1) TNF-mediated 26kDa protein induction in cells was related
   to the control of mitogenic action of TNF (15)
2) TNF-mediated 2-5A synthetase mRNA induction in some cell
   lines was correlated to TNF antiviral activity by Wong et
   Goeddel (personal communication).

REFERENCES

1. Content J, De Wit L, Pierard D, Derynck R, De Clercq E ,
   and Fiers W . Proc. Natl. Acad. Sci. USA, 79, 2768-2772
   (1982).
2. Weissenbach J, Chernajovsky T, Zeevi M, Shulman L, Soreq H,
   Nir U, Wallach D, Perricaudet M, Tiollais P and Revel M.
   Proc. Natl. Acad. Sci. USA, 77, 7152-7156 (1980).
3. Vaquero C, Sanceau J, Weissenbach J, Beranger F and Falcoff
   R. J. of Interferon Res. 6, 161-170 (1986).
4. Poupart P, De Wit L and Content J. Eur. J. Biochem., 143,
   15-21 (1984).
5. Haegeman G, Content J, Volckaert G, Derynck R, Tavernier J
   and Fiers W. Eur. J. Biochem (1986, in press).
6. Content J, De Wit L, Poupart P, Opdenakker G, Van Damme J
   and Billiau A. Eur. J. Biochem., 152, 253-257 (1985).
7. Zilberstein A, Ruggieri R and Revel M. in : "The Interferon
   System", Serono Symposia. Rossi GB and Dianzani E, eds.
   Raven Press N.Y., 73-83 (1985).
8. Fiers W. and al. Lymphokines and monokines in therapy.
   Cold Spring Harbor (in press).

9. Fiers W. and al. in : "The biology of the Interferon System", W.E. Stewart II and M. Schellekens, eds., Elsevier Sci. Publ. Amsterdam., 241-248 (1985).

10. White BA and Bancroft FC. J. Biol. Chemistry, 257, n° 15, 8569-8572 (1982).

11. Content J, De Wit L, Tavernier J and Fiers W. Nucl. Acid Res., 7, 1541-1552 (1983).

12. Amasino RM. Analytical Biochemistry, 152, 304-307 (1986).

13. Fransen L, Van der Heyden J, Ruysschaert MR and Fiers W. Eur. J. Canc., 22, 419-426 (1986).

14. Zullo JN, Cochran BH, Huang AS and Stiles CD. Cell, 43, 793-800 (1985).

15. Kohase M, Henriksen-De Stefano D, May LT, Vilcek J and Sehgal PB. Cell, 45, 659-666 (1986).

16. Wathelet M, Moutschen S., Defilippi P, Cravador A, Collet M. Huez G and Content J. Eur. J. Biochem, 155, 11-17 (1986).

17. Moutschen S, Wathelet M, Cravador A, Defilippi, P, Huez G, and Content J. in "The 2-5A system", B.R.G. Williams and R.H. Silverman, eds., Alan R. Liss, New York, Vol. 202, 183-192 (1985).

INTERACTIONS OF TUMOR NECROSIS FACTOR, INTERFERON AND INTERLEUKIN-1 IN CELL KILLING

D. WALLACH, H. HOLTMANN, T. HAHN and S. ISRAEL
Department of Virology, The Weizmann Institute of Science, Rehovot, Israel

1. SUMMARY

Modulation of cell vulnerability to tumor necrosis factor (TNF) and lymphotoxin (LT) by leukocyte-produced cytokines was examined. Inhibitors of RNA and protein synthesis sensitize cells to the cytotoxic effect of the two cytotoxins (CTXs). The effectivity of killing of such sensitized cells is, however, largely decreased in cells which have been treated for a few hours with crude preparations of the CTXs in absence of the inhibitors. Fractionation of the crude preparations revealed that the cytokines inducing resistance to cytotoxicity of the CTXs are the CTXs themselves and interleukin-1 (IL-1). Effective resistance can also be induced with phorbol diesters. On the other hand, IFN-γ increases the effectiveness of cell killing by the CTXs. Modulation of sensitivity to TNF by these agents may in part be explained by their effect on TNF receptor expression: IL-1, the CTXs and phorbol-diesters down-regulate the receptors to TNF while IFN-γ induces an increase in the TNF receptors. However, besides receptor down-regulation additional mechanisms seem to contribute to the resistance since following removal of IL-1, CTXs or phorbol-diesters from treated cells and recovery of the receptors to TNF the cells still exhibit decreased vulnerability to killing by the CTXs.

2. INTRODUCTION

Tumor necrosis factor (TNF) and lymphotoxin (LT) affect the function of cells in a rather complex manner. The two closely related "cytotoxins" (CTXs) are found to alter the pattern of cell growth and viability, metabolic functions and immune activities in multiple, sometimes contrasting ways (1). Some of their effects are likely to be beneficial to the organism, yet others can be detrimental (2). Knowledge on the mechanisms which determine the nature of response to the proteins in differing situations is quite limited.

Among the many possible effects of the CTXs no doubt the one with the most far-reaching bearing on functioning of the cell is cell death, resulting from the cytotoxic activity of the CTXs. There is particular interest in this activity from the point of view of IFN research since cells treated by IFN are found to exhibit a significant increase in vulnerability to killing by the CTXs (reviewed in 1). There are several indications that vulnerability to the cytotoxic effect is subjected to regulation by mechanisms which, at least in part, are independent to those controlling the response to other effects of CTXs. Thus, comparison of effects of the CTXs on cells of differing cultured lines has revealed marked differences in vulnerability from one cell line to another; tumor cells being in general more vulnerable than normal ones. These differences did not correlate to the level of receptors to the CTXs nor to the effectiveness at which non-cytotoxic effects could be induced in the cells. Cell killing by the CTXs is enhanced not only by IFN but also by inhibitors of RNA and protein synthesis or of energy metabolism; it is also enhanced consequently to infection of the cells with certain viruses (reviewed in 1). The findings presented below suggest that resistance of cells to killing by the CTXs involves mechanisms of desensitization - namely mechanisms whose activation by the CTXs themselves and by other cytokines, of related function, results in decreased responsiveness to the cytotoxic effect. That decrease in response is affected on more than one mechanistic level: it involves down-regulation of the receptors to the CTXs and, to a greater extent, some other induced change(s) of yet unclear nature.

3. MATERIALS AND METHODS

Quantitation of the cytotoxic and protective activities of preparations of cytokines: Both activities were determined with the use of SV80 or HeLa cells as previously described (3). Cytotoxicity was quantitated by applying the tested samples on cells at serial dilutions, in the presence of cycloheximide (CHI)(50 µg/ml) and, 12 hr later, determining viability of the cells by measuring the uptake of neutral red. Protective activity was quantitated by applying the tested sample on the cells, in serial dilutions, in absence of CHI and 3 hr later, rinsing the cells and incubating for a further 12 hr with CTXs (usually - 125 cytotoxicity units/ml) in the presence of CHI and then determining cell viability as above. Units of cytotoxic and protective activities are defined as the concentrations of cytokines at which 50% of the cells in the culture were killed or protected from killing (as compared to cells incubated in absence of the tested cytokine).

Quantitation of the receptors for TNF: rTNF, radioiodinated with the chloramine-T reagent to specific radioactivity of 42 Ci/gr, as previously described (12), was applied on cultures of SV80 cells or FS11 foreskin fibroblasts at a concentration of 3.6 ng/ml, either alone or in presence of 1000-fold excess of unlabelled TNF, 14 hr after seeding the cells into 18 mm tissue culture wells at a density of 2.5×10^5 cells/well. The radio-labelled TNF was applied at 4°C in 150 µl growth medium also containing 20 mM Hepes buffer and 15 mM sodium azide. Following 2 h incubation the cells were rinsed and the amount of label specifically bound, calculated by subtracting the amount bound in the presence of excess unlabelled protein from that bound in its absence, was determined.

Materials: rTNF, produced by Genentech Corp., San Francisco was kindly provided by Dr. G. Adolf from the Boehringer Institute, Vienna, Austria . rIL-1α was a gift from Dr. A. Stern and Dr. P.T. Lomedico, the Roche Research Center, Nutley, N.J, and rIFN-γ a gift from Dr. M. Rubinstein, The Weizmann Institute of Science, Rehovot, Israel. Antiserum against IL-1, raised by immunizing rabbits with rIL-1β was purchased from Cistron, Pine Brook, N.J.

4. RESULTS

Antagonistic effects of IFN and of crude preparations of cytotoxins on vulnerability of cells to their killing by the cytotoxins. IFN potentiates the cytotoxic effect of CTXs, yet cells of certain tumor lines (such as HeLa or SV80) will not be killed by the CTXs, even when treated with IFN. Such resistant cells can, nevertheless, be sensitized to the cytotoxic effect with inhibitors of protein synthesis, such as cycloheximide (CHI). Treatment with IFN, prior to applying the CTXs and CHI together, will augment killing of the cells. However, if rather than IFN, crude preparations of CTXs are applied on the cells the opposite effect can be observed: even though such preparations contain significant amounts of IFN they are found to induce, within a few hours of application, a marked reduction in vulnerability to the cytotoxic effect of a subsequent challenge with the CTXs and CHI (Fig. 1). That protective effect of CTX-preparations could be induced with minute concentrations of those preparations – similar to those at which, in vulnerable cells, the cytotoxic effect is observed (3) suggesting that the protective effect, like the cytotoxic one is induced by some effective regulatory factors present in those cytokine preparations.

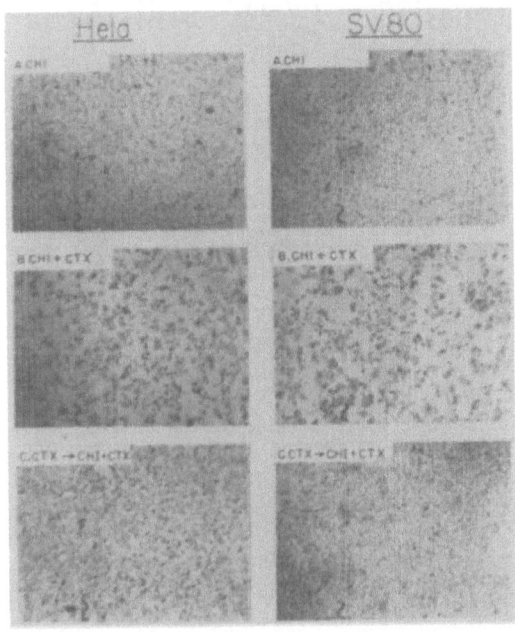

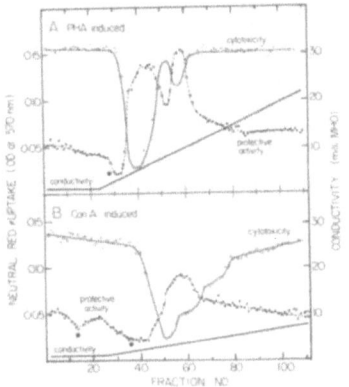

FIGURE 2. Pattern of fractionation of cytotoxic and protective activities of phytohemagglutinin (PHA) and Concanavalin A-induced CTX preparations, on DEAE cellulose. For details see (3).

FIGURE 1. Cytotoxic and protective effects of crude preparations of CTXs. Death of HeLa and of SV80 cells following 12 h incubation with crude preparations of CTXs (8 cytotoxic U/ml) together with CHI (B) is shown in comparison to the normal morphology observed in incubation with CHI alone (A), and to the decreased extent of death in cells which were treated for 3 hr with the CTXs in absence of CHI (80 cytotoxic U/ml), then rinsed and incubated for a further period of 12 h with the CTXs (8 U/ml), this time in the presence of CHI. For further details see (3).

Identification of the cytokines inducing resistance to the cytotoxic effect of the CTXs as the CTXs themselves and as IL-1.

Subjecting crude preparations of CTXs to fractionation, aimed at elucidating the nature of the cytokines which induce in cells resistance to killing by CTXs, we observed much similarity between the pattern of fractionation of that protective activity (determined by applying the proteins on the cells in absence of CHI and, a few hours later, challenging these cells with CTXs in presence of CHI) and the cytotoxic activity (observed in applying the proteins in cells simultaneously with CHI)(Fig. 2) suggesting that the CTXs themselves are the proteins which induce in cells resistance to their cytotoxicity. Indeed, after purifying to homogeneity the major CTX present in these preparations – TNF – we found the protein to induce, effectively, resistance to its own cytotoxicity (4). Similarly, we found the other CTX present in these preparations – LT – to induce resistance to its own cytotoxicity. The two CTXs also induced resistance to the cytotoxicity of each other (unpublished results).

TABLE I. Presence of pH2.0-resistant cytokine(s) with protective activity in crude preparations of CTXs

	Cytotoxic Activity	(U/ml)	Protective Activity
Purified TNF	125		60
Purified TNF/pH2-treated	1		< 1
Crude CTXs	125		125
Crude CTXs/pH2-treated	1		250

Crude PHA-induced CTXs and native TNF, purifed from the crude preparations to apparent homogeneity, both produced as described in (4) as well as samples of these preparations which were treated for 12 h at pH 2.0 and then brought back to pH 7.0 were examined for cytotoxic and protective activities as described under Materials and Methods (5).

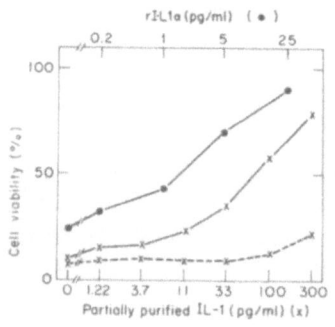

FIGURE 3. Protective effect of IL-1. The extent of protection of SV80 cells from the cytotoxicity of TNF by differing amounts of a preparation of CTXs, induced in U937 cells (14) and treated at pH 2.0 (x——x) is shown in comparison to the protection induced by this same preparation following its incubation with monospecific antiserum to IL-1 (applied at a dilution of 1:100)(x---x). Also shown is the protective effect observed in treating the cells with various concentrations of rIL-1 (●). (5).

However, the resistance-inducing-activity of TNF and LT could not fully account for the protective activity of crude preparations of those proteins. As shown in Table I, treating such prearations at pH2.0, which resulted in inactivation of TNF and LT, did not decrease their protective effect. Actually, an increase in that activity could be observed. That increase could be related to the inactivation, at pH2.0, also of IFN-γ present in these preparations, which sensitizes cells to the cytotoxicity of the CTXs. Subjecting such pH2.0-treated-CTX preparations to a series of fractionation steps we found their protective activity to fully copurifiy with interleukin-1 (IL-1)(5). Furthermore, the protective activity of such preparations could be neutralized with specific antiserum to IL-1 and could be reproduced with recombinant IL-1 (Fig. 3). We thus concluded that the resistance-inducing activity of crude preparations of the CTXs is mediated primarily by three cytokines; TNF, LT and IL-1.

Mechanisms involved:

A likely mechanism for the decrease in responsiveness to a cytokine, consequently to treatment by that cytokine itself is depletion of its receptors. Indeed, like many other polypeptide cytokines, TNF is found to be taken up into the cell following binding to its receptors (6,7), which results in a significant decrease of these receptors (not shown). Marked reduction in the receptors to TNF can also be induced by IL-1 (Fig. 4), although competition experiments indicate that IL-1 and TNF bind to differing and independent receptors (8). This IL-1 induced decrease in TNF receptors is rapid, reaching maximal extent at 1 h following IL-1 application and, while being temperature dependent, it can not be blocked with the protein synthesis inhibitor CHI (not shown). In contrast to the effect of IL-1, IFN is found to induce some increase in TNF receptors (Fig. 4 and refs. 9-12), which is rather slow and is dependent on de-novo synthesis of protein(s)(11). Cells in which the receptors to TNF have been increased by treatment with IFN respond to IL-1 in decrease of TNF receptors although not to as low a level as in cells which have not been treated with IFN (Fig. 4).

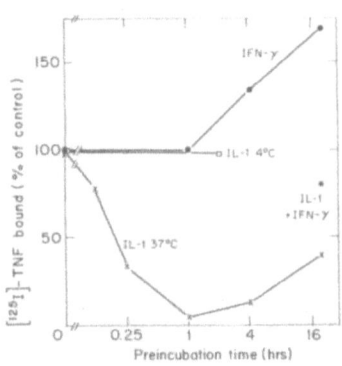

FIGURE 4. Down-regulation of the receptors for TNF by IL-1 and increase of the receptors by IFN-γ. The level of TNF receptors at various times following application of rIL-1α (60 pg/ml) at 37°C (x), on SV80 cells as opposed to the lack of any decrease in TNF binding in treatment by IL-1 at 4°C (☐). Also shown is the increase in TNF receptors at various times following application of rIFN-γ (12.5 ng/ml) at 37°C (●), and the level of TNF receptors in cells treated for 16 hr with IFN-γ (12.5 ng/ml) and then for a further 4 hr with IL-1 (60 pg/ml)(*). Specific binding to the control cells (100% - not treated by IL-1 or IFN-γ was 1200 CPM) (8).

There are several other known examples for down-regulation of receptors to one agonist consequently to binding of another agonist to its own receptor. In a number of these, the decrease of receptors could be correlated with induced phosphorylation of the down-regulated receptor and could be initiated with phorbol-diesters which stimulate the activity of protein kinase-C. Examining the effects of phorbol myristate acetate (TPA) and of phorbol dibutyrate on the response of cells to TNF we found these phorbol-diesters to induce, just like the CTXs themselves and like IL-1, a marked decrease in the level of the receptors to TNF (Fig. 5) as well as resistance to the cytotoxicity of TNF (13).

To determine the extent to which the decrease in TNF receptors, in response to TNF, IL-1 and phorbol-diesters, contribute to the resistance to cytotoxicity of TNF, which they induce, we followed, kinetically, the recovery of TNF receptors following removal of TNF, IL-1 and phorbol-diesters and its correlation to the extent of responsiveness of the cells to the cytotoxic effect. Within a few hours of removal of IL-1, TNF or the water soluble phorbol-diester phorbol-dibutyrate from cells, TNF receptors increased, almost back to the normal level. Yet, in spite of the recovery of the receptors, the cells remained resistant to killing by TNF (12,13 and Fig. 6) indicating that the resistance to killing is not due solely to the lack of receptors to TNF but, even to a greater extent, to some other more lasting change(s) which these agents induce.

228

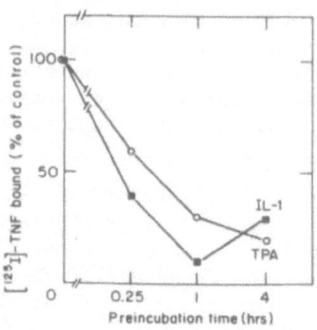 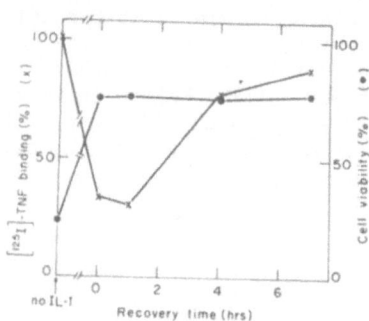

Figure 5 (left): Down-regulation of the receptors to TNF by TPA and IL-1. Binding of radiolabelled TNF to receptors on FS11 foreskin fibroblasts at various time points following application, at 37°C, of IL-1 (60 pg/ml) or TPA (5 ng/ml). Specific binding to control cells was 1000 CPM (13).

FIGURE 6 (right): Maintenance of resistance to killing by TNF, following removal of IL-1, in spite of recovery of TNF receptors. Human SV80 cells were incubated for 4 h with rIL-1 α (60 pg/ml), then rinsed and, at various times afterwards, examined for specific binding of radiolabelled TNF (40 U/ml) and for vulnerability to the cytotoxic effect of TNF; the latter determined by applying the protein (at 1000 U/ml) in presence of CHI for a period of 12 hrs. Specific TNF binding prior to application of IL-1 (100%) was 1900 CPM.

5. DISCUSSION

By what mechanisms do cells resist killing by the CTXs? Since resistant cells can be rendered vulnerable by inhibitory agents, such as CHI, they are probably not devoid of those activities which take part in the cytotoxic effect but rather express some additional activities which counteract the cytotoxic ones and thus protect them from killing. The findings presented above suggest that resistance to the cytotoxicity of the CTXs involves mechanisms of desensitization, namely mechanisms whose activation upon exposure of the cell to prearations of the CTXs results in decreased responsiveness to their own cytotoxic effect. The fact that IL-1 and the CTXs function similarly in inducing that desensitization is not surprising in view of quite a number of other, recently revealed similarities in the function of the CTXs and IL-1. Actually, in certain cells IL-1 functions as a CTX, exerting cytotoxic effects which can be potentiated by inhibitors of RNA/protein synthesis (5,15,16). That similarity in function of the CTXs and IL-1 and the ability of IL-1 to modulate the response to the CTXs - reflected in the induction, by IL-1, of resistance to their cytotoxicity - seems to make physiological sense: By the nature of their producing cells and inducing stimuli the CTXs are likely to be accompanied, under physiological situations, by IL-1. However, it is of interest to note that cytokines whose mode of production is completely unrelated to that of the CTXs can also induce resistance to the cytotoxicity of the CTXs. Thus, we have recently observed that, in certain cells, effective resistance to cytotoxicity can be induced also by insulin.

An idea as to the molecular basis for the induction of resistance and of the physiological role of this phenomenon can perhaps be derived from the finding that such resistance can also be induced with phorbol-diesters, which are known to affect cells by stimulating protein kinase-C, thus enhancing phosphorylation, primarily of tyrosine residues, in specific target proteins. Stimulation of protein-kinase-C is a growth stimulatory signal. Similarily, receptors to several growth stimulatory cytokines are known to possess tyrosine-kinase activity which is enhanced upon their occupation by the cytokine and which is believed to have a central role in their growth promoting activity. It seems possible that TNF, LT and IL-1, which have been found to function as growth promoting agents for certain cells, also stimulate tyrosine kinase activity, mediated by their receptors and that the resistance induced by these cytokines reflects enhanced phosphorylation, by these kinase(s) of certain proteins. One of the substrates for this putative phosphorylation process may be the receptor to TNF which is found to decrease in amount in response to IL-1, TNF and phorbol-diesters. Indeed, receptors for several hormones are down-regulated as a consequence of their induced phosphorylation. Besides that decrease in receptors some other, more lasting change(s) in the cell take part in the induced resistance to cytotoxity of the CTXs. The nature of those changes is unkown. One may speculate that phosphorylation, perhaps of a protein of a longer half-life than that of the receptor, is involved in this effect as well. But it also seems possible that the resistance involves the function of some proteins whose synthesis is enhanced by IL-1, TNF and phorbol-diesters. The existence of such inducible "protective proteins" would be consistent with the effective potentiation of the cytoxicity of the CTXs by inhibitors of RNA and protein synthesis.

Similarly to the opposing effects that the CTXs and IL-1 can have on cell viability - cytotoxic effect or induced resistance to killing - these proteins also have diametrical effects of cell growth - being growth stimulatory to certain cells and cytostatic to others. The finding that growth stimulatory agents such as phorbol-diesters and insulin, can also induce the resistance to cytotoxicity raises the possibility that the mechanism by which the CTXs and IL-1 induce such resistance is related to their growth promoting activity - perhaps the consequence of the same induced biochemical changes. It can be further speculated that suppression of such growth and viability-related mechanisms by IFN, as opposed to stimulation of those mechanisms by growth factors, may underly the marked potentiation both of the cytotoxic and of the cytostatic effects of the CTXs in cells which were treated by IFN.

6. ACKNOWLEDGEMENTS
This study was supported in part by grants from Inter-Yeda Ltd.,(Israel) and the NCRD (Israel). We thank Dr. G. Adolf from the Boehringer Institute, Vienna, Austria for the gift of rTNF; Dr. A. Stern and Dr. P.T. Lomedico from the Roche Research Center, Nutley, N.J. for the gift of human rIL-1α and Professor Menachem Rubinstein from the Weizmann Institute of Science for the gift of rIFN-γ.

REFERENCES
1. Wallach D: in Interferon 7 (I. Gresser, ed.) Academic Press (London) pp. 89-122, 1986.
2. Beutler B and Cerami A: Nature 320:584-588 (1986).

3. Wallach D: J Immunol 132:2464-2469, 1984.
4. Hahn T, Toker L, Budilovsky S, Aderka D, Eshhar Z and Wallach D: Proc Natl Acad Sci USA 82:3814-3818, 1985.
5. Holtmann H, Hahn T and Wallach D: Submitted.
6. Tsujimoto M, Yip YK and Vilcek J: Proc Natl Acad Sci USA 82:7626-7630, 1985.
7. Baglioni C, McCandless S, Tavernier J and Fiers W: J Biol Chem 260: 13395-13397, 1985.
8. Holtmann H and Wallach D: Submitted.
9. Aggarwal BB, Essalu TE and Hass PE: Nature 318:665-667, 1985.
10. Tsujimoto M, Yip YK and Vilcek J: J Immunol 136:2441-2444, 1986.
11. Ruggiero V, Tavernier J, Fiers W and Baglioni C: J Immunol 136:2445-2450, 1986.
12. Israel S, Hahn T, Holtmann H and Wallach D: Immunol Lett 12:217-224, 1986.
13. Holtmann H and Wallach D. Submitted.
14. Aderka D, Holtmann H, Toker L, Hahn T and Wallach D: J Immunol 136: 2938-2942, 1986.
15. Onozaki K, Matsushima K, Aggarwal BB and Oppenheim JJ: J Immunol 135: 3962-3968, 1985.
16. Lachman LB, Dinarello CA, Llansa ND and Fidler IJ: J Immunol 136: 3098-3102, 1986.

NEW RELATIONSHIPS AMONG IFN-GAMMA, TNF AND IL-1: CELLULAR AND BIOCHEMICAL EFFECTS[1]

LOIS B. EPSTEIN, RAMILA PHILIP, MAUREEN H. BERESINI AND MARCIA J. LEMPERT
Cancer Research Institute and Department of Pediatrics
University of California, San Francisco, CA 94143 U.S.A.

INTRODUCTION

In recent studies we found differences between the mechanism of enhanced monocyte cytotoxicity observed with IFN-gamma and IFN-alpha and demonstrated new interactions between the monokines, interleukin-1 (IL-1) and tumor necrosis factor (TNF) (1,2). Specifically we observed that pure recombinant human IFN-alpha, IFN-gamma, TNF and natural IL-1 each enhance human monocyte cytotoxicity as determined by a short term ^{51}Cr release assay using actinomycin D-treated WEHI cells as tumor targets. The ability of TNF and IL-1 to enhance cytotoxicity is specific for monocytes, as TNF and IL-1 do not enhance natural killer (NK) cell activity. This is in contrast to IFN-alpha and IFN-gamma which enhance both monocyte and NK cytotoxicity. We also found that IFN-alpha, IFN-gamma and TNF enhance IL-1 production by the monocytes. Enhanced production of IL-1 is not the mechanism whereby IFN-alpha, IFN-gamma, TNF and IL-1 itself enhance monocyte cytotoxicity. Of the four agents (IFN-alpha, IFN-gamma, IL-1, and TNF), only TNF has direct cytotoxic effects on the tumor targets. In experiments in which antibodies to each of the four agents were used during the cytotoxic phase of the monocyte-tumor cell interaction we then went on to prove that TNF alone mediates the enhanced monocyte cytotoxicity observed with IFN-gamma, IL-1 and TNF itself, but not that observed with IFN-alpha. The means by which monocyte cytotoxicity occurs spontaneously or is enhanced by IFN-alpha are yet to be determined.

The natural human IL-1 that we employed in the studies described above is composed of 2 forms, IL-1-alpha and IL-1-beta, in a ratio of 1:9. The recent availability of pure recombinant IL-1-alpha and beta made it possible for us to determine in the present studies if both forms of IL-1 are capable of enhancing monocyte cytotoxicity. Furthermore, like IL-1, IFN-gamma and TNF also share in common the ability to enhance monocyte cytotoxicity. Therefore, we have examined the induction of peptides by IFN-gamma and TNF in human fibroblasts, a well standardized system previously used for the comparison of the effects of IFN-gamma and IFN-alpha (3-5).

MATERIALS AND METHODS

Reagents and Cell Lines

Pure human recombinant IL-1-alpha and IL-1-beta, each having a specific activity of 10^8 u/mg were purchased from Genzyme. Both were negative in the Limulus assay for endotoxin. Pure human recombinant TNF, specific activity 10^8 u/mg and pure human recombinant IFN-gamma, specific activity 3.4×10^7 I.U./mg, were gifts from Dr. H.M. Shepard, Genentech. Both had negligible amounts of endotoxin (<0.8 and <0.125 ng/ml, respectively).

[1]This work was supported by NIH grants CA27903 and HD17001.

WEHI-164, a murine fibrosarcoma cell line obtained from Dr. Noel Warner, Becton Dickinson, was employed (after pretreatment with Actinomycin D) as the targets in the short term ^{51}Cr release monocyte cytotoxicity assays. Fetal lung fibroblasts, strain 153, were derived from a 20 week twin abortus trisomic for human chromosome 21 (6).

Methods

Cytotoxicity assay: Monocytes with >90% purity were isolated by adherence to plastic from Ficoll-Hypaque gradient-separated peripheral blood mononuclear cells (1,2). They were pretreated with either medium, 10 u/ml pure natural IL-1 or recombinant IL-1-alpha or beta for 18 hr, washed and combined for 8 hr with actinomycin D treated WEHI 164 cells in a short term ^{51}Cr cytotoxicity assay with slight modifications (1,2) of a previously described technique (7). Spontaneous, maximum, and specific ^{51}Cr release were determined after counting supernatant aliquots in a gamma counter (1,2).

2-D-gel analyses: Strain 153 human fibroblasts were seeded in microtest culture wells and incubated at 37° for 24 hr in the presence of 125 u Ci/ml L-[U-^{14}C] amino acids (Amersham) in the presence of either medium, 100 or 1000 u/ml TNF, or 100 I.U./ml IFN-gamma. Cell lysates were obtained (6) and 2-D-gel electrophoresis was performed by Protein Data Bases, Inc. according to the procedure of Garrels (8). Autoradiograms were prepared and subjected to computer based analyses by the QUEST System (9).

Antiproliferative assays: Strain 153 human fibroblasts were seeded in microtest culture wells at 10^4 cells/well. After 24 hr at 37°, either medium, TNF (100 or 1000 u/ml) or IFN-gamma (100 I.U./ml) was added. The cultures were incubated for 72 hrs at 37°, washed twice, and stained with Armstrong's dye-fixative, and eluted with ethylene glycol monomethyl ether. The optical density in each well was quantitated at 550 nm. in a Titertek Multiskan, a vertical light path photometer, as described previously (10).

RESULTS AND DISCUSSION

Enhancement of monocyte cytotoxicity by IL-1 alpha and beta

Table 1 depicts the results of 3 experiments in which human monocytes were pretreated for 18 hrs with either medium, or 10 u/ml pure natural IL-1 or recombinant IL-1-alpha or beta, washed and then combined with ^{51}Cr labelled, actinomycin D pretreated WEHI-164 cells for the 8 hr cytotoxic phase of the response. As can be seen from the data, natural IL-1 and both recombinant IL-1-alpha and beta were effective in enhancing monocyte cytotoxicity, all to approximately the same extent.

TABLE 1
Enhancement of Monocyte Cytotoxicity by Natural Human IL-1 and by Recombinant IL-1-Alpha and IL-1-Beta*

Agent Used for the Pretreatment of Monocytes	Cytotoxicity (% specific release of ^{51}Cr)
Medium	17.3 + 1.8 (SEM) **
Natural IL-1	42.0 + 1.2
Recombinant IL-1-alpha	34.7 + 2.6
Recombinant IL-1-beta	38.7 + 1.5

* All reagents used at a concentration of 10 u/ml
** The results depicted represent the mean + SEM of triplicate samples from each of 3 experiments.

We previously observed that natural IL-1 (a mixture of one part IL-1-alpha and 9 parts IL-1-beta) could enhance the cytotoxicity of human monocytes for murine tumor target cells. The present studies, using pure recombinant IL-1-alpha and beta, demonstrate that each has the capacity to enhance monocyte cytotoxicity. Although these molecules have only 27% homology in their amino acid sequence (11,12), they bind to the same receptor (13,14) and in accordance with our findings on enhancement of monocyte cytotoxicity, they appear to subserve the same functions (15).

Effects of TNF and IFN-gamma on polypeptide synthesis and growth of human fibroblasts

Table 2 depicts the overall results of 2 experiments in which the effects of a 24 hr exposure to TNF and IFN-gamma on polypeptide synthesis by strain 153 fibroblasts were assessed, as described in Materials and Methods.

TABLE 2

Effects of TNF and Interferon-Gamma on Polypeptide Synthesis in Human Fibroblasts (Strain 153)

Presence in Cells

Polypeptide Spot Number	Mr	pI	Untreated	Treatment TNF 100u/ml	TNF 1000u/ml	IFN-gamma 100 u/ml
5605(80a)*	79	6.0	−	+	+	++
6408(54)	52	6.3	++	+++	+++	++++
8412	51	6.8	+	+	+	++++
4204(44)	42	5.7	+	++++	++++	++
4105(30)	28	5.7	+	+	+	++
5103	28	6.0	+	+	+	++
8004	22	N.D.	+	++++	++++	++

*Polypeptide spot number designation from our previous publications (Ref. 3-5) are in parenthesis.

Only one polypeptide (5605) was induced de novo by both TNF and IFN-gamma whereas the synthesis of 6 additional polypeptides was enhanced by both of these agents. Peptides 5605, 6408, 8412, 4105, and 5103 were enhanced to a greater extent by IFN-gamma, while synthesis of peptides 4204 and 8004 was increased to a much higher level by TNF. Four of the 24 peptides we described previously (3) to be induced or enhanced in these same fibroblasts by IFN-gamma (i.e., p 80a,54,44,30) are also induced by TNF. This is particularly intriguing in view of the fact that IFN-gamma and TNF differ in their binding sites (16,17), cell of origin (18) and their primary amino acid sequence (19,20).

In parallel with the studies on polypeptide induction in strain 153 fibroblasts by TNF and IFN-gamma we also examined the effects of these agents on fibroblast growth, and the data are depicted in Table 3.

TABLE 3

Effects of TNF and IFN-Gamma on the Growth of Human Fibroblasts (Strain 153)

Agent Used	Concentration (units/ml)	Cell Growth (% of Control)
Medium	-	100
TNF	100	109 + 2 *
TNF	1000	105 + 3
IFN-gamma	100	81 + 2

*Mean + S.E.M. of six replicates. Cultures harvested at 72 hours after the addition of either TNF or IFN-gamma.

We found that growth of the fibroblasts is slightly increased in the presence of TNF, and suppressed in the presence of IFN-gamma.

What then can we say about the significance of polypeptides induced in common by TNF and IFN-gamma in fibroblasts in which these same agents have disparate effects on the growth of the cells? Two hypotheses could be considered. The first is that the peptides induced in common by TNF and IFN-gamma are not involved in the control of growth by these agents, but are involved in as yet unknown, other actions.

The second, and perhaps more tenable, hypothesis takes into account our observation of the non-coordinate induction of each peptide by the 2 agents (i.e., that some proteins are induced to a greater degree by IFN-gamma than TNF, and vice-versa). This suggests the possibility of a threshold effect in growth suppression.

Our studies thus highlight common functional effects of IL-1, TNF and IFN-gamma in enhancing human monocyte cytotoxicity, and common biochemical effects of TNF and IFN-gamma on the induction of polypeptides in human fibroblasts. The latter studies should also prove useful in determining those peptides induced by TNF and IFN-gamma that are responsible for their recently described synergy (21,22). Such studies are now underway in our laboratory.

ACKNOWLEDGEMENTS
    The authors thank Elissa Doty for technical assistance and Eileen Hunt for editorial assistance.

REFERENCES
1. Philip R and Epstein LB: Nature, 323, 86, 1986.
2. Epstein LB and Philip R: in Interferons as Cell Growth Inhibitors and Antitumor Factors (Friedman RF, Merigan T and Sreevalsan T eds.), UCLA Symposia on Molecular and Cellular Biology, New Series, Vol. 50 (Fox CF, ed.) p. 197, Alan R. Liss, Inc., New York, 1986.
3. Weil J, Epstein CJ, Epstein LB et al: Nature 301, 437, 1983.
4. Weil J, Epstein CJ, Epstein LB et al: Antiviral Res. 3, 303,1983.

5. Weil J, Epstein CJ, Epstein LB: Nat. Immun. and Cell Growth Reg. 3, 51, 1984.
6. Weil J, Epstein LB and Epstein CJ: J. Interferon Res. 1, 111, 1980.
7. Chen AR, McKinnon KP and Koren HS: J. Immunol. 135, 3978, 1985.
8. Garrels JI: Meth. Enzymol. 100, 411, 1983.
9. Garrels JI, Farrar JT, Burwell CB: in Two Dimensional Gel Electrophoresis of Proteins (Celis JE and Bravo R, ed.) p.37, Academic Press, Florida, 1984.
10. Epstein LB, McManus NH, Hebert SJ et al: in Methods for Studying Mononuclear Phagocytes (Adams DO, Edelson PJ and Koren H, eds.) p. 619, Academic Press, New York, 1981.
11. Gubler U, Chua AO, Stern AS et al: J. Immunol. 136, 2492, 1986.
12. Auron PE, Webb AC, Rosenwasser LJ et al: Proc. Nat. Acad. Sci. 81, 7907, 1984.
13. Matsushima K, Akahoshi T, Yamada M et al: J. Immunol 136, 4496, 1986.
14. Kilian PL, Kaffka KL, Stern AS et al: J Immonol. 136, 4509, 1986.
15. Perlmutter DH, Goldberger G, Dinarello CA et al: Science 232, 850, 1986.
16. Tsujimoto M, Yip YK and Vilcek J: Proc. Nat. Acad. Sci. 82, 7626, 1985.
17. Baglioni C, McCandless S, Tavernier J et al: J. Biol. Chem. 260, 13395, 1985.
18. Kelker HC, Oppenheim JJ, Stone-Wolff DS et al:Int. J. Cancer 36, 69, 1985.
19. Pennica D, Hayflick J, Bringman TS et al: Proc. Nat. Acad. Sci. 82, 6060, 1985.
20. Gray, PW, Leung DW, Pennica D et al: Nature 295, 503, 1982.
21. Soehnlen B., Liu R and Salmon SE: Proc. Amer. Assoc. Cancer Res. 26, 303, 1985.
22. Mazumder, A: Proc. Amer. Assoc. Cancer Res. 27, 342, 1986.

ANTAGONISTIC EFFECT OF TYPE I INTERFERONS ON HUMAN MACROPHAGE ACTIVATION BY INTERFERON-γ

G. GAROTTA°, K.W. TALMADGE°, J.R.L. PINK°, B. DEWALD+,
M. BAGGIOLINI+
°Central Research Units, F.Hoffmann-La Roche & Co. Ltd. Basle,
+Theodor-Kocher-Institut,University of Bern, Bern, Switzerland

1. INTRODUCTION

IFNγ (type II interferon) has been reported to activate macrophages and to enhance functions related to host defense, i.e. the production of hydrogen peroxide and plasminogen activator, the expression of class I and class II MHC molecules and of receptors for C3b and Fc, and the killing of intracellular parasites and tumor cells [1-4].
IFNsα and IFNß (type I interferons) have a different profile of activity on macrophages, but share some effects with IFNγ, e.g. the induction of class I MHC molecules, Fc receptor expression and plasminogen activator release, and activation for tumor cell lysis [2,3]. With respect to these actions, type I and type II IFNs are additive or even synergistic [3,5-7].
We report here the effects of IFN combinations on one of the key functions of human monocytes and macrophages, H_2O_2 production during the respiratory burst. We found that type I interferons, which by themselves have little influence on H_2O_2 production, strongly antagonize the enhancing effect of IFNγ.

2. METHODS

Monocytes from single donors were purified from blood mononuclear leukocytes by counterflow elutriation [8]. Adherence-selected monocytes were obtained by incubating blood mononuclear cells on plastic and removing non-adherent cells by washing. Monocyte-derived macrophages were obtained by culturing elutriated monocytes for 3 days in medium with 20% of human serum in the absence of IFNs. Monocytes or macrophages were dispensed in flat-bottom 96-well plates (10^5 cells/well) and were cultured for 3 days in the presence of IFNγ alone or in combination with either IFNsα or ß. H_2O_2 production following stimulation with PMA (phorbol myristate acetate) or opsonized zymosan was then determined [9,10]. Human recombinant IFNγ (9.5 x 10^6 U/mg), IFNα2A (2.7 x 10^8 U/mg), IFNß (10^8 U/mg) were produced and purified by F. Hoffmann-La Roche & Co. Ltd., Basle, Switzerland. Human recombinant IFNαA/D (7 x 10^7 U/mg) was produced and purified by Hoffmann-La Roche Inc., Nutley, N.J., USA. Human recombinant IFNαB (2 x 10^8 U/mg), IFNαD (3 x 10^7 U/mg), IFNαF (5 x 10^7 U/mg) were provided by Ciba-Geigy Ltd., Basle, Switzerland. Natural human IFNγ (2.8 x 10^6 U/mg) and IFNα (2.7 x 10^8 U/mg) were

from Interferon Science Inc., New Brunswick, N.J., USA. All
media, sera and IFN preparations were negative for endotoxins
according to the limulus test. Toxic effects on the
mononuclear phagocytes were excluded on the basis of vital
staining and absence of lactate dehydrogenase leakage during
the culture and treatment periods.

3. RESULTS AND DISCUSSION

Pretreatment with IFNγ (10-100 U/ml) markedly increased
(180-316%) the capacity of the mononuclear phagocytes to
produce H_2O_2. The enhancement by IFNγ pretreatment was
similar for H_2O_2 production induced with PMA, which
directly activates protein kinase C [11], or opsonized zymosan
which interacts with surface receptors. This suggests that the
pretreatment with IFNγ affects the H_2O_2-producing
oxidase rather than the process of stimulus-receptor
interaction or stimulus transduction. In contrast, no
enhancement of PMA- or zymosan-induced H_2O_2 production was
observed in any of the mononuclear cell populations following
treatment with IFNsα or IFNβ between 1 and 10^4 U/ml.
Mononuclear phagocytes treated with IFNγ spread and adhered
faster and more firmly than untreated cells, while those
exposed to type I interferons tended to round up.
When IFNγ was supplied to monocytes or macrophages together
with type I IFNs, the enhancement of H_2O_2 production was
prevented (Fig. 1). High concentrations of type I interferons
decreased the yield of PMA-induced H_2O_2 release even below
the control level. Similar results were obtained when
opsonized zymosan was used as stimulus (data not shown). With
ratios of type II to type I IFNs between 1 and 0.1, a
reduction of 50% was observed. Subsequent experiments showed
that type I interferons must be present together with IFNγ
over the whole 3-day period in order to exert maximum
antagonism. When added one or two days after IFNγ, type I
interferons had little influence on the H_2O_2 response of
the phagocytes, and no effect was observed when they were
supplied only a few hrs before PMA stimulation. By contrast,
the enhancement of H_2O_2 production was nearly prevented
when type I IFNs were added together with or one day before
IFNγ. The type I IFNs studied were: natural IFNα,
rIFNα2A, rIFNαB, rIFNαD, rIFNαF, rIFNαA/D and rIFNβ.
All of them lowered stimulus-dependent H_2O_2 release by
human macrophages activated with recombinant or natural
IFNγ. However, rIFNαD and rIFNαF showed less
antagonistic activity than the other type I IFNs tested.
Since type I and type II IFNs bind to different receptors, it
is unlikely that the antagonism shown here is due to a
competition for binding sites [12]. Preliminary experiments
indicate that IFNα does not reduce the expression of IFNγ
receptors. The fact that IFNα or IFNβ had to be added to the
cultures before or together with IFNγ in order to be
effective suggests that type I interferons inhibit the
inductive process by which IFNγ enhances the activity of the
respiratory burst oxidase.
As the mononuclear phagocytes may not represent a homogeneous

cell populations, it is conceivable that the observed
antagonism could be due to complex mechanisms involving
different subsets of macrophages. Two reasons argue against
this possibility. The different mononuclear phagocyte
preparations (monocytes selected by adherence, the total
monocyte population purified by elutriation and macrophages
derived from the latter by culturing) responded in a similar
fashion to IFNs. Secondly, after IFNγ treatment all
mononuclear phagocytes expressed class II MHC determinants and
all these cells were able to produce superoxide as judged by
NBT staining.
Other examples of antagonism between type I and type II IFNs
have been reported recently. IFNα and IFNß counteracted the
induction of mouse macrophage class II antigens by IFNγ,
while IFNγ antagonized type I IFNs in their capacity to
induce mannose-fucose receptor expression [2,7,13]. Our own
investigations on the monocytic cell line U937 showed
enhancement of the expression of class I and class II MHC
molecules with IFNγ, but only of class I molecules with
IFNα. When IFNα and IFNγ were applied together, the
expression of class I molecules was increased while the
density of class II molecules per cell was decreased. In the
same experiments, 77% of the cells treated with IFNγ were
positive for the C3b receptor as compared to 38% in the
controls. When the cells were exposed to combinations of IFNs,
no increase of C3b receptor expression was observed.
This and other reports on antagonism or synergism between type
I and type II IFNs suggest that the relative local
concentration of IFNs could control some host defense
functions [14]. The effects discussed are summarized in
Table I. A prevalence of INFγ would be expected to enhance
macrophage activity, while a prevalence of IFNα or IFNß
appears to enhance the capacity of macrophages to kill tumor
cells at the expense of antimicrobial functions. Phagocytic
activity (C3b receptor expression), microbicidal capacity
(H_2O_2 production) and antigen presentation (class II MHC
molecule expression), which are enhanced by IFNγ, are
depressed by type I IFNs. On the other hand, the anti-cellular
activities of macrophages, which are related at least in part
to class I MHC molecule expression, Fc receptor expression and
plasminogen activator production, are enhanced by type I
interferons in synergism with IFNγ.

REFERENCES

1. Svedersky LP, Benton CV, Berger WH, Rinderknecht E,
   Harkins RN and Palladino MA. J.Exp.Med. 159: 812, 1984.
2. Alan R, Ezekowitz B, Hill M and Gordon S.
   Biochem.Biophys.Res.Commun. 136: 737, 1986.
3. Vogel SN and Friedman RM. Interferon, Vol. 2:
   Interferons and the immune system, J Vilcek and
   E DeMayer, Eds, Elsevier Publishers , 1984.
4. Nathan CF, Murray HW, Wiebe ME and Rubin BY.
   J.Exp.Med. 158: 670, 1983.

240

5. Oleszak E and Stewart WE. J.Interferon Res. 5: 361, 1985.
6. Pace JL, Russel SW, Le Blanc PA and Murasko DM.
   J.Immunol. 134: 977, 1985
7. Ling PD, Warren MK and Vogel SN. J.Immunol. 135: 1857,
   1985.
8. Leigh PCJ, van den Barselaar MT, Daha MR and
   van Furth R. J.Immunol. 129: 332, 1982.
9. Pick E and Mizel D. J.Immunol.Methods 46: 211, 1981.
10. Ruch W, Cooper PH and Baggiolini M. J.Immunol.Methods
    63: 347, 1983.
11. Castagna M, Takai Y, Kaibuchi K, Sano K, Kikkawa U
    and Nishizuka Y. J.Biol.Chem. 257: 7847, 1982.
12. Orchansky P, Novick D, Fischer DG and Rubinstein M.
    J.Interferon Res. 4: 275, 1984.
13. Inaba K, Kitaura M, Kato T, Watanabe Y, Kawade Y and
    Muramatsu S. J.Exp.Med. 163: 1030, 1986.
14. Bocci V. Immunology Today 6: 7, 1985.

Figure 1

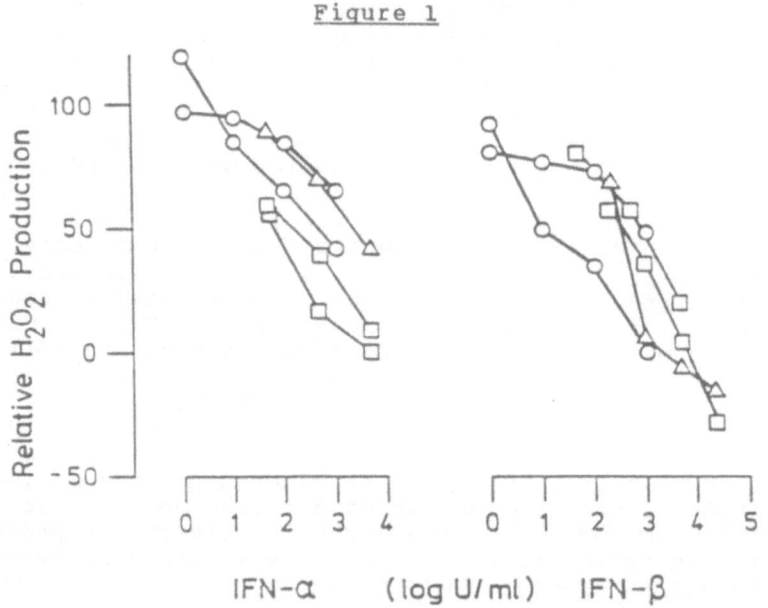

IFN-α   (log U/ml)   IFN-β

The curves represent the relative production of H_2O_2 on a
scale extending between 0 (value in absence of IFNs) and 100
(value with IFNγ alone) at different concentrations of
IFNα or IFNß. H_2O_2 production (pmol/hr per 10^3 cells,
mean ± SD) of control cultures stimulated with 50 nM PMA was
7 ± 1; in cultures treated with 50 U/ml IFNγ it was
57 ± 4. Each curve refers to cells from different donors.
Adherent monocytes treated with 50 (△) or 150 (□) U/ml of
IFNγ, elutriated macrophages treated with 50 U/ml IFNγ (○).

Table 1

MODULATION OF MACROPHAGE FUNCTIONS BY IFNs

Macrophage Treatment	Release of			Tumor Killing	Expression of				
	O_2^- (1)	H_2O_2 (1)	Plasminogen Activator		class I	class II	RC3b	RFc	R man-fuc
IFNγ	+	+	+	+	+	+	+	+	0
IFNα/β	0	0	+	+	+	0	0	+	+
IFNγ+IFNα/β	-	-	++	++	++	-	-	++	-
References:		2		6	3	7,13	3	3	2

(1) after PMA or zymosan stimulation

+ induction; ++ synergism or additivity; - antagonism; 0 no effect.

IFNγ and IFNα/β show antiviral and antiproliferative activity. They synergize when given in combination to fibroblasts or tumor cell lines.

GAMMA-INTERFERON REGULATES SECRETION OF G-CSF IN HUMAN
MONOCYTES ON THE TRANSCRIPTIONAL LEVEL.

W. Oster, A. Lindemann, R. Mertelsmann and F. Herrmann.
Dept. of Hematology, University of Mainz, FRG.

The production of colony stimulating factors (CSF) for granulocytes and monocytes is integrated into a network of communi-
cating soluble messenger molecules resulting from T-cell/mono-
cyte interactions. We assessed the capactiy of gamma-Interferon to modulate monocyte secretion of CSF by colony assays
and Northern blot analysis to hybridize monocyte RNA with cDNA
probes of different CSF-types. Whereas mRNA for GM-CSF was undetectable in untreated and gamma-IFN treated peripheral blood
monocytes, the constitutive expression of mRNA for G-CSF and
subsequent production of a CSF with biological activities
similar to G-CSF could highly be enhanced by exposure of monocytes to gamma-IFN.

INTRODUCTION

In vitro survival, proliferation, and differentiation of
myeloid progenitor cells into mature granulocytes and macrophages is dependant on the presence in culture of growth
factors, termed colony-stimulating factors (CSF) (1).
Colony stimulating factors are glycoproteins produced by T-
cells, macrophages, endothelial cells and fibroblasts in response to activation stimuli (2-5). In man, several CSF-species
have now been purified, sequenced and cDNA cloned including
CSF for granulocytes/macrophages (GM-CSF), for granulocytes
(G-CSF) and macrophages (M-CSF) (6-10). However, little is
known about the respective cellular origin of these CSF-
species. In a previous report by our group, gamma-Interferon
(gamma-IFN) was shown to recruite macrophages to produce CSF
(11). The present work extends our earlier observation to show
that macrophage derived CSF, induced by gamma-IFN has the
quality of a G-CSF. Unlike GM-CSF and M-CSF and similar to G-
CSF (9), CSF derived from macrophages in response to gamma-IFN
failed to support eosinophil colony growth of highly purified
myeloid progenitor cells but induced terminal differentiation
of murine myelomonocytic leukemia line WEHI3BD+. Whereas mRNA
for GM-CSF was undetectable in uninduced and gamma-IFN induced
peripheral blood derived monocytes (unpublished observations),
the constitutive expression of mRNA for G-CSF in monocytes
could be highly amplified by exposure of monocytes to gamma-
IFN. This effect of gamma-IFN was restricted to monocytes, as
secretion and gene expression of G-CSF from T cells was not
inducible with gamma-IFN.
These results point out a previously unknown function of

gamma-IFN in regulating macrophage G-CSF release on the trans-
scriptional level.

MATERIALS AND METHODS

<u>Preparation of monocytes and T-cells:</u> Peripheral blood derived
mononuclear cells were isolated from leukapharesis residues of
healthy volunteer donors by Ficoll-Hypaque gradient centrifugation. T-cells were recovered by rosetting with AET-treated
SRBC (5% v/v solution); monocytes were separated by repeated
adherence steps of the E rosette negative fraction. Purity of
the individual cell fractions, assessed by morphology, cytochemistry and immunofluorescence analysis with monoclonal
antibodies to the T11, MO2, and B1 antigens revealed preparations consisting of > 98% monocytes and T-cells respectively.
<u>Preparation of monocyte- and T-cell-conditioned medium:</u> Mono-
cytes or T-cells were plated at 10^5 cells/ml in RPMI1640 supplemented with 10% FCS, penicillin/streptomycin, L-glutamine,
and sodium pyruvate in plastic petri dishes (35 mm/well). Recombinant gamma-IFN (Genentech, San Francisco, CA, kindly
provided by Dr. Flener, Böhringer, Vienna, Austria) was added
to the cultures at 100 U/ml. In control cultures the addition
of gamma-IFN was omitted. Cellfree supernatants collected at
day 5 of cultures were treated at room temperature for 2 hr
with monoclonal antibody to gamma-IFN previously described
(12) at 300 neutralyzing U/ml to exclude any effect of inhibition of growth by gamma-IFN in target assay for colony formation. Numbers of monocytes cultured, length of the culture
period, and concentration of gamma-IFN used were selected for
optimal monocyte CSF secretion based on our previously published experiments (11).
<u>Northern blot analysis:</u> Monocytes or T-cells (10^5 /ml) were
cultured as described above. Prior to addition of gamma-IFN to
the culture and 6 hr or 12 hr later each cell fraction was
harvested and total RNA was extracted using standard procedures (13) followed by glyoxilation and fractionation in a
0,8% autoclaved agarose gel (10ug/well). RNA was transferred
by capillary blotting to a nytran membrane and baked for 2 hr
at 80 ° C. Hybridization of prehybridized filters was performed
for 4 hr at 58° C. in 6 x NET, 10 x Denhardt`s 0,1% SDS con-
taining a complementary G-CSF oligonucleotide probe (kindly
provided by Dr. Blohm, BASF, Ludwigshafen, FRG) labeled
with p 32-gammaATP using T4 oligonucleotide kinase.
The filters were washed three times for 10 min with 6 x SSC at
room temperature and exposed to Kodak XAR-5 films with 2
Dupont lightning plus intensifier screens at -70° C. In additional experiments, filters were washed in 5mM EDTA at 65 C.
for 15 min and at 80° C. for 10 min followed by rehybridiza-
tion with a cDNA-probe for the constant region of the alpha-
chain of the T-cell antigen receptor (kindly provided by Dr.
Royer, DFCI, Boston, MA) labeled with p32 alpha d CTP using
previously described procedures (14).
<u>Purification of hematopoietic progenitor cells:</u> Normal bone
marrow was obtained by aspiration from the iliac crest into
sterile heparin after obtaining informed consent. The mono-

nuclear cell fraction was separated by Ficoll-Hypaque gradient centrifugation. Progenitor cells were enriched 80 - 100 fold by an immune rosette technique previously described (15). Monoclonal antibodies used in the purification of progenitor cells included anti-MY8 (granulocytic cells as immature as promyelocytes), -MY4, -MO2 (monocytes), -39 (nucleated red blood cells), -T3, -T11 (T-cells), -B1, -B4 (B-cells), NKH1 (NK-cells), -N901 (NK-cells, basophils), and -J5 (common ALL antigen). The mononuclear cell fraction was simultaneously depleted of cells positive for MY8, MY4, MO2, 39, T3, T11, B1, B4, NKH1, N901 and J5 by forming rosettes with SRBC covalently coupled to affinity-purified rabbit anti-mouse immunoglobulin. Rosetted cells were separated from unrosetted cells by density gradient centrifugation. Unrosetted cells (containing the progenitor cells) were then subjected to a second rosetting step involving selection of HLA-DR positive cells with I2 monoclonal antibody to improve the enrichment of myeloid progenitor cells. All antibodies used in this study were extensively characterized before (for ref. see 16).

Assay for myeloid progenitor derived colonies: Colonies derived from myeloid progenitor cells purified as described above were assayed in agar (Agar Noble, Difco, Detroit, MI) by a standard double layer method (15). Underlayers (0,5 ml) were composed of 0,5% agar in Iscoves modified Dulbeccos MEM (IMDM) supplemented with penicillin/streptomycin, L-glutamine and 20% FCS. As a source of colony stimulating activity either recombinant GM-CSF (2 ug/ml; Immunex, Seattle, WA, kindly provided by Dr. Krumwieh, Behring, Marburg, FRG) HPLC-puri-fied G-CSF (500 U/ml; kindly provided by Dr. K. Welte, Sloan-Kettering Cancer Center New York, NY) or 15% media conditioned by monocytes or T-cells as described above, was added to the underlayer. The overlayer (0,5 ml) consisted of 0,3% agar in the same medium and contained $2,5 \times 10^3$ purified myeloid progenitor cells. The cultures were set up in quadruplicates in 24 well plastic culture plates (Costar, Cambridge, MA) and were incubated at $37°$ C. in 5% CO2 and humidified air. After 14 days of culture, overlayers were removed from underlayers by agitation and were dried onto glass slides under filter paper. To determine colony lineage, the dried cultures were fixed in acetone-methanol-sodium citrate fixative and stained for naphthol AS-D chloracetate esterase (CAE; granulocytic colonies), alpha naphthyl acetate esterase (ANAE; monocytic colonies) and luxol fast blue (LFB; eosinophil colonies).

Assay for WEHI3BD+ colonies (leukemia differentiation induction: 3×10^2 WEHI3BD+ cells (kindly provided by Dr. Williams, Sloan-Kettering Cancer Center, New York, NY) were incorporated into an identical clonal assay system as described above for purified normal bone myeloid progenitor cells. Cultures were harvested at day 7 and scored for induction of desperse, differentiated colonies or tight, blast cell colonies by recombinant GM-CSF, HPLC purified G-CSF and gamma-IFN induced and uninduced monocyte and T-cell conditioned medium.

RESULTS

Effect of gamma-IFN on monocytes and T-cells to release CSF:

To assess the effect of gamma-IFN on release of CSF from monocytes and T-cells, purified population of each fraction (10^5
/ml) were allowed to condition media for 5 days in the presence of absence of 100 U/ml of recombinant gamma-IFN. Gamma-
IFN induced monocyte and T-cell conditioned media were pretreated with a monoclonal antibody to gamma-IFN at neutralyzing concentrations to avoid the effects of carry-over of
gamma-IFN on growth of myeloid progenitor cells. Control
cultures (not shown) containing appropriate amounts of the
anti gamma-IFN monoclonal antibody were indistinguishable from
control cultures without these reagents. Media contitioned by
monocytes were effective in promoting colony growth of purified myeloid progenitor cells at a final concentration of 15%,
when gamma-IFN was present during the culture period, whereas
CSF release of monocytes cultured in the absence of gamma-IFN
was neglegible (Table 1A). To determine if gamma-IFN induced
CSF-release was derived from monocytes or possibly contamina-
ting-T-cells, experiments were performed in which the effects
of gamma-IFN on CSF-induction in resting T-cells was investigated. Under the experimental conditions described, resting T-
cells were unable to release CSF in their culture supernatants
neither constitutively nor induced by gamma-IFN (Table 1A).

Table 1: Presence of biological active CSF with G-CSF properties in gamma-IFN induced mono-
cyte-(gamma-IFNMOCM) and T-cell condition medium (gamma-IFNTCM).

A)   Failure of induction of eosinophil colony growth by gamma-IFNMOCM and HPLC purified
G-CSF.

Source of CSF	Colony number at day 14/2,5 x 10^3 NBM myeloid progenitor cells		
	CAE + (gran.)	ANAE + (mono.)	LFB + (eosinophil)
HPLC p G-CSF (500 U/ml)	150 + 9 *	18 + 2	0
rGM-CSF (2 μg/ml)	88 ∓ 6	45 ∓ 3	16 + 3
gamma-IFNMOCM (15 % v/v)	98 ∓ 5	20 ∓ 3	0
MOCM (15 % v/v)	6 ∓ 2	2 ∓ 1	0
gamma-IFNTCM (15 % v/v)	0	0	0
TCM (15 % v/v)	0	0	0
Medium	0	0	0

B)   Leukemia differentiation induction by gamma-IFNMOCM and HPLC p G-CSF

Source of CSF	Colony number at day 7/3 x 10^2 WEHI3BD+ cells	
	blast colonies	differentiated colonies
HPLC p G-CSF (500 U/ml)	8 + 2	163 + 4
rGM-CSF (2 μg/ml)	169 ∓ 4	10 ∓ 1
gamma-IFNMOCM (15 % v/v)	7 ∓ 1	159 ∓ 4
MOCM (15 % v/v)	159 ∓ 4	19 ∓ 3
gamma-IFNTCM (15 % v/v)	170 ∓ 4	11 ∓ 1
TCM (15 % v/v)	170 ∓ 3	12 ∓ 2

* Values are expressed as colony number of quadruplicate culture ± SD of one representative experiment.
Three additional experiments gave comparable results.

CSF produced by gamma-IFN treated monocytes has properties of
G-CSF: To identify the species of CSF present in gamma-IFN
induced monocyte conditioned medium (gamma-IFN MOCM) we
compared the effects of gamma-IFN MOCM with the known
biological activities of GM-CSF, i. e. the induction of
eosinophil colonies, and those of G-CSF, i. e. the capacity to
induce differentiation of the murine myelomonocytic line WEHI
3BD+ (9). To this end, lineages of normal bone marrow derived
myeloid colonies (Table 1A) and numbers of blast colonies and
differentiated colonies of the WEHI 3BD+ line (Table 1A)
induced by gamma-IFN MOCM, recombinant GM-CSF and G-CSF purified using sequential ammonium sulfate precipitation, anion
exchange dermatography, gel filtration and RP-HPLC (17)
(HPLCpG-CSF) were determined by cytochemistry and Wright
Giemsa morphology. Unlike GM-CSF but similar to G-CSF, gamma-
IFN MOCM failed to induce eosinophil colonies (Table 1A), but
was able to promote differentiation of the WEHI 3BD+ leukemia
(Table 1A).
Northern blots experiments confirmed the presence of a G-CSF
message in monocytes being highly amplified after gamma-IFN
treatment for 6 hr (Fig. 1), whereas resting T-cells failed to
express a G-CSF mRNA (Fig. 2). Consequently neither colony
stimulating activity (Table 1A) nor leukemia differentiation
activity (Table 1B) could be detected in media conditioned by
untreated and gamma-IFN induced resting T-cells.

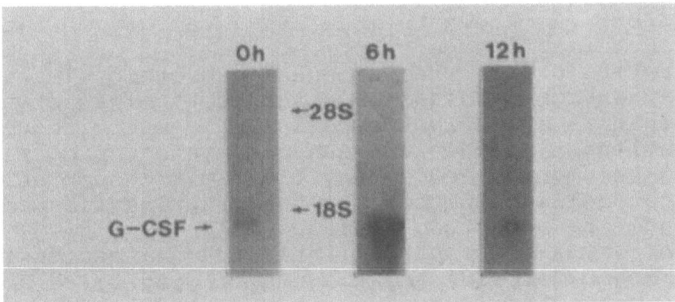

Fig. 1: Kinetics of G-CSF mRNA expression in monocytes:
        RNA was obtained from purified monocytes incubated
        with gamma-IFN (100 U/ml) for different time periods.
        Maximal G-CSF transscription was detectable at the 6 h
        time point.

248

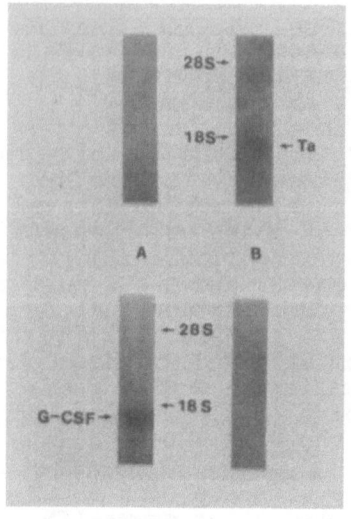

Fig. 2:
Absence of G-CSF m RNA expression in resting T-cells: RNA from purified monocytes that were incubated with gamma-IFN (100 U/ml) for 6 hr are lacking a signal for T alpha whereas a G-CSF messenger RNA of 1,6 KB is detectable (lanes A). RNA from identically treated T-cells characterized by the T alpha mRNA of 1,7 KB does not contain transscribed G-CSF-RNA (lanes B).

DISCUSSION

Gamma-Interferon is known to modulate a variety of functions of the monocyte/macrophage system.
Included are its capacity a) to enhance protein synthesis and membrane expression of class II HLA-DR determinants on monocytic progenitor- and effector cells (18); b) to induce differentiation of normal and leukemic myeloid progenitor cells along the monocytic pathway (19); c) to suppress growth of progenitor cells committed for the granulocytic and (to a lower extend) the monocytic lineage (15).
In addition, gamma-IFN enhances interleukin-1 secretion of LPS challenged monocytes (20) and induces binding sites for interleukin 2 in monocytes (19).
The present report provides evidence that T-cells may recruit monocytes to cooperate in the secretion of G-CSF via gamma-IFN as a messenger molecule. Although detection of CSF-activity was neglegible (albeit present) in MOCM generated in the absence of gamma-IFN, release of G-CSF by monocytes could be highly enhanced by gamma-IFN. Enhanced expression of mRNA for G-CSF in gamma-IFN treated peripheral blood monocytes suggests

that the biological activity detectable in gamma-IFN MOCM is
G-CSF itself, rather than an activity potentiating colony
formation and leukemia differentiation that is being released
from other accessory cells present in the marrow preparations
or the leukemic cells themselfes in response to gamma-IFN.
Particularly T-cells known as a mayor source of CSF under
conditions of T-cell activation (2) failed to express G-CSF
mRNA and consequently to release G-CSF in the presence or absence of gamma-IFN.
Thus, the data presented here, point out a previously unknown
function of gamma-IFN to modulate myeloid blood cell production by transcriptional regulation of monocyte G-CSF release.

Supported by a grant from the Deutsche Forschungsgemeinschaft
(He 1380-2/1).

REFERENCES

1. METCALF D.: The granulocyte-macrophage colony stimula-
    ting factors: Science 229:16, 1985
2. GRIFFIN J.D., MEUER S.C., SCHLOSSMAN S.F., REINHERZ E.L.
    T-cell regulation of myelopoiesis: analysis at a clonal
    level. J. Immuno. 133:1863, 1984
3. BAGBY G.C., RIGAS V.D., BENNETT R.M., VANDENBARK A.A.,
    GAREWAL H.S. Interaction of lactoferrin, monocytes and T
    lymphocyte subsets in the regulation of steady-state granu-
    lopoiesis in vitro. J. Clin. Invest. 68:56, 1981
4. BAGBY G.C., McCALL E., BERGSTROM K.A., BURGER D. A mono-
    kine regulates colony stimulating activity production by
    vascular endothelial cells. Blood 62:663, 1983
5. BAGBY G.C., McCALL E., LAYMAN D.L. Regulation of colony
    stimulating activity production. Interaction of fibro-
    blasts, mononuclear phagocytes, and lactoferrin. J. Clin.
    Invest. 71:340, 1982
6. CANTRELL M.A., ANDERSON D., CERETTI D.P. et al. Cloning,
    sequence, and expression of a human granulocyte/macrophage
    colony-stimulating factor. Proc. Natl. Acad. Sci. USA
    82:6250, 1985
7. WONG G.G., WITEK J.S., TEMPLE P.A., et al. Human GM-CSF:
    Molecular cloning of the cDNA and purification of the na-
    tural and recombinant proteins. Science 228:810, 1985
8. NAGATA S., TSKCHIYA M., ASANO S., et al: Molecular cloning
    and expression of cDNA for human granulocyte colony
    stimulating factor. Nature 319:415, 1986
9. SOUZA L.M., BOONE T.C., GABRILOVE J., et al: Recombinant
    human granulocyte colony-stimulating factor: Effects on
    normal and leukemic myeloid cells. Science 232:61, 1986
10. KAWASAKI E.S., LADNER M.B., WANG A.M., et al. Molecular
    cloning of a cDNA encoding human macrophage-specific
    colony stimulating factor. Science 230:291, 1985.
11. HERRMANN F., CANNISTRA S.A., GRIFFIN J.D., T cell-monocyte
    interactions in the production of humoral factors regula-
    ting human granulopoiesis in vitro. J. Immunol. 136:2856,
    1986.

12. HERRMANN F., SCHMIDT R.E., RITZ J. GRIFFIN J.D. Effect of human NK cells on regulation of hematopoiesis in vitro: Analysis at the clonal level. Blood, in press, 1986
13. MANIATIS T., FRITSCH G.F., SAMBROOK J. Molecular cloning, a laboratory mannal. Cold Spring Harbor laboratory publication. 1982.
14. FEINBERG A.P., VOGELSTEIN B. A technique for radiolabeling DNA restriction endonuclease fragments to high specific activity. Anal. Biochem. 132:6, 1983
15. HERRMANN F., GRIFFIN J.D., MEUER S.C., MEYER ZUM BÜSCHENFELDE K.-H. Establishment of an interleukin-2 dependent T cell line derived from a patient with aplastic anemia, which inhibits in vitro hematopoiesis. J. Immunol. 136:1629, 1986.
16. Leukocyte Typing. Human Leukocyte Differentiation antigens detected by monoclonal antibodies. A. Bernard, L. Noumsell, J. Dausset, C. Milstein. S. Schlossman (eds.), Springer, Berlin, 1984.
17. WELTE K., PLATZER E., LU L., et al. Purification and biochemical characterization of human pluripotent hematopoietic colony-stimulating factors. Proc. Natl. Acad. Sci. USA. 82:1526, 1985.
18. GRIFFIN J.D., SABBATH K.D., HERRMANN F., et al. Differential expression of HLA-DR antigens in subsets of human CFU-GM. Blood 66:788, 1985
19. HERRMANN F., CANNISTRA S.A., LEVINE H., GRIFFIN J.D. Expression of interleukin-2 receptors and binding of interleukin-2 by gamma-Interferon induced leukemic and normal monocytic cells. J. Ex. Med. 162:1111, 1985.
20. ARENZANA-SEISDEDOS F., VIRELIZIER J.L., FIERS W. Interferons as macrophage activating factors. III. Preferential effects of Interferon-gamma on the IL-1 secretory potential of fresh or aged human monocytes. J. Immunol. 134:2444, 1985.

PRESENCE OF TUMOR NECROSIS FACTOR AND LYMPHOTOXIN IN CLINICAL INTERFERON
PREPARATIONS DERIVED FROM LEUKOCYTES

D. WALLACH[1], K. CANTELL[2], S. HIRVONEN[2], L. TOKER[1], D. ADERKA[1], and H.
HOLTMANN[1]. [1]The Weizmann Institute of Science, Rehovot, Israel, [2]National
Public Health Institute, Helsinki, Finland.

1. SUMMARY

Crude preparations of IFN-γ, produced by stimulating human leukocytes
for 3 days with lentil-lectin were found to contain significant amounts of
CTXs, at the range of 10^5-10^6 cytotoxic units/10^6 U of IFN-antiviral-activ-
ity. About 60 to 80% of that activity could be neutralized with antiserum
to TNF and the rest with antiserum to LT. The amounts of TNF, as estimated
by immunoassay were at the range of micrograms/10^6 IFN U. Following enrichment of the IFN by CM-Sephadex ion-exchange chromatography, the preparations still contained TNF at the range of 10^3-10^4 U/10^6 IFN U. However, on
further purification of the IFN with the use of a monoclonal antibody to
the protein the CTXs were effectively eliminated. Crude preparations of
IFN-α, induced with Sendai virus, were found to contain CTXs at the range
of 10^2-10^4 U/10^6 IFN U. Most (>90%) of that activity could be neutralized
with antiserum to TNF. Concentrated, non-purified preparations of the IFN,
used in initial clinical trials (Strander et al. J. Nat. Cancer Inst. 1973:
51,722-742) also contained TNF at levels reaching more than 10 ng per 10^6
IFN U, while in partially purified preparations of the IFN (Cantell et al.
Method. Enzymol 1981:78,499-505) CTXs could not be detected by bioassay nor
by the immunoassay for TNF.

2. INTRODUCTION

Studies on the therapeutic potential of IFN have, so far, centered on
elucidating the activity of IFN in its purified state: Isolating the
various IFN species, developing the technology for producing large amounts
of the proteins and testing the effect of such purified preparations in
various diseases. Yet, in exploring, to date, ways to extend the use of
IFN to kinds of applications at which, until now, it has proved only partially effective, it seems necessary to go back to the crude preparations
where the IFNs are "contaminated" with various other, simultaneously induced, cytokines. Since the formation of such cytokines is linked to the
production of IFN it is likely that their functions are also related. Use
of IFN together with such cytokines may, therefore, prove more physiologically relevant and perhaps also more effective than the use of IFN alone.

Studies in recent years on the cytotoxines (CTXs)-tumor-necrosis-factor
(TNF) and lymphotoxin (LT)- revealed close interrelations between the formation and function of these cytokines and those of IFN. IFN was found to
enhance production of the CTXs, and to potentiate their cytostatic effects
as well as their cytotoxic effect on tumor and virus infected cells
(reviewed in 1). Furthermore, the CTXs induce in certain cells the production of IFN (2). We therefore found it of interest to determine whether
CTXs are present in clinical IFN preparations, induced in leukocytes.

In crude preparations of both IFN-α and IFN-γ significant amounts of CTXs
could be detected: preparations of IFN-α containing primarily TNF, and

preparations of IFN-γ both TNF and LT. In the purification procedures, presently in use, those CTXs were effectively eliminated. However, TNF could still be detected in concentrated preparations of IFN-α used for clinical trials in the past.

3. MATERIALS AND METHODS

Quantitation of cytotoxic activity. Cytotoxic activity of the IFN preparations was determined with SV80 cells (3) as previously described (4). Samples to be tested were applied in serial dilutions together with cyclo-heximide (CHI, 50 μg/ml) on confluent monolayers of the cells grown in 9 mm microwells. Following 20 hr incubation, viability of the cells was deter-mined by measuring uptake of neutral-red. A unit of cytotoxic activity is defined as the amount which, under conditions of the assay, caused killing of 50% of the cells.

Antibody neutralization of the CTXs. Rabbit antiserum to TNF, raised in our laboratory, which neutralized 10 U of TNF/ml at a dilution of 1:1000 and rabbit antiserum to LT (kindly provided by Dr. G.R. Adolf of the Ernst-Boehringer Institut fur Arzneimittelforsch. Vienna, Austria),found to neutralize the cytotoxic activity of 1000 ng/ml LT at a dilution of 1:1000, were both used at a dilution of 1:100. Samples of the tested CTXs were incubated with either one of the antisera as well as with the two antisera together, first for 2 h at 37°C and then for a further 12 hr at 4°C and their cytotoxicity was then titrated on the SV80 cells and compared to that of a sample of CTXs which was incubated under the same conditions in the absence of antisera.

Immunoassay for TNF. An ELISA for TNF, based on the combined use of a high affinity mouse monoclonal antibody and of rabbit polyclonal antiserum against the protein was carried out as will be described in detail else-where. The results were standardized by comparison to a purified prepara-tion of rTNF, kindly provided by Dr. Leo Lin of the Cetus Corporation, Emersville, Calif.

IFN preparations. Samples of clinical IFN preparations of the following kinds were examined: Preparations of IFN-α induced with Sendai virus (5). Concentrated (6) and partially-purified (7) preparations of IFN-α. Prepara-tions of IFN-γ induced by 3 day stimulation with lentil-lectin (8). Samples obtained from various steps of purification of IFN-γ, carried out as described in (8) as well as samples of the IFN, further purified by immuno-affinity on a monoclonal antibody to IFN-γ.

4. RESULTS

Presence of CTXs in cytokine preparations which contain also IFN can be sensitively quantitated by measuring cytotoxic effects of such preparations on cells treated with inhibitors of RNA or protein synthesis. Such inhibi-tors block the response to IFN and, simultaneously, potentiate the cytoto-xicity of the CTXs. Thus for determining the level of CTXs in crude prepa-rations of IFN-α and γ and the amounts of CTXs remaining in various steps of purification of the IFNs, cytotoxic effects of these preparations were titrated in presence of cycloheximide. Representative results from testing several different preparations of each kind are presented in Table I.

Crude prearations of IFN-α, induced with Sendai virus in human peripheral blood leukocytes (5), were found to contain significant amounts of CTXs, ranging from a few to several hundreds of units per ml. Practically all that cytotoxic activity could be neutralized with antiserum to TNF nad none with antiserum to LT. Presence of TNF in such preparations was further

TABLE I.   Content of Cytotoxins (CTXs) in Preparations of IFN-α and IFN-γ

	IFN antiviral U/ml	CTXs cytotoxic U/ml	CTXs/IFN cytotoxic U/10^6 antiviral U	TNF ng/ml	Relative contribution of TNF and LT to cytotoxic activity
IFN-α					
Crude preparations (5)	8.10^4	300	3750	N.D.	TNF ⩾ 90%
	8.10^4	15	300	4.5	
	8.10^4	10	125	4.4	
Concentrated preparations produced as described in (6)	2.10^6	65	32	20	TNF ⩾ 90%
	$1.5 \cdot 10^6$	22	15	15	
	2.10^6	15	7.5	30	
Partially-purified preparations (7)	6.10^6	< 1	< 1	< 0.1	
	6.10^6	< 1	< 1	< 0.1	
IFN-γ					
Crude preparations	7.10^3	6400	910,000	20	TNF 65% LT 20%
	6.10^3	2800	470,000	11	TNF 65% LT 30%
Preparations concentrated with PEG and $(NH_4)_2SO_4$ (8)	6.10^4	180	3,000	N.D.	
	9.10^4	180	2,000	21	
Partially-purified preparations (on CM-Sephadex (8))	2.10^6	18,000	9,000	N.D.	
	2.10^6	43,000	21,000	N.D.	
Purified (on a monoclonal antibody)	1.10^6	< 1	< 1	< 0.1	

confirmed with an enzyme-linked immunosorbent assay (ELISA) for this protein.

Some amounts of TNF at the range of tens of units of cytotoxic activity and tens of picograms of immunologically detectable TNF per ml, were also found in concentrated preparations of IFN-α, produced as described by Strander et al. (6). Six of these IFN preparations, which in the past were in use for clinical trials, were examined and were all found to contain TNF at similar levels. On the other hand, in preparations of IFN-α partially-purified as described by Cantell et al. (7) no CTX could be detected by the cytotoxicity assay nor could TNF be detected in those preparations by the ELISA for the protein.

Crude preparations of IFN-γ, induced in human peripheral-blood leuko-cytes by three-days stimulation with lentil-lectin (8), were found to con-tain significant amounts of CTXs, at the range of thousands of units per ml. About three quarters of that cytotoxicity could be neutralized with anti-serum to TNF and the rest with antiserum to LT. The concentration of immunologically detectable TNF in these preparations, as determined by the ELISA, were at the range of tens of picograms per ml. Isolation of IFN from these preparations was carried out essentially in three steps: Con-centration with the use of polyethylenlglycol (PFG) and ammonium sulphate, followed by fractionation on carboxy methyl cellulose (8) and then by affinity purification on a monoclonal antibody to the IFN. The first two steps of purification resulted in only partial removal of the CTXs. Inter-estingly, in concentrating IFN-γ with PEG and $(NH_4)_2SO4$, which involved precipitation of the IFN in PEG followed by partitioning the proteins between two phases formed in presence of PEG and $(NH_4)_2SO4$, the CTXs fractionated somewhat differently than the IFN. Like IFN, the CTXs were effectively precipitated in 30% PEG, however in the ensuing partitioning step (8) a major part of the CTXs (about 75%) partitioned to the upper phase while IFN and most other proteins, present in the preparations, could be quantitatively recovered from the lower phase. Still, the final product of this con-centration-procedure was found to contain a significant amount of CTXs; about 10% of that initially present in the crude IFN preparations. Most of that CTX was found to elute together with IFN in its purification by a step-wise fractionation on carboxy methyl cellulose (8). However, in further purifying the IFN, on immunosorbent constructed from a monoclonal antibody to the protein, the CTXs were effectively eliminated.

5. DISCUSSION

Preparations of IFN-α and preparations of IFN-γ are both found to con-tain CTXs. Indeed, agents applied as inducers of these IFNs are known to induce also the production of CTXs. Sendai virus, used as inducer of IFN-α is a potent inducer of TNF in mononuclear phagocytes (9), while T cell mito-genic agents, such as lentil-lectin, which are used as inducers of IFN-γ, induce also CTXs in preparations of mononuclear leukocytes which contain lymphocytes (10). Even though lectin-induced CTX preparations have initially been termed "lymphotoxins" it is now known that such preparations contain significant amounts of TNF. Actually, in the first day of lectin-stimulation it is primarily TNF which is induced and very little of the CTX now called lymphotoxin (LT) can be detected (9,11). Production of LT does increase, however, at longer stimulation (11) and in preparations induced by stimulation for a few days, like those of the IFN-γ examined in this study, LT may indeed constitute a significant portion of the CTXs (12).

Presence of CTXs in the crude preparations of IFN is a challenge for those purifying IFN from such preparations. The CTXs are potent regulatory agonists which, at even minute concentrations, can elicit a variety of biological responses, several of which are potentiated by IFN, so that in order to examine the function of IFN alone one has to assure effective depletion of the CTXs from the preparations.

It should be noted that bioactivity of the CTXs may give only a partial indication for the actual level of these proteins in a cytokine preparation. Comparison of the level of TNF, as determined by the immunoassay, in the various preparations examined in this study, and of the amounts of biologically active CTXs in these preparations, as measured by the cytotoxicity assay, suggests that at the time of the assay some of the preparations contained a significant proportion of inactivated TNF. The specific activity of purified, fully active, TNF is, under the conditions of the bioassay, applied in this study, between 10^7 to 10^8 units of cytotoxicity per mg protein. Among the various preparations of IFN which were examined, only those of crude IFN-γ exhibited cytotoxic activity whose quantitative proportion to the level of immunologically detectable TNF seems consistent with such a high specific activity. In all other preparations, particularly in the concentrated preparations of IFN-α, the cytotoxic activity was significantly lower than that expected from the amounts of TNF, determined by the immunoassay, suggesting that much of the immunologically detectable TNF in these preparations was biologically inactive.

The fact that IFNs and the CTXs are formed together, in response to the same stimuli, implies not only the need to monitor carefully effectiveness of separation of the two, when their purity is sought, but also the possibility that function of these cytokines together may be more physiologically relevant and more effective in therapy than their function alone. What amounts of the CTXs should be used when applying the CTXs for therapy in combination with IFN. At which situations such combinations may indeed become helpful, and at which may the adverse effects which CTXs can have, interfer with such applications, is, unfortunately, to be determined primarily by trial and error. Therefore, it seems of particular interest that, unknowingly, TNF has already been applied, in low amounts, for therapy in combination with IFN. Preparations of concentrated IFN-α which have been in use for clinical trials throughout the period of 1971-1978 (6) are shown here to contain TNF. The amounts of TNF in these preparations are quite low, at the range of tens of units of TNF per ml, as compared to millions of IFN. Yet, as TNF can have effects at even minute concentrations and these effects may greatly be potentiated by IFN, some contribution of TNF to the clinical effects observed with these preparations cannot be excluded.

6. ACKNOWLEDGEMENTS

This study was supported in part by grants from InterYeda Ltd., Ness-Ziona (Israel) and the NCRD (Israel). We thank Dr. Leo Lin from the Cetus Corp., Emersville, California for the gift of rTNF and Dr. G. Adolf from the Boehringer Institute, Vienna, Austria for the gift of antiserum against LT.

REFERENCES

1. Wallach D: in: Interferon 7 (I. Gresser, ed). Academic Press (London) pp. 89-122, 1986.
2. Kohase M, Henriksen-De Stefano D, May LT, Vilcek J and Sehgal PB: Cell 45:659-666, 1986.
3. Todaro JG, Green H and Swifl MR: Science 153:1252-1254, 1966.

4. Wallach D: J Immunol 132:2464-2469, 1984.
5. Cantell K, Hirvonen S, Kauppinen H-L and Myllylä G: Meth Enzymol 78: 29-38, 1981.
6. Strander H, Cantell K, Carlstrom G and Jakobsson PÅ: J Nat Cancer Inst 51:733-742, 1973.
7. Cantell K, Hirvonen S and Koistinen V: Meth Enzymol 78:499-505, 1981.
8. Cantell K, Hirvonen S and Kauppinen H-L: Meth Enzymol 119:54-63, 1986.
9. Aderka D, Holtmann H, Toker L, Hahn T and Wallach D: J Immunol 136: 2938-2942, 1986.
10. Granger GA and Kolb WP: J Immunol 101:111-120, 1968.
11. Nedwin GE, Sverdersky PL, Bringman ST, Palladino AM and Goeddel DV: J Immunol 135:2492-2497, 1985.
12. Stone-Wolff DS, Yip YK, Chroboczek Kelker H, Le Y, Henriksen-De Stefano D, Rinderknecht E, Aggarval BB and Vilcek J: J Exp Med 159:828-843, 1984.
13. Aderka D, Novick D, Hahn T, Fischer DG and Wallach D: Cell Immunol 92:218-225, 1985.

PHENOTYPE OF γ-INTERFERON PRODUCING LEUKOCYTES

Ulf Andersson, Tamas Laskay, Jan Andersson, Rolf Kiessling and Marc DeLey

I. INTRODUCTION

Although human gamma interferon (IFNγ) is a well-defined lymphokine of central importance for the immune system, the cellular origin of its production still remains to be determined. This question has until now been studied either by enriching or depleting IFNγ-producing cells in unfractionated cell cultures, as well as by scrutinizing IFNγ synthesizing cell lines and hybridomas. These approaches have, however resulted in incomplete information.

In the present study we demonstrate a method for cytoplasmic detection of IFNγ in individual human cells in suspension using specific mAbs. In order to visualize cellular proteins using immunofluorescent microscopy, the cells must be exposed to a fixation procedure which immobilizes the intracytoplasmic protein molecules and makes them accessible to specific antibodies. Fixatives, such as acetone and alcohols preserve the antigenicity of macromolecules, but damage the cell morphology. Other fixatives, such as aldehydes are better preservatives of cell morphology but another fixation step is required to make the membranes permeable for antibody penetration. In the present study a paraformaldehyde fixation was combined with an acetone treatment. This procedure prevented the IFNγ molecules to be dissolved from the cytoplasm and preserved their antigenicity to be recognized by monoclonal antibodies. This method enables us to detect the presence of cytoplasmic IFNγ by an indirect immunofluorescence technique which can be used on cells in suspension (1).

Mononuclear blood cells (MNC) have been stimulated to produce IFNγ in vitro by OKT3 antibodies, which induce polyclonal T cell activation by interacting with the constant region of the T cell antigen receptor (2). The phenotype of the IFNγ-producing cells was characterized by double-staining immunofluorescence. The activated MNC were first stained with a panel of mAbs against cell surface antigens, then were fixed, permeabelized in suspension and stained for cytoplasmic IFNγ. We have thus been able to study the phenotypes of the IFNγ producing cells in a direct manner, as outlined in this communication.

2. PROCEDURE
2.1. Material and Methods
2.1.1. Cells and Culture Conditions.
MNC were isolated from buffy coats from healthy adult blood donors or from umbilical cord blood by centrifugation on a Ficoll Isopaque gradient. MNC were cultured for 20 hours at a cell density of 2×10^6/ml in serum-free RPMI 1640 supplemented with L-glutamin (2mM) and 0,5% human serum albumin (Kabi-Vitrum, Stockholm, Sweden) in 250 ml plastic flasks (Nunc AS, Kamstrup, Denmark). The cultures were stimulated with 12,5 ng/ml of purified OKT3 antibody (Ortho Pharmaceutical, Raritan, N.J.).

2.1.2. mAbs. The following antibodies were used to identify cell surface antigens: OKT3 (pan T cell) OKT4 (helper/inducer T cell), OKT8 (suppressor/cytotoxic T cell), OKT11 (sheep erythrocyte receptor) OKIa1 (HLA-D frame work), OKM1 (complement 3bi receptor) and OKB7 (pan B cell) were purchased from Ortho Pharmaceutical, Raritan, N.J., whereas Leu M3 (macrophage/monocyte), and Leu 7 (NK-cell) were purchased from Becton Dickinson, Mountain View, CA. Anti-Tac antibody (IL-2 receptor) (3) and B73.1 (reactive with a structure closely associated with the Fc receptor of large granular lymphocytes) (4) were gifts from Dr. T. Waldmann and from Dr. G. Trinchieri, respectively.

2.1.3. mAb IFN γ. Eight different mouse mAbs (all IgG1, obtained from 5 different laboratories) with specificity for natural human IFNγ have been used: Balb/c mice were immunized with FPLC-purified natural human IFN γ . Spleen cells were then fused, cloned and hybridoma supernatants were screened for anti-IFN γ antibodies in an ELISA against purified human natural IFN γ or E.coli produced rIFNγ . Three subclones D9D10, D1G2 and D13C8 had considerable neutralizing activity. One ml of the D9D10 mAb in ascitic fluid neutralized 5×10^6 units of IFNγ. The D9D10 mAb (ascites) was used at a dilution of 1:500; the D1G2 (ascites) and D13C8 (ascites) at a dilution of 1:50.

The B3 antibody is a well-characterized mAb and has already been successfully used in a specific RIA (5). The B3 mAb (used at dilution 1:2) was purchased from Centocor Inc., Malvern, PA, as provided in the IMRX Interferon-Gamma RIA kit. The 1-DIK-1 (Affi-Gel-Purified ascites, 0.3 mg prot/ml, diluted to 1:10) and 7-B6-1 (Affi-Gel Protein-A MAPS-purified ascites, 1 mg prot/ml, diluted to 1:10) mAbs recognize different epitopes on the IFNγ molecule, and are useful in IFNγ specific ELISA (6). These antibodies were gifts of Dr. G. Andersson, Kabi-Vitrum, Stockholm, Sweden. The GZ-4 (PEG precipitated ascites, 4 mg/ml, used at 1:20 dilution) which binds and neutralizes natural IFNγ and which has been used in specific ELISA (7) was kindly provided by Dr. W. Berthold, Dr. Karl Thomae, GMBH, Biberach, Germany. Monoclonal antibody S1-1 (8) was obtained from Dr. S. Stefanos, Paris, France.

Monoclonal mouse antibodies (IgG1) with specifficity for human IFNα , 9-1-1 and 2-2-1 (precipitated ascites, 8 mg prot/ml diluted to 1:3, or 1:20) and 7-AZ specific for recombinant human IFN γ (Dr. G. Andersson personal communication) (Affi-Gel Protein A MAPS-purified supernatant, 0.3 mg prot/ml, diluted to 1:2) were kindly supplied by Dr. G. Andersson, Kabi-Vitrum, Stockholm, Sweden. The C-15 anti-recombinant human IFNγ mouse monoclonal IgG1 antibody (7) (PEG-precipitated ascites 4 mg prot/ml, diluted to 1:3), and anti-human IFNα EBI-1 mouse IgG1 mAb were obtained from Dr. W. Berthold, Dr. Karl Thomae, GMBH, Biberach, Germany.

2.1.4. Fluorescent Antibodies. For indirect immunofluorescent staining of cell surface antigens we used a FITC-rabbit F(ab')$_2$ anti-mouse Ig (Dakopatts, Glostrup, Denmark) whereas for cytoplasmic IFNγ a rhodaminated (TRITC)-goat anti-mouse IgG1 (Southern Biotechnology, Birmingham, AL) was used. To visualize the Golgi apparatus, purified WGA, supplied by Dr. H. Wigzell, was coupled to FITC (Sigma, St. Louis, MO), and used at the concentration of 10 ug/ml. All antibodies for immunofluorescent staining were deaggregated prior to use by ultracentrifugation at 100.000 xg.

2.1.5. IFNγ . Natural human IFNγ which was affinity-purified by mAb was a gift from Dr. K. Cantell.

2.1.6. Fixation and Immunofluorescent Staining of the MNC. The cultured MNC were washed in PBS without calcium and magnesium and aliquoted (5-10x10^6 MNC) into 5 ml plastic tubes. The cells were then pelleted to a volume of 50 μl and incubated with 10 ul of the appropriate monoclonal reagent (Ortho or Becton-Dickinson) for 30 minutes. Anti-Tac antibodies were used at a final dilution 1:5000 and mAb B73.1. at 1:100. After one wash in PBS a second incubation with the FITC-F(ab')$_2$ α mouse Ig at a final dilution of 1:50 was performed for 30 min. . After another wash an additional incubation with the first antibody was carried out in order to block remaining free sites of the anti-mouse Ig. All procedures were done on ice. These surface-stained MNC (5-10x10^6/tube) were then fixed and stained for the cytoplasmic presence of IFNγ (1). Briefly, the cells were fixed in phosphate-buffered 4% formaldehyde (Paraform-aldehyde 40 g/l, NaH$_2$PO$_4$xH$_2$O 16,883 g/l, NaOH 3.856 g/l and glucose 5,4 g/l, pH 7.4) for 5 min at room temperature. After one wash in PBS the cells were carefully resuspended in 0.5 ml PBS per tube and added to 10 ml ice-cold acetone in round bottomed glass-tubes, which were immediately centrifuged at 400 xg for 5 minutes at +4°C. The acetone was removed and 10 ml cold PBS was added to each glass tube followed by the addition of 0.2 ml 30% bovine serum albumin. The MNC were centrifuged to a pellet and transferred to 5 ml plastic tubes for cytoplasmic IFNγ staining, which was performed in PBS supplemented with 0.2% bovine serum albumin. A sequential incubation with an αIFNγ-mAb (D9D10) at a final dilution 1:500 was followed by one wash and the addition of a TRITC α mouse IgG1 antibody (final dilution 1:60).

2.1.7. Fluorescent Microscopy. The double-stained MNC were examined with a Zeiss fluorescent microscope using a Ploem-pac with filter sets 44-77-10 for FITC and 48-77-15 for TRITC. Buffered glycerol containing 2% diazo-bicyclooctane was used as the mounting medium to reduce UV quenching of FITC. In order to determine the percentage of FITC-mAb-surface positive cells, 200 cells of each sample were counted. To study the phenotype of the IFNγ-containing MNC 200 consecutive TRITC-αIFNγ stained cells were then examined for co-expression of positive FITC surface staining for each mAb used. When the stained MNC were suspended in PBS with 0.02% sodium azide, we found that the samples could be stored for weeks at +4° without any observable change.

3. RESULTS AND DISCUSSION
In the present study OKT3-activated cells in suspension were fixed with paraformaldehyde and the cell membranes were made permeable to antibodies by acetone treatment. Using eight different αIFNγ-mAbs, separately, and a second fluorochrome conjugated antibody, uniform staining results were obtained regardless of the mAb used. The morphology of the immunofluorescent IFNγ staining was very characteristic with a local, polar accumulation in the cytoplasm in a juxtanuclear position (Fig. 1), consistent with the appearance of the Golgi apparatus (9).

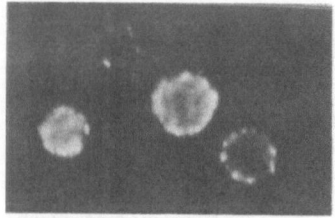
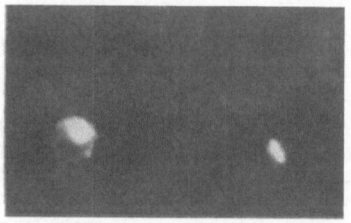

FIGURES 1a and b. Right: Two IFNγ-positively stained MNC showing typical cytoplasmic fluorescence. Left: The same cells are shown after OKT11 antibody surface staining.

In order to substantiate this finding, fixed MNC were stained sequentially with TRITC-anti IFNγ mAb and FITC-WGA. This was based on the observation that certain lectins, such as WGA, bind preferentially to the Golgi apparatus (9). Our finding of totally overlapping TRITC and FITC staining indicated that the majority of the cytoplasmic IFN was localized in the Golgi organelle (1). The cytoplasmic presence of IFNγ in a minority (range 6-11% of the cultured cells from adult blood donors was demonstrated. Control studies with irrelevant murine IgG1 antibodies, were in contrast, negative. Additional evidence for the specific staining of IFNγ with the αIFNγ mAbs came from experiments in which it was possible to block the reaction by incubating the mAbs with purified natural IFNγ.

It is conceivable that this cytoplasmic staining of IFNγ represents its production, rather than an uptake, since the incubation of cultured MNC with high concentrations of natural IFN did not lead to any detectable IFNγ staining (1). The compartmentalization of the IFNγ to the Golgi apparatus gives additional support for this view. Moreover, in other experiments we have noted that exposure of MNC cultures to other stimuli known to induce IFNγ production, such as PHA, ConA and Epstein-Barr virus, resulted in IFNγ producing cells of identical morphology to those seen after OKT3 activation.

When umbilical cord blood MNC were stimulated with OKT3 antibodies, we found much fewer IFN positive cells (range 0.2-0.6% in four experiments) than in control cultures from adults. This is in agreement with reports on the IFNγ levels in supernatants from such cultures (10). MNC which were cultured in medium alone, from adults as well as newborns contained maximally 0.1% IFNγ producing cells. We also found that OKT3-induced IFNγ production in MNC from adults could be detected by cytoplasmic staining as early as 2,5 hours after initiation of cultures and increased up to 20 hours. This finding is also in good accordance with a previous report (11).

To study the phenotypes of the IFNγ producing MNC the cells were stained with a panel of mAbs specific for cell surface antigens followed by the addition of a FITC conjugated F(ab')2-antibody before fixation. They were then stained for cytoplasmic IFNγ. The results of four experiments with OKT3-activated MNC from adult donors are reported in Table I. The percentage of IFN positive MNC co-expressing one particular surface antigen at a time is presented.

TABLE 1. Surface markers on IFN γ producing cells after OKT3 activation.

Phenotypic marker	Frequency on IFNγ+-MNC (%)				Frequency on total MNC (%)
	Exp. 1	Exp. 2	Exp. 3	Exp. 4	range
T3	36	50	47	26	50-66
T4	14	10	29	7	38-55
T8	16	39	63	22	21-30
T11	90	92	92	93	58-79
Tac	98	97	89	88	35-58
OKIa1	12	9	0	0	12-17
OKB7	0	0	0	0	5-10
OKM1	82	55	26	75	16-21
LeuM3	0	0	0	0	2- 4
Leu 7	13	12	4	8	7-24
B73.1	ND	7	0	0	3- 6

6-11% of the total MNC were IFNγ producing cells in the OKT3 stimulated cultures, while the corresponding frequency in unstimulated cultures was <0.1%.

The majority of the IFN γ producing MNC were T cells since they displayed the T11 antigen. They also expressed receptors for IL-2 (Tac+). The presence of IL-2 receptors is of great interest, since it has been shown that IL-2 is a major regulator of IFNγ production (12). The absence of the T3 antigen on approximately 50% of the IFNγ positive MNC is not surprising, since the cells were stimulated by OKT3 antibody and there is a rapid capping and shedding phenomena of the OKT3-T3 complex during culture, followed by a slow re-expression of the T3 antigen, as reported by Reinherz et al. (2). An alternative explanation, that Fc receptors on monocytes cause OKT3-treated lymphocytes to internalize the T3 antigen, has also recently been presented (13).

Only a minority of the IFN γ-producing cells was OKT4+, which is in contrast to a previous report by Chang and coworkers (14). However, these authors did not use a direct approach in determining the phenotype of the IFN γ-producing cells. In the present study the frequency of OKT8+ IFN γ-producing cells varied considerably between experiments. In three out of four experiments the sum of the OKT4+ and OKT8+ IFN γ-producing MNC was less than 50%, thus suggesting that also an OKT4-,8- T cell subset produces IFNγ after OKT3 stimulation.

The high number of IFNγ producing cells expressing receptors for complement factor 3bi (OKM1) was unexpected to us, but this finding is consistent with those obtained by O'Malley and coworkers made after PHA stimulation of peripheral blood MNC (15). Our experiments also further suggest the existence of a reciprocal relationship between the percentage of OKT8+ and OKMI+ IFN γ-producing cells. However, further data are needed to substantiate this finding.

Another unexpected finding was the absence of class II-molecules (OKIal) on the IFNγ-producing cells, since they are frequently expressed on activated T cells. However, it cannot at present be excluded that our culture period of 20 hours was too short for the class II-molecules to appear. NK cells, as defined by the Leu 7 and B73.1. antibodies did not play a significant role in the OKT3-induced IFN γ production. B lymphocytes (OKB7) and Leu M3+ mono-cytes/macrophages were negative for cytoplasmic IFNγ. This latter finding indicates that the OKMI+ IFNγ-producing cells are probably not of the monocyte/macrophage lineage. Taken together our data strongly suggest that the IFN γ-producing cells after OKT3 antibody stimulation are T cells of heterogeneous phenotypes. One such major subset is anti-Tac+, OKT4-, OKT8-, OKM1+ and OKIal-.

Our more recent observation is that the cells stained for cytoplasmic IFN γ in suspension can be analysed by flow cytofluorometry. This therefore makes it possible to carry out a more detailed and accurate characterization of the nature of the IFN γ -producing cells.

Since monoclonal antibodies now have been produced to a variety of other well defined lymphokines, the possible usefulness of this assay in the charac-terization of the cells responsible for their production should be further investigated.

REFERENCES

1. Laskay T, Andersson U, Andersson J, Kiessling R, DeLey M: An immuno-fluorescent method for identifying individual IFN -producing lymphocytes. J Immunol Meth eds. (in press).
2. Reinherz EL, Meuer S, Fitzgerald KA, Hussey RE, Levine H, Schlossman SF: Antigen recognition by human T lymphocyytes is linked to surface expression of the T3 molecular complex. Cell 30:735, 1982.
3. Uchiyama T, Broder S, Waldmann TA: A monoclonal antibody (anti-TAC) reactive with activated and functionally mature human T cells. J Immunol 126:1393, 1981.
4. Perussia B, Starr S, Abraham S, Fanning V,Trinchieri G: Human natural killer cells analyzed by B73.1, a monoclonal antibody blocking Fc-receptor functions. J. Immunol 130:2133, 1983.
5. Chang TW, McKinney S, Liu V, Kung PC, Vilcek J, Le J: Use of monoclonal antibodies as sensitive and specific probes for biologically active human gamma-interferon. Proc Natl Acad Sci USA 81:5219, 1984.
6. Troje-Blomberg M, Andersson G, Stoczkowska M, Shabo R, Romero P, Patarrayo E, Wigzell H, Perlmann P: Production of IL2 and IFN-gamma by T cells from malaria patients in response to Plasmodium falciparum or erythrocyte antigens in vitro. J Immunol 135:3498, 1985.
7. Berthold W, Zahn G: Antiviral Res ABSTR 1:69 (meeting abstract).
8. Stefanos S, Wietzezkin J, Catinot L, Devos R, Jalcoff R: Characterization of antibodies against recombinant Hu IFN-gamma produced by hybridoma cells. J Interferon Res 5:455, 1985.
9. Virtanen I, Ekblom P, Lauria P: Subcellular compartmentalization of saccharide moieties in cultured normal and malignant cells. J Cell Biol 85:429, 1980.
10. Wakasugi N, Virelizier JL: Defective IFNγ production - the human neonate. J Immunol 134:67, 1985.

11. von Wussow P, Platsoucos CD, Wiranowska-Stewart M, Stewart WE: Human γ-interferon production by leukocytes induced with monoclonal antibodies recognizing T cells. J Immunol 127:1197, 1981.
12. Farrar WL, Johnson HM, Farrar JJ: Regulation of the production of immune interferon and cytotoxic T lymphocytes by interleukin 2. J Immunol 126:1120, 1981.
13. Rinnoy-Kan EA, Wright SD, Welte K, Wang CY: Fc receptors on Monocytes cause OKT3-treated lymphocytes to internalize T3 and secrete IL2. Cell Immunol 98:181, 1986.
14. Chang T-W, Testa D, Kung PC, Perry L, Dreskin HJ, Goldstein G: Cellular origin and interactions involved in γ-interferon production induced by OKT3 monoclonal antibody. J. Immunol. 128:585, 1982.
15. O'Malley JA, Nussbaum-Blumenson A, Sheedy D, Grossmeyer BJ, Ozer H: Identification of the T cell subset that produces human γ-interferon. J Immunol 128:2522, 1982.

INTERFERON REGULATES MAJOR HISTOCOMPATIBILITY CLASS I GENE EXPRESSION
THROUGH A 5' UPSTREAM REGULATORY REGION

KENJI SUGITA, JUN-ICHI MIYAZAKI, ETTORE APPELLA* AND KEIKO OZATO,
LABORATORY OF DEVELOPMENTAL AND MOLECULAR IMMUNITY, NICHD, AND *LABORATORY
OF CELL BIOLOGY, NCI, NIH, BETHESDA, MD, 20892, USA.

1. INTRODUCTION

Major Histocompatibility complex (MHC) class I genes encode membrane
glycoproteins involved in tissue rejection and T cell recognition of viral
and foreign antigens. MHC class I gene expression is induced in a variety
of cells by interferons (IFN) (both by IFN-α/β and by IFN-γ) (1-4). It is
suggested that the increased MHC class I gene expression by IFNs plays a
role in enhanced T cell immunity. However, the mechanisms of the IFN
action in the induction of MHC class I gene expression have not been
elucidated so far. Recently, Friedman and Stark reported the presence of
~30-bp IFN consensus sequence in the 5' flanking region of IFN inducible
genes which may be involved in IFN induced gene regulation (5). To study
the molecular mechanism of MHC class I gene induction by IFNs, we tested
promoter activity of the 5' flanking region of a mouse MHC class I gene,
H-2L^d. Various portions of the upstream region of the H-2L^d gene were
connected to the chloramphenicol acetyltransferase (CAT) gene, and CAT
activity was examined after transient and stable transfection of the genes
into NIH 3T3 cells. We show that IFN treatment enhances promoter activity
of the MHC class I gene, and that the IFN consensus sequence is responsible
for the increased transcription of the gene.

2. MATERIALS AND METHODS
2.1. Plasmids

A H-2L^d genomic clone, pL^{d}4, which contains ~4.5-kilobase (kb) 5'
flanking region (6) was used. Fragments containing various length of the
H-2L^d 5' flanking region were placed in front of the cat gene after removal
of simian virus 40 (SV 40) promoter from the pSV2-cat plasmid (7) (see
Fig. 1).

Plasmid pAZ 1037 (8), in which the chicken actin promoter is connected
to the cat gene, was a gift from Dr. Crombrugghe.

2.2. DNA Transfection

DNA transfection was performed according to the DNA calcium phosphate
coprecipitation method (9). For transient transfection, a mixture of
the cat hybrid gene (1 µg) and carrier DNA, pUC19 (19 µg), was added to
NIH 3T3 cells incubated overnight after seeding at 5 x 10^5 cells/plate.
After 6 hr incubation with the DNA precipites, cells were rinsed, and
incubated with fresh medium containing IFNs for a further 44 hrs. For
stable transformation, pL^d-cat 392 (10 µg) and pSV2-neo (0.5 µg) used as a
selectable marker (10) were added to the cells followed by selection with
G418 for the subsequent 10-14 days.

2.3. Interferons

Mouse IFN-α/β was purchased from Lee Biomolecular. Mouse recombinant
IFN-γ was donated by Shionogi Research Laboratories (Osaka, Japan). The

and were incubated with or without IFN-α/β for 72 hr. IFN treatment caused a 4-fold enhancement in CAT activity. These results indicate that IFNs increases promoter activity of the MHC class I gene regardless of whether the cat gene is expressed transiently or after integration into the host chromosomes.

3.2. Analyses of Deletion Constructs

To localize the sequence that is involved in the response to IFN signals within the xbaI-BamHI fragment, various deletion constructs were prepared and CAT activity was measured (Table 1).

Table 1

Plasmid	Deletion*	CAT Enhancement by IFN***
pL^d-cat 392	None	4.0
pL^d-cat 237	(-392 to -238)	4.3
pL^d-cat 210	(-392 to -211)	4.2
pL^d-cat del5	(-341 to -135)	0.6
pL^d-cat 123 ICS	**	1.9
pL^d-cat 123 CRE	**	0.7
pL^d-cat 123	(-392 to -124)	0.8

*The H-2L^d 5' flanking sequences deleted from the cat constructs (the number represents nucleotide position relative to the cap site (1))

**ICS or CRE was added to the 123 bp 5' flanking region of the H-2L^d gene.

***The ratio of CAT activity obtained by IFN treated cells vs that obtained by untreated cells.

Two deletion constructs, pL^d-cat 237 and pL^d-cat 210, showed elevated CAT activities on IFN treatment with the same efficiency as pL^d-cat 392. However, no CAT enhancement was detected with pL^d-cat del 5. These results suggested that the sequence responsive to IFN signals may be localized within the sequence from -210 to -134. This region from -210 to -134 is highly conserved among MHC class I genes (11), and contains two salient structural features; i.e., class I regulatory element (CRE) (12) and IFN consensus sequence (ICS) (Fig. 2). The CRE is located from -195 to -161, which elicits an enhancer like function in cells expressing MHC class I genes constitutively. In addition, the CRE represses MHC class I gene transcription in undifferentiated embryonal carcinoma cells which do not express the MHC class I genes (12). The ICS is located between position -167 and -139, which shares high homology with the IFN consensus sequence (11). To determine if either of the two sequences is involved in responding to the IFN signals, two additional constructs were prepared: oligonucleotides correspondig to the CRE or the ICS were joined to the AvaII-BamHI fragment (from -123 to +14) of the 5' flanking region of the H-2L^d gene. pL^d-cat 123 did not show enhanced CAT activity after IFN treatment. Similarly, pL^d cat 123 CRE did not respond to the IFN

titer of these IFNs were determined by detection of the cytopathic effects after infecting L929 cells with Vesicular Stomatitis Virus.

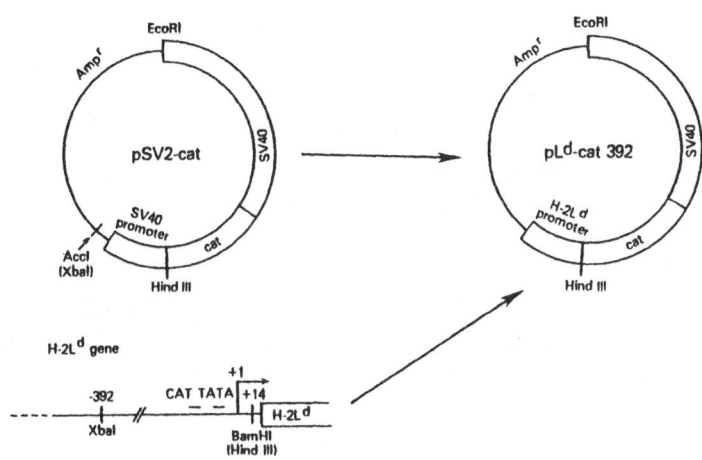

Fig. 1: Construction of pL^d-cat plasmid. See detailed explanation in Materials and Methods and in the text.

2.4. CAT Assay
CAT assay was performed according to the method of Gorman et al. (7). Cell extracts were assayed for the CAT activity by incubation with [^{14}C]-chloramphenicol and acetoacetyl CoA for 30 min at 37°C. Radioactive spots were excised from the thin layer chromatograph silica gel plate, and radioactivity was counted.

3. RESULTS
3.1. IFNs Enhance Promoter Activity of the H-2L^d cat Gene Introduced into NIH 3T3 cells.
To examine a possible contribution of the 5' flanking region of the H-2L^d gene to the IFN-induced enhancement of MHC class I gene expression, we tested CAT activity of pL^d-cat 392, in which transcription of the cat gene is driven by the XbaI-BamHI fragment of 5' flanking region of the H-2L^d gene (-392-bp to +14-bp from the cap site, see Fig. 1).
After the constructs were introduced transiently into NIH 3T3 cells, cells were treated with various concentrations of IFN-α/β or IFN-γ for 44 hrs. IFN-α/β enhanced CAT activity in the cells transfected with pL^d-cat 392 in a dose dependent manner: a slight increase was detected even at 1 u/ml of IFN-α/β, and the maximum elevation was seen at 1000 u/ml, where CAT activity was about 6 times higher than the untreated control. IFN-γ increased CAT activity modestly (~2 fold), and the dose dependence was not found between 10 u/ml and 1000 u/ml of IFN-γ. The augmentation of CAT activity by IFNs was specific for pL^d-cat 392, since the chicken actin-cat construct did not show enhanced activity even at 1000 u/ml of IFN-α/β.
We also tested the effect of IFNs in the stable transformants. NIH 3T3 cells were transformed with pL^d-cat 392 using pSV2-neo as a selectable marker. All the colonies that arose after selection with G418 were pooled

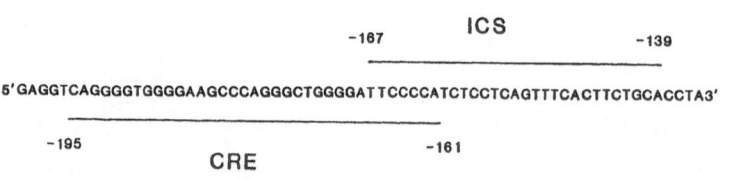

Fig. 2:   The sequence of interferon consensus sequence (ICS) and class I
regulatory element (CRE) of the H-2L^d gene. See details in the
text.

Table 2

Plasmid	Addition of H-2L^d sequence*	CAT Enhancement by IFN***
pSV2-cat	None	0.8
pLS-cat C1	(-392 to -121)	3.2
pLS-cat C2	(-210 to -121)	2.5
pLS-cat Y2	(-121 to -210)**	2.4
pLS-cat ICS	ICS	2.5
pLS-cat CRE	CRE	0.6
pSV-cat delE	None	0.7

*The H-2L^d 5' sequence was joined to the pSV-cat delE.

**This portion was connected in the reverse orientation to pSV-cat delE.

***The ratio of CAT activity obtained by IFN treated cells vs that obtained
by untreated cells.

treatment. However, pL^d-cat 123 ICS exhibited a two-fold increase in CAT
activity (Table 1). The augmentation by pL^d-cat 123 ICS was highly
reproducible, although its degree was lower than that seen by pL^d-cat 237
or pL^d-cat 210.

3.3. The ICS Confers the Responsiveness to IFNs to a Heterlogous Promoter

To examine the functions of the ICS further, H-2L^d fragments
containing the ICS were connected to a SV40 promoter fused with the cat
gene,and CAT activity was measured after IFN treatment (Table 2 and Fig.3).
The 72-bp repeat enhancer sequence was removed from the original pSV2-cat,
and the H-2L^d fragments were placed either in the native or in the reverse
orientation. No response to IFN treatment was seen in the cells
transfected with pSV2-cat or the enhancerless pSV-cat delE. Placement of
the H-2L^d fragment from -392 to -121 in front of SV40 promoter in the
native orientation (pLS-cat C1) enhanced CAT activity by 3.2 fold after IFN

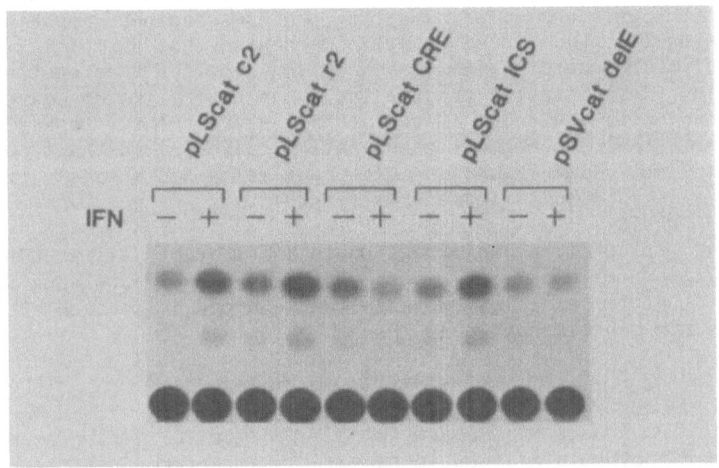

Fig. 3: Autoradiograph of CAT activity by H-2L^d and SV40 promoter cat hybrid genes enhanced by the interferon treatment.

treatment. With the shorter fragment (from -210 to -121) connected to the SV40 promoter, IFN treatment also gave about 2.5 fold increase in CAT activity, regardless of the orientation (pLS-cat c2 andpLS-cat r2). The ability of the ICS to increase promoter activity was also examined in the SV40 promoter (pLS-cat ICS vs pLS-cat CRE). Similar to the results of the H-2L^d promoter, the ICS, but not the CRE, showed significant augmentation in CAT activity after IFN treatment. The degree of enhancement by pLS-cat ICS was the same as those by pLS-cat cl or by pLS-cat Y1. These results indicate that ICS alone confers the responsiveness to the IFN to a heterologous promoter and that it acts in an orientation-independent manner.

4. DISCUSSION

In this study, we show that IFNs elevate promoter activity of a mouse NHC class I gene, H-2L^d and that the ICS present in the 5' flanking region of the gene is capable of conferring IFN induced transcription enhancement. The presence of the IFN consensus sequence was first reported by Friedman and Stark (5) after a homology search of IFN-inducible genes. However, its functional significance has not been verified. The present paper demonstrates a function of the ICS in the IFN-induced gene regulation. We show that the ICS acts in a manner similar to an inducible enhancer (13,14), since it exhibits the activity even in the reverse orentation and in the heterologous SV40 promoter. The enhanced promoter activity seen with the cat constructs by IFNs most likely reflects the enhanced endogenous MHC class I gene expression by IFNs, as the enhanced CAT activity was observed both in transient and stable transformation.

Although the ICS plays a predominant role in enhancing the transcription of the MHC class I gene, neighboring sequences are also involved in the regulation, because the ICS alone elicited lower CAT enhancement than do larger fragment (Table 1). Both constructs, however, exhibited equivalent augmentation in the SV40 promoter (Table 2).

How does the ICS enhance promoter activity of the class I gene upon

IFN treatment? The simplest explanation would be that a diffusible
trans-acting factor(s) that binds to the ICS are induced by IFN treatment.
Evidence for the presence of trans-acting proteins has been reported for a
number of cases (15,16). An alternative mechanism that postulates
derepression of MHC class I gene transcription upon IFN treatment would be
equally valid. In the latter hypothesis, a repressor like negative
regulatory factor(s) would be removed after IFN treatment. Regardless of
the mechanisms, the hypothetical trans-acting factor(s) must act across the
species, since MHC class I genes of different mammalian species introduced
into mouse cells respond to mouse IFNs (17,18).

5. ACKNOWLEDGMENTS

We thank Drs. Crombrugghe, Long, Dawid and Okayama (all at NIH) for
providing chicken actin-cat, pSV2-cat, pSVO-cat genes, mycoplasma free NIH
3T3 cells and helpful advices. We also thank Shionogi Research
Laboratories for mouse recombinant IFN-γ.

6. REFERENCES

1. Basham TY, Bourgeade MF, Creasey AA and Merigan TC: Interferon
increase HLA synthesis in melanoma cells: Interferon-resistent and
-sensitive cell lines. Proc. Natl. Acad. Sci. (USA) 79:3265-3269,
1982.
2. Fellous M, Nir U, Wallach D, Merlin G, Rubinstein M and Revel M:
Interferon-dependent induction of mRNA for the major histocompatibility
antigens in human fibroblasts and lymphoid cells. Proc. Natl. Acad.
Sci. (USA) 79:3082-3086, 1982.
3. Friedman RL, Manly SP, McMahon M, Kerr IM and Stark GR:
Transcriptional and post-transcriptional regulation of
interferon-induced gene expression in human cells. Cell 38:745-755,
1984.
4. Rosa FM, Cochet MM and Fellous M: Interferon and major
histocompatibility complex gene: A model to analyze eukaryotic gene
regulation? (ed.) Gresser I, Academic Press (London) (in press).
5. Friedman RL and Stark GR: α-interferon-induced transcription of HLA
and metallothionein genes containing homologous upstream sequences.
Nature 314:637-639, 1985.
6. Evans GA, Margulies DH, Shykind B, Seidman JG and Ozato K: Exon
shuffling: Mapping polymorphic determinants on hybrid mouse
transplantation antigens. Nature 300:755-757, 1982.
7. Gorman C, Moffat L and Howard BH: Recombinant genomes which express
chloramphenicol acetyltransferase in mammalian cells. Mol. Cell. Biol.
2:1044-1054, 1982.
8. Schmidt A, Setoyama C, and deCrombrugghe B: Regulation of a collagen
gene promoter by the product of viral mos oncogene. Nature
314:286-289, 1985.
9. Graham FL and van der Eb AJ: A new technique for the assay of
infectivity of human adenovirus 5 DNA. Virology 52:456-467, 1973.
10. Southern PJ and Berg P: Transformation of mammalian cells to
antibiotic resistance with a bacterial gene under control of an SV40
early region promoter. J. Mol. Appl. Genet. 1:327-341, 1982.
11. Kimura A, Israel A, Bail OL and Kourilsky P: Detailed analysis of the
mouse H-2K^b promoter: Enhancer-like sequences and their role in the
regulation of class I gene expression. Cell 44:261-272, 1986.
12. Miyazaki J, Appella E and Ozato K: Negative regulation of the major
histocompatibility class I gene in undifferentiated embryonal carcinoma

cells. Proc. Natl. Acad. Sci. USA (in press).

13. Seguin C, Felber BK, Carter AD and Hamer DH: Competition for cellular factors that activate metallothionein gene transcription. Nature 312:781-785, 1984.

14. Bienz M and Pelham HRB: Heat shock regulatory elements function as an inducible enhancer in the Xenopus hsp 70 gene and when linked to a heterologous promoter. Cell 45:753-760, 1986.

15. Emerson BM and Felsenfeld G: Specific factor conferring nuclease hypersensitivity at the 5' end of the chicken adult β-globin gene. Proc. Natl. Acad. Sci. (USA) 81:95-99, 1984.

16. Kovesdi I, Reichel R and Nevins JR: Identification of a cellular transcription factor involved in EIA trans-activation. Cell 45:219-228, 1986.

17. Yoshie O, Helmuth S, Reddy ESP, Weissman S and Lengyel P: Mouse interferons enhance the accumulation of a human HLA RNA and protein in transfected mouse and hamster cells. J. Biol. Chem. 257:13169-13172, 1982.

18. Satz ML and Singer SD: Effect of mouse interferon on the expression of a procine major histocompatibility gene introduced into mouse L cells. J. Immunol. 132:496-501, 1984.

THE ANTIVIRAL EFFECTS OF TUMOR NECROSIS FACTORS

GRACE H.W. WONG and DAVID V. GOEDDEL, Department of Molecular Biology,
Genentech, Inc., 460 Point San Bruno Blvd., South San Francisco, CA 94080

Introduction

Tumor necrosis factor (TNF) and lymphotoxin were originally described
as antitumor proteins released by activated macrophages and lymphocytes,
respectively (1-3). Both proteins have been purified and sequenced, and
their cDNA clones have been isolated and the expressed proteins purified
(4,5). Because they share similar biological activities, bind to the
same cell surface receptor (6), and their genes are closely linked (7),
TNF and lymphotoxin have been renamed TNF-α and TNF-β, respectively (8).
In addition to their known antitumor effects, TNFs have been shown to
mediate pleiotropic biological activities (9-16). Here we summarize our
data showing that TNFs alone have intrinsic antiviral activity and
potentiate the antiviral activity of IFNs.

Characterization of the antiviral activity of TNFs

The antiviral effect of TNF-α and TNF-β was tested on various cell
lines which were not susceptible to the cytotoxicity of the TNFs. In
these experiments, no difference in the antiviral activities of TNF-α or
TNF-β was observed. The results showing the effect of TNF-α on vesicular
stomatitis virus (VSV) replication are summarized in Table I. Although
TNF-α had little anti-VSV activity on most cell lines tested, it was by
itself effective against VSV replication in the human 7860 renal
carcinoma, HT-29 colon carcinoma, RPMI-8226 lymphoblastoid, and U87MG
glioblastoma cell line, murine C127 epithelial and Raw-264 macrophage
cell lines, and Rat-1, a rat fibroblast cell line.

In addition to inhibiting virus replication, both TNF-α and TNF-β
prevented virus-mediated cytopathic effects on cells from several species.
The antiviral effects of the TNFs, like their antitumor activity, are not
species-specific. This contrasts with the antiviral activity of IFN-γ
which exhibits a considerable level of species specificity.

Complete protection against VSV-mediated cytopathic effects required
at least 12 hr to develop after the onset of exposure to TNFs. However,
a one-hour exposure to TNF-α or TNF-β followed by a 12-hour incubation
was sufficient to induce complete resistance to VSV infection. The
antiviral activity of TNFs was abolished only by monoclonal antibodies
against the respective TNF but not by antibodies against IFN-α, -β or -γ,
thus confirming that the effects observed are indeed mediated by TNFs.

TNFs did not induce the expression of IFN-α, -β or -γ mRNA but did
induce "IFN-β2" mRNA in human T24 bladder carcinoma and U87MG
glioblastoma cells. However, there was a lack of correlation between the
antiviral activity of TNFs and the induction of "IFN-β2" mRNA (17). The
results suggest that the effects of TNFs are direct rather than mediated
through the production of IFNs (17).

As with the IFNs, TNFs induce mRNA for 2'-5' oligoadenylate synthetase (17), an enzyme associated with the IFN-mediated antiviral state. However, the antiviral activity of the TNFs are not solely due to induction of this enzyme, since TNFs induce high levels of this mRNA in A549 and T24 cells, yet show no detectable antiviral activity in these cells (Table I).

Table I. TNF-α potentiates the ability of IFN-γ to inhibit virus replication in diverse cell types.

	VSV yield ($\times 10^6$ PFU/ml)			
	Virus Control	TNF-α	IFN-γ	TNF-α+IFN-γ
Human Cell Lines				
HT-1080 (lung fibrosarcoma)	36	27	31	0.64
HeLa (cervical carcinoma)	18	14	4.6	0.73
T24 (bladder carcinoma)	56	39	2.8	0.59
HT-29 (colon carcinoma)	49	8.5	2.3	0.38
7860 (renal carcinoma)	120	16	85	0.45
ST-486 (Burkitt lymphoma)	340	180	230	8.6
RPMI-8226 (myeloma)	290	21	18	0.33
U87MG (glioblastoma)	180	23	120	0.031
Non-human Cell Lines				
MDBK (bovine kidney fibroblast)	130	110	3.1	0.12
Rat-1 (rat fibroblast)	89	0.52	2.3	0.019
C127 (mouse epithelial cells)	48	0.93	0.81	0.038
RAW-264 (mouse macrophage)	160	14	8.7	0.57

Cell lines were obtained from the American Type Culture Collection. Confluent cells were seeded in 24-well tissue culture plates, pretreated with TNF-α (0.1 µg/ml) and/or IFN-γ (10 ng/ml) for 18 hr before adding VSV. After 24 hr, the virus yield was determined as described (18). Purified recombinant bovine IFN-γ (specific activity of 10^6 units/mg; provided by Dr. G. Burton) was used in the bovine cell line, whereas recombinant murine IFN-γ (specific activity = 3×10^7 units/mg) was used for rat and murine cells. Human IFN-γ was used for all human cell lines.

Synergism between IFN-γ and TNFs

Although TNFs alone have antiviral activity, TNFs together with IFNs result in a much greater antiviral response. This synergistic effect is discussed below. In some cells, no antiviral activity is apparent unless both are added together. With the human lung carcinoma (A549) cells, TNFs or IFN-γ alone does not protect against VSV infection. However, when A549 cells were treated with both TNFs and IFN-γ, VSV-mediated cytopathic effect was completely inhibited (Fig. 1). Other experiments showed that TNFs not only synergize with IFN-γ but also enhance the antiviral activities of IFN-α or IFN-β.

In the presence of human TNF-α, IFN-γ became more potent and effective in inhibiting VSV replication on various cell lines (Table I). With human U87MG glioblastoma cells, the combination resulted in inhibition about 4,000 times greater than IFN-γ alone in these cells (Table I). Human TNF-α enhances the antiviral effect of bovine or murine IFN-γ (Table I). Conversely, the antiviral activity of human IFN-γ is also enhanced by murine TNF-α. Thus, the potentiating activity of TNFs is also not species-specific.

The antiviral or antiviral enhancing activity of TNFs was not limited to VSV but could be demonstrated with encephalomyocarditis virus, adeno-virus 2 and Herpes Simplex virus type II (17). The antiviral activity of TNFs alone and the synergistic activity of TNFs and IFN-γ could only be

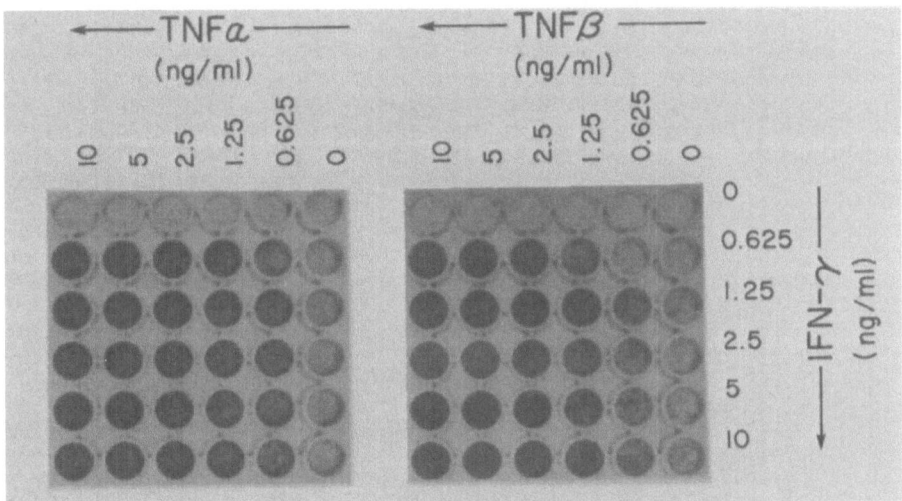

FIGURE 1. IFN-γ protects A549 cells from VSV infection only in the presence of TNF-α or TNF-β. A549 cells were treated with two-fold serially diluted samples giving the indicated concentrations of TNFs in combination with IFN-γ. After 24 hr, the cells were challenged with VSV and the cytopathic effect was determined by staining the cells with crystal violet. The specific activities for purified recombinant human TNF-α and TNF-β were $3-4 \times 10^7$ units/mg assayed in LM cells (19). Using human A549 cells and EMC virus, the specific activities for purified recombinant human IFN-αA, IFN-β and IFN-γ were 2×10^8, 3×10^7 and 4×10^7 units/mg, respectively.

276

demonstrated in confluent cells. TNFs do not protect against VSV infection in subconfluent 7860 cells. Likewise, TNF's ability to potentiate the activity of IFNs could not be detected in subconfluent A549 cells. It will be of interest to examine the importance of confluency for the induction of antiviral resistance by TNFs.

Other cytokines such as interleukin-1, interleukin-2, nerve growth factor, platelet-derived growth factor, and transforming growth factor -α or -β had no antiviral activity and did not enhance the antiviral activity of IFNs in A549 cells.

In addition to inducing antiviral resistance in uninfected cells, TNFs selectively kills VSV and Ad-2 infected cells. Destruction of TNF-treated virus-infected cells is enhanced with IFN-α, IFN-β and IFN-γ (17) and the selective killing is also dependent on the state of confluency.

Virus infection resulted not only in the production of IFNs but also synthesis of TNFs. Hybridization of mRNA extracted from virus infected RPMI-1788 cells with specific cDNA probes showed that both TNF-α and TNF-β mRNA were induced (Fig. 2). Similar results were obtained using normal human peripheral blood leukocytes (17).

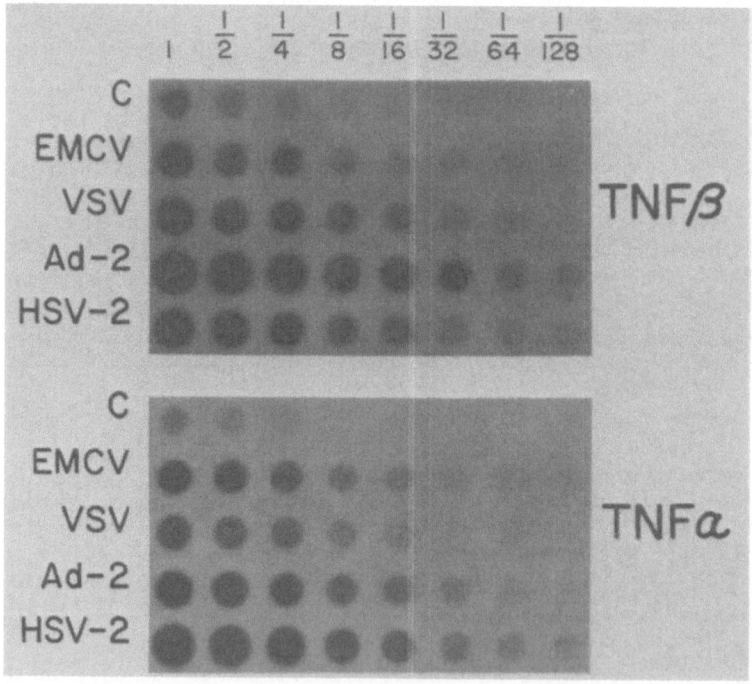

FIGURE 2. Dot blot hybridization of RNA isolated from RPMI-1788 cells (2x10^6 cells/ml) infected with indicated viruses. RNA isolated at 5 and 12 hr was used for hybridization with TNF-α and -β probes, respectively. TNF-α or -β probes were ^{32}P-labeled AvaI-HindIII restriction fragment (578 bp) from human TNF cDNA (4) and an EcoRI restriction fragment (~650 bp) from the coding region of human TNF-β cDNA (5), respectively.

Conclusion

Both TNF-α and TNF-β have intrinsic antiviral activity on certain cells, demonstrating that these cytokines represent a new class of virus inhibitory proteins. Both TNF-α and TNF-β synergize with IFNs in inhibiting virus-induced cytopathic effects, reducing viral yield and selectively killing virus-infected cells. In some circumstances (depending on virus and cell types), IFNs have little or no effect except in the presence of TNFs. Thus, TNFs increase the effectiveness of the IFNs as antiviral mediators. The antiviral synergistic activity of TNFs appears to be general because such activity was observed with several types of viruses on a range of cell types. The regulation of production of TNFs and IFNs by virus infection and their ability to interact with a variety of cells suggests that these cytokines are essential components of the host antiviral defense system. Our findings strongly support the view that one of the major physiological roles of TNFs may be to protect against infectious pathogens.

References

1. Carswell, E.A. et al. Proc. natn. Acad. Sci. USA 72, 3666–3670 (1975).
2. Ruddle, N.H. and Waxsman, B.H. Science 157, 1060–1062 (1967).
3. Granger, G.A. and Kolb, W.P. J. Immunol. 101, 111–120 (1968).
4. Pennica, D. et al. Nature 312, 724–729 (1984).
5. Gray, P.W. et al. Nature 312, 721–724 (1984).
6. Aggarwal, B.B., Essalu, T., and Hass, P.E. Nature 318, 665–667 (1985).
7. Goeddel, D.V. et al. Cold Spring Harbor Symposium, N.Y. (in press).
8. Shalaby, M.R. et al. J. Immunol. 135, 2069–2073 (1985).
9. Sugarman, B.J. et al. Science 230, 943–945 (1985).
10. Gamble, J.R. et al. Proc. natn. Acad. Sci. USA 82, 8667–8671 (1985).
11. Silberstein, D.S. and David, J.R. Proc. natn. Acad. Sci. USA 83, 1055–1059 (1986).
12. Beutler, B. et al. Nature 316, 552–554 (1985).
13. Torti, F.M. et al. Science 229, 867–869 (1985).
14. Bertolini, D.R. et al. Nature 319, 516–518 (1986).
15. Dayer, J.M., Beutler, B., and Cerami, A. J. Exp. Med. 162, 2163–2168 (1985).
16. Broudy, V.C. et al. Proc. natn. Acad. Sci. USA (in press).
17. Wong, G.H.W. and Goeddel, D.V. Nature (in press).
18. Rager-Zisman, B. and Merigan, T.C. Proc. Soc. Exp. Med. 142, 1174–1179 (1973).
19. Kramer, S.M. and Carver, M.E. J. Immunol. Methods (in press).

ANIMAL MODELS

INTERFERON AND TUMOR NECROSIS FACTOR ALTER THE PHOSPHOLIPID METABOLISM IN FRIEND LEUKEMIA CELL TUMORS IN MICE

E. PROIETTI[1], F. BELARDELLI[1], G. CARPINELLI[1], M. DI VITO[1], I. GRESSER[2], W. FIERS[3], J. TAVERNIER[4] AND F. PODO[1].
[1]Istituto Superiore di Sanità, Rome, Italy. [2]Institut de Recherches Scientifiques sur le Cancer, Villejuif, France. [3]Laboratory of Molecular Biology, Ghent, Belgium. [4]Biogent, Ghent, Belgium.

INTRODUCTION

Interferon (IFN) has been shown to inhibit the growth of different tumors in experimental animals (reviewed in 1 and 2) and it is currently used to treat some patients with cancer in clinical trials. However the mechanisms of antitumor effects of IFN are still unknown (2). We have previously shown that administration of highly purified α/β IFN was equally effective in inhibiting tumor growth in mice injected with either IFN-sensitive or IFN-resistant Friend leukemia cells (3,4). These data suggested that the antitumor effects of IFN in this system were not due to a direct effect of IFN on tumor cells but that these effects were in some way host-mediated (4). Daily injection of IFN into FLC tumors implanted subcutaneously (s.c.) caused an arrest of tumor growth and tumor necrosis and resulted in a complete tumor regression in some mice transplanted with either IFN-sensitive or IFN-resistant FLC (5).

Tumor necrosis factor (TNF) is toxic for some tumor cells in vitro (6) and induces necrosis of tumors in experimental animals (7-9). We have recently demonstrated that intratumoral treatment with TNF of mice injected s.c. with FLC resulted in a marked inhibition of tumor growth and in the rapid appearance of tumor cell necrosis, even though these cells appeared to be resistant in vitro to the cytotoxic effect of TNF (10).

High resolution ^{31}P NMR spectroscopy offers a powerful non-invasive approach to study metabolic alterations in intact tumor cells either cultured in vitro or transplanted into syngeneic mice (11-14). We recently demonstrated that intratumoral injection of either IFN-sensitive 745 or IFN-resistant 3C18 FLC tumors with mouse IFN α/β resulted in a marked decrease in the levels of some phospholipid metabolites such as glycerophosphorylcholine (GroPCho), glycerophosphorylethanolamine (GroPEtn) and phosphorylcholine (PCho) (11). These metabolic alterations occurred concomitantly with a significant increase in the average intratumoral pH and were detectable after 2 days of intratumoral injection of mouse IFN α/β. Very similar metabolic alterations have been also observed very recently in TNF-treated FLC tumors (10).

The interest of further investigating these metabolic alterations during in vivo tumor regression is supported by the growing experimental evidence that changes in pool sizes of these phospholipid metabolites may represent well defined biochemical responses of cells to factors regulating their conditions of growth, differentiation and proliferative capabilities (11-14).

By using NMR spectroscopy techniques we further describe herein some similarties in the metabolic alterations induced by either IFN or TNF at early stages of tumor regression.

MATERIALS AND METHODS

Mice: 5-8 week old male or female DBA/2 and C3H/HeN mice were obtained from Charles River ITALIA S.p.A. (Milan).

Tumor cells: IFN-sensitive (745) or resistant (3C18) Friend leukemia cells (FLC) were passaged by weekly intraperitoneal (i.p.) inoculation of 5-6 week-old DBA/2 mice.Methylcholanthrene-induced fibrosarcoma cells (C3H) were kindly provided by Dr. A. Mantovani (Milan) and were passaged by s.c. inoculation in 5-6 week old C3H/HeN mice.

Interferon preparations: The methods of production, partial purification and assay of IFN α/β have been previously described (11).One of our units is the equivalent of 4 international mouse reference units. In some experiments we used highly purified α/β mouse IFN (specific activity: $5x10^8-10^9$ IU/mg of protein). Mice were injected daily into the tumor with 0.2 ml of either IFN (10^6 U/ml) or BSA (100 μg/ml).

Tumor necrosis factor preparations:
in E. Coli, was purified to apparent homogeneity (Tavernier at al., unpublished results) and had a specific activity of $8x10^7$ U/mg of protein (assayed in L929 cells in the presence of 1 μg of Actinomycin D for 18 h). In the experiments TNF was diluted in PBS containing BSA (Sigma) 100 μg/ml.

Tumor preparation: FLC and fibrosarcoma cell tumors were rapidly dissected, washed with physiological solution containing 20% D_2O and inserted into 10 mm $\emptyset$ NMR tubes. Several tumors were packed at the bottom of the NMR tube, up to a height of at least 2.5 cm, in order to optimize the filling factor of the radio frequency coil in the probe.

Tumor extracts: Tumor extracts were prepared by addition of a 60% solution of ethanol, homogeneization and centrifugation at 27,000 g for 30 min. The supernatant was evaporated to   dryness by rotovapor and the residue suspended in H_2O-EDTA (100 mM), reading pH 7 for NMR measurements.

Nuclear magnetic resonance: ^{31}P NMR spectra of freshly dissected tumors were recorded at 4°C by means of a Bruker AM 400 WB spectrometer working at 162 MHz, interfaced to an ASPECT 3000 computer, under conditions of broad-band proton decoupling. ^1H NMR spectra of ethanol tissue extracts were recorded at 27°C by using the same spectrometer at 400 MHz, under conditions of selective residual HDO signal.

RESULTS AND DISCUSSION

Fig. 1 shows typical ^{31}P NMR spectra of freshly dissected FLC tumors in vivo treated with BSA (control tumors), IFN or TNF. As shown in the spectra of control BSA-treated tumors, the following main resonances were detected: two rather broad peaks, a and b, arising from phosphomonoester (PME) compounds (a: phosphorylethanolamine (PEtn) and, to a much lower extent,α-glycerophosphate; b: phosphorylcholine (PCho) with a minor contribution from AMP); a peak, c, due to inorganic phosphate (Pi), two narrow signals (d and e) in the spectral region of the phosphodiesters (PDE),

assigned to glycerophosphorylcholine (GroPCho) and glycerophosphorylethanolamine (GroPEtn) respectively. These resonances with similar peak areas
were found also in ^{31}P NMR spectra of C3H fibrosarcoma tumors implanted
s.c. in C3H/HeN mice. These assignments were confirmed by ^{31}P NMR analysis
of the tumor ethanolic extracts.

Daily intratumoral administration of IFN induced a decrease of PCho,
GroPCho and GroPEtn levels and an increase of the average intratumoral
pH. These modulations were observed in both IFN-sensitive (745) and IFN-
resistant (3C18) FLC tumors after at least 2 days of treatment (Fig.1,
panel A and ref. 11). Treatment of FLC tumors with X-rays did not result
in any comparable metabolic change 2 days after irradiation (11). In vitro
cultivation of 745 or 3C18 FLC with IFN α/β for 4 days did not result in
any significant ^{31}P NMR spectral difference with respect to untreated cells
(11).

As also shown in Fig. 1,(panel B), treatment of FLC tumors with highly
purified murine recombinant TNF also resulted in the reduction of PDE
(GroPCho and GroPEtn) and PME levels as well as in the increase of the
intratumoral pH. These metabolic alterations occurred as early as 1 h
after intratumoral injection of TNF, even though they were more marked 3-6
h after treatment (table 1). Similar changes in the PME and PDE levels and
in the intracellular pH were also observed in the TNF treated fibrosarcoma
cell tumors transplanted in syngeneic C3H/HeN mice. Moreover ^{31}P NMR analysis of ethanolic extracts of TNF-treated FLC tumors showed a marked increase in the levels of glycerophosphate with respect to control BSA-
treated tumors (data not shown). Such conspicuous increases were not observed in IFN-treated FLC tumors.

Recent studies in our laboratories (17) demonstrated that glucose-6-
phosphate (G-6-P) injected in FLC tumors in vivo is actively converted
into products of the Embden-Meyerhoff pathway whose concentration levels
can easily be monitored by ^{31}P NMR of tissue extracts. Combined ^{31}P and ^{1}H
NMR analyses of tissue extracts indicated that 6 h after injection, TNF
did not reduce the glycolytic capacity of these tumors in terms of formation of fructose-6-phosphate and lactic acid. Therefore the alkalinization
observed in FLC tumors after treatment with TNF does not seem to be due to
a reduced production of lactic acid.

Although the biochemical events underlying the metabolic modulations
described above are still under investigation,these results indicate that
both IFN and TNF may alter in vivo the phospholipid metabolism of tumor
cells by acting primarily on the host. The finding that these changes in
the phospholipid metabolism and in the intracellular pH can be revealed
before the appearance of tumor necrotic areas suggests that these phenomena do not simply reflect the necrotic process but are in some way related
to the antitumor effects of IFN and TNF. Moreover, these results further
underline the interest of monitoring by in vivo NMR spectroscopy metabolic
alterations induced by antitumor agents and suggest that variations in the
intracellular pH as well as in signal ares of PME and PDE can represent
early markers of tumor regression.

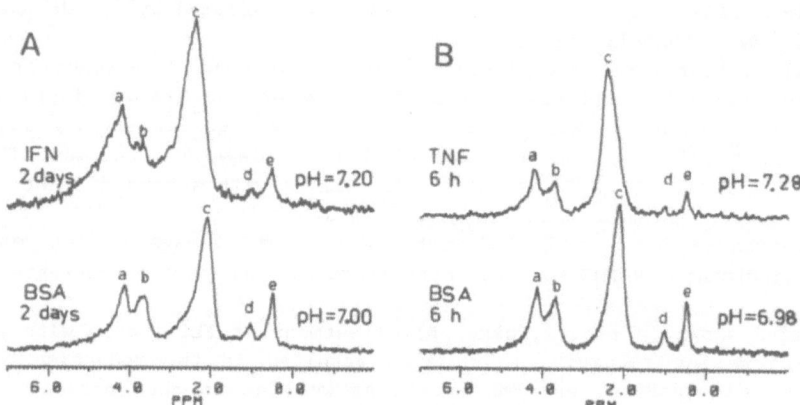

Fig. 1 : Typical ^{31}P NMR spectra of freshly dissected FLC tumors after in vivo treatment with BSA, IFN or TNF.

Panel A: 5-7 week-old male DBA/2 mice were inoculated s.c. with 5×10^6 745 cells. On day 11 of tumor growth, mice were injected daily with either BSA or IFN as described in Methods.

Panel B: 5-7 week-old male DBA/2 mice were injected s.c. with 5×10^6 745 cells. On day 8 of tumor growth mice were treated with either BSA or TNF (4 μg per mouse).

a) PEtn and αglycerophosphate, b) AMP and PCho, c) Pi, d) GroPEtn, e) Gro-PCho. The chemical shift of the peak c indicates the intracellular pH variation.

Table 1 : ^{31}P NMR determination of the effects of intratumoral injection of TNF on pH and some phosphorylated compounds in FLC (clone 745) and C3H fibrosarcoma tumors implanted s.c. in syngeneic mice.

TIME AFTER TNF ADMINISTRATION

	745 tumors			C3H fibrosarcoma tumors	
	1h	3h	6h	6h	12h
ΔpH *	+0.10	+0.12	+0.30	+0.12	+0.24
Pi	108%	114%	115%	132%	117%
PME (a+b) *	96%	89%	93%	69%	82%
GroPEtn (d) *	98%	74%	41%	<5%	<5%
GroPCho (e) *	79%	70%	49%	<5%	58%
PDE (d+e)	85%	71%	46%	<5%	40%

* The values represent the percentage of the relative signal area of each peak in the spectra of TNF-treated tumors with respect to the corresponding peaks of BSA- treated tumors.

REFERENCES

1) Gresser, I. and Tovey, M. G. (1978), Biochem., Biophys. Acta. 516: 231- 247.

2) Gresser, I. How does interferon inhibit tumor growth?. In "Interferon 6" (ed. I. Gresser). Academic Press, London, pp. 93-126, 1985

3) Belardelli, F., Gresser, I., Maury, C., and Maunoury, M., T., (1982a), Int. J. Cancer, 30: 813-820.

4) Belardelli, F., Gresser, I., Maury, C., and Maunoury, M., T.,(1982b), Int. J. Cancer, 30: 821-825.

5) Belardelli, F., Gresser, I., Maury, C., Duvillard, P., Prade, M., and Maunoury, M., T., (1983), Int. J. Cancer, 31: 649-653.

6) Fransen, L., Van Der Heyden, J., Ruysschaert, M., R., and Fiers, W., (1986), Eur. J. Cancer Clin. Oncol., in press.

7) Carswell, E., A., Old, L., J., Kassel, R., L., Green, S., Fiore, N., and Williamson, B., (1975), Proc. Natl. Acad. Sci. USA, 72, 3666-3670.

8) Haranaka, K., Satomi, N., and Sakurai, A., (1984), Int. J. Cancer,34: 263-267.

9) Balkwill, F., R., Lee, A., Aldam, G., Moodie, E., Thomas, J., A., Tavernier, J., Fiers, W., (1986), Int. J. Cancer, in press.

10) Gresser, I., Belardelli, F., Tavernier, J., Fiers, W., Podo, F., Federico, M., Carpinelli, G., Duvillard, P., Prade, M., Maury, C., Bandu, M., T., Maunoury, M., T., (1986), Int. J. Cancer, in press.

11) Proietti, E., Carpinelli, G., Di Vito, M., Belardelli, F., Gresser, I., and Podo, F., (1986),Cancer Res., 46: 2849-2857.

12) Warden, C., H., Friedkin, M., and Geiger, P., J., (1980), Biochem. Biophys. Res. Commun., 94: 690-696.

13) Carpinelli, G., Podo, F., Di Vito, M., Proietti, E., Gessani, S., and Belardelli, F., (1984), FEBS Lett., 176: 88-92.

14) Maris, J., M., Evans, A., Mc Laughlin, A., Bolinger, L., and Chance, B. In: Program of the Third Annual Meeting of the Society for Magnetic Resonance in Medicine, New York, August 13-17, 1984, pp. 501-502.

15) Fransen, L., Mueller, R., Marmenout, A., Tavernier, J., Van Der Heyden, J., Kawashima, E., Chollet, A., Tizard, R., Van Heuverswyn, H., Van Vliet, A., Ruysschaert, M., R., and Fiers, W., (1985), Nucl. Acid. Res., 4417-4429.

16) Carpinelli, G., Podo, F., Di Vito, M., Gresser, I., Proietti, E., and Belardelli, F., In: Book of Abstracts of the Fourth Annual Meeting of the Society for Magnetic Resonance in Medicine, London, August 19-23 1985, pp. 454.

ANTITUMOR EFFECTS OF RAT REC.IFN-γ IN RATS BEARING TRANSPLANTABLE RAT TUMORS

K. NOOTER, P.H. VAN DER MEIDE, J. DEURLOO, D.W. VAN BEKKUM and H. SCHELLEKENS

TNO Radiobiological Research Institute and TNO Primate Center, P.O. Box 5815, 2280 HV Rijswijk, The Netherlands

1. INTRODUCTION

Clinical studies have shown that IFNs can be effective towards solid tumors and hematological malignancies (4). However, analysis of these clinical results is difficult due to variations in the type of IFN and in the specific activity of the IFN preparations, the dose and schedule used in the trials, the route of administration, the duration of the treatment, the use of IFN alone or in combination with other treatments and the kind and stage of the disease. Evaluation of the results of animal studies on antitumor activity of IFNs has been hindered by the lack of sufficient pure species-specific IFNs. Recently, Schellekens and co-workers have been able to clone and express rat IFN-γ in E. coli and CHO cells. In contrast with natural T-cell derived IFN-γ this recombinant preparation is free of other lymphokines and biological effects can be unequivocally attributed to IFN-γ.

Preclinical and clinical studies suggest that when large tumor masses are present IFNs are generally not very active in decreasing the tumor volumes. We think that IFNs can best be tested for antitumor activity in tumor models at a stage of the disease comparable with complete remission in man, the so-called minimal residual disease stage. This situation can be induced by inoculation of low numbers of tumor cells. The aim of the study presented here was to assess optimal treatment modalities for rat recombinant IFN-γ, in rats bearing small numbers of tumor cells.

2. PROCEDURE

2.1 Materials and methods

2.1.1 Rats. Female rats of the inbred Brown Norway strain were used. The animals were bred under specific pathogen-free conditions and were 14 to 16 weeks old.

2.1.2 Tumor. The antitumor studies were carried out with a well-differentiated bronchioalveolar adenocarcinoma (LTR-27), originally induced with 40 rad total body irradiation in BN female rats. A tumor developed about two years after irradiation and was transplantable in BN rats. The subcutaneously transplanted tumor metastasizes to the lungs. The tumor doubling time in vivo is dependent on the number of cells inoculated and varies between 4 and 10 days.

2.1.3 Interferon. The molecular cloning and expression of rat IFN-γ by Dijkema et al. (2) made it possible to produce pure IFN-γ in sufficient amounts for pharmacokinetic and antitumor studies. The IFN preparation used in our studies was ± 98% pure and had a specific activity of ± 4 x 10^6 U/mg protein. The IFN-γ was purified from tissue culture fluid by affinity chromatography as described by Van der Meide et al. (5). IFN concentrations (U/ml) were determined by a biological antiviral assay employing a laboratory standard IFN-preparation as a comparative standard.

2.2 Experimental procedure

2.2.1 Pharmacokinetic studies. The pharmacokinetic behaviour of rat rec.IFN-γ was studied upon administration by i.v. and i.p. route. The i.p. IFN administrations were carried out at a single dose as well as at repeated doses. At specific time points (0.5, 1, 2, 4, 6, 8, 16 and 24 h) after IFN injection blood was taken by aortic punction under light ether anesthesia and thereafter the IFN titers in serum were determined.

2.2.2 Antitumor studies The antitumor studies with IFN-γ were designed to study the effects of IFN-γ on the outgrowth of s.c. transplanted LTR-27 tumor cells. The IFN-γ preparation was diluted in phosphate buffer (10 mM, pH 8.5), supplemented with 0.005% rat serum albumin and 0.5 ml aliquots were injected either i.v. or i.p. The controls received buffer plus serum albumin alone. We have chosen for a rather high pH of the inoculum because it has been shown that at lower pH values rat IFN-γ is rapidly degraded (2).

3. RESULTS AND DISCUSSION

3.1 Pharmacokinetic studies

The aim of the pharmacokinetic studies was to find out what route of administration is the most appropriate one for the maintenance of circulating IFN-γ levels over a prolonged period of time (e.g., one week). In rats the most appropriate (in terms of animal handling) routes for repeated injections are either the i.m. or i.p. one. Since it has been reported (3) that upon i.m. injection of HuIFN-γ in man only very low circulating antiviral titers are found, we only have studied repeated i.p. administration and have compared a single i.p. injection with a single i.v. push injection.

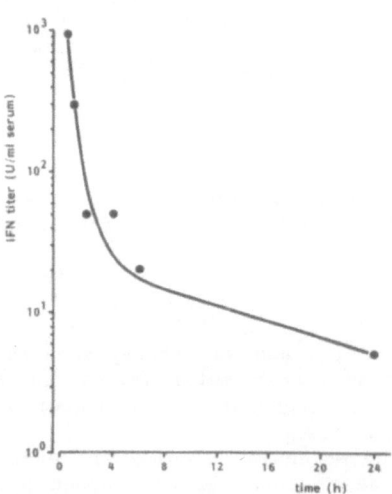

FIGURE 1. IFN serum titers in rats after an i.v. bolus dose of rat rec. IFN-γ (10^6 U/kg).

After an i.v. bolus dose of rat rec.IFN-γ (10^6 U/kg) into the tail vein of BN rats the serum disappearance curve can be fitted with equations describing an open two-compartment model with drug excretions from the central compartment only (Fig. 1). The $t\frac{1}{2}$ value of the distribution (α) phase is 18 min and of the elimination (β) phase 14 h. The area under the plasma curve $(AUC)_{0 \quad 24 \ h}$ was 1510 U/ml.h and the plasma clearance 1.6 ml/min. HuIFN-γ appeared to have a one-compartment distribution in man (3) but a two-compartment distribution in rabbits and rhesus monkeys (1).

After an i.p. injection of the same dose of IFN the $AUC_{0 \quad 24 \ h}$ appeared to be much smaller (975 U/ml.h) and subsequently the plasma clearance was quicker (2.5 ml/min) (Fig. 2). The bioavailability after i.p. administration is about 65%. Fig. 3 shows the IFN-γ serum titers in rats after repeated i.p. injections of rat rec.IFN-γ. The IFN was injected i.p. 3 times a day at a 4 h interval for 5 days (10^6 U/kg per injection). Hardly any accumulation of IFN-γ was seen in the serum. After 5 days of treatment about the same serum titers were detected as obtained after a single i.p. injection. Only the 24 h serum titer seems to be slightly increased.

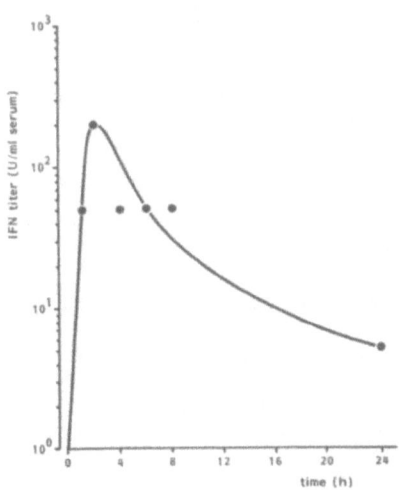

FIGURE 2. IFN serum titers in rats after an i.p. injection of rat rec. IFN-γ (10^6 U/kg).

Our idea was that IFN-γ would only be active as an antitumor agent when circulating IFN molecules are present during a (yet unknown) prolonged period of time. For that reason the antitumor experiments were carried out with an i.p. scheme of 7 days, 10^6 U/kg, i.p., 3 times a day.

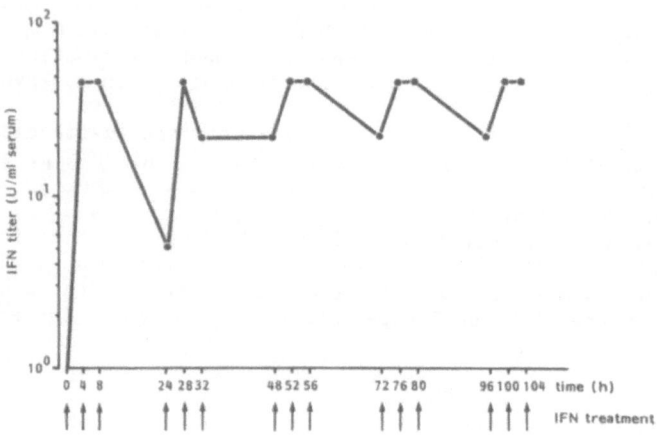

FIGURE 3. IFN serum titers in rats after repeated i.p. injections of rat rec. IFN-γ (10^6 U/kg per injection).

3.2 Antitumor studies

Fig. 4 shows the tumor take rate versus number of LTR-27 tumor cells inoculated s.c. Upon inoculation of 10^4 tumor cells in about half the cases palpable tumors will develop. Inoculation of 10^6 tumor cells will lead to about 100% take rate. For testing the antitumor activity of IFN-γ rats were injected s.c. with 10^3, 10^4, 10^5, 10^6, 10^7 or 10^8 LTR-27 cells. One day later the IFN-treatment was started (7 days, 10^6 U/kg, i.p., 3 times a day). It appeared that this IFN-treatment caused about a two log shift to the right of the tumor take rate curve. IFN treatment completely inhibited the outgrowth of 10^5 s.c. transplanted tumor cells. However, 10^8 cells transplanted s.c. and thereafter treated with IFN for one week still developed palpable tumors in 100% of the cases. These data suggest that IFN-γ has an antitumor activity of limited potential. The next experiments were set up to find out whether we could increase the antitumor activity of IFN-γ by increasing the treatment time.

Inoculation of 10^6 tumor cells and subsequently one-week IFN-treatment resulted in a 30% tumor take (Table I). Doubling the IFN-treatment time did not lead to better results. A 7 day-treatment of 10^5 s.c. transplanted cells was effective in preventing the outgrowth of tumors. However, a reduction of this treatment time to 3 days is not effective anymore.

In conclusion:
1.   The minimal duration for an IFN-γ treatment to be effective in our system is between 3 and 7 days.
2.   The antitumor effect of IFN-γ is rather limited. IFN-γ treatment causes about a two log cell kill irrespective of the number of cells

inoculated suggesting a direct effect of IFN-γ on the tumor cells.
3.   However, doubling the IFN-γ treatment time after s.c. inoculation of
     10⁶ tumor cells does not improve the results. This finding is not in
     favor of a direct effect of IFN-γ on the tumor cells, but rather
     suggest an (indirect) host mediated phenomenon.

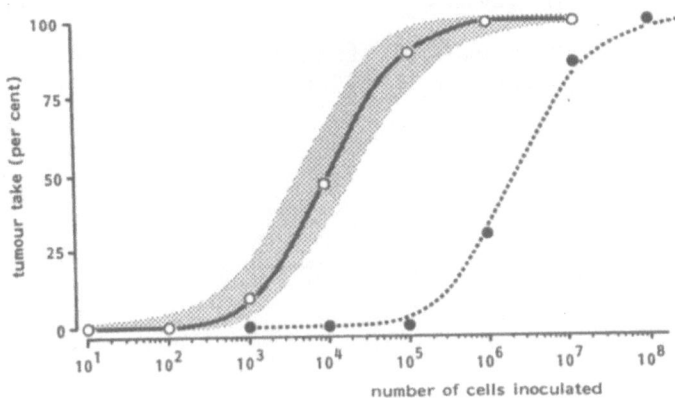

FIGUUR 4. The effect of rat rec.IFN-γ treatment on the take rate of LTR-27
tumor cells. Solid line: take rate in untreated rats; dotted line: take
rate in IFN-treated rats (10⁶ U/kg, i.p., 3 times a day).

TABLE 1. Treatment of LTR 27 tumor bearing rats with IFN-γ.

no. of LTR 27 tumor cells inoculated s.c.	duration of IFN-γ treatment	tumor take rate
10^5	controls	90%
	3 days	100%
	7 days	0%
10^6	controls	100%
	7 days	30%
	14 days	50%

4. ACKNOWLEDGEMENTS
   This work was supported, in part, by a grant from the Dutch Organization
for the Fight against Cancer, the "Koningin Wilhelmina Fonds".

5. REFERENCES

1. Cantell K, Hirvonen S, Pyhälä L, De Reus A and Schellekens H (1983). J. Gen. Virol. <u>64</u>, 1823-1826.
2. Dijkema R, Van der Meide PH, Pouwels PH, Caspers M, Dubbeld M and Schellekens H (1985). The EMBO Journal <u>4</u>, 761-767.
3. Kurzrock R, Rosenblum MG, Sherwin SA, Rios A, Talpaz M, Quesada JR and Gutterman JU (1985). Cancer Res. <u>45</u>, 2866-2872.
4. Strander H (1986). Advances in Cancer Research Vol. <u>46</u>, Acad. Press, New York.
5. Van der Meide PH, Dubbeld M, Vijverberg K, Kos T and Schellekens H (1986). J. Gen. Virol. <u>67</u>, 1059-1071.

MECHANISMS OF ANTICANCER ACTION OF INTERFERONS ON HUMAN TUMOUR XENOGRAFTS IN NUDE MICE

BALKWILL F R, WARD B G, FIERS W F *, RAMANI P

Imperial Cancer Research Fund
Lincoln's Inn Fields
London WC2A 3PX

*Laboratory of Molecular Biology,
State University of Ghent
Ledeganckstraat 35,
9000,
Ghent   Belgium,

INTRODUCTION

For several years we have been studying the effects of biological therapies such as interferons, IFNs, on subcutaneous xenografts of human bowel, breast, lung and ovarian cancers in nude mice (1,2). We have found that human IFN-αs have direct effects on sensitive human tumours which range from a slowing down of tumour growth or tumour stasis to complete regression. Other tumours are insensitive to therapy. The response of individual tumours remains similar over several tumour passages in the nude mouse, but after prolonged passage, some changes in sensitivity have been seen (Balkwill, unpublished data).

The tumour inhibition by human IFN-αs is due to a direct effect on the tumour cells because the human IFN therapy does not alter activity of the nude mouse immune system, increase levels of IFN induced enzymes in the mouse tissues, or give any signs of systemmic or haematological toxicity (1,2). The direct effects of IFN-αs on the human tumours may not of themselves be important in a clinical setting but when combined with other putative effects of IFNs on the host, they may contribute to an anticancer effect.

In contrast to the effects of IFN-α, human IFNs-β and γ had little or no anticancer activity on subcutaneous xenografts in this model system (2). However, human IFN-γ induced HLA class II expression on 2 of 3 human tumour xenografts in nude mice (3). Thus, human IFN-γ can have other direct effects on tumours which may affect a host immune response to them.

IFNs may also cause tumour stasis in this model system via host mediated mechanisms. Murine IFNα/β inhibited the growth of 3 subcutaneous xenografts irrespective of their sensitivity to human IFN-α (4). Extensive studies showed that the murine IFN had no activity on the human tumours per se (4) and the possibility that the tumour stasis was mediated by some non-immunological host effect is currently under investigation. Therefore, although this model system employs an immune deficient host, several components of the activities of biological response modifiers can still be investigated.

To further our studies in nude mouse model systems   we have looked at human tumours grown in 2 other sites. First we have studied the effects of IFNs and another biological response modifier, tumour necrosis factor, TNF, on intraperitoneal xenografts of human ovarian cancer. The biological behaviour, and anatomical site of these xenografts bears close resemblance to the disease in man (Ward et al, submitted for publication). Second, we have studied the effects of IFNs on the metastatic potential of human cells in nude mice.

MATERIALS AND METHODS
Mice

Female outbred six to ten week old nu/nu mice were used in all experiments with the intraperitoneal tumours. Six to ten week old female Balb-c nu/nu mice were used in the experimental metastases studies with DX3 melanoma cells. Mice were housed in negative pressure isolators as described previously (1,2).

Tumours

The OS line was derived from ascites from a 51 year old woman with Stage III moderately differentiated serous cystadenocarcinoma and was passaged by sterile removal of ascites from tumour bearing animals. The approximate time of passage was 6 weeks. This line grew solely as ascites (Ward et al, submitted for publication).

The La line was derived from ascites from a 72 year old woman with Stage IV poorly differentiated mucinous cystadenocarcinoma. This line grew as ascites and i.p. tumour and was passaged every 6 weeks.

The DX3 AzaC cells were derived from DX3 human melanoma cells by exposure to 5 azacytidine (5).

Reagents

Recombinant human interferon gamma, rHuIFN-γ was a gift from Biogen S.A. It was over 99% pure and had a specific activity of 2×10^7 U/mg. Its activity was tested in a biological assay and it was calibrated against reference standard Gg-23-901 530 (National Institute of Allergy and Infectious Disease, National Institute of Health, Bethesda, M.D.). Recombinant human tumour necrosis factor was greater than 99% pure and had a specific activity of 2×10^7 U/mg. Human lymphoblastoid interferon derived from Namalwa cells was a gift from Dr G Lewis, Wellcome Research Laboratories, Langley Court, Beckenham, Kent, UK. It had a specific activity of 2×10^8 U/mg. Recombinant murine IFN-γ, MuIFN-γ was a gift from Dr G Adolph, (Ernst-Boehringer Institut fur Areneimittel-Forshung, Bender and Co, Ges, MbH, Vienna).

All reagents were aliquoted in single dose amounts in PBS + 3mg/ml Bovine serum albumin, (Sigma Chemical Company, Dorset, UK), and stored at -70°C until use.

Experimental Procedures

0.3ml of a 1+1 dilution in tissue culture medium of OS or La cells was injected i.p. into each mouse and treatment started the next day, or on day 7 post injection. Treatment continued until all control mice died. Mice were observed weekly for signs of disease and sacrificed when the ascites was advanced and wasting was obvious. 5×10^5 DX3 AzaC cells were injected intravenously as described previously (5).

RESULTS

The effect of interferons and tumour necrosis factor on the development of intraperitoneal xenografts of ovarian cancer

After implantation of 0.3ml of tumour cell suspension mice remained well for 3-5 weeks. Thereafter, signs of disease, generally wasting and abdominal swelling, were obvious as the ascites and intraperitoneal tumours developed. Mice were treated with daily i.p. therapy immediately or 1 week after implantation of tumour cells. As shown in Fig 1 & 2 the OS tumour line was very sensitive to recombinant IFN-γ, rHuIFN-γ, or recombinant human TNF, rHuTNF, given daily from 24 hrs after tumour cell injection. Doses as low as 1ug/day rHuTNF protected the majority of mice from disease. In one experiment mice treated with 5ug TNF are still surviving 5 months after cessation of therapy and death of all control mice.

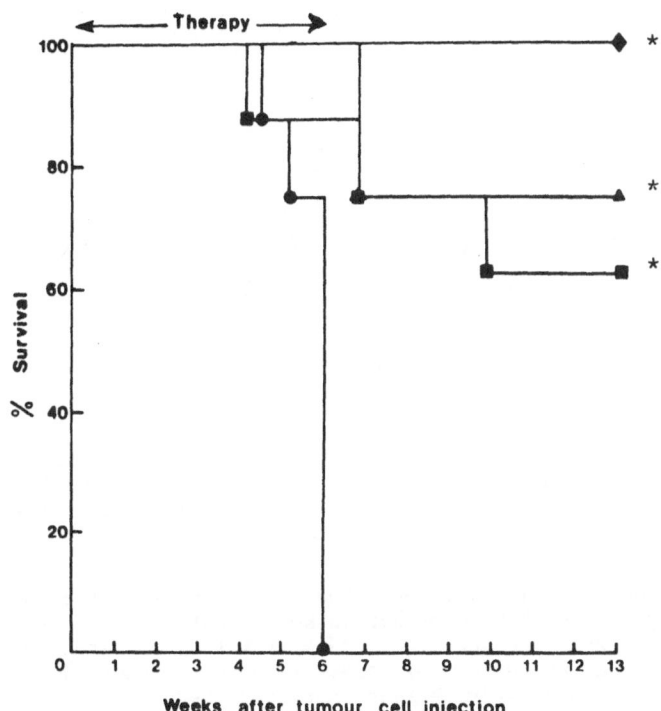

Fig. 1 The effects of rHuIFN-γ on the development of OS
ascites in vivo

Mice were treated daily i.p. with IFN from day 1 to 6 weeks
◆ 5X10⁴U rHuIFN-γ daily   ■ 1X10⁴U rHuIFN-γ daily
▲ 2X10⁴U rHuIFN-γ daily   ● Control diluent

*Mice still surviving

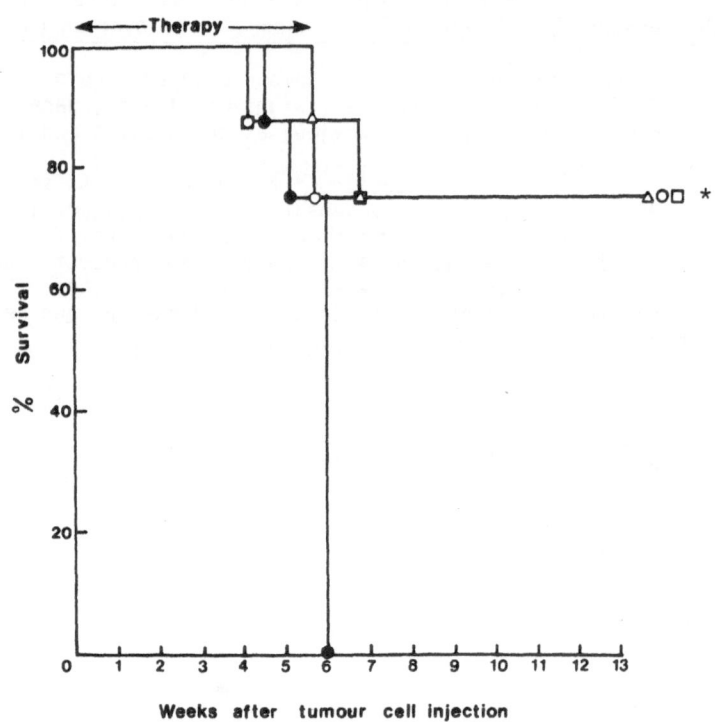

Fig. 2 The effect of rHuTNF on the development of OS ascites
in vivo

Mice were treated daily i.p. with TNF from day 1 to 6 weeks
—□— 10ug rHuTNF daily   —○— 5ug rHuTNF daily
—▽— 1ug rHuTNF daily   —●— Control diluent
*Mice still surviving

        In contrast, if mice were treated 1 week after the tumour implant-
ation, therapy with either agent was markedly ineffective. It can be seen
from Fig. 3 only 25% mice survived treatment with the highest dose of IFN-γ
5X10⁴u daily, when treated at + 7 days. With lower doses all mice died
(data not shown). Similarly, no mice survived when treated with 5ug or
1ug of TNF i.p. daily (see Fig 3 for an example of the results with 1ug.)
        Synergy between IFN-γ and TNF has been reported in several
in vitro studies (6,7). In the next series of experiments, we attempted
to see if a combination of these two agents could protect mice from
disease 1 week after tumour cell injection. Fig 3 shows that neither
5X10⁴U daily i.p. rHuIFN-γ or 1ug rHuTNF could prevent the majority of
mice from dying of ascitic disease 6-8 weeks after tumour injection. However, in the combination group all mice survived and were free of disease
13 weeks after the start of the experiment. A similar survival advantage
was seen when this combination was tested on a second ovarian cancer

xenograft, La. All control mice were dead 4 weeks after the start of the experiment but all in the TNF/IFN-γ combination group were alive and well at 6 weeks. As rHuTNF has activity in the mouse, we investigated whether the host played a role in the synergistic interactions reported here, by using a combination of rHuTNF and murine IFN-γ . In two experiments mice were not protected from disease by this therapy and rMuIFN-γ had no appreciable antitumour activity per se (data not shown). Thus we concluded that the therapeutic activity of the rHuIFN-γ and rHuTNF was due to a direct effect on the human tumour cells.

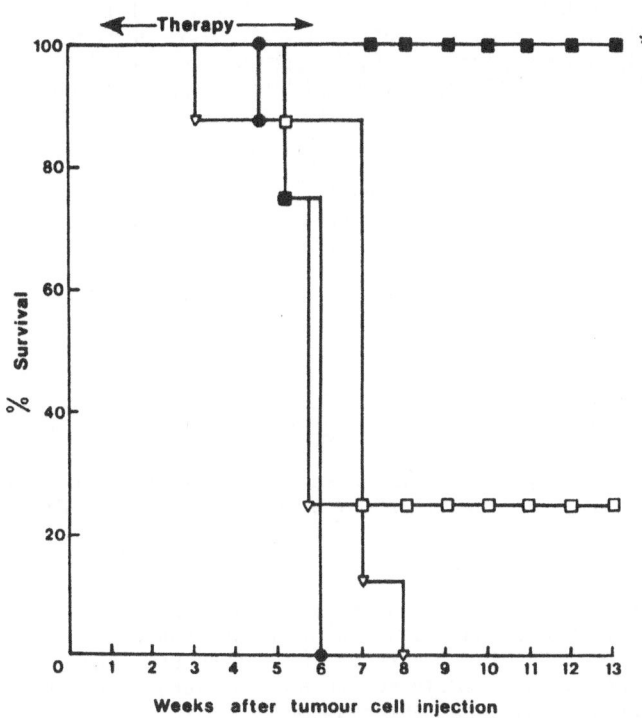

Fig 3   The effect of combinations of rHuIFN-γ and rHuTNF on
the development of OS ascites

Mice were treated daily i.p. from day 7 to day 42

- 5X10⁴U rHuIFN-γ + 1ug rHuTNF daily
- 5X10⁴U rHuIFN-γ   — 1ug rHuTNF
- Control diluent

* Mice still surviving

The effect of interferons on experimental metastases of
DX3 AzaC melanoma cells

When 5X10⁵ DX3 AzaC cells were injected i.v. into the Balb

c nu/nu mice, numerous lung nodules developed 4-6 weeks later (9). Table 1 shows the results of a typical experiment in which the mice were treated with human IFNs from the time of tumour cell injection. Inhibition of the development of lung colonies is seen with both HuIFN-γ and HuIFN- α

The doses of human IFNs used in this study had no effect on the activity of NK cells in the nude mouse lung. Control values for NK cell activity were 18% specific cytolysis of YAC lymphoma cells by control lung mononuclear cells, and 16% and 17% for rHuIFN-γ or HuIFN-α treated lungs respectively.

Moreover, the human IFN therapy did not increase the clearance of DX3 AzaC cells from the lungs of the Balb c nu/nu mice. When^{125}IUdR labelled tumour cells were injected i.v. all injected cpm were localised in the lungs by one hour. Thereafter, the clearance of the counts from thelungs was identical in mice treated with HuIFN-α or control mice.

In a typical experiment 50,000 cpm were recorded in the lungs two hours after tumour cell injection and by 72 hours the control group mean cpm/lung was 694, treated group 927.

Table 1
The effect of daily administration of
Human IFN-α or γ on the development of
DX3 AzaC lung nodules

		wts of lungs median(range)	'p'value	lung nodules median(range)	'p'value
	Control	200(140-390)		75(2-200)	
2X10^5 U	HuIFN-α	140(120-160)	0.0298	1(0-3)	0.0040
2X10^{5}U	rHuIFN-γ	145(130-150)	0.05	6(0-40)	0.108

Balb/c nu/nu mice were injected with 0.2ml s.c. IFN daily 5 times a week starting on the day of tumour cell injection. Mice were killed after 6 weeks.

Discussion

In previous experiments we have studied the mechanism of anti-tumour activity of interferon using subcutaneous xenografts of human tumours. We report here experiments where we have used two other anatomical sites for the tumour, namely peritoneum and lung. The use of these 2 sites has given us additional information.

First, in contrast to its lack of effect on subcutaneous tumours, we have found that HuIFN-γ can be an effective antitumour agent when given directly into the site of a peritoneal tumour and also when given i.p. to mice with experimental lung metastases. The reason for this difference is not known and may well be due to different distribution in the body of IFN-α and γ . We found that rHuIFN-γ was cleared from the circulation more rapidly than HuIFN-α (2). As the blood supply of subcutaneous xenografts is comparable to that of normal skin (F Balkwill, unpublished results), it is likely that insufficient levels of rHuIFN-γ can be sustained on the subcutaneous xenografts. Confirmation of this comes from preliminary studies in which we have found that culture of cell colonies derived from xenografts insensitive to rHuIFN-γ in vivo are sensitive to rHuIFN-γ in vitro (Balkwill et al, unpublished results). However, rHuIFN-γ can reach such tumours because it switched on or enhanced tumour cell HLA-D antigens in vivo (3). In the peritoneum it is probable that there is, at least for a short period, a high concentration of rHuIFN-γ which is sufficient for an antiproliferative effect.

DX3 AzaC cells were also sensitive to rHuIFN-γ when growing in the lungs. In this instance it is possible that microcolonies of cells would be more sensitive to the IFN than established subcutaneous xenografts which have a diameter of approximately 0.5cm when first treated.

Synergy between IFN-γ and TNF has been reported in vitro in many different cell systems (6,7). This, to our knowledge, is the first report of such synergy in vivo. The experiments using rMuIFN-γ show that a direct interaction between TNF & HuIFN-γ is responsible for the survival of mice in the combination group. This difference between the groups was dramatic. 8/8 mice in the combination group survived, compared with 0/8 in the TNF and 2/8 in the rHuIFN-γ groups. Preliminary experiments in a second xenograft that grew as ascites and i.p. tumour have indicated a similar survival advantage. These results indicate that such therapy with a combination of biologicals may have a role in the management of ovarian cancer in man.

Several studies have shown that IFNs or IFN inducers can have strong inhibitory effects on the development of experimental metastases (8,9,10). In these experiments IFNs enhancement of host immune defences, particularly NK cells, has been implicated. However, in the DX3 AzaC metastases model the human IFNs had no effect on the nude mouse lung or spleen NK cell activity, or white blood cell counts. Previous experiments have shown that human IFNs do not bind to nude mouse tissues as evidenced by a lack of elevation of the IFN induced enzyme 2-5A synthetase (1,2). The DX3 AzaC cells were very sensitive to the growth inhibitory effects of IFNs in vitro and our preliminary conclusion is that the antimetastatic effect of IFNs seen here was due to direct antiproliferative effect of the IFNs.

There is much controversy concerning the mechanisms of the anti-cancer activity of IFNs. Experimental evidence is available that in some model systems host mediated immune or non immune effects are important whereas in other systems, direct effects on the tumours predominate. Our evidence so far would suggest that in all the xenograft model systems we have investigated direct effects of IFNs on the tumour cells predominate and can have strong therapeutic activity.

REFERENCES

1. Balkwill F R, Moodie E M, Freedman V, Fantes K H. Human interferon inhibits the growth of established human breast tumours in the nude mouse. Int J Cancer 1982 30: 231-235

2. Balkwill F R, Goldstein L, Stebbing N. Differential action of six human interferons against two human carcinomas growing in nude mice. Int J Cancer 1985 35: 613-617

3. Balkwill F R, Stevens M H, Griffin D B, Thomas J S, Bodmer J G, Interferon gamma regulates HLA-D expression on solid tumours in vivo Europ J Cancer and Clin. Oncol. 1986 in press

4. Balkwill F R, Proietti E. The effects of mouse interferons on human tumour xenografts in the nude mouse host Int J Cancer 1986 in press

5. Ormerod E J, Everett C A, Hart I R.   Enhanced metastatic capacity of
a human tumour line following treatment with 5 azacytidine.   Cancer Res
1986   46: 884

6.Fransen L, Van Der Heyden J, Ruysschaert R, Fiers W.   Recombinant
tumour necrosis factor ; Its effect and its synergism with interferon on
a variety of normal and transformed human cell lines.   Eurp J Cancer,
Clin. Oncol.   1986   22: 419-426

7. Sugarman B J, Agaarwal B B, Hass P E, Girari I S, Palladino M A jnr,
Shepard H M.   Recombinant human tumour necrosis factor - : effects of
proliferation of normal and transformed cells in vitro. Science   1985
230: 943-945

8. Hanna N   The role of natural killer cells in the control of tumour
growth and metastasis.   Biochemica et Biophysica ACTA   1985   780: 213-226

9. Brunda M, Rosebaum D, Stern L   Inhibition of experimentally induced
metastases by recombinant interferon:correlation between the modulatory
effect of interferon treatment on natural killer cell activity and
inhibition of metastases.   Int J Cancer   1984   34: 421-426

10. Ramani P, Hart R, Balkwill F R,   The effect of interferon on
experimental metastases in immunocompetent and immunodeficient mice.
Int J Cancer   1986   37: 563-568

IFN-GAMMA INHIBITS OSTEOSARCOMA XENOGRAFTS IN NUDE MICE

Otte Brosjö, Henrik C.F. Bauer, Olle S. Nilsson, Hans Strander.
Departments of Orthopaedics and Oncology, Karolinska Institute and
Hospital, S-104 01 Stockholm, Sweden.

Since 1981, we have transplanted fresh osteosarcoma tissue to nude
mice and succeeded in establishing 13 osteosarcoma xenografts in serial
passage. Studies on these human xenografts have not revealed any
significant changes in tumor characteristics compared to the original
tumor (1). The sensitivity of 11 xenografts to nIFN-alpha has been studied
(2). All xenografts could be growth inhibited but the lowest growth
inhibitory daily dose ranged from 1×10^5 IU for the most sensitive xenograft
to 1×10^6 IU for the most insensitive. No IFN resistent tumors were found.
The induced growth inhibition has mainly been attributed to a direct
anti-tumor effect since human IFN does not effect mouse cells.

In this study, we have investigated the effect of natural and
recombinant IFN-gamma on osteosarcoma xenografts, both alone and in
combination with natural or recombinant IFN-alpha.

MATERIALS AND METHODS

The tumors have originally been obtained at surgery and directly
transplanted to female nu/nu BALB/c nude mice. Four different osteosarcoma
xenografts (labelled T2, T7, T8, and T11), maintained in serial passage,
were used. The histological characteristics and sensitivity to nIFN-alpha
is presented in Table 1.

There was one tumor per mouse and treatment was started when the mean
tumor volume was more than 80mm^3. Tumor bearing mice were divided into one
control group and one or more IFN treatment groups with 4-8 animals in
each group.

The human nIFN-alpha (spec. activity 3.3×10^6 IU/mg protein) and human
nIFN gamma (1×10^6 IU/mg protein) were produced and purified at the Finnish
Red Cross Blood Transfusion Service and provided by Dr. Kari Cantell. The

rIFN-alpha (2×10^8 IU/mg protein) and rIFN-gamma (1×10^7 IU/mg protein) were provided by Boehringer Ingelheim. The IFNs were diluted to proper concentration with saline and given by daily subcutaneous injections at a distance from the tumor. The dose given was expressed in IU. The effect on growth was evaluated by measurements of tumor volumes at regular intervals.

Table 1. Histologic data and nIFN-alpha sensitivity of 4 osteosarcoma xenografts.

Tumor	Type	Grade	Lowest growth inhibitory dose of nIFN-alpha
T2	OBL	IV	2×10^5
T7	CHO	III	1×10^6
T8	CHO	III	1×10^6
T11	FIB	IV	2×10^5

OBL; Osteoblastic, CHO; Chondroblastic, FIB; Fibroblastic.

RESULTS
Recombinant IFN-alpha and -gamma.

Daily injections of 2×10^5 IU of rIFN-gamma had no effect on the growth of T7 and T8 neither alone, nor in combination with 2×10^5 IU of nIFN-alpha. A weak but not significant growth inhibition was seen with 2×10^5 IU of rIFN-gamma to T2.

The daily dose of rIFN-gamma was therefor increased in a subsequent experiment with T11. 5×10^5 IU of rIFN-gamma had still no effect on the growth but 1×10^6 IU significantly inhibited the growth of this tumor. In the same experiment, 2×10^5 IU of rIFN-alpha almost arrested the growth and a combination of 1×10^6 IU rIFN-gamma and 2×10^5 IU rIFN-alpha actually resulted in a reduction of mean tumor volume.

Natural IFN-alpha and -gamma.

2×10^5 IU nIFN-alpha almost completely arrested the growth of T11 and the same dose of nIFN-gamma resulted in tumor size reduction. A combination of 2×10^5 IU of nIFN-alpha and nIFN-gamma induced an even stronger antitumor effect since almost no tumors remained after 30 days of treatment. (Fig.1)

Fig. 1 Growth of T11. Treatment was started at arrow.

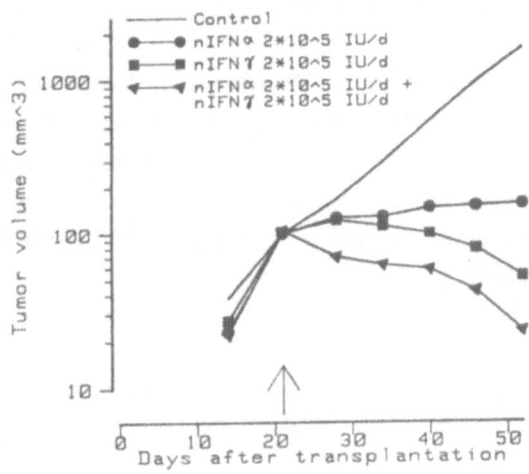

A stronger growth inhibition has earlier been found when nIFN-alpha was injected every day compared to every third day, with the same total amount of IFN. A comparison between nIFN-gamma in a dosage of $2x10^5$IU given daily and $1x10^6$IU of nIFN-gamma given twice daily was thus performed with T11. No difference in tumor growth inhibition was seen after 27 days of treatment.

DISCUSSION

These experimets have demonstrated that IFN-gamma, both recombinant and natural, can inhibit the growth of osteosarcoma xenografts. The necessary daily dose, to obtain growth inhibition of T11, was $2x10^5$IU of nIFN-gamma, nIFN-alpha and rIFN-alpha. However, it seems that rIFN-gamma was less effective since this dose had no effect at all on T11 and also no significant effect on the other 3 xenografts. Abscence of growth inhibition with $2-4x10^5$IU/d of rIFN-gamma to nude mice xenografts have also been reported by other authors (3,4). The difference in effectiveness between recombinant and natural IFN-gamma can be explained by differences in _in vivo_ absorption, stability and clearance since Balkwill et al. (3)

reported a higher antitumor effect of rIFN-gamma given i.p. compared to
the s.c. route. This assumption is supported by the study on Tll where the
extremely high rIFN-gamma dose of $1x10^6$IU/day did have a significant
growth inhibitory effect.

The experiments with combinations of IFN-alpha and IFN-gamma shows that
a definite additive anti-tumor effect was present. The greatest reduction
of tumor volumes was seen with IFN combinations, which may be suggestive
of a synergistic anti-tumor effect, but may also be explained by the fact
that the more IFN that is given, regardless of type, the stronger
anti-tumor effect is evoked.

REFERENCES

1. Bauer HCF, Brosjö O, Broström L-Å, Nilsson OS, Reinholt FP, Tribukait
B: Eur J Cancer Clin Oncol 22:821-830, 1986.

2. Brosjö O, Bauer HCF, Broström L-Å, Nilsson OS, Reinholt FP, Tribukait
B: Cancer Res In press, 1986.

3. Balkwill FR, Goldstein L, Stebbing N: Int J Cancer 35:613-617, 1985.

4. Twentyman PR, Workman P, Wright KA, Bleehen NM: Br J Cancer 52:21-29,
1985.

TRANSGENIC MICE CARRYING EXOGENOUS MOUSE INTERFERON GENES

Yoshimi KAWADE, Masahide ASANO, Hitoshi NAGASHIMA, and Yoichiro IWAKURA*
Institute for Virus Research, Kyoto University, Kyoto, and *Institute of
Medical Science, University of Tokyo, Tokyo, Japan

1. INTRODUCTION

Interferon is well known to be highly pleiotropic in its action. In cells it acts on, it induces or modulates expression of various genes (1,2), including major histocompatibility antigen genes (3,4), and cellular onc genes (5), besides antiviral protein genes. It seems natural, therefore, to presume that IFN plays important roles in immune functions, cell growth and differentiation, embryonic development, and so on. However, most investigations so far were concerned with events in cells in culture or with effects of artificial administration of IFN to animals; questions concerning the physiological roles of IFN in the body remain largely unanswered (6,7), such as whether endogenously produced IFN has any role in normal immune responses, and whether IFN has anything to do with normal embryonic development.

One possible approach to these questions will be to produce transgenic mice that harbor extra mouse IFN genes, and see the effects of the endogenously produced IFN on the animal. It will also be of much interest to see whether such mice are highly resistant to virus infections. Various mouse IFN genes have been cloned (8), and also, various constructions of plasmid consisting of an appropriate enhancer-promoter and the IFN-coding sequence are possible that would allow either constitutive or induced production of IFN. The use of a single IFN gene in this way might highlight the role of that particular IFN species.

In this report, our initial attempts to introduce mouse IFN-β cDNA (9) into fertilized mouse eggs are described. At first, we tried to introduce the IFN gene by using a SV40 early gene-replacement vector (10), but no integration of the gene into the chromosome was observed. Then, two constructions of DNA, in which the same IFN cDNA was ligated downstream to either the long terminal repeat (LTR) of murine leukemia virus or to the metallothionein enhancer-promoter region, were injected to fertilized mouse eggs. So far, 55 pups were born, and 5 of them carried intact exogenous IFN gene. One of them carried only one extra IFN gene, but others had multiple copies, and three of them transmitted the genes to the offspring. Though various tissues were examined, we did not detect any IFN mRNA, nor did we find IFN activity in the serum. Therefore, expression of these exogenous genes appeared to be repressed by some mechanism in the transgenic mice we could so far obtain.

306

2. MATERIALS AND METHODS

2.1. Plasmids
DNA constructions used in this study are illustrated in Fig. 1.
pSV40-Muß is a transducing viral DNA in which mouse IFN-β cDNA (9) is ligated downstream to the early gene promoter of SV40.   Upon transfection of the DNA to COS cells, infectious virus particles (called SV40-Muß) are produced without involvement of helper virus (10); this was used previously to infect various cells in culture and transduce the mouse IFN-β cDNA (10).
The second construction, pSVX-Muß, was prepared by inserting IFN-β cDNA downstream to the LTR at the Bam HI site of pZIP-NeoSV(X)I, a Moloney murine leukemia virus vector (11) (kindly provided by Dr. R.C. Mulligan, together with ψ2 cells).   This also can produce infectious virus particles (called SVX-Muß) when transfected to 2 cells.
In the third construction, pMK-Muß, IFN-β cDNA was ligated to the enhancer-promoter region of the metallothionein I gene at the Bgl II site, upstream to the thymidine kinase gene (12; pMK and m$_1$pEE$_{3.8}$ kindly supplied by Dr. R.D. Palmiter).
For microinjection, DNA was digested with EcoRI, and dissolved in 10 mM Tris-HCl (pH 7.5)-0.5 mM EDTA at a concentration of 10 µg/ml.

2.2. Embryos
Mice (CF1) were superovulated with injection of pregnant mare serum (Sankyo-zoki) at 17:00, followed by human chorionic gonadotropin 48 hr later, and mated immediately.   For virus infection, embryos harvested at 2-cell stage were stripped of zona pellucidae by pronase treatment, and cocultured with the transducing viral particles in a Terasaki plate under mineral oil until they developed into the blastocyst stage.   Then, they were brought back to the uteri of a pseudo-pregnant foster mother (ICR) that had been mated one day later than the embryonic donor.   For microinjection of DNA, embryos at one-cell stage were collected at 14:00 on the day vaginal plugs were observed.   DNA was injected to the male pronuclei of the embryos, which were brought back to the oviduct of foster mothers mated

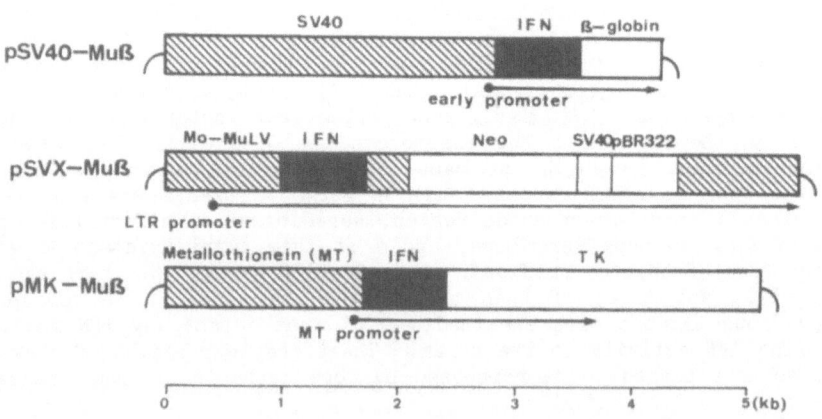

FIGURE 1. DNA constructions

on the same day.   Culture conditions of embryos were described previously
(13).

2.3. Hybridization

For Southern hybridization, DNA was prepared from the tail or liver,
digested with restriction enzymes, and after electophoresis, transferred to
a nitrocellulose filter (14).   Hybridization was carried out as described
(9,15).   For Northern hybridization, total RNA from various tissues was
prepared by guanidium thiocyanate-CsCl centrifugation method   (16).   RNA
was electrophoresed in 2.2 M formaldehyde, and transferred to a nitrocellulose filter.   Hybridization was done overnight at 42°C in 5 x SSC, 0.1%
bovine serum albumin, 0.1% Ficoll 400, 0.1% polyvinyl pyrrolidone, 50 mM
sodium phosphate (pH 6.5), 50% formamide, 0.1% sodium dodecylsulfate (SDS),
100 µg/ml denatured salmon sperm DNA, 10% dextran sulfate and ^{32}P-labeled
probe DNA (17,18).   After hybridization, the filter was washed 3 times in 2
x SSC, 0.1% SDS at room temperature for 30 min each time, and then 3 times
in 0.2 x SSC, 0.1% SDS at 45°C for 30 min.   The filter was autoradiographed
at -80°C for 2 days.

3. RESULTS AND DISCUSSION

First, we attempted to use the transducing virus, SV40-Muβ, to introduce
mouse IFN-β cDNA into the mouse chromosome.   Earlier, it was reported that,
when wild type SV40 infected mouse embryos at the morula stage, the viral
genome was integrated into the host chromosome (19).   Our virus, SV40-Muβ,
had been shown to be able to transduce the IFN gene efficiently into cultured cells (10); it is a defective virus free of helper, and therefore we
expected that the target cells would not be disturbed by the virus multiplication processes.   We allowed the virus to infect 2-cell stage embryos,
and after incubation for 48 h,   the blastocyst stage embryos were brought
back to the uteri of foster mothers.

As shown in Table 1, 72 pups were born from 303 blastocysts which were
brought back to foster mothers, although the high doses of the virus used
were toxic for the embryo (at 45 TCID$_{50}$/ml, the virus input is approximately 1,000 plaque-forming units/embryo).   DNA from these mice was
examined for integration of the extra IFN cDNA by Southern hybridization
using IFN cDNA as the probe, but none of the mice was found positive.
Thus, the viral vector did not appear to be as efficient in transferring
IFN gene to fertilized eggs as to cultured cells.   The difference between
our virus SV40-Muβ and the wild type SV40 could be explained by the fact
that wild type genome can multiply to some extent in embryonic cells, in
contrast to the transducing virus which is totally defective.

Next, we tried to introduce DNA by microinjection to the pronuclei of
fertilized eggs. pSVX-Muβ, in which the IFN gene was ligated to the LTR
structure of murine leukemia virus, and pMK-Muβ, in which the DNA was
ligated to an inducible promoter, were used.   Both of the constructions
were confirmed to be functional by transfecting the DNA to cultured cells,
and pMK-Muβ was shown to be responsive to Cd^{2+} induction (Asano et al.,
manuscript in preparation).   After linearization, 500 to 1,000 copies of
the DNA were injected into the pronuclei of fertilized eggs.   Table 2 shows
the frequency of integration of the DNA to the host chromosome.

Integration of the DNA was investigated by Southern hybridization of
liver or tail DNA using mouse IFN-β cDNA as the probe (Fig. 2).   So far,
2,756 eggs received injection of DNAs, and 1,614 eggs survived it.   Fifty-

TABLE 1. Frequency of offspring born from the embryos treated with SV40-Muβ

| | Control | Input dose of SV40-Muβ ($TCID_{50}/ml$) | | |
		4,500	450	45
Blastocyst formation*	35% (138/391)	0.5% (2/414)	19% (197/1,048)	25% (160/637
Offspring born**	42% (40/95)	-	17% (28/163)	31% (44/140

*Number of embryos that developed to blastocysts at 48h of cultivation/total number of embryos treated.
**Pups born/embryos implanted to the uteri.

TABLE 2. Frequency of transgenic mice produced.

DNA	Eggs injected	Eggs that survived	Pups born	Pups with extra IFN gene
pSVX-Muβ	1,097	450	26	1
pMK-Muβ	1,659	1,164	29	4
Total	2,756	1,614	55	5

five pups were born from the 1,538 eggs which were brought back to oviducts, and five of them were shown to have integrated the DNA. The frequency of integration of pSVX-Muβ was not higher, but rather lower, than that of pMK-Muβ, although we had expected that the presence of the LTR structure in the former would help integration of the DNA.

Patterns of Southern hybridization of DNA from the mice with integrated pMK-Muβ are shown in Fig. 2. All the mice showed extra bands other than those of endogenous IFN gene (3.1 kb band in the PvuII digest, 2.2 and 10 kb in the Bam HI digest, and 3.0 kb in the EcoRI digest). Judging from the intensity of the bands, mouse T124 contained approximately one copy of the extra IFN gene, T126 had about 9 copies, T161 about 6 copies, and T162 about 3 copies. The way of integration was deduced from the data shown in Fig. 2 and also from those obtained using pBR322 and TK gene fragments as the probes (not shown);the most likely structures are shown in Fig. 3, though other structures may also be conceivable for T126 and T162.

Interestingly, in spite of the fact that the IFN gene was cleaved from the pBR322 sequence before injection, all the four mice contained the pBR322 sequence next to the IFN gene. The pattern of integration was very complicated in most cases; some pBR322 sequences and IFN genes were ligated head to tail, but in some part, there were head to head or tail to tail structures. These patterns of integration suggest that the Eco RI fragments of the MT-IFN-TK and pBR322 portions were ligated randomly before integration into the host chromosome. Three mice out of the four which carried pMK-Muβ (except T124) were shown to transmit the DNA to the offspring in a Mendelian manner without rearrangement (data not shown). T124 is considered to be a mosaic mouse, consisting of cells that carry the extra IFN DNA and those that do not; the germ line belongs to the latter.

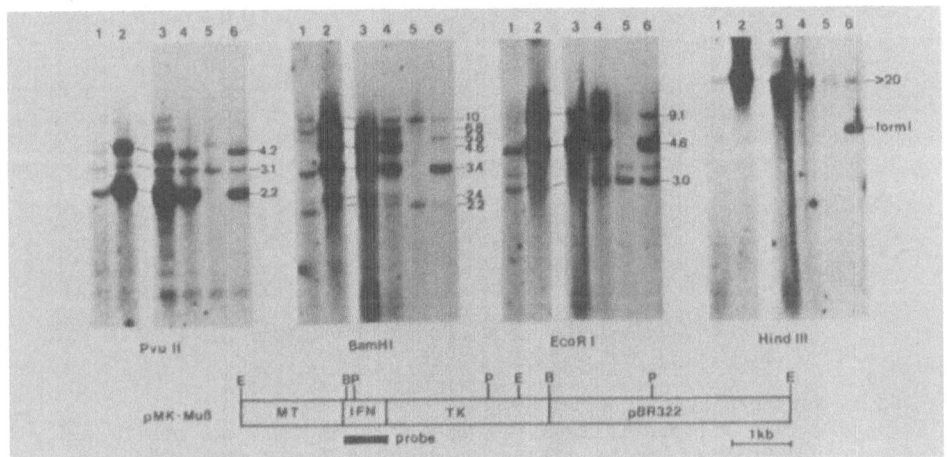

FIGURE 2. Southern hybridization
Mouse tail DNA (10 μg) was digested with Eco RI (E), Bam HI (B), Pvu II (P), or Hind III (H), and after electrophoresis, the DNA was blotted to nitrocellulose filters and hybridized with ^{32}P-labeled mouse IFN-β cDNA. 1: mouse T124, 2: T126, 3: T161, 4: T162, 5: control mouse, 6: control plus one copy equivalent amount of pMK-Muβ (30 pg). Form I means closed circle form of the plasmid.

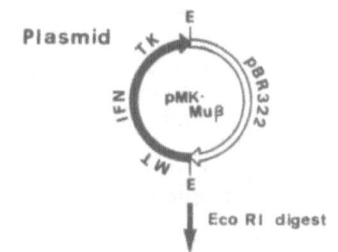

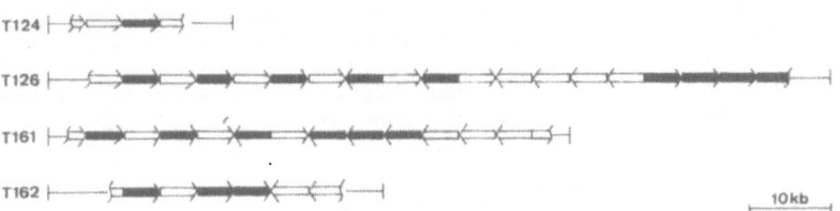

FIGURE 3. Possible patterns of integration of extra IFN genes into the host chromosome.
Filled and open arrows indicate structures related to the metallothionein-IFN-thymidine kinase gene and pBR322, respectively.

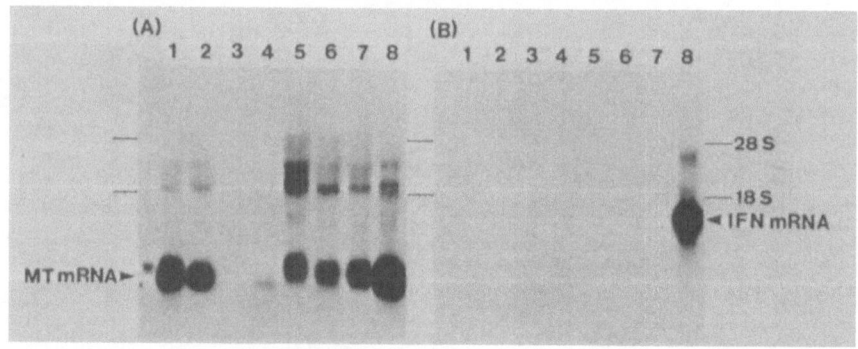

FIGURE 4. Northern hybridization.
Total RNA from various tissues (20 µg) of mouse T126 was hybridized with mouse metallothionein probe (A) and mouse IFN probe (B). 1: liver, 2: kidney, 3: spleen, 4: small intestine, 5: testis, 6: heart, 7: brain, 8: L cells infected with Newcastle disease virus.

Expression of the integrated DNA was investigated by dot hybridization and Northern hybridization using IFN cDNA as the probe. No RNA preparation from liver cells of any mouse which carries pMK-Muβ hybridized to the IFN probe, even after induction with $CdSO_4$ (1 mg/kg body weight, 18 h before sampling; data not shown). To examine the possibility that the introduced gene is expressed in unexpected tissues other than the liver, we examined RNA from various tissues after Cd^{2+} induction. As shown in Fig. 4, no IFN mRNA was detected in any tissue examined, whereas the expression of endogenous metallothionein gene was amply detected. Furthermore, we assayed the serum of all the transgenic mice for IFN activity, but we could detect none (less than 15 IU/ml). Therefore, expression of the IFN gene appeared to be repressed in these transgenic mice. Since no deletion or rearrangement in the IFN gene was observed, this repression could be due to inhibitory effects of the flanking sequences at the integration sites (including the pBR322 sequences), or due to a mechanism of the host to inactivate foreign genes, such as DNA methylation. However, it may also be possible that only nonproducing embryos can survive and are selected for, if endogenously produced IFN is harmful to the embryonic development. In this connection, it should be noted that the metallothionein promoter was not totally inert, and significant amounts of metallothionein mRNA was detected in the liver or kidney under uninduced conditions. We are now trying to reduce the background expression during the embryonic development, and also trying to introduce IFN genes with another inducible promoter, that of heat shock protein.

REFERENCES

1. Lengyel P: Biochemistry of interferons and their actions. Annu. Rev. Biochem. 51, 251-282, 1982.
2. Revel M and Chebath J: Interferon-activated genes. Trends in Biochem. Sci. 11, 166-170, 1986.

3. Friedman RL, Manly SP, McMahon M, Kerr IM and Stark GR: Transcriptional and posttranscriptional regulation of interferon-induced gene expression in human cells. Cell 38, 745-755, 1984.
4. Ozato K, Wan Y-J and Orrison BM: Mouse major histocompatibility class I gene expression begins at midsomite stage and is inducible in earlier-stage embryos by interferon. Proc. Natl. Acad. Sci. USA 82, 2427-2431, 1985.
5. Einat M, Resnitzky D and Kimchi A: Close link between reduction of c-myc expression by interferon and G_0/G_1 arrest. Nature 313, 597-600, 1985.
6. Bocci V: The physiological interferon response. Immunol. Today 6, 7-9, 1985.
7. Kawade Y: The interferon system in the mouse. In "The Interferon System: A Current Review". Baron S et al., eds. Texas Univ. Press, in press.
8. Weissmann C and Weber H: The interferon genes. Prog. Nucleic Acid Res. Mol. Biol., in press.
9. Higashi Y, Sokawa Y, Watanabe Y, Kawade Y, Ohno S, Takaoka C and Taniguchi T: Structure and expression of a cloned cDNA for mouse interferon-β. J. Biol. Chem. 258, 9522-9529, 1983.
10. Asano M, Iwakura Y and Kawade Y: SV40 vector with early gene replacement efficient in transducing exogenous DNA into mammalian cells. Nucl. Acids Res. 13, 8573-8586, 1985.
11. Cepko CL, Roberts B and Mulligan RC: Construction and applications of highly transmissible murine retrovirus shuttle vector. Cell 37, 1053-1062, 1984.
12. Brinster RL, Chen HY, Trumbauer M, Senear AW, Warren R and Palmiter RD: Somatic expression of herpes thymidine kinase in mice following injection of a fusion gene into eggs. Cell 27, 223-231, 1981.
13. Iwakura Y and Nozaki M: Effects of tunicamycin on preimplantation mouse embryos: prevention of molecular differentiation during blastocyst formation. Dev. Biol. 112, 135-144, 1985.
14. Southern EM: Detection of specific sequences among DNA fragments separated by gel electrophoresis. J. Mol. Biol. 98, 503-517, 1975.
15. Roop DR, Nordstrom JL, Tsai SY, Tsai M-J and O'Malley BW: Transcription of structural and intervening sequences in the ovalbumin gene and identification of potential ovalbumin mRNA precursors. Cell 15, 671-685, 1978.
16. Chirgwin JM, Przybyla AE, MacDonald RJ and Rutter W: Isolation of biologically active rebonucleic acid from sources enriched in ribonuclease. Biochem. 18, 5294-5299, 1975.
17. Thomas PS: Hybridization of denatured RNA and small DNA fragments transferred to nitrocellulose. Proc. Natl. Acad. Sci. USA 77, 5201-5205, 1980.
18. Wahl GM, Stern M and Stark GR: Efficient transfer of large DNA fragments from agarose gels to diazobenzyloxymethyl-paper and rapid hybridization by using dextran sulfate. Proc. Natl. Acad. Sci. USA 76, 3683-3687, 1979.
19. Willison K, Babinet C, Boccara M and Kelly F: Infection of preimplantation mouse embryos with simian virus 40. Cold Spring Harbor Symp. Quant. Biol. 46, 307-317, 1982.

STUDIES ON INTERFERON IN THE BOVINE SPECIES : BIOLOGICAL AND BIOCHEMICAL EFFECTS.

C. VANDENBROECKE*, A. SCHWERS°, J.E. SEVERIN*, M. BUBLOT°,K. HAUZEUR*,S. PISCITELLI*, E. THIRY°, P. ZILIMWABAGABO*, M. MAENHOUDT°, L. DEVOS*, P.P. PASTORET ° and J.J. WERENNE.
*. Faculty of Sciences, Université Libre de Bruxelles, Belgium.
°. Faculty of Veterinary Medicine, Université de Liège, Belgium.

1. INTRODUCTION

In 1982, our group obtained the first evidence for the efficiency of interferon as an antiviral in cattle (1). In view of its importance for the veterinary field, this successful achievement, using on the basis of the rational short cut we proposed (2) the only cloned interferon available in sufficient quantity at this moment (Hu IFN α2), prompted us to further investigate this model on a fully quantitative basis (3,4) and to evaluate the action of interferon on other viral infections of economical importance for breeders.

As we showed *in vitro* that Rotavirus (5) and Bovine Herpes Virus I (6) respectively involved in newborn calf diarrhoea and in bovine rhinotracheitis, one of the manifestations of the shipping fever symptoms, were very differently susceptible to interfferon we concentrated first our efforts on these viruses, to correlate further the *in vivo* situation to the *in vitro* effect.

It appeared that interferon response in colostrum deprived newborn calves infected with bovine rotavirus exerted a regulating role on the pathogenicity (7,8) and that exogenous interferon was active in the animals experimentally infected with rotavirus alone (9) or with the virus and enterotoxigenic Escherichia coli (10). The data we have obtained with Bovine Herpes virus 1 (BHV1) demonstrated that even if interferon does not fully block viral multiplication, it exerts clear biological effects on the animal (11).

In this paper we describe the biological effects of Human interferon in newborn calves and in older animals, whether infected or not by these viruses, in comparison to its biochemical effects, namely the induction of 2'5' A synthetase or Protein Kinase. Before doing this we wish to mention here that the first evidence for an antiviral action of interferon in cattle that we have obtained five years ago, using Human interferon against an experimental infection with Vaccinia virus, at a time when sufficient bovine interferon was not available, has been now confirmed using interferon from the homologous species.Two groups Genentech (12) and the Israel Institute for Biological Research (13) ,are indeed able now to produce efficiently Bovine interferon by genetic engeneering in E. coli. With the collaboration of Medical Research Limited (MRL Australia) and IIBR, we had recently the opportunity to show that intramuscular injection of Bovine interferon (Bo IFN a c) is very efficient in protecting young calves against a vaccinia infection. We showed also that this interferon induces a high level of 2'5' A synthetase in the lymphocytes of those animals. These yet unpublished results show that the data obtained previously with Human interferon are also valid for Bovine interferon. Our approach (14) is therefore meaningful even on the therapeutical point of view.

2. PROCEDURE
2.1. Materials and methods.
2.1.1. Interferons. The interferons used were supplid either by Dr. C. Weissman from Zürich University (HuIFN α 2) or by Dr. P. Swetly (Boehringer Ingelheim, HuIFN α2 Arg). Their titration in the collected plasma, was made using the CPE technique with VSV. In some experiments IFN was quantified also using an ELISA assay.

2.1.2. Animals and viruses. For the experiments involving rotavirus infection, newborn calves strictly deprived of colostrum, were treated within the few hours following calving with interferon and/or the virus strain 81/36F as we have described previously (9). When indicated, the animals were also orally infected 24 hours later with E. coli 126 A at $2,5 \times 10^9$ bacteria The experimental procedure used with older animals (6-8 months for the study with BHV 1, infection, was described in detail elsewhere

2.1.3. Determination of the 2'5' A synthetase activity. The method for preparation of the lymphocytes as used in the 2'5' Asynthetase assay was described before (11, 15). For determination of the enzyme in the different tissue extracts studied, the procedure was adpated accordingly (16).

2.1.4. Determination of the protein kinase. The assay was derived from an assay recently described by C. Buffet-Janvresse et al (17). The enzyme was measured in unfractioned plasma obtained after low speed centrifugation of heparinized blood samples, using calf thymus histone as substrate.

3. RESULTS
3.1. Inhibition of interferon induction in virus infected animals by treatment with exogenous interferon
When heifers were intranasally infected with BHV1, an important amount of interferon was induced and appeared in the blood circulation as the bovine rhinotracheitis symptoms developped. Fig. 1 showed that this induction was

totally suppressed by intramuscular interferon treatment (10^6 U/kg for 6 consecutive days starting 24 hours before viral infection).

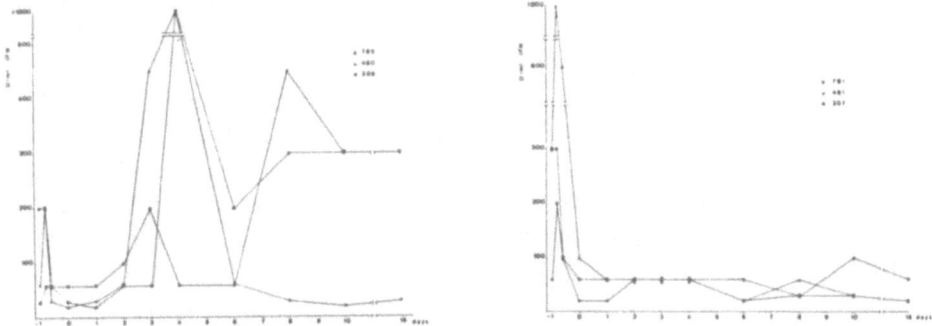

Fig. 1 <u>Interferon in BHV1 infected heifers</u> : in the different animals mentioned, interferon was titrated daily in the plasma (LEFT : infected animals – RIGHT : infected heifers treated with exogenous interferon).

In the newborn calves infected with Rotavirus this effect was surely not as striking. The total interferon dose we used in this case was however less important; we treated the animal only 3 times on 3 consecutive days and the first injection was made the same day as the viral infection. Despite those different conditions, when using a low Rotavirus infectious dose (Fig. 2.) exogenous interferon treatment reduced significantly the first peak of endogenous interferon.

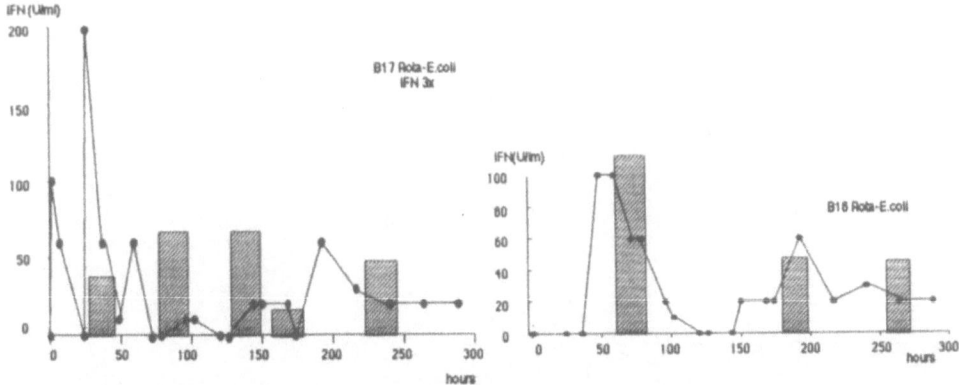

Fig. 2. <u>Interferon in Rotavirus infected newborn calves :</u> in the animals infected with the virus and also with E. coli IFN was titrated at intervals in plasma. Dashed bars represents kinase activity (the highest level reached an activity of 800 p moles ATP incorporated /mg added histone) (left : infected calf – right : infected calf treated 3 times with interferon).

It did not abolished however tthe subsequent waves of interferon production.
Exogenous interferon treatment seemend to reduce the induction by endogenous
interferon of kinase activity. This effect was partly masked as the injected
interferon itself induced a kinase activity. The kinase activity followed the
interferon peaks much more later in this case, than what was observed with
the endogenous interferon, and none of those peaks showed a kinase activity as
high as it would do in absence of interferon treatment before infection.As time
elapsed the full kinase inducing capacity of endogenous IFN is progressively
recovered. On the contrary in experiments where we used a much higher
Rotavirus infectious dose we did not observe such a phenomenon (Fig. 3). In
this case, endogenous interferon appeared much earlier in the circulation after
viral infection.

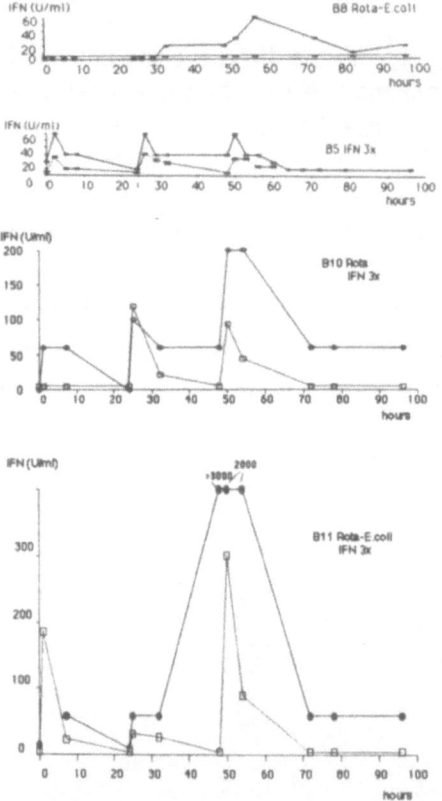

Fig. 3. Endogenous interferon production and effect of exogenous interferon in
rotavirus infected newborn calves : interferon titration by CPE (closed circles)
or by ELISA test showing only human exogenous interferon (open squares).
When exogenous interferon was also used,it was the second infection and not
anymore the last one which did coïncide with the increase of the endogenous

interferon level. In the experiments reported, we titrated in this situation higher amount of circulating interferon than expected (Calf B10 and B11). In one instance (Calf B11) we observed about 10-20 times more antiviral activity after the third interferon injection than in the previous experiment (Fig. 2.). Most of this activity is of an unknown nature, but is not closely related to the human interferon since only a minor part was recognized by an ELISA test used in parallel. We may not exclude that this observation was due to the presence of a mixture of interferons of different origins that may show in the *in vitro* test a synergistic activity. Whether this bears any important significance for the physiological role of interferon or not , is not known for the moment.We do not know for exemple if this enormous increase of endogenous interferon was caused by the experimental Rotavirus infection or if it was due to another factor, since each cow had its own historical background before calving.

There was however as described later (Table I) a good correlation between the observed circulating interferon level and the level of induced enzyme.

Animal treatment	B9 control	B5 IFN 3x *	B6 IFN 3x *	B7 natural infection	B14 E.coli	B8 rota E.coli	B12 rota E.coli IFN 1x	B10 rota E.coli IFN 3x	B11 rota E.coli IFN 3x
IFN (U/ml)									
Day 0	0	60	60	10	0	0	30	60	60
1	0	60	60	10	15	0	0	100	60
2	0	60	20	60	0	60	0	200	>200
3	0	10	0	20	0	20	60	60	60
2'-5'A synthetase (nmol ATP/h/DO$_{260}$)									
LIVER	0.08	4.04	nd	3.11	0.39	27.35	1.9	8.36	14.78
SPLEEN	1.4	4.05	17.5	10	1.84	17.25	5.22	42	23.16
LUNG	2.13	6.55	17.95	10.35	0.39	16.3	3.86	36.13	63.3
KIDNEY	0.09	0.97	1.7	2.09	<n.mol	4.55	0.72	4.6	4.7
INTESTINE IG1	0.7	6.2	10	10.85	5.28	17.45	10.66	55.38	64.07
IG2	1.5	2.92	1.95	8.4	nd	5.45	11.29	35.04	37.03
IG3	1.61	4.1	2.88	3.35	1.06	3.8	7.38	10.23	21.04
COLON	2.9	2	4.12	4.1	1.1	1.08	4.76	25.17	47.01

* : B5 and B6 were slaughtered 48h and 7h after the last IFN injection respectively
IG1 : 1 meter after the pylore
IG2 : mid-jejunum
IG3 : 30 cm ahead of the ileo-caecale valve
nd : not done

Table I : Interferon and 2'-5'A synthetase levels in newborn calves

3.2. Induction of 2'5' A Synthetase by interferon.

Fig. 4 showed that a single injection of interferon (Hu IFN α2 Arg) at 10^6U/Kg in a six month old heifer induced efficiently the 2'5' A Synthetase in the circulating lymphocytes. While interferon reached its peak of activity within an hour after injection the induced 2'5' A Synthetase activity increased progressively up to 24 hours after interferon treatment . It leveled off after about doubling the basal level, and then slowly reduced towards the initial value which was not yet completely reached 72 hours after the injection of IFN.

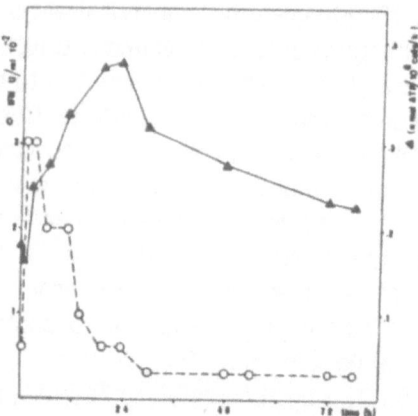

Fig. 4. <u>Induction of 2'5' A Synthetase by Hu IFN α2 Arg</u> : the pharmacokinetics of interferon is shown by the dashed line on parallel to 2'5' A synthetase activity (closed triangles).

In the calves intranasally infected with BHV1, endogenous interferon was induced and appeared in the circulation soon after the virus was excreted (titrated in samples collected on ansal swabs). Fig.5. showed that the sequence of events was as followed : first, virus excretion, then interferon production and finally 2'5' A Synthetase induction. The amount of interferon produced may vary considerably from one animal to the other for unknown reasons. We showed in Fig. 5. the two extreme situations that we have encountered (heifer IBR 460 and IBR 785). This example showed also that the level of 2'5' A Synthetase reached in the lymphocytes is a direct function of the level of circulating interferon indicating, that a cause to effect relationship exists between those two properties. However a direct correlation between the amount of virus excreted and the amount of interferon cannot be drawn. This observation coul be due to the fact that different animals may be either good or poor interferon producers but many other explanations could be also possible. In heifer infected with BHV and treated starting the day before, for 6 consecutive days, with a high dose of Hu IFN α2 (10^6 U/kg) no endogenous interferon could be detected. The exogenous interferon (which does not show up in Fig. 5. – IFN IBR 307 since blood samples were collected just before the daily interferon treatment 24 hours after the previous injection, a time after which the injected interferon was completely cleared out from the circulation) suppressed completely the normal process of induction of interferon by BHV1. As already mentioned, in Rotavirus infected calves, the 2'5' A Synthetase detected in the different tissue extracts studied, was in direct proportion to

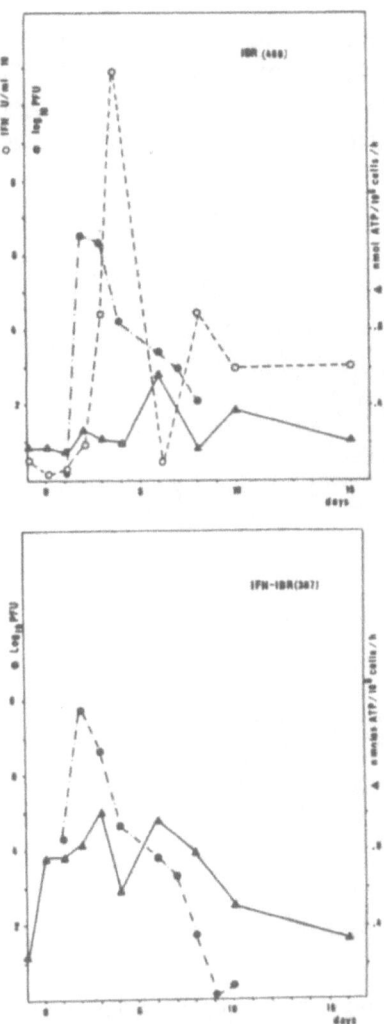
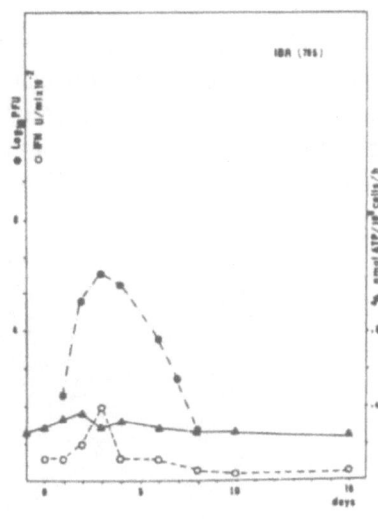

Fig. 5. <u>2'5' A synthetase in BHV1 infected animals</u> : the level of 2'5' A Synthetase (closed triangles) is shown in parallel to interferon induced (open circles)by BHV 1 infection (IBR 460 et 785). The closed circles indicates the virus titers excreted.

the interferon showing up in the circulation, whether the animals were treated by interferon, infected by the virus of both.

Table I showed moreover that when detected, interferon was quite abundant in the spleen, the lungs and particularly in the the intestine, which is the site of replication of the rotavirus. We know that interferon inhibits virus excretion, but never completely. The residual viral replication occuring in presence of interferon was sufficient to permit normal seroconversion in the infected calves.

On the other hand, this low amount of virus produced under interferon treatment exerted enough cell damage to permit colonisation of the intestine by superinfecting coli. The size of the intestinal microvillosites were indeed reduced to similar extent in the mixed infection whether the animals were treated or not with interferon (Fig. 6.).

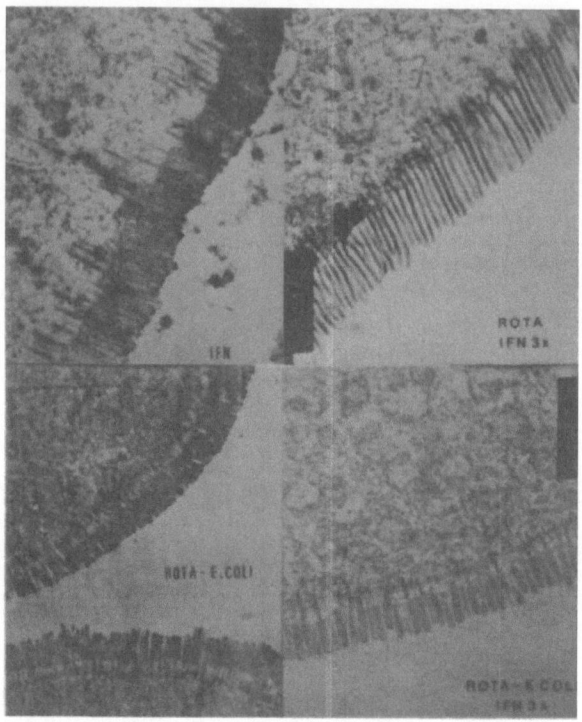

Fig. 6 _Morphological changes in microvillosities_ : The lenght to diameter ratio of the microvillosities in calves B5 (IFN), B7 (ROTA - not shown in the Fig.); B8 (ROTA - E. coli), B11 (ROTA - coli - IFN 3x) or B10 (ROTA - IFN 3x) are respectively 6-8,5-4-4,5 and 10.
Interferon did not appear to affect this pathophysiological modification. As no sign of diarrhoea was observed in these calves we may conclude that the extent of the intestine damage was not sufficient in the treated animal to cause an extensive pathogenecity. Alternatively this could also indicate that the alteration of the intestine morphology is not a process which by itself might cause the disease.

4. CONCLUSIONS

We have been able to show that experimental infections of young calves with some viruses could be inhibited by interferon while infections by others could not. The susceptibility of the virus as shown in *in vitro* systems is not the only criteria to explain those differences.

2'5' A Synthetase is elevated by interferon in circulating lymphocytes collected from treated animals. While the importance of this observed increase has been correlated with the antiviral effect shown against vaccinia virus by human interferon (15) or bovine interferon (unpublished data) its role in the protective action is still not clearly established.

Moreover, an elevation of this enzymeactivity in the lymphocytes does not confer a resistancce to all viruses. Replication of BHV1, for example is not affected by interferon despite important 2'5' A Synthetase induction in the lymphocytes rotavirus which is more susceptible to IFN is indeed inhibited, but not totally, by interferon. Again, here we showed a correlation between the level of circulating interferon and the activity of 2'5' A synthetase in selected. organs.

Therefore the bovine system, as we have proposed to approach it, constitutes a fascinating tool to study the mechanism of action of interferon *in vivo*. It is not only a very good animal model system in which both human and bovine interferons could be investigated ,but it is also for this reason an excellent mean to test simultaniously the potentialities of interferon as an antiviral substance in veterinary medicine. Others have indeed followed us recently to this direction using the shipping fever model and have obtained essentially the same results confirming our data, showing marginal (18) or no (19) effects of interferon on BHV1, excretion in experimentally infected animals. Despite this limitation concerning a direct antiviral effect against BHV1, striking biological actions of interferon on this system were detected.

We should therefore investigate this system further; the efforts invested will undoubtly be soon rewarding.

5. REFERENCES

1. Goossens A., Schwers A., Vandenbroecke C., Maenhoudt M. ,Bugyaki L., Pastoret PP, Werenne J. : Arch. internat. Physiol. Biochim., 90 (4) , B 193. Antiviral efficiency of interferon in the bovine species : preliminary data on the activity in calves of bacterially produced human interferon, 1982.
2. Werenne J. : Ann. Méd. Vét. 126, 93-122. L'interféron : perspective actuelles, 1982.
3. Wérenne J., Pastoret J.J. , Vandenbroecke C., Schwers A., Gossens A., Bugyaki L., Maenhoudt M., : The Biology of the Interferon System, Elsevier Science Publishers B.V. E. De Maeyer and H. Schellekens, editors, 419-424. Bacterially produced interferon as an antiviral in the bovine species, 1983.

322

4. Werenne J., Vandenbroecke C., Schwers A., Goossens A., Bugyaki L., Maenhoudt M., Pastoret P.P., : J. IFN Research 5, 129-136. Antiviral effect of bacterially produced human interferon (Hu-IFN α2) against experimental vaccinia infection in calves, 1985.

5. Dagenais L., Pastoret P.P., Vandenbroecke C., Werenne J.J. : Archives of Virology, 70, 377-379. Susceptibility of bovine rotavirus to interferon, 1981.

6. Goossens A., Schwers A., Vandenbroecke C., Dagenais L., Maenhoudt M., Duwijn R., Van Camp E., Pastoret P.P., Werenne J.J., : Ann. Méd. Vét. 127, 135-139. Sensibilité des virus de la rhinotrachéite infectieuse bovine (BHV1) et de la maldie d'Aujeszky (SHV1) à l'interféron humain produit par des bactéries (Hu-IFN α2), 1983.

7. Schwers A., Vandenbroecke C., Pastoret P.P., Werenne J., Dagenais L., Maenhoudt M. : The Veterinary Record, 112, 250. Dose effect on experimental reproduction of rotavirus diarrhoea in colostrum deprived newborn calves, 1983.

8. Vandenbroecke, C. Schwers A., Dagenais L. Goossens A. Maenhoudt M., Pastoret P.P., Werenne J.J. : Ann. Rech. Vét.15, (1) 29-34, Interferon response in colostrum-deprived newborn calves infected with bovine rotavirus : its possible role in the control of the pathogenicity, 1984.

9. Schwers A., Vandenbroecke C., Maenhoudt M., Beduin J.M., Werenne J., Pastoret J.J., : Ann. Rech. Vet. 16, 213-218. Experimental rotavirus diarrhoea in colostrum deprived newbron calves : assay of treatment by administration of bacterially produced human interferon, 1985.

10. Vandenbroecke C., Schwers A., Plasman P.O., Pirak M., Piscitelli S., Thiry E. Koch K., Thein P., Pastoret P.P., Werenne J.J. : TNO-ISIR Meeting on the Interferon system, Clearwater Beach, Florida, U.S.A., . 2'5'A Synthetase as a marker for the biological activity of interferon in the bovine species, 13-18 october 1985.

11. Goossens A., Vandenbroecke C., Schwers A., Dagenais L. , Maenhoudt M., Pastoret P.P., Werenne J.J. : IN VITRO antiviral efficiency of bacterially produced Human interferon (HU-IFN α2) against several bovine viruses.
In Adjuvants, Interferon and non-specific immunity. Eds. F.M. Cancellotti and D. Gallassi, CEE, Luxembourg, 1984.

12. Capon DJ., Shepard HM., Goeddel DV., : Mol. Cell. Biol., 5, (4) 768-779. Two distinct families of human and bovine IFN genes are coordinately expressed and encode functional polypetides.

13. Velan B., Cohen S., Grosfeld H., Leitner M. and Shafferman A. : Bovine interferons and genes structures and expression. J. Biol. Chem. 260, 5498, 1985.

14. Werenne J. : Interferon in veterinary medicine : the present position and future prospects. Outlook on Agriculture, 14, 3, 1985.

15. Vandenbroecke C. , Plasman P.O., Pirak M., Thiry E., Schwers A., Piscitelli S., Ligongo G., Pastoret P.P. and Werenne J.J. : Human recombinant interferon in the bovine species : induction of 2'5' A Synthetase and antiviral activity. Progress in clinical biological research, 202, 332-338, Eds : Br. R.G. Williams & R.H. Silverman, Alan R. Liss, Inc, N-Y, 1985.

16. Vandenbroecke C., Plasman P.O., Hauzeur K., Goossens A., Ligongo G., Piscitelli S., Werenne J. : The 2'5' A Synthetase in the bovine system : preliminary characterization. Progress in clinical biological research, 202, 338-344, Edt : Br. R;G; Williams I R;H; Silverman, Alan R. Liss, Inc, N-Y, 1985.

17. Buffet-Janvresse C., Vannier J.P., Laurent A.G., Robert N. and Hovanessian A.G. : Enhanced Level of Double-Stranded RNA- Dependent Protein Kinase in Peripheral Blood Mononuclear Cells of Patients with viral infections. Journal of Interferon Research, 6, 85-96, 1986.

18. Roney C. S., Rossi C. R. ,Smith P.C. Lauerman L. C., Spano J.S. Hanrahan L.A. William J.C. : Effect of human leukocyte A interferon on prevention of infectious bovine rhinotracheitis virus infection of cattle. Am J Vet Res, 46, N° 6, June 1985.

19. Babiuk L.A., Bielefeldt-Ohmann H., Gifford G., Czarniecki C.W., Scialli V.T. and Hamilton E.B. : Effect of Bovine α1 Interferon on Bovine Herpesvirus Type 1 induced Respiratory Disease. J. Gen. Virol. , 66, 2383-2394, 1985.

CLINICAL STUDIES

TREATMENT OF MALIGNANT ENDOCRINE PANCREATIC TUMORS WITH HUMAN LEUKOCYTE
INTERFERON

B. Eriksson[1], K. Öberg[1], G. Alm[2], A. Karlsson[3], G. Lundqvist[4], T.Andersson[5]
and E. Wilander[6]. Ludwig Institute for Cancer Research and Department of
internal Medicine[1], Department of Internal Medicine[3], Clinical Chemistry[4],
Radiology[5] and Pathology[6], University Hospital, Uppsala, and Interferon Laboratory and Department of Immunology[2], Biomedical Centre, Uppsala.

SUMMARY

Twenty-two patients with malignant endocrine pancreatic tumors were treated with human leukocyte interferon (IFN) at doses of 3-6 x 10^6 IU per day.
Eighteen of the twenty-two patients had earlier been treated with chemotherapy, mainly streptozocin plus 5-fluorouracil or adriamycin, but showed
progressive disease. Two patients had received somatostatin analogue, SMS
201-995. Seventeen of twenty-two patients (77%) showed an objective response with a mean duration of 11.2 months, median 8.5 months (range 2-36+).
7/7 with the WDHA-syndrome, 3/4 with the Zollinger-Ellison syndrome, 6/9
with "non-functioning" tumors and one patient with a mixed tumor mainly
producing somatostatin responded. Fifteen of 22 patients had a significant
reduction of tumor markers, <50% of pretreatment levels. Totally a significant reduction of tumor size was seen in six patients, out of which two had
complete remissions. Improvement of clinical manifestations was seen in all
patients with an objective response. Adverse effects, including influenza-
like syndrome, reduction of blood cells, chemically signs of liver dysfunction and disturbed lipid metabolism occurred but were reversible or could
be circumvented by dose reduction. Liver steatosis was demonstrated radiologically in three patients. Autoimmune disease was also noted such as development of thyroid autoantibodies, antinuclear and rheumatoid factors.
IFN therapy seems to be as potent as cytotoxic treatment in reducing hormone levels and clinically alleviating symptoms. It also controls tumor growth
for extendes time periods (mean 11.2 months). The adverse effects are more
tolerable than with cytotoxic treatment.

INTRODUCTION

Pancreatic endocrine tumors are rather rare and the prevalence in unselec-
ted autopsy cases has been estimated as 0.5-1.5 per cent[1]. The clinical
course is dominated by serious and potentially fatal complications caused
by tumor derived products. Surgery is the primary form of treatment[2] but
unfortunately many of the patients have metastatic disease at diagnosis and
alternative treatment is required. Sometimes symptomatic treatment can be
satisfactory, but the majority of patients requires causal therapy. Chemotherapy (streptozocin plus 5-fluoracil) has been shown to produce tumor responses in more than 60% of the patients[3,4]. Cytotoxic treatment, however,
rarely cures the patient and eventually other alternatives are needed. Promising results in a number of patients with carcinoid tumors[5], closely related to endocrine pancreatic tumors, encouraged us to try human leukocyte
interferon (IFN) in twenty-two patients with advanced endocrine pancreatic
tumors.

PATIENTS AND METHODS

Twenty-two patients, 9 women and 13 men, with a mean age of 58,5 years were included in the study. Seven patients had the WDHA-syndrome, whereas four had the Zollinger-Ellison syndrome. One patient had a malignant insulinoma and another patient had a mixed tumor mainly secreting somatostatin. In the remaining nine cases, no specific hormone-related symptoms were present and they were considered as "non-functioning" tumors. Six patients had a MEN-1 trait (two had WDHA-syndrome, four "non-functioning" tumors). Fourteen had a remaining pancreatic tumor, twenty patients had liver metastases. One patient had metastases in the lungs and two had skeletal spread.

Twelve patients had earlier undergone pancreatic surgery and two patients had been subjected to debulking liver surgery. Embolisation and/or ligation of tumor vessel had been undertaken in six cases. One patient had received irradiation therapy. Eighteen patients had been treated with chemotherapy, mainly streptozocin plus 5-fluorouracil or adriamycin but were refractory to that treatment. Two patients had been treated with somatostatin analogue SMS 201-995, without any effect. All patients showed progressive disease, when IFN therapy was started.

Methods

Peripheral blood samples were collected before and during IFN therapy. Serum insulin, pro-insulin, c-peptide, gastrin, HGC-alpha and -betasubunits, pancreatic polypeptide (PP), calcitonin, plasma vasoactive polypeptide (VIP) glucagon, somatostatin and neurotensin were analyzed by radioimmunoassays described earlier. Routine hematology, liver enzymes, blood lipids, thyroid antibodies, antinuclear factor, rheumatoid factor, serum T_4 and TSH were checked before start of IFN therapy and regularly during the treatment. Histopathological diagnosis was established in twenty cases on tissues taken at surgery or by coarse-needle biopsy. All tumors were argyrophil with the Grimelius silver nitrate stain. Two patients had not been operated on or biopsied; both had MEN-1 and biochemically and radiologically verified tumors. Routinely fixed, paraffin embedded sections were immunohistochemically investigated with a number of antibodies against gastrin, insulin, PP, glucagon, somatostatin, HCG-alpha, -beta, calcitonin, VIP and neurotensin.

Interferon

The human leukocyte interferon was produced according to Cantell et al (6) at the Interferon Laboratory, Uppsala and Pharmaceutical Department, Karolinska Hospital, Stockholm. The IFN was administered as daily intramuscular or subcutaneous doses of $3-6 \times 10^6$ IU. An escalating schedule was used to avoid side effects, starting the treatment with 3×10^6 per day during three days, and then increasing it to 6×10^6 IU per day.

Evaluation of tumor responses

In the evaluation of the therapeutic effect, response was defined as 1) a reduction of a specific tumor marker (e.g. gastrin in Zollinger-Ellison) and 2) a reduction of a general marker (PP, HCG-alpha and -beta in "non functioning" tumors) by more than 50% and/or reduction of tumor mass by more than 50% measured as the product of the maximum and its perpendicular diameter on CT, ultrasound or angiography. Stable disease was defined as a reduction of tumor markers and/or tumor mass by less than 50% and no appearance of new metastases. Progressive disease was accordingly defined as an increase in tumor markers and/or tumor mass by more than 25% or appearance of nes metastases.

Results

Objective responses were seen in seventeen of twenty-two patients (77%) treated with IFN. The mean duration of response was 11.2 months, median 8.5 months (range 2 - 36+). Ten patients are at the time of last follow up, still on IFN treatment and still responding. Fifteen of the seventeen responders showed decreasing tumor markers (Fig 1a, 1b), whereas to responders had no detectable marker. These two patients, however, had a significant reduction of tumor size (one complete remission, one partial remission). Totally, six out of seventeen responders had decreasing tumor masses on CT-scans by more than 50% (Fig 2a, 2b), two had complete remissions. All seven cases with the WDHA-syndrome, three of four with the Zollinger-Ellison syndrome, six of nine with "non functioning" tumors and a somatostatinoma patient responded. Stable disease was seen in one patient with the Zollinger-Ellison syndrome. She showed stable disease for 27 months. The remaining four cases, one insulinoma patient and three cases with "non-functioning" tumors, showed progressive disease according to the criteria. Two of them progressed in parallell in both "markers" an tumor size, whereas two increased in peptide levels but the tumor size remained unchanged. The parallell decrease in tumor markers and tumor size are illustrated in Table 1. Subjective responses were noted in all cases with the WDHA-syndrome, where diarrhoeas and flushes disappeared. The three responding Zollinger-Ellison patients had relief of dyspeptic/ulcer symptoms. In the patients with "non functioning" tumors, the symptomatic or subjective responses were not as easily discernible but in one case analgesics (morphine) could be withdrawn because of diminishing tumor size and abdominal pain. In the case with a somatostatinoma, insulin therapy could be withdrawn.
Totally are sixteen patients still living (mean observation time 15 months, median 9.5 months) and ten patients are still on IFN therapy. Three of seventeen initial responders subsequently died after two, three and ten months of response in progressive tumor disease.

330

Legends

Fig 1.a

Serum gastrin le-
vels before and
during interferon
treatment in a
patient with the
Zollinger-Ellison
syndrome.

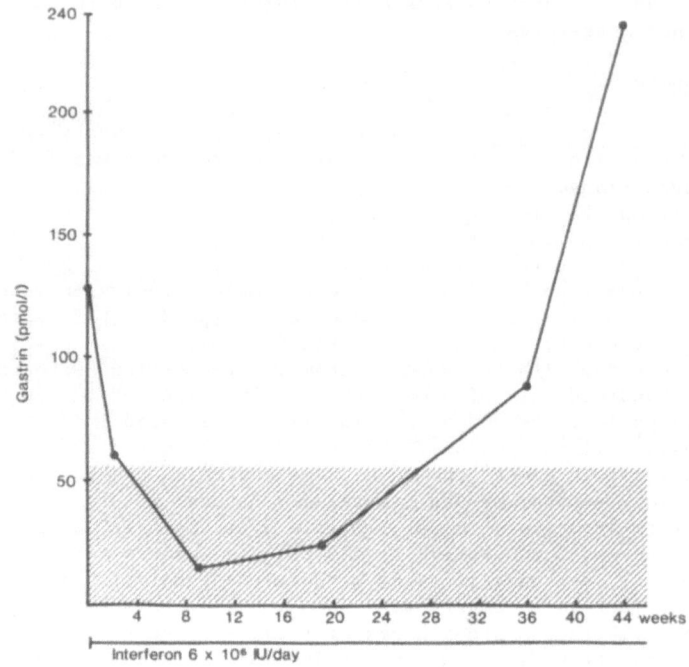

Fig 1.b   Plasma VIP before and during interferon treatment in a patient
with the WDHA-syndrome.

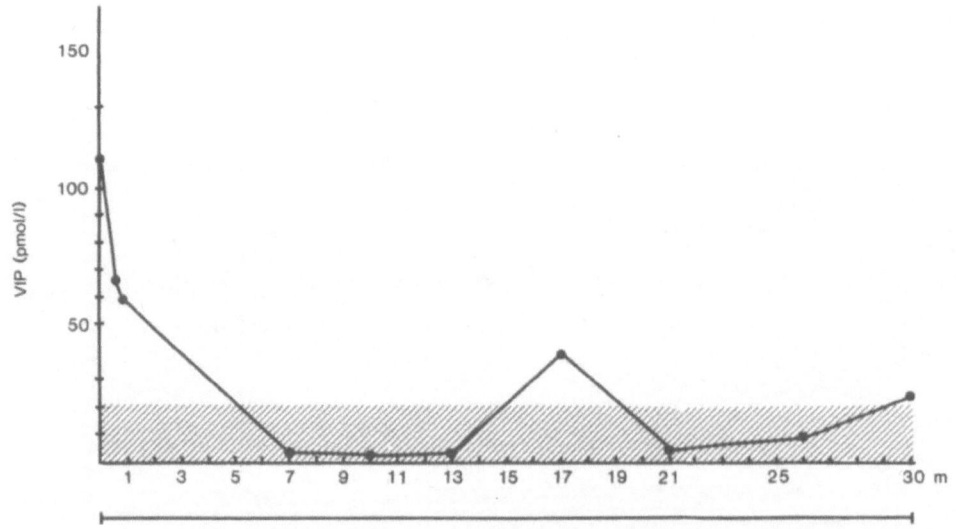

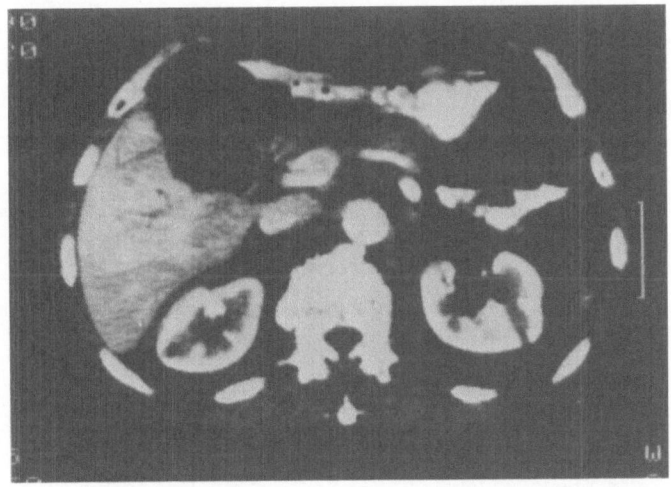

Fig 2a. CT-scan showing a liver metastase in a patient
with the WDHA-syndrome.

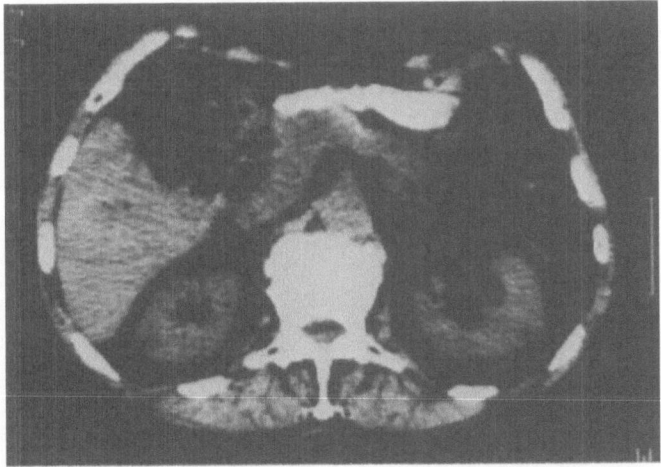

Fig 2b. CT-scan showing the complete disappearance of
the liver metastase in the patient with the
WDHA-syndrome.

Table I

Clinical syndrome	Total no of patients	Liver met	Objective tumor responses * tumor size	markers	Total
WDHA-syndrome	7	6	S=5 R=2 P=0	S=1 R=6	S=1 R=7
Zollinger-Ellison	4	4	S=2 R=2	S=2 R=2	S=1 R=3
insulinoma	1	1	P=1	P=1	P=1
somatostatinoma	1	1	S=1	R=1	R=1
non functioning tumors	9	8	S=6 R=2 P=1	R=6 P=3	R=6 P=3

* R=objective response > 50% reduction of tumor size or tumor marker
S=stable disease=unchanged tumor size and < 50% reduction of tumor of tumor marker
P=progressive disease > 25% increase of tumor size or tumor marker

Adverse effects

The side effects are summarized in Table II. The most frequent side effect was "influenza-like" symptoms, observed in 77% of the patients at initiation of IFN therapy, rarely lasting more than a week. Anorexia and a slight weight loss were noted in 59% of the cases.

Routine hematology and clinical chemistry data were available in 21 patients. A mild reduction of hemoglobin levels occurred in 8/21 patients (38%), the reductions being transitory in three cases, the other received iron suplementation. Fourteen patients (66%) had a moderate decrease in the number of leukocytes. One patient went below 2.0×10^9 g/l and the IFN dose was reduced by 50%, whereafter the counts normalized. A moderate reduction of platelets was seen in 38% of the patients, the decrease was transitory in three. Increased levels of the liver enzymes ASAT, ALAT and alkaline phosphatases were observed in 8/21 patients (38%) and in one patient the IFN dose had to be reduced. Blood lipids increased in eight patients (38%), seven had elevated serum triglycerides and one elevated serum cholesterol. Dose reduction was performed in one patient because of this and serum triglyceride levels normalized. A liver steatosis was noted in three cases (13%).

Thyroid autoantibodies were found in one patient after 20 months of treatment and she developed hypothyroidism, which was easily substituted. The same patient also developed antinuclear factors and a rheumatoid factor already after seven months of treatment without any clinical symptoms. Two other patients also developed antinuclear factors after 6 and 24 months respectively and one of them who also had a rheumatoid factor developed joint involvement. The joint symptoms disappeared spontaneously after one month without interruption of IFN therapy but the elevated serum factors persisted. One of the responders, who has had an insulin-dependent diabetes for many years, developed a myocardial infarction after 22 months of IFN treatment. She was one of the patients, who had an increase in serum triglycerides.

Table II

Adverse effects

Influenza-like symptoms	17/22	77%
Weight loss	13/22	59%
Decreased content of hemo-globulin below 110 g/l	8/21	38%
Decreased of no of leukocytes below 4.0×10^9/l	14/21	66%
Decreased no of platelets below 150×10^9/l	8/21	38%
Increased liver enzymes		
ASAT	8/21	38%
ALAT	6/21	28%
alkaline phosphatase	5/21	23%
Increased blood lipids		
s-triglycerides	7/21	33%
s-cholesterol	1/21	4%
Liver steatosis	3/22	13%
Thyroid autoantibodies	1/21	4%
Antinuclear factor	2/21	9%
Myocardial infarction	1/22	4%

Discussion

The optimal treatment in malignant endocrine pancreatic tumors has not yet
been established. Streptozocin plus 5-FU has earlier been described to produce response rates of 63% (3,4) with a median duration of 17 months. Eighteen of our 22 patients had earlier received chemotherapy but were refractory to that treatment at the start of interferon treatment. Two patients
had also received a new somatostatin analogue SMS 201-995. One had a subjective improvement but no objective response. The other patient, who had the
WDHA-syndrome, showed no improvement at all, probably because of endogenously excessive somatostatin secretion and hence down regulation of somatostatin receptors.
Our result indicate objective response rates as high as with cytotoxic treatment. Obviously, this comparison may be biased in favor of chemotherapy,
which started at a less advanced stage of the disease. The duration of response is shorter, which could be explained by a shorter observation time.
Ten patients are still on IFN and responding.
The majority of objective tumor responses were seen as rapid reductions of
tumor markers with concomittant alleviation of clinical symptoms. In contrast, only six patients displayed an objective reduction of tumor size judged by CT-scans. Therapy-induced liver steatosis seen in three cases made
the radiological interpretation difficult.
The precise mechanism of action of IFN on endocrine tumors is not yet clear
but we favor a direct inhibitory effect on tumor cell proliferation and hormone synthesis, causing inhibition of hormone secretion. This might explain
the pronounced and rapid IFN-induced reduction or hormone levels, but pro-
portionately smaller and slower effects on tumor size determined by CT-scans
During IFN therapy, the tumor cells might stop to divide but remain viable
for extended time periods. Clinically these observations might be supported
by the finding that during a mean observation period of 11.2 months, 20 of
the 22 (90%) did not increase in tumors size measured on CT-scans.
It is possible that tumor derived hormones also may serve as autocrine growth
factors and that IFN may control cell growth via inhibition of such factors.
Interrestingly, bombesin-like peptides have recently been shown to be autocrine growth factors for small-cell lung cancer (7) closely related to endocrine pancreatic tumors. Recently, antagonistic effects of interferon on
various growth factors have been demonstrated, suggesting that cellular proliferation may be regulated by opposing actions between growth promoting
factors and IFN (8).
Since the effect of IFN may be primarily noncytotoxic and controlling, it
might be of value to combine IFN with conventional cytotoxic treatment to
produce additive and even synergistic effects on these tumors.
In the present study, most adverse effects of IFN were less severe than with
cytotoxic treatment and corresponded to those described by others (9). Flu-
like symptoms were transitory in all but two patients, in whom they were
controlled by dose reduction. The reduction of blood counts produced no
symptoms. The treatment was discontinued in one patient because of anorexia,
although she belonged to the responders.
Two more serious side effects occurred, a disturbed lipid metabolism and
autoimmune disease. The disturbed lipid metabolism resulted in increased serum triglyceride levels and liver steatosis and might be due to an interferon mediated inhibition of lipolytic enzymes (10). One patient developed
thyroid autoantibodies and hypothyroidism and required substitution with
thyroxin. Three patients developed antinuclear factors, but only one of them,
who also had a rheumatoid factor, developed joint involvement. The symptoms,
however, disappeared spontaneously without withdrawal of IFN therapy. The

ausal relationship between IFN and autoimmunity seems obvious but the
echanism is not yet fully understood. These side effects may be due to
FN-alpha itself or contaminants of human leukocyte interferon including
FN-gamma.

n conclusion, our results indicate that IFN is a new promising alternati-
e in the management of malignant endocrine pancreatic tumors. It appears
ffective in the amelioration of severe clinical symptoms caused by secre-
ory products, an important goal for any therapy in this patient group. In
his respect it seems as potent as cytotoxic treatment. Leukocyte IFN may
lso control the growth of the tumor but rarely eradicate them. In the fu-
ure, the combination of IFN and cytotoxic treatment may render even bet-
er results.

EFERENCES

1. Creuzfeldt W: Endocrine tumors of the pancreas, clinical, chemical and
   morphological findings. Monograph Pathol 1980;21:208-230.
2. Wellbourn RB, Wood SM, Polak JM and Bloom SR: Pancreatic endocrine tumors. In: Bloom SR, Polak.
3. Moertel CG, Hanley JA and Johnson LA: Streptozocin alone compared with
   streptozocin plus fluorouracil in the treatment of advanced islet cell
   carcinoma. N Engl J Med 1980;303:1189-1194.
4. Öberg K and Boström H: Endocrine gastrointestinal tumors. Diagnosis
   and management. Uppsala J Med Sci 1984; suppl 39:183-189.
5. Öberg K, Funa K and Alm G: Effects of leukocyte interferon on clinical
   symptoms and hormone levels in patients with mid-gut carcinoid tumors
   and carcinoid syndrome. N Engl J Med 1983;309:129-133.
6. Cantell K, Hirvonen S and Koistinen V: Partial purification of human
   leukocyte interferon on a large scale. Methods in Enzymology 1981;78:
   499-505.
7. Cuttitta F, Carney DN, Mulshine J, Moody TW, Fedorko J, Fischler A and
   Minna JD: Bombesin-like peptides can function as autocrine growth factors in human small cell lung cancer. Nature 1985;316:823-826.
8. Inglot A: The hormonal concept of interferon. Arch Virol 1983;76:1-13.
9. Ingimarsson S, Bergström K, Broström LÅ, Cantell K and Strander H: Effect of longterm treatment with human leukocyte interferon on various
   laboratory parameters. Acta Med Scand 1980;208:155-159.
0. Ekenholm C, Aho K, Huttunen JK, Kostianen E, Mattila K, Pikkavainen J
   and Cantell K: Effect of interferon on plasma lipoproteins and on the
   activity of postheparin plasma lipases. Arteriosclerosis 1982;2:68-73.

EFFICACY OF ALPHA INTERFERONS IN HAIRY CELL LEUKEMIA (HCL) AND STATUS OF THE INTERFERON SYSTEM OF PATIENTS IN REMISSION

J.R. Quesada[1], J.L. Lepe-Zuniga[1] and J.U. Gutterman[1]
[1]Department of Clinical Immunology and Biological Therapy,
University of Texas M.D. Anderson Hospital & Tumor
Institute, Houston, TX 77030

Hairy cell leukemia (HCL) is a malignant disease of mononuclear cells which infiltrate predominantly the liver, spleen and the bone marrow, resulting in pancytopenia and immunologic deficiencies.[1] The latter two abnormalities result in a high incidence of infections, which has been the main cause of morbidity and mortality in these patients. The great majority of patient's hairy cells express predominantly B-cell markers. Hairy cells can also express monocyte-like characteristicsand the Tac antigen (the putative receptor for interleukin 2) which is predominantly expressed by T-cells, suggesting that hairy cells are in a unique stage of differentiation.[2] The T cell variety of HCL is rare. The etiology of the disease or the precise mechanisms(s) responsible for the pancytopenia have not been elucidated. A deficient bone marrow hematopoiesis secondary to infiltration by leukemic cells is partly responsible. However, because hematopoiesis is regulated by cells of the immune system, the immunologic deficits described in HCL may play a role in the defective bone marrow function.[3]

Splenectomy has been the traditional front-line therapy of HCL. Long-term control of the disease is achieved only in a few instances. In a substantial number of patients, the response to splenectomy is inadequate or short-lived. Chemotherapy can effectively induce bone marrow remissions in HCL, but it is complicated by severe and prolonged myelosuppression.[1] Other approaches to the treatment of HCL have included leukapheresis or transfusions of allogeneic mononuclear cells which resulted in improvement of hematologic indexes in some patients.[3]

We began in July 1982 treatment of selected HCL patients with a partially purified preparation of interferon alpha (IFNα), based on the results which have been obtained in other malignancies including those of B-cell lineage.[4] Interferon was also of interest because of its potentiating activities in several functions of the immune system.[5] Of 7 patients treated with IFNα (Finnish Red Cross Blood Center) at a dose of 3 million units daily, 3 achieved complete remissions and 4 partial remissions.[6] The criteria of response used in that study was stringent including response in the bone marrow aspirates and biopsies. Complete remission (CR) was defined as disappearance of hairy cells from at least two consecutive bone marrow biopsies and aspirates and reduction in the size of the spleen or other organomegaly to normal. Partial remission (PR) was defined as a 50% reduction in the bone marrow leukemic index (% cellularity x % hairy cells). All patients must have a recovered hemoglobin ≥ 12 gm 1 dl; ≥ 100,000 platelets/mm³ and ≥ 1500 neutrophiles/mm³.

The results of that preliminary study have been confirmed using purified recombinant alpha interferons (rIFNα). The results obtained with rIFNα excluded the involvement of lymphokines other than IFNα which are contained in the partially purified preparation. rIFNα in contrast to IFNα contains a single purified species of IFNα. The results in the first 30 patients treated at M.D. Anderson Hospital with rIFNα have been recently published.[7] Nine CR and 17 PR were obtained. The other four patients had hematologic improvement of clinical significance but did not achieve remission status as defined. In this study, we included 7 patients with splenomegaly (and therefore candidates for splenectomy). The incidence of CR was significantly higher (p = < 0.01) in previously untreated patients (5 of 7) than in those in whom splenectomy had been performed (4 of 23).

During the course of these studies other investigators have reported similar results as shown in Table 1. The paucicity of CR in some of these studies is perhaps explained by the less frequent schedule of administration, shorter duration of therapy, or the selection of patients with more advanced disease. Altogether these studies confirmed that interferon alpha is highly effective therapy for all stages of the disease; IFNα is probably the treatment of choice for patients progressing after splenectomy and an adequate initial treatment for selected untreated patients.

None of 23 patients treated with IFNα have relapsed while on treatment. Median duration of remission is exceeding 20 mo. It has become evident that treatment for less than 6 months may be insufficient because there are patients who continue to improve beyond this point and remissions have been obtained as late as 10 or 12 months after initiation of treatment. A small number of patients have relapsed while receiving rIFNα treatment, coincidentally with the development of antibodies to this molecule (unpublished).

In order to determine the duration of remissions without further treatment, interferon was discontinued at 24 months in our first study[6] and at 12 months in the second study[7]. Clinical relapse, which required reinitiation of treatment with interferon, was defined as progressive anemia, granulocytopenia or thrombocytopenia with a minimum of three consecutive counts showing either a hemoglobin of less than 10 gm/dl; a granulocyte count of less than 1,000/mm³ or platelets less than 100,000/mm³; progressive increase in the WBC or reappearance of circulating hairy cells in the peripheral blood. Bone marrow relapse was defined as a doubling of the absolute leukemic index from the end of study values. Bone marrow relapse was not considered an indication to restart treatment.

Table 2 shows the comparison of both studies in regards to the number of relapses observed following discontinuation of therapy. A larger number of both clinical and bone marrow relapses has occurred in patients in whom treatment was discontinued at 12 months. Analysis of the blood and bone marrow indexes prior to treatment, at the end of the study and at regular intervals following discontinuation of treatment indicate a statistically significant increase in the leukemic indexes and a decrease in the granulocyte counts from patients in the 12 month study. It is still unclear what should be the optimal duration of treatment, but our data suggest that treatment beyond one year may be superior. Notably, five of 13 patients who attained CR after 24 mo. of treatment have sustained their remission status for periods ranging from 4+ to 21+ months (median 8+ months) after discontinuation of treatment. In patients in whom a clinical relapse was diagnosed, treatment with interferon was reinstituted. PR have been successfully reinduced in all patients who have received reinduction treatment for longer than 3 months. The ability to reinduce remissions suggests that periodic administration of interferon may prove satisfactory in the management of some patients.

The toxicity of low doses of alpha IFNs is minor except for unusually sensitive individuals in whom moderate to severe fatigue or fever may require reduction from daily to thrice weekly administration. Caution must be exerted in patients with severe neutropenia because IFN may decrease the WBC and neutrophil count in the first 2-4 weeks of treatment. Approximately 20% of our patients have complicated their induction treatment with infections or with fever of unclear origin that merited prescription of antibiotics. Platelets rarely decreased in individuals with pretherapy counts above $50,000/mm^3$. A recent study has suggested that initial lower doses of IFNα (0.5-1.5 MU/d, s.c.) may be preferable and for patients with severe thrombocytopenia, equally effective.[8]

The exquisite sensitivity of HCL to alpha interferons remains unexplained. We have previously speculated on the possible mechanisms responsible for the antitumor activity and the ensuing recovery of the bone marrow function.[6] Among several possibilities we proposed that a deficiency in IFNα production in these patients could be related to the physiopathology of the disease and explain the response to the exogenous administration of this protein. The effects of IFNα in patients with HCL are somehow reminiscent of the results obtained when a deficient hormone or vitamin is replaced with exogenous products. Recently we have studied 19 patients in different stages of disease activity and

remission for production of IFNα by peripheral blood
mononuclear cells (MNC). Among the different groups of
patients, a severe deficiency in IFNα production was
identified in eight of eight patients with active disease
and in six of six patients who had achieved PR after
treatment with IFNα or rIFNα, and had treatment discontinued
for a minimum of 6 months. These patients were deficient
despite normal numbers and normal immunophenotype of
circulating MNC. In contrast, 5 patients who achieved CR
after IFNα treatment, splenectomy or infection produced
amounts of IFNα similar to those of the control population.
These data suggest that a deficiency of IFNα production may
be involved in the physiopathogenesis of HCL; that
endogenous production of IFNα may bear relevance to the
induction and sustenance of remission after splenectomy, or
"spontaneous" remissions and, that relapse after induction
of PR may be associated to an incomplete restoration of
endogenous production of IFNα. The nature of the defect
requires further study.

TABLE 1

THERAPY OF HAIRY CELL LEUKEMIA WITH ALPHA INTERFERONS

TYPE	DOSE	NO. PTS	CR	PR	MR	STUDY CENTER
IFNα	3 MU/d	22	5	13	4	M.D. Anderson (USA)
rIFNα	3-12 MU/d	30	9	27	4	M.D. Anderson (USA)
rIFNα	2 MU/m²/tiw	8	0	7	1	Univ. Chicago (USA)
rIFNα	2 MU/m²/tiw	14*	0	6	3	UCLA (USA)
rIFNα	2 MU/m²/tiw	8	1	6	0	Germany
rIFNα	2 MU/m²/tiw	14	3	10	1	France
rIFNα	3 MU/d	17	7	10	0	Seattle (USA)
rIFNα	3-6 MU/d	14	1	12	0	NCI (USA)
rIFNα	5 MU/d	26	3	15	4	Austria

MU = million units; CR = complete remission; PR = partial
remission; MR = minor response.
* criteria of response not specified; results are approxi-

TABLE 2

ALPHA INTERFERONS IN HAIRY CELL LEUKEMIA
Relapses After Treatment Discontinuation
M.D. Anderson Hospital Studies

	IFNα			rIFNα		
RESPONSE STATUS	NO. PTS	%	Median Duration (mo.)	PTS	%	Median Duration (mo.)
SUSTAINED REMISSIONS	11	52	21+	6	25	10+
CLINICAL RELAPSES*	4	19		10	40	
BONE MARROW RELAPSES*	6	28		9	36	

* see text for definitions.

REFERENCES

1. Flandrin G, Sigaux F, Sebahoun G and Bouffette P:
   Hairy Cell Leukemia: clinical presentation and follow
   up of 211 patients. Sem Oncol 1984; 11 (Suppl. 2):
   458-471.

2. Korsmeyer SJ, Green WC, Cossman J, et al:
   Rearrangement and expression of immunoglobulin genes
   and expression of Tac antigen in hairy cell leukemia.
   Proc Natl Acad Sci 1983; 80:4522-4526.

3. Quesada JR, Hersh EM and Gutterman JU: Biologic
   therapy of hairy cell leukemia. Sem Oncol 1984; 11
   (Suppl. 2): 507-510.

4. Gutterman JU, Blumenschein GR, Alexanian R, et al.
   Leukocyte interferon-induced tumor regression in
   human metastatic breast cancer, multiple myeloma, and
   malignant lymphoma. Ann Intern Med 1980; 93:399.

5. DeMaeyer-Guignard J and DeMaeyer E: Immunomodulation
   by interferons: recent developments. In: Gresser I,
   ed. Interferon, vol. 6. London: Academic Press, 1985;
   69-92.

6. Quesada JR, Reuben J, Manning JT, et al: Alpha
   interferon for induction of remission in hairy cell
   leukemia. New Engl J Med 1984; 310: 15-18.

342

7. Quesada JR, Hersh EM, Manning J, et al: Treatment of
    hairy cell leukemia with recombinant alpha
    interferon. <u>Blood</u> 1986; 68:493-97.

8. Porzsolt F, Thoma J, Unsold M, et al.
    Platelet-adjusted IFN dosage in the treatment of
    advanced hairy cell leukemia. <u>Blut</u> 1985; 51:73-82.

WELLFERON[R] EFFICACY IN THE TREATMENT OF HAIRY CELL LEUKAEMIA (HCL)

J.M. Bottomley, A. Nethersell, J.C. Cawley, D. Catovsky, P.C. Bevan, C. Burke, N.B. Finter and others.
Wellcome Research Laboratories, Langley Court, Beckenham, Kent, BR3 3BS

Hairy Cell Leukaemia (HCL) is a lymphoproliferative disorder characterised by splenomegaly, pancytopenia and the presence of typical malignant cells in the bone marrow, spleen, liver and peripheral blood(1). HCL is a B-cell neoplasm as shown by the presence of rearranged immunoglobulin genes(2). More rarely, T-cell forms of HCL have been reported (3).

HCL accounts for about 2% of all leukaemias (4) and diagnosis depends on the recognition of characteristic hairy cells in the peripheral blood or bone marrow in association with clinical features of the disease. A bone marrow trephine is usually mandatory to confirm diagnosis, particularly as aspiration commonly results in a "dry tap". Cytochemical staining is used to support the diagnosis; in particular a strongly positive tartrate resistant acid phosphatase activity in the cytoplasmic granules of hairy cells (HC'S).

Median survival is about 50 months (4) with main features of the disease relating to progressive bone marrow infiltration and splenomegaly which occurs in the majority of patients. The resultant anaemia, leucopenia and thrombocytopenia in part caused by marrow failure and in part by hypersplenism are associated with symptoms including increased incidence of infections, dyspnoea, bleeding or bruising tendency, fever and general debility.

Patients with splenomegaly have traditionally been treated by splenectomy, although most patients ultimately deteriorate as marrow infiltration worsens. Prognosis for most patients is then poor. Treatment at this stage is difficult with aggressive combination chemotherapy poorly tolerated whilst low dose chlorambucil reduces the HC burden at the expense of further depression of normal leucocytes.

Quesada et al (5) described favourable responses in HCL patients treated with partially purified human leucocyte interferon. Following this, the Wellcome Research Laboratories were asked to supply Wellferon (Wellcome Human Lymphoblastoid Interferon) for 53 patients who were considered to show a deterioration in their haematological or clinical status. Some of these patients have been described in previous publications (6).

Patients

53 patients with progressive HCL began Wellferon therapy between January 1984 and February 1985. Three patients have been excluded; one patient had concurrent malignant melanoma, an absence of circulating HC's and hepatomegaly as only marker of disease. Two patients were subsequently considered to be suffering from non-Hodgkin's Lymphoma. One of these patients died with end stage disease and the other failed to respond to prolonged Wellferon therapy.

There were 50 evaluable patients (39 men, 11 women, median age 50 years). Patients were treated in 31 different centres; informed consent was obtained in all cases.

All patients were cytopenic or leukaemic prior to Wellferon therapy. 39 patients had been splenectomised and subsequently deteriorated; in 16 of these Wellferon was the only treatment apart from splenectomy. 27 patients received prior chemotherapy and/or steroids prior to Wellferon. 6 patients had non palpable spleens, and 5 patients had splenomegaly, of whom 4 and 3 patients, respectively, received Wellferon as first treatment option.

The disease duration prior to starting Wellferon therapy was variable, ranging from 1 week to 168 months (median 15 months). Circulating HC's were present in 42 patients (17 leukaemic). 45 patients were neutropenic (median 0.6×10^9/l), 37 patients were thrombocytopenic (median 66×10^9/l) and 48 patients were anaemic (median 10.1g/dl). Bone marrow infiltration was considered heavy (>80%) in 41 patients, moderate (40-80%) in 8 and low (10%) in 1 patient. 31 patients had definite evidence of transfusion dependency prior to Wellferon. 29 patients had clear evidence of previous infections.

Methods

After initial full haematological and biochemical investigations, bone marrow aspirate and trephine, most patients received 3 Mu daily Wellferon, either s.c. or i.m. The subcutaneous route of administration was found to be preferable and effective, many patients were taught to administer the dose themselves. Exceptions to this included 3 patients receiving 3Mu thrice weekly and seven patients 6Mu thrice weekly. In about one half of the patients dose reductions to 3Mu thrice weekly were common after 12-16 weeks. Individual physicians were also responsible for providing an assessment of the side-effect profile for each patient. Side-effects were graded 0-3 according to WHO criteria.

Results

As of August 1986, 50 patients had been studied for between 200 and 894 days (median 570). 32 of these patients have received one course, 13 have received 2 courses and 5 patients have received >3 courses of Wellferon respectively. Currently 35 patients are off therapy and 15 remain on therapy. Therapy was interrupted primarily following achievement of a satisfactory peripheral blood response, or occasionally because of intercurrent illness or intolerance. The total cumulative Wellferon dose per patient had a median value of 594Mu (range 42-1317Mu).

Follow-up data was not available for one patient, and three patients died on study. One patient died after 57 weeks of treatment and an excellent response which almost certainly prolonged his survival. The cause of death remains unknown and no post mortem was performed. The second patient received 3 months Wellferon with complete normalisation of the peripheral blood. He died unexpectedly 3 months after stopping therapy, of bronchopneumonia superimposed on longstanding emphysema. The third patient, having a known psychiatric history of depression, tolerated 3 weeks Wellferon and experienced worsening of his depression. He died over one year later.

Haematological Responses

Median HC's fell to zero by 3 months. There seemed to be no difference in the rate at which leukaemic and non-leukaemic patients cleared their peripheral HC's. Platelets reached a median value of 200×10^9/l by 3 months and haemoglobin rose to median 12g/dl by 3 months. Median neutrophils exceeded 1.5×10^9/l by 3 months while monocytes rose from 0 to 0.14×10^9/l by 6 months.

CRITERIA OF RESPONSE

Complete response:

(CR)   Normalisation of all blood indices; Haemoglobin > 12g/dl, Neutrophils >2.5×10^9/l, Platelets >150×10^9/l; absence or ≈5% HC's in bone marrow trephine.

Partial response:

(PR)   Significant improvement and/or normalisation of haematological indices, no HC's; greater than 50% reduction in the bone marrow HC infiltration.

Minor response:

(MR)   Improvement in peripheral blood, HC's persist; Less than 50% reduction in HC infiltration.

No response:

(NR)   No improvement in peripheral blood or bone marrow.

Overall peripheral blood responses included 25 complete responses, 14 partial responses, 10 minor response and 1 no response (this patient only tolerated 3 weeks of treatment because of worsening of pre-existing depression).

Repeat trephine biopsies were not available for 4 patients (2 patients stopped therapy due to CNS toxicities, see later, one gentleman went abroad after 3 months Wellferon and one trephine awaited). 27 patients achieved ≈5% HC infiltration and in 11 of these no HC's could be detected. All but 7 of these 27 patients had heavy infiltration initially. A further 13 patients showed significant reduction (>50%) in HC infiltration while 2 patients achieved only modest improvement. (<50% reduction in HC infiltration). Bone marrow infiltration was unchanged in 4 patients; in 1 patient treatment had been intermittent; in the second 3 months therapy resulted in reduction in size to normal of an extremely large spleen; in the third improvement in all other parameters occurred despite persistence of marrow infiltration. The fourth patient received 3 months therapy with peripheral blood remission. This patient, mentioned earlier, died unexpectedly 3 months later and post mortem revealed HC infiltration of the marrow. Clearance of HC's from the marrow was accompanied in some patients by a completely normal marrow appearance, and in others by a somewhat hypocellular marrow.

Duration of response

19 of 24 patients (79%) in whom Wellferon was stopped (at median 430 day) having achieved bone marrow complete remission sustained a normal peripheral blood count for a median 161 + days off Wellferon (range 1-654+day; 2 patients just stopped therapy). 6 of 8 partial responders (75%) sustained a normal peripheral blood count for a median of 282+ days off Wellferon (range 153+-655+ days), therapy was stopped at median 339 day. Clinical relapses were observed in 7 patients, and further Wellferon therapy resulted in reinduction of remission in all cases. Patients with lesser responses in the bone marrow or peripheral blood responses only, relapsed more quickly off therapy. These results suggest longer term Wellferon therapy may be superior.

Side Effects

α Interferons are known to suppress bone marrow function. During the first four weeks of therapy most patients showed a fall in neutrophil and platelet counts. Median neutrophil levels fell from $0.7 \times 10^9/l$ to $0.3 \times 10^9/l$, whilst median platelets fell from $111 \times 10^9/l$ to $73 \times 10^9/l$. Both these occurred at a median time of nadirs 14 days.

A detailed analysis of the side-effects was performed. Fever was the most common side-effect during the first week of treatment, but after the first month only mild or moderate fever was experienced transiently. Side-effects could be lessened by concurrent administration of paracetemol and by evening administration of Wellferon. Fatigue tended to be more of a problem with continued therapy, but was rarely severe. Cardiovascular abnormalities and local reaction at injection site were rare. Mental disturbance was severe in 3 patients. One elderly female patient became apathetic and withdrawn after 3 weeks therapy and lapsed into a coma. Upon cessation of therapy she recovered completely and remains well, having resumed therapy at a lower dose. One patient with a history of depression tolerated only 3 weeks therapy experiencing severe worsening of depression. He died over one year after stopping Wellferon. A 79 year male epileptic patient experienced severe confusion, necessitating withdrawal of therapy at 2 weeks. Thereafter his spleen continued to reduce in size over the next 6 weeks and his mental condition improved. It was apparent that elderly patients or those with pre-existing neurological or psychiatric disease may be more susceptible to mental change while receiving interferon.

Conclusions

Wellferon is extremely active in inducing haematological improvement and in some cases complete remission in patients with progressive HCL. It also appears effective in patients with splenomegaly, inducing rapid shrinkage of the spleen. Long term remissions have been sustained off therapy (longer term therapy appears to be superior) although some patients have relapsed slowly off treatment. Reinduction with Wellferon is effective in all cases. The side-effects of treatment compare favourably with those associated with cytotoxic therapy.

The role of Wellferon in patients with splenomegaly and the role of maintenance therapy is currently being evaluated; 157 patients have been entered into this second multicentre randomised trial.

References

1.   Bouroncle,B.A., et al. Leukemic reticuloendothelosis. Blood (1958) 13, 609-630.

2.   Korsmeyer,S.J. et al.   Rearrangement and expression of immunoglobulin genes and expression of Tac antigens in hairy cell leukemia. Proc. Natl. Acad. Sci. (USA) (1983) 80, 4522-6.

3.   Cawley,J.C. et al. Hairy Cell Leukaemia with T cell features (1978) Blood. 51 61-69.

4.   Cawley,J.C. et al. Hairy Cell Leukaemia. Recent Results in Cancer Research. 72 Springer-Verlag (1980).

5.   Quesada,J.R. et al. Alpha Interferon for induction of remission in hairy cell leukaemia. N. Engl. J. Med. (1984) 310, 15-18.

6.   Worman, C.P. et al. Interferon is effective in hairy cell leukaemia. Br. J. Haem. (1985) 60, 759-763.

DETECTION OF URINARY ENZYME EXCRETION AS A PARAMETER OF THE NEPHROTOXICITY OF HUMAN INTERFERON ALPHA-2b IN PATIENTS WITH HAIRY CELL LEUKEMIA (HCL)

E. KURSCHEL, U. METZ-KURSCHEL*, N. NIEDERLE, O. KLOKE AND
C.G. SCHMIDT
WEST GERMAN TUMOR CENTER AND DEPT. OF RENAL AND HYPERTENSIVE DISEASES*
UNIVERSITY OF ESSEN, HUFELANDSTR:55, D-4300 ESSEN, FRG

Introduction
Interferons have antiviral, antiproliferative and immunomodulating properties. Their therapeutic value has been established in the treatment of
hairy cell leukemia (1-3), whereas in other malignancies their role is
still in debate.
A broad spectrum of clinical adverse reactions and laboratory abnormalities has been reported in numerous clinical trials including fever,
headache, flu-like symptoms, myalgia, arthralgia, leucocytopenia and
others (Tab. 1)

SIDE EFFECTS DURING TREATMENT WITH INTERFERON ALPHA-2b

Headache	Fever
Myalgia	Hypotension
Arthralgia	Leucocytopenia
Abnormal Taste	Thrombocytopenia
Confusion	Anemia
Anorexia	Abnormal Liver Function Tests

Renal function disturbances have been observed in several studies. They
comprised interstitial nephritis with the nephrotic syndrome (4) as well
as nephrotic syndromes and proteinuria of unknown origin (5-7). During a
previous study of the application of interferon alpha-2b in 13 patients
with cancer of the colon or rectum, renal cancer and carcinoid tumors we
did not find signs of acute or chronic renal toxicity (8).
In a study on the effect of human interferon alpha-2b in patients with
hairy cell leukemia we investigated 14 patients in order to define the
nephrotoxic potential of this new drug by using sensitive non-invasive
techniques which are capable to detect subclinical renal damage.

Patients and Methods

So far fourteen patients with hairy cell leukemia entered the study
(11 male, 3 female), median age 50,0 ± 6,4 years. Prior to therapy all
patients had normal renal function tests.
Treatment consisted in subcutaneous injections of interferon alpha-2b
(Schering Corp., Kenilworth, N.J., USA) of a specific activity of
$1 - 2 \times 10^8$ U/mg protein. Induction therapy was 4×10^6 U/m^2 s.c. daily,
dose adjustment depending on leucocyte count and remission status.
Besides the determination of serum creatinine, BUN, protein excretion and

the analysis of the urinary sediment, the urinary excretion of the following enzymes : Lactate Dehydrogenase (LDH, EC 1.1.1.27), Leucine Arylami-
dase (LAP, EC 3.4.11.2), Gamma-Glutamyltransferase (GGT, EC 2.3.2.2),
N-Acetyl-beta-Glucosaminidase (NAG, EC 3.2.1.30) and Beta-Galactosidase
(GAL, EC 3.2.1.23).
Urine for the enzyme and protein determination was collected in 3 hours
morning periods from 6 to 9 a.m. to obviate any influence of circadian
variation of urinary enzyme excretion. Immediatedly after the collection
period urines were prepared by gel filtration of Sephadex G 50 fine.
Determinations of enzyme activities were carried out as previously described (9). Urinary enzyme output was calculated as mU/3h, corrected for
the average adult body surface ($1,73$ m^2).
Measurement of protein excretion was made by using the Biuret method and
in two patients SDS-polyacrylamide gel electrophoresis was done for the
qualitative differentiation of selective and unselective glomerular and
tubular proteinuria (10,11).

Results

No patients had a rise in serum creatinine or BUN. 6 patients developped
proteinuria of up to 2 g/l (Tab. 2).

Tab. 2
  Renal Function During Treatment With Interferon Alpha-2b

Proteinuria < 2 g/l	6/14
Proteinuria > 2 g/l	0/14
Microhematuria	1/14
S-Creatinine mg/ml	$0,8 \pm 0,2$

In 11/14 patients there was a rise in the excretion of LAP, GGT and NAG,
while only in 4 patients GAL and in another 2 patients LDH was excreted in
abnormal amounts (Tab. 3)

Tab. 3
  Enzymuria During Treatment With Interferon Alpha-2b

Lactate Dehydrogenase (LDH)	2/14
Leucine Arylamidase (LAP)	3/14
Gamma-Glutamyltransferase (GGT)	9/14
N-Acetyl-Beta-Glucosaminidase (NAG)	9/14
Beta-Galactosidase (GAL)	4/14

The kinetics of enzyme excretion show that elevated enzymuria occurred
during the first week of treatment and reached its maximum between the

fourth and sixth week. A slight decrease could be observed following dose of interferon alpha-2b (Fig. 1 - 3).

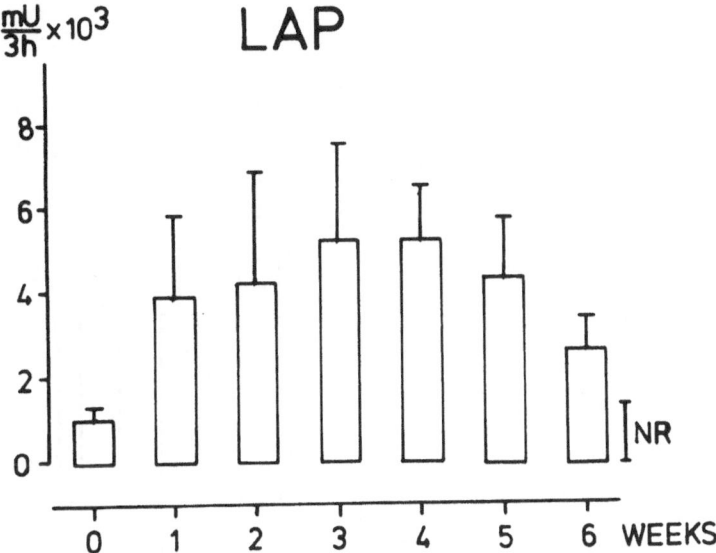

Fig. 1: Urinary excretion of LAP during treatment with interferon alpha-2b. NR: Normal range

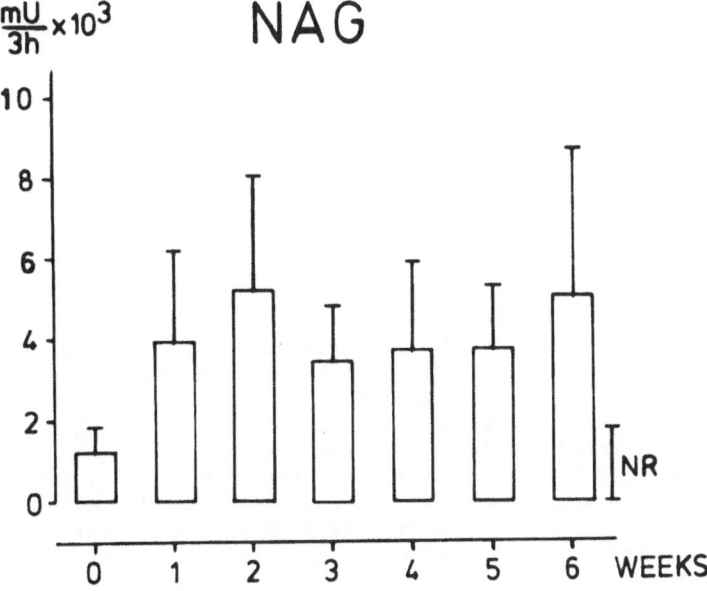

Fig. 2: Urinary excretion of NAG during treatment with interferon alpha-2b. NR: Normal range

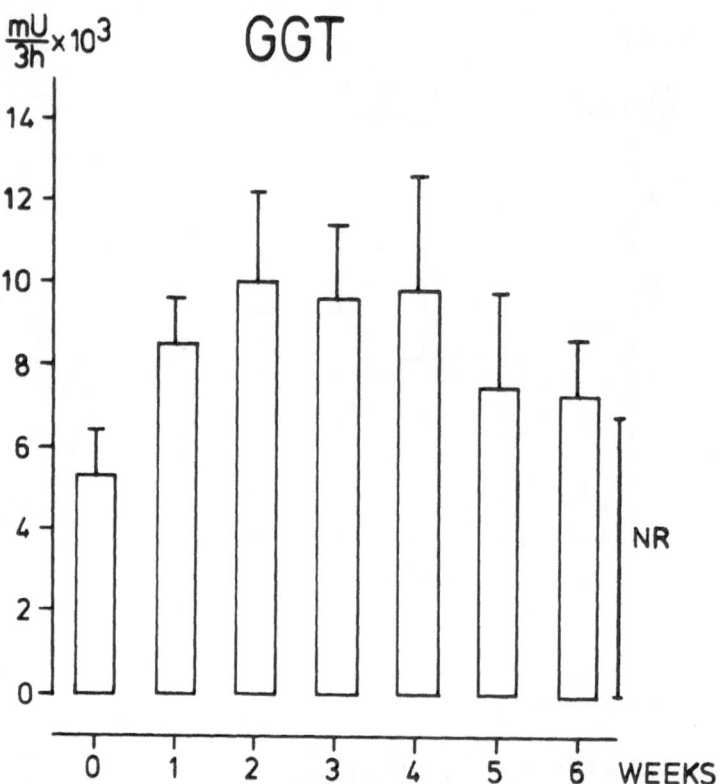

Fig. 3: Urinary excretion of GGT during treatment with interferon alpha-2b.
NR: Normal range

SDS-polyacrylamide gel electrophoresis demonstrated unselective glomerular
and incomplete tubular proteinuria in 2 patients.

Discussion

Since its introduction into clinical research only a few reports on renal
damage during treatment with human interferons have been published. In
these studies a causative relation between interferon therapy and impairment of renal function was likely because cessation of therapy led to the
disappeareance of the symptoms, whereas reexposition with the drug deteriorated renal function again (4,5) . Renal biopsy in one patient with
Mycosis fungoides revealed acute interstitial nephritis, further pathological studies have not been carried out to our knowledge.
In our study we tried to confirm data from previous trials by using sensitive techniques. For these purposes, analyses of the excretion of tubular
macroproteins are well established in the clinical and preclinical evaluation of potentially nephrotoxic substances. The exactly defined localization of the enzymes to certain compartments of the tubular cells is very
helpful in order to correlate specific excreted enzymes with putative

renal injury. LDH is a cytoplasm enzyme with activity in the proximal and
distal tubule, whereas LAP and GGT are localized in the brush border membrane of the proximal tubule. NAG and GAL are found in the lysosomes of
the proximal tubules. A basal excretion of these enzymes is due to the
physiological turn over of the tubular epithelia.
As a consequence of disturbances of the metabolism or the cellular integrity of tubular cells increased amounts of cellular enzymes appear in the
urine. Alterations of tubular epithelia cause excretion of LAP and GGT.
Moreover, lesions are accompanied by additional changes in tubular functions, i.e. renal glucosuria or microglobulinuria. Necrosis of tubular
epithelia is followed by an output of intracellular enzymes (LDH,NAG,GAL)
and tubular dysfunction.
Our data show that during treatment with interferon alpha-2b renal damage
occurs in a high incidence. The constant increase of enzymuria during
induction therapy and the reduction of enzymuria following dose adjustment
are hints to a dose depending relation. This hypothesis is supported by
proteinuria in 6 patients and the findings in SDS-polyacrylamide gel electrophoresis. So far, the exact localization of the injury cannot be exactly definde , but in the majority of our patients the results are consistent
with a combined glomerular and tubular lesion.
The pathophysiological basis of these findings is unclear. An influence of
the initial febrile episodes can be excluded, because increase of enzymuria continued in all patients after termination of the febrile reactions.
Concommitantly used drugs have only been applied during the first treatment week for the controll of fever and then withdrawn, therefore their
causative role is rather unlikely. On the other hand, immunological
mechanisms may contribute to renal damage, because in NZB/W mice a dramatic deterioration of the spontaneously occurring glomerulonephritis took
place after application of exogeneous interferon(12 - 14). Furthermore,
renal transplant recipients had severe rejection periods after prophylactic treatment with interferon(15). An expression of HLA-DR antigens on
tubular epithelia may be another factor to render renal parenchyma
susceptible for immunological attacks. Another factor might be a substance
specific effect of interferon on those renal cells which are responsible
for the synthesis of the basal mambrane.

Conclusion

Human interferon alpha-2b is capable to induce clinical and subclinical
renal injury in a high percentage of patients. Urinary enzyme determination is a useful tool in the detection of even subclinical nephrotoxic
effects. The pathophysiological basis of the combined glomerular and
tubular lesion remains unclear, but immunological mechanisms are likely to
be a causative factor.

Literature:
1. Foon KA (1986)Recombinant leucocyte A interferon therapy for advanced
   hairy cell leukemia. Am.J Med 80: 351 - 356
2. Niederle, N., Kloke; O., Nowrousian MR, Kruse, H., Schmidt, CG (1986)
   Efficacy of Gamma- and Alpha-Interferon in the treatment of hairy cell
   leukemia. J Cancer Res Clin Oncol, Supp. 111, 40
3. Quesda JR,Reuben, J. Manning, JT, Hersch, EM, Guttermann JU (1984)
   Alpha Interferon for induction of remission in hairy cell leukemia.
   N Eng J Med 310: 15 - 18

4. Averbuch, SD, Austin, HA, Sherwin, SA, Antonovych, T, Bunn, PA, Longo,DL (1984) Acute interstitial nephritis with the nephrotic syndrome following recombinant leucocyte A interferon therapy for mycosis fungoid N Eng J Med 310: 32 - 35

5. Selby, P,Kohn, J., Raymond, J, Judson, I. McElwain, T (1985) Nephrotic syndrome during treatment with interferon. Br Med J Apr. 20: 1180

6. Skriskandan K, Tee DEH, Pettingale KW (1985) Nephrotic syndrome during treatment with interferon Br Med J Vol 290: 1590 - 1591

7. Spiegel RJ.(1985) Phase II study of recombinant alpha-2-interferon in resistant multiple myeloma. J Clin Oncolo 5: 654 - 659

8. Metz U, Kurschel E, Niederle N, Oehl S (1985) Urinary enzymes in the detection of the nephrotoxic potential of recombinant human-alpha-interferon. Bach HP, Lock EA (EDS:): Renal Heterogeneity and Target Ell Toxicity. Chicester New York Brisbane Toronto Singapure: Wiley & Sons, 267 - 270

9. Maruhn D, Fuchs I, Mues G, Bock KD (1976) Normal limits of urinary excretion of eleven enzymes. Clin Chim Acta 22: 1567 - 1574

10. Weber MH, Scholz P (1985) Mikro-Flachgelelektrophorese und Alpha-1-Mikroglobulin - Neue Begrife in der Proteinuriediagnostik. Mitt.Klin. Nephrologie XIV, 117 - 122

11. Guder WG, Heidland A (1986) Urine Analysis. J Clin Chem Clin Biochem 24: 611 - 620

12. Morel-Maroger L, Verroust P, Woodrow D (1980) Role of interferon in murine glomerulonephritis. Renal Pathophysiol. Ed. ALeaf: 9 - 16

13. Woodrow D, Ronco P, Riviere Y, Moss J, Gresser I, Guillon JC, Morel-Maroger L, Sloper JC, Verroust P (1982) Severity of glomerulonephritis induced in different strains of suckling mice by infection with lymphocytic choriomeningitis virus: Correlation with amounts of endogenous interferon and circulating immune complexes. J Pathol 138:325 - 336

14. The effect of exogenous interferon:accelaration of autoimmune and renal diseases in (NZB/W) F_1 mice. Clin exp Immunol 40 : 373 - 382.

15. Kramer P, Bijnen AB, Ten Kate FWJ, Jeekel J, Weimar W (1984) Recombinant leucocyte interferon A induces steroid-resistant acute vascular rejection episodes in renal transplant recipents. Lancet May 5, 895

INTERFERON ALFA-N1 WELLFERON[R] IN KAPOSI'S SARCOMA: SINGLE AGENT OR COMBINATION WITH VINBLASTINE

M.A. FISCHL, M.D., E. GOROWSKI, R.N., G. KOCH, R.N., L.S. LUCAS, R.PH., S. RICHMAN, M.D., AND M.A. FLETCHER, PH.D.

INTRODUCTION

Kaposi's sarcoma is the most common malignancy associated with the acquired immunodeficiency syndrome (AIDS). AIDS-associated Kaposi's sarcoma frequently involves lymph nodes and the gastrointestinal tract. In some patients, visceral disease and edema secondary to venous or lymphatic obstruction can be seen and is often associated with significant morbitity and mortality. Since the pathogenesis of AIDS-associated Kaposi's sarcoma is believed to be related to the underlying immunodeficiency associated with AIDS, treatment modalities with both antitumor and immunomodulatory effects have been sought.

Several studies have shown that high-dose alpha-interferon is useful in the treatment of AIDS-associated Kaposi's sarcoma (1-4). Patients with prior opportunistic infections, depressed immunologic studies, and visceral disease, however, are less likely to respond (3,5,6). Cytotoxic chemotherapy has also been shown to be effective in treating AIDS-associated Kaposi's sarcoma (7-11). However, its use has often been limited because of underlying immunosuppression and occurrence of opportunistic infections. The possibility of combining cytotoxic chemotherapeutic agents and alpha-interferon to achieve a better antitumor response seemed warranted. Therefore, we undertook a phase II study to evaluate both the long-term use of high-dose alpha-interferon alone and in combination with vinblastine in the treatment of AIDS-associated Kaposi's sarcoma.

PATIENT POPULATION

All patients had measurable, biopsy-proven Kaposi's sarcoma associated with AIDS. No prior history of immunosuppression or second primary malignancy was elicited. Eligibility criteria included age $\geq$ 18 years, a Karnofsky performance status $\geq$ 80%, a hemoglobin $\geq$ 10 g/dl, an absolute granulocyte count $\geq$ 1500 cells/mm^3, a platelet count $\geq$ 100,000 per mm^3, a serum bilirubin $\geq$ 1.5 mg%, serum glutamic oxaloacetic transaminase less than 3 times normal, a serum creatinine $\leq$ 1.5 mg%, and no prior therapy with any interferon or antineoplastic agents. Patients with known sensitivity to Neomycin or Polymixin B were excluded. Patients were instructed not to take salicylate compounds or nonsteroidal antiinflammatory agents for at least two weeks before entry into the study and throughout the duration of interferon treatment.

METHODS

Laboratory Data

Pretreatment evaluation included a medical history, physical examination, tumor assessment (including photographs) and laboratory tests. All patients were evaluated with a computerized tomographic scan of the abdomen, gastroscopy and colonoscopy. Other tests, scans, or biopsies were done, if needed, to document the presence of tumor or opportunistic infection.

Immunologic Data

Immunocompetence was assessed by T-lymphocyte markers, blastogenic responses to mitogens and antigens and serum immunoglobulins. T-lymphocytes and T-lymphocyte markers were quantitated as previously reported (12). Proliferative assays were done using the mitogens phytohemagglutinin concanavalin A and pokeweed and the antigen Candida albicans. Each test was done in triplicate. Immunoglobulin determinations were done by nephelometry (Beckman ICS and Beckman reagents, Norcross, Georgia). Immunocompetence was assessed before therapy and every three months thereafter for the duration of interferon therapy.

Response Criteria

Response was assessed by standard oncologic criteria. A complete response was defined as absence of detectable disease and normalization of previously abnormal radiographic or endoscopic studies for at least 1 month confirmed by repeat biopsy, endoscopy, and/or lymph node biopsy when appropriate. A partial response was defined as greater than a 50% reduction in the sum of the product of the perpendicular diameters of indicator lesions for at least 1 month without the appearance of new lesions. A minor response was regarded as less than a 50% decrease in measured lesions lasting for at least 1 month or as a greater regression of lesions that persisted for less than 1 month. Stabilization of disease was defined as less than a 25% increase in the size of measured lesions without the appearance of new lesions. Progression of disease was defined as at least a 25% increase in the size of any lesion or the appearance of any new lesions. Patients were considered evaluable for toxicity after one dose of interferon and evaluable for response after 4 weeks of therapy.

Clinical Staging Criteria

The clinical staging criteria used were a modification of that proposed by Krigel et al. (13). Stage I was defined as limited cutaneous disease ($\leq$ 10 cutaneous lesions); stage II, as lymph node involvement alone; stage III, as generalized mucocutaneous disease with or without lymph node involvement. Patients with minimal gastrointestinal disease defined as fewer than 5 lesions and less than 2 cms in combined area, and patients with minimal oral or penile disease defined as fewer than 3 lesions and less than 2 cms in combined area were included in Stage III. Stage IV was defined as visceral disease other than minimal gastrointestinal, oral or penile disease. Subtype A included no systemic symptoms. Subtype B included fevers greater than 39.0°C, drenching night sweats and a greater than 10% weight loss without an identifiable infection.

Patients were followed regularly with repeat histories, physical examinations, tumor assessments (including photographs) and laboratory tests. All patients recorded fevers, weight and subjective symptoms during interferon treatment. Toxicity was measured using the Standard World Health Organization criteria (14).

Treatment Regime

After written informed consent was obtained and all baseline studies completed, patients with Stage I, II and IIIa disease were treated with interferon alone, and patients with stage IIIb or IV disease were treated with interferon and vinblastine. Human lymphoblastoid interferon, Wellferon, an electrophoretically pure preparation of eight species of alpha-interferon purified from culture supernatants of the Namalva lymphoblastoid cell line was used. Interferon was given intramuscularly at a dose of 20 Mu/m^2 once per day for 8 weeks after a 3 day escalation dose regime of 3 Mu/m^2, 5Mu/m^2 and 10Mu/m^2. Vinblastine was given intravenously at a dose of 5mg/m^2 once per day every 2 weeks for 8 weeks. Patients achieving a complete response after 8 weeks of therapy were placed on maintenance therapy. Patients achieving less than a complete response without progressive disease were placed on consolidation therapy. During consolidation therapy, interferon was administered intramuscularly at a dose of 20 Mu/m^2 once per day three times per week, and vinblastine was administered intravenously at a dose of 5mg/m^2 once per day every 2 weeks for 8 weeks. Patients with a complete or partial response following consolidation therapy were placed on maintenance therapy. During maintenance therapy, interferon was administered intramuscularly at a dose of 10 mu/m^2 once per day three times per week for 24 weeks. Vinblastine was administered intravenously at a dose of 5 mg/m^2 once per day every three weeks for a total of 24 weeks.

RESULTS

Patient Population

Sixty-four patients with biopsy-proven Kaposi's sarcoma associated with AIDS were enrolled. All were men between 23 and 55 years of age with a median age of 31.0 years. Sixty of the patients were homosexual or bisexual men, 3 were of Haitian ancestry, and 1 used intravenous drugs. None had skin disease alone, 3 had lymphadenopathy alone, 34 had skin disease with lymphadenopathy and/or minimal gastrointestinal, oral or penile disease, and 26 had visceral disease. Four patients were withdrawn from protocol during the first 3 weeks of therapy because of toxicity.

After 8 weeks of induction therapy; 1 patient had a complete response; 6 had a partial response; and 10 had progressive disease. After 20 weeks of therapy including both induction and consolidation, 5 patients had a complete response; 14 had a partial response; and 19 had progressive disease for an overall antitumor response of 32%. Remission durations ranged from 1 to 16 months with a median of 8 months. In evaluating unmaintained remission durations, it was noted that patients with a partial response had unmaintained remission durations of less than 1 to 4 months with a median of 1.2 months as compared with patients who had a complete response and unmaintained remission durations of 12 to 16 months with a median of 14 months.

Antitumor responses relative to the stage of disease and therapy (interferon alone versus interferon and vinblastine therapy) are outlined in Table 1. Eleven of 37 patients treated with interferon alone had a complete response or partial response for an overall antitumor response rate of 30% compared with 8 of 27 patients treated with interferon and vinblastine for an overall antitumor response rate of 37%. No difference in remission durations were noted.

TABLE 1 - ANTI-TUMOR RESPONSE ACCORDING TO THE STAGE OF KAPOSI'S SARCOMA

Stage	I	II	IIIa	IIIb/IV	TOTAL
Number of cases	0	3	34	27	64
Complete response	-	-	3	2	5
Partial response	-	1	7	6	14
Min-partial response	-	-	6	4	10
Stable disease	-	-	8	5	13
Progression	-	2	8	9	19
Inevaluable	-	1	2	1	4
Overall anti-tumor response*	-	1/3 (33%)	10/34 (29%)	8/27 (30%)	19/60 (32%)

Antitumor responses were seen at all stages of disease, including skin, lymph nodes, oral cavity and gastrointestinal tract. Although no correlation between the stage of disease and response to therapy was noted, several subgroups could be identified. Five patients with conjunctival or orbital Kaposi's sarcoma lesions had complete resolution of eye lesions after 8 weeks of therapy. Five patients with bulky oral cavity lesions had no response or rapidly progressive disease after 8 weeks of therapy. Three patients with pulmonary Kaposi's sarcoma had rapidly progressive disease on therapy.

Fifteen patients developed opportunistic infections while on therapy. Opportunistic infections occurred throughout the study and among patients with all stages of Kaposi's sarcoma. Opportunistic infections occurred more commonly among patients with opportunistic infection prior to entry into the study which also correlated negatively with response. The most common infection noted was Pneumocystis carinii pneumonia. Two patients developed life-threatening gastrointestinal hemorrhages secondary to an initially unrecognized or late developing B-cell lymphoma.

We also evaluated survival of patients in comparison with varying response to therapy. Among patients with a complete or partial response, 12 of 19 (63%) are still alive. The mean ($\pm$ SD) length of survival was 17.6 $\pm$ 6.8 months., ranging from 6 to 28 months. Thirty-three per cent survived beyond 22 months. Among patients with progressive disease, 1 of 19 patients (5%) is still alive. The mean ($\pm$ SD) length of survival was 8.1 $\pm$ 5.7 months. Two patients had survivals of 19 and 23 months, the remainder ranged between 1 to 10 months.

Immunologic Studies

Immunologic abnormalities characterized by skin test anergy, decreased T-helper cells, an inverted T-lymphocyte subset ratio, depressed blastogenic responses to both mitogens and antigens and depressed natural-killer cell activity were noted in the majority of patients. Variable responses in immunologic studies were noted during therapy. No correlation between immunologic studies and antitumor response was noted. However, patients with T-helper cells less than 100 cells/mm^3 were more likely to have visceral disease, opportunistic infections and no antitumor response to therapy. Patients who had a complete response had a statistically higher number of T-helper cells and were more likely to react to one or more skin test antigens compared with patients who had progressive disease (p<0.05, analysis variance).

359

Adverse Reaction

Adverse reactions to interferon are listed in Table 2. Severe fatigue with mylaise, anorexia, and weight loss averaging 4.5 kg was common. Fevers as high as 39.9°C associated with chills were prominent during the first 2 to 6 weeks of therapy and typically occurred 2 to 3 hours after injections of interferon. Cardiac toxicity was minimal. Two patients developed chest pain without electrocardiogram changes. Neurological toxicity was substantial and included headaches, lethargy, confusion, difficulty in concentration, forgetfullness, spacial mispreception and paresthesias. Two patients had an acute encepathopathy, and 1 had grand mal seizures. Patients who had underlying central nervous system disease secondary to human T-lymphotropic virus, type III/lymphadenopathy-associated virus (HTLV-III/LAV) had a higher incidence of neurologic toxicity. Two patients developed a cellulitis at injection sites, and 1 developed a muscle abscess requiring surgical drainage. Six patients were removed from protocol secondary to adverse reactions, including severe fatigue, weight loss, and encephalopathy.

TABLE 2 - TOXICITY/CLINICAL TOXICITY OF HIGH DOSE ALPHA INTERFERON THERAPY

fatigue with myalgias	100%	fevers	83%
anorexia/weight loss	89%	headaches	58%
nausea	72%	paresthesias	42%
diarrhea	58%	confusion	18%
hypotension	49%	hair loss	47%

Pretreatment and mean nadir laboratory values measured during therapy are listed in Table 3. All patients had myelosuppression. The lowest granulocyte count was 100 cells/mm^3 with a mean of 620 cells/mm^3. More than 50% of patients had thrombocytopenia. Two patients had platelet counts less than 45,000/mm^3, necessitating interruption of therapy. All patients had mild to moderate hepatotoxicity. Seven patients had a greater than ten-fold increase in transaminases. Hepatotoxicity was more commonly noted in patients with chronic hepatitis B infection. Both myelosuppression and hepatotoxicity were reversible. One patient died of a fulminant hepatitis secondary to a herpes virus infection shortly after recovery from moderately severe hepatotoxicity secondary to interferon therapy.

TABLE 3 - LABORATORY DATA TOXICITY OF HIGH DOSE ALPHA INTERFERON THERAPY

	Pretreatment		Nadir	
	Range	Mean	Range	Mean
Hematocrit	30 - 49	40.0 $\pm$ 6.0	20 - 38	30.9 $\pm$ 5.3
WBC count	2100 - 6400	4033 $\pm$ 1232	300 - 3800	1738 $\pm$ 861
Granulocyte ct	1500 - 400	2296 $\pm$ 791	100 - 1806	620 $\pm$ 274
platelet count (x10^3)	140 - 326	203 $\pm$ 69.3	47 - 120	88.3 $\pm$ 48.4
SGPT	8 - 56	30 $\pm$ 7.3	49 - 628	200 $\pm$ 167

Patients on interferon and vinblastine had significant myelosuppression with marked decreases in both the total white blood cell

count and granulocyte count, frequently necessitating interruption of
therapy. In 23% of the patients on combination therapy, administration of
vinblastine was interrupted because of myelosuppression. Decreases in
white blood cell count and granulocyte count occurred within 1 week of
vinblastine administration, and counts typically remained depressed for 4
to 6 weeks before vinblastine therapy could be restarted. Even with a 50%
reduction in the dose of both interferon and vinblastine significant
myelosuppression reoccurred. Patients with T-helper cell < 200 cells/cm^3,
prior opportunistic infection and subtype B symptoms exhibited the most
significant and persistant myelosuppression. Otherwise, no significant
difference in adverse reactions were noted between patients on interferon
alone versus combination therapy.

DISCUSSION

The present study demonstrates that high-dose human-lymphoblastoid
interferon can induce regression of AIDS-associated Kaposi's sarcoma.
Antitumor response durations were long with a median of 8 months. In
particular, patients with a complete antitumor response had remission
durations in excess of 12 months. Patients with a complete response also
had a decreased incidence of opportunistic infections associated with
improvement in immunologic studies and had prolonged survival compared to
patients with progressive disease. Based on these findings and those of
others (1-3), it appears that Kaposi's sarcoma is responsive to high-dose
alpha-interferon and may be particularly benefical to patients with AIDS.

As shown by Krown et al. high-dose alpha-interferon therapy is needed
to achieve any significant antitumor response (3). It also appears from
the present study, that the length of therapy is also important with
reference to both response rates and durable responses. It was noted that
after 8 weeks of daily high-dose interferon therapy, response did occur
but the overall antitumor response rate was low (10%). During
consolidation therapy, antitumor responses were still seen and approached
32%, suggesting the need for long-term therapy to achieve maximal
response. This finding may account for some of the lower response rates
reported previously with short term interferon therapy in AIDS-associated
Kaposi's sarcoma (6). Patients that had a complete response or partial
response during induction therapy continued to respond during
consolidation therapy, and responses were not lost with lower dose of
interferon. In addition to high-dose interferon therapy, long-term
therapy should be considered when treating patients with AIDS-associated
Kaposi's sarcoma. A consolidation course of lower doses of interferon,
thereby decreasing toxicity, appears acceptable.

In evaluating durability of responses, a considerable difference was
noted between those patients with a partial response and a complete
response. In all but one patient, patients with a partial response had
unmaintained remission durations of less than 1 month compared with
greater than 12 months in patients with a complete response. These data
suggest that patients with any evidence of residual tumor should be
continued on maintenance therapy.

Antitumor responses to interferon therapy did not necessarily
correlate with the stage of Kaposi's sarcoma as noted in prior studies
(1,3). However, patients with conjuctival or orbital lesions appeared to
be particularly sensitive to interferon therapy. In all patients that had
ocular lesions a dramatic decrease in lesions was seen within a relatively
short period of time. In two of these patients, ocular lesions
disappeared in the face of progressive disease elsewhere. Patients with

bulky oral disease or lung disease, however, all had progressive disease on therapy.

Immunologic parameters did not necessarily correlate with response to interferon therapy. However, certain subgroups of patients could be identified. In particular, patients with low T-helper cells ($<$ 100 cells/mm^3) had a poor prognosis and were less responsive to interferon therapy. All patients with a complete response had a statistically higher number of T-helper cells prior to therapy and were more likely to react to one or more skin test antigens, similar to what has been noted previously with high-dose recombinant alpha interferon therapy (3). This was not true for patients with a partial response.

Adverse reactions associated with high-dose human-lymphoblastoid interferon were considerable. Although adverse reactions occurred during consolidation therapy and maintenance therapy, they were most prominent during induction therapy. Most patients had constitutional symptoms with severe fatigue, decreased Kornofsky performance, anorexia, and weight loss. Neurotoxicity with progressive peripheral neuropathy, dementia or seizures were noted and were more prominent in patients with underlying neurological disease secondary HTLV-III/LAV. In considering future use of high-dose interferon therapy, caution should be used in treating patients with underlying neurologic disease. The majority of patients also had mild to moderate evidence of hepatotoxicity. In 12% of patients, a significant increase in liver function studies were noted, necessitating interruption of therapy. These data suggest that caution should be used when evaluating patients with underlying liver disease for interferon therapy. In addition, many drugs already shown to have activity in Kaposi's sarcoma are metabolised by the liver. Therefore, caution should again be used when selecting drugs for combination regimens which include interferon.

Mild to moderately severe hematologic toxicity with depression in white blood cell counts, granulocyte counts, and occassionally platelet counts was seen. This was particularly prominant in patients on combination therapy. In more than 20% of patients on combination therapy, significant and sustained myelosuppression was seen, necessitating interruption of vinblastine therapy for up to 4 to 6 weeks. Even with reduction in doses of both interferon and vinblastine, myelosuppression reoccurred. The reasons for this were unclear. The possibility of altered pharmakinetics of vinblastine secondary to interferon therapy needs to be considered especially in the face of hepatotoxicity. Both myelosuppression and hepatotoxicity were reversible with discontinuance of therapy. Three patients developed either a cellulitis in the soft tissue or muscle abscess secondary to injections of interferon. Carefull sterile techniques in injecting interferon and rotation of injection sites need to be considered when using long-term interferon.

In evaluating interferon therapy alone and in combination with vinblastine, several points were noted. Antitumor responses with combination therapy were slightly better (37%) compared with interferon therapy alone (30%). This was surprising, since combination therapy was used in patients with more advanced disease and subtype B symptoms. These patients were also more likely to have had opportunistic infection and significant depression of immunologic studies; all poor prognostic indicators. Further, myelosuppression with combination was significant and often limited the use of vinblastine. Despite this limitation, responses were still seen. Further studies evaluating alternate doses and/or frequency of vinblastine in combination therapy for Kaposi's sarcoma should be considered.

In summary, interferon alone and in combination with vinblastine appears to have activity in AIDS-related Kaposi's sarcoma. Although combination therapy appears as effective or perhaps slightly better than interferon alone, combination therapy was limited secondary to myelosuppression, and the possibility of lower doses of vinblastine or the use of other cytotoxic chemotherapeutic agents should be considered.

REFERENCES

1. Krown SE, Real FX, Cunningham-Rundles, et al. Preliminary observations on the effect of recombinant leukocyte interferon in homosexual men with Kaposi's sarcoma. N Engl J Med. 1983; 308:1071-6.

2. Groopman JE, Gottlieb MS, Goodman J, et al. Recombinant alpha-2 interferon therapy of Kaposi's sarcoma associated with the acquired immunodeficiency syndrome. Ann Intern Med. 1984; 100:671-77.

3. Krown SE, Real FX, Vadhan-Raj S, et al. Kaposi's sarcoma and the acquired immunodeficiency syndrome: treatment with recombinant interferon and analysis of prognostic factors. Cancer. 1986; 57:1662-5.

4. Fischl MA, Gorowski E, Koch G, et al. Human lymphoblastoid interferon in the treatment of Kaposi's sarcoma. Intern Conf on AIDS. 1986; 2:39.

5. Vnadham-Raj S, Wong G, Gnecco C, et al. Immunologic variables as predictors of prognosis in patients with Kaposi's arcoma and the acquired immunodeficiency syndrome. Cancer Res. 1986; 46:417-25.

6. Gelmann EP, Preble OT, Steis R, et al. Human lymphoblastoid interferon treatment of Kaposi's sarcoma in the acquired immunodeficiency syndrome. Ann Intern Med. 1985; 78:737-41.

7. Volberding P, Conant MA, Stricker RB, et al. Chemotherapy in advanced Kaposi's sarcoma. Ann Intern Med. 1983; 74:652-6.

8. Laubenstein LJ, Krigel RL, Odajnkl CM, et al. Treatment of Epidemic Kaposi' sarcoma with VP-16 (etoposide) and a combination of doxorubicin, bleomycin, and vinblastine. J Clin Oncol. 1984; 2:1115-20.

9. Mintzer DM, Real FX, Jovinol L, et al. Treatment of Kaposi's sarcoma and thrombocytopenia with vincristine in patients with the acquired immunodeficiency syndrome. Ann Intern Med. 1985; 102:200-2.

10. Lewis BJ, Abrams DI, Zigler J, et al. Single agent or combination chemotherapy of Kaposi's sarcoma in AIDS. 1983; Proc AM Soc Clin Oncol. 2:59.

11. Volberding P, Abrams D, Kaplan L, et al. Therapy of AIDS-related Kaposi's sarcoma with ICRF-159. Proc Am Soc Clin Oncol. 1985; 4:4.

12. Scott GB, Fischl MA, Klimas N, et al. Mothers of infants with the acquired immunodeficiency syndrome (AIDS): evidence for both symptomatic and asymptomatic carriers. JAMA. 1985; 253:363-6.

13. Krigel RL, Laubenstein LJ, Muggia FM. Kaposi's sarcoma: a new staging classification. Cancer Tret Rep. 1983; 67:670-6.

COMBINED INTERFERON-POLYCHEMOTHERAPY VERSUS POLYCHEMOTHERAPY
IN MULTIPLE MYELOMA: A PHASE-III STUDY

H. LUDWIG[1], A. CORTELEZZI[2], E. FRITZ[1], I. KÜHRER[1], E. POLLI[2],
W. SCHEITHAUER[1], and R. FLENER[3]

[1]II. Med. Dept. Univ. Vienna, Austria; [2]Inst. Clin. Med. I,
Milan, Italy; [3]Ernst Boehringer Res. Inst., Vienna, Austria.

1. INTRODUCTION

The discovery of inhibitory effects of human leukocyte
interferon on in vitro cellular proliferation stimulated the
first clinical trials in patients with multiple myeloma (1,2).
After the recent advances in recombinant DNA-technology,
highly purified interferon has become available in ample
quantities to render adequately controlled clinical studies
feasible (3,4,5).

The "European Myeloma Study Group for Interferon" (EMSI)
conducted a clinical trial which confirmed the therapeutic
effect of recombinant interferon alpha-2C (rec-IFN) in the
treatment of patients with multiple myeloma. Polychemotherapy,
however, induced significantly more responses than the rec-IFN
monotherapy (6). These results suggested a possible improvement of both treatment modalities by combining the standard
VMCP-polychemotherapy treatment regimen with rec-IFN. Therefore, a second prospective randomized study has been organized
to compare the combined treatment of rec-IFN plus VMCP-polychemotherapy to VMCP-polychemotherapy alone.

2. PATIENTS AND METHODS
2.1 Patients:

So far, 32 patients with multiple myeloma have been entered
into the study. Diagnosis of multiple myeloma required the
presence of two out of the three following findings:
(1) serum or urinary M-component,
(2) bone marrow plasma cell infiltration >15 %,
(3) osteolytic bone lesions.

In accordance with the legal regulations on clinical
trials, informed consent has been obtained from all participating patients. Details on patients' characteristics are
given in table 1.

Of the 15 patients randomized to the rec-IFN+VMCP subgroup
and the 17 patients randomly assigned to the VMCP chemotherapy
arm, one patient in each group has not yet become eligible for
evaluation. Another patient had to be excluded from evaluation
in the VMCP chemotherapy arm because of early death. Thus, up

to date, 14 patients have become eligible for evaluation of
rec-IFN+VMCP treatment and 15 patients have remained for
evaluation of VMCP-polychemotherapy.

Table 1

PATIENTS' CHARACTERISTICS

	rec-IFN+VMCP	VMCP
Number of cases:	15	17
Age (years):		
median	69	63
range	43-76	45-82
Sex:		
male	10	5
female	5	12
Myeloma stage (Durie & Salmon):		
I	1	3
II	7	7
III A	6	6
III B	1	1
Paraprotein class:		
IgA	5	4
IgG	8	13
light chain	2	0
kappa/lambda	9/6	9/8
Performance status before therapy (WHO):		
median	1	1
range	0-2	0-2
Number of treatment cycles:		
median	3	4
range	1-8	1-6

2.2 Response Criteria:

Response to treatment has been evaluated according to the following criteria of the Chronic Leukemia and Myeloma Task Force:

Response: more than 50% reduction of serum or urinary M-component.

Minor response: reduction of M-component between 25% and 50%.

Disease progression: at least 25% increase of serum M-component or, in light chain myeloma, 100% increase of urinary protein excretion.

Stable disease: neither response nor progression criteria met.

2.3 Treatment Regimens:

Table 2 shows the therapy schedules and the doses administered. Rec-IFN alfa-2C (Boehringer Ingelheim International, FRG; specific activity 3.2×10^8 U/mg protein) was administered intramuscularly; polychemotherapy consisted of intermittent courses at 4-6 weeks' intervals.

2.4 Statistical Evaluation:

Comparisons were performed using Fisher's exact test, Student's t test, and the two sample test by Wilcoxon. All statements about time intervals refer to censored data, i.e., they distinguish between completed and current observations.

Table 2

TREATMENT REGIMENS

rec-IFN+VMCP					VMCP		
rec-IFN alfa-2C 2×10^6 U/d i.m.		L-PAM	15	mg/m^2	i.v.	day 1
5x/week		CTX	450	mg/m^2	i.v.	day 1
		VCR	1.4	mg/m^2	i.v.	day 1
+ VMCP		PRED	40	mg/m^2	p.o.	day 1-7
			20	mg/m^2	p.o.	day 8-14

L-PAM: melphalan, CTX: cyclophosphamide, VCR: vincristine, PRED: prednisolone

3. RESULTS

3.1 Treatment Responses:

Up to now, the following response rates have been observed: in the rec-IFN+VMCP subgroup 50 % of the patients achieved complete responses and 21 % showed minor responses giving a total of 71 % responses. The respective values in the VMCP arm were 33 % complete responses and 33 % minor responses, adding up to a total of 67 % responses. For detailed comparisons between the two treatment arms consult table 3.

Responses or minor responses were observed after a median time span of 0.9 months (range: 0.5-4.4) under rec-IFN+VMCP and after 1.3 months (range: 0.4-5.0) under VMCP treatment.

Out of the patients who had achieved responses or minor responses, all patients under rec-IFN+VMCP and all but one under VMCP treatment are still in remission. These observation periods cover a median of 4.9 months (range: 0.7-13.8) and 3.8 months (range: 1.0-13.8) for the rec-IFN+VMCP and the VMCP treatment arm, respectively.

Table 3

R E S P O N S E   R A T E S

| | rec-IFN+VMCP | | VMCP | |
	N	Percentage	N	Percentage
Number of patients	15		17	
not evaluable	1		2	
evaluable	14	100 %	15	100 %
Response	7	50 %	5	33 %
Minor Response	3	21 %	5	33 %
Stable course of disease	3	21 %	4	27 %
Progression	1	7 %	1	7 %
Response + minor response	10	71 %	10	67 %

Table 4

T R E A T M E N T   R E S U L T S

	rec-IFN+VMCP	VMCP
Number of cases:	13	13
Serum M-component before treatment (g/l):		
median	3.63	3.07
range	2.9-6.6	0.8-5.3
Nadir serum M-component (g/l):		
median	2.40	1.60
range	0.7-3.4	0.6-3.1
Per cent reduction serum M-component:		
median	46.9	33.3
range	7.1-76.7	2.3-89.0
Number of cases:	1	2
Urinary M-component before treatment (g/24h):		
median	0.83	1.86
range		1.4-2.3
Nadir urinary M-component (g/24h):		
median	0.10	0.11
range		0.04-0.2
Per cent reduction urinary M-component:		
median	87.7	94.6
range		92.0-97.1

Table 4 shows the changes in serum and urinary M-component concentrations of the entire patient population, i.e., the results of treatment regardless whether it induced responses or ultimately had to be considered inadequate. Both rec-IFN+VMCP treatment and VMCP-polychemotherapy effected significant reductions in M-component levels. Similar to response rates, the combined rec-IFN+VMCP therapy appears to be slightly superior to treatment with VMCP alone, but those differences remain below the statistical limit of significance.

The addition of rec-IFN to the VMCP-polychemotherapy regimen did not enhance the side effects of chemotherapy. Clinical observation yielded more symptoms in the VMCP subgroup than in the rec-IFN+VMCP treatment arm (table 5).

Table 5

S I D E   E F F E C T S

| | rec-IFN+VMCP | | VMCP | |
	N	Percentage	N	Percentage
Hematol. toxicity I+II	9	60 %	15	88 %
Hematol. toxicity III+IV	6	40 %	8	47 %
Fever	5	33 %	8	47 %
Chills	–		1	6 %
Infection	3	20 %	2	12 %
Nausea	2	13 %	6	35 %
Vomiting	–		7	41 %
Diarrhoea	–		1	6 %
Fatigue	2	13 %	3	18 %
Weakness	–		1	6 %
Vertigo	1	7 %	3	18 %
ARDS	–		1	6 %
Pulmonary oedema	1	7 %	–	
Oedema	–		1	6 %
Alopecia	–		1	6 %
Neuralgia	–		1	6 %
None	0		0	

4. DISCUSSION

Since the introduction of polychemotherapy (7), treatment results in multiple myeloma have considerably improved. Nevertheless, the current standard therapy regimens are far from being optimal and much effort is exerted in search of better therapeutic agents or novel drug combinations. When pilot studies applying human leukocyte interferon to patients with multiple myeloma showed the first encouraging results (1,2), it was, therefore, appropriate to publish those reports rapidly and share the hope for a superior treatment mode with the scientific community. Subsequent investigations on a larger scale confirmed some beneficial effect of interferon in the treatment of multiple myeloma (5), but they could not support the early proposals of interferon surpassing the benefit of an established polychemotherapy regimen (6).

Our current study in which rec-IFN is combined with VMCP-polychemotherapy has, so far, not rendered significantly improved treatment results. There may be marginal benefits in adding a "biological response modifier" to the standard cytostatic drug regimen, as observed in the earlier achievement of complete responses, the larger reduction of median serum M-component or the fewer side effects, but none of those slight differences proved to be statistically significant and they may well be due to random variations.

Since this paper constitutes only an interim report on the second cooperative clinical EMSI trial, definite conclusions have to await the completion of the study. If the final results will confirm the observed marginal benefits of a combined rec-IFN+VMCP-polychemotherapy treatment, one will have to carefully weigh the additional clinical expenditure of interferon application against its possible gain in therapeutic efficacy.

REFERENCES
1. Mellstedt H, Ahre A, Björkholm M, et al.: Interferon therapy in myelomatosis. Lancet i:245-247, 1979.
2. Gutterman JU, Blumenschein GR, Alexanian R, et al.: Leukocyte interferon-induced tumor regression in human metastatic breast cancer, multiple myeloma, and malignant lymphoma. Ann Intern Med 93:399-406, 1980.
3. Ahre A, Björkholm M, Mellstedt H, et al.: Human leucocyte interferon and intermittent high-dose melphalan-prednisone administration in the treatment of multiple myeloma: A randomized clinical trial from the Myeloma Group of Central Sweden. Cancer Treat Rep 68:1331-1338, 1984.

4. Foon KA, Sherwin ST, Abrams PG, et al.: Treatment of advanced non-Hodgkin's lymphoma with recombinant leukocyte A interferon. N Engl J Med 311:1148-52, 1984.
5. Ohno R, Kimura K, Amaki I, et al.: Treatment of multiple myeloma with recombinant human leukocyte A interferon. Cancer Treat Rep 60:1433-1435, 1985.
6. Ludwig H, Cortelezzi A, Scheithauer W, et al.: Recombinant interferon-alpha-2C versus polychemotherapy (VMCP) for treatment of multiple myeloma: A prospective randomized trial. Eur J Cancer Clin Oncol (in press).
7. Alexanian R, Bonnet J, Gehan E, et al.: Combination chemotherapy for multiple myeloma. Cancer 30:382-389, 1972.

Supported in part by the Austrian Research Grant:P4999 and the Ludwig Boltzmann Institute for Gerontology, Vienna

RECOMBINANT INTERFERON ALPHA (2A) COMBINED WITH DACARBAZINE IN TREATMENT
OF METASTATIC MELANOMA.

P. HERSEY, G.R.C. McLEOD[t] AND D.B. THOMSON [t]
Oncology and Immunology Unit, Royal Newcastle Hospital, Newcastle.
N.S.W.;[t]Department of Surgery and Medical Oncology, Princess Alexandra
Hospital, Brisbane, Queensland, AUSTRALIA.

INTRODUCTION
Systemic therapy for patients with unresectable malignant melanoma is
extremely poor and a variety of regimes have not altered the median
survival of the disease[1,10]. Several recent studies have shown that
systemic treatment with interferon alpha 2A (rIFN-α2A) produced by
genetic engineering may induce complete or partial disease remission in
from 11 to 23% of patients with malignant melanoma [7,8] (see also
reviews [11,14]). Responses were seen mainly in patients with skin and
or pulmonary disease. These results compared favourably with those from
"standard" chemotherapy with dacarbazine (dimethyl triazeno-imidazole-4-
carboxamide, DTIC) where response rates of 14-18% were reported in
several studies[3,6]. See reviews [4,5,9,15]. Complete responses (CR)
were uncommon[5,9] and occurred in less than 5% of the patients. Response
duration was also usually short and in the vicinity of 14 weeks[6].

Interferon has been reported to synergize with a number of
chemotherapeutic agents in vitro[17]. This together with the reported
differences in antiproliferative effects of dacarbazine and IFN-α on the
cell cycle (IFN active in G1, DTIC in G2) suggested that this combination
may have complimentary antitumour effects. The present report describes
the preliminary results in the first 44 patients treated with this
combination.

MATERIALS AND METHODS
Patients studied
Fifty two patients have so far been entered into the study from the
Melanoma Units at Newcastle and Brisbane. This report describes the
results only in those where the follow up time was adequate to assess
treatment response (i.e. 3 months or greater). Details of the subjects
were as follows. Nineteen were from Newcastle and 25 from Brisbane.
Only 5 patients had been treated with prior chemotherapy and or
irradiation and 38 had been treated with surgery. Metastases were in
lymph nodes alone(7), skin alone(7), lung plus other sites(16), liver
plus other sites(6) and other sites(8). Twenty seven were males and 17
females. Performance status by the ECOG scale[16] was grade 0 or 1 in 36
patients and grade 2 in 8 patients. Average age was 52 (range 21-73).

Study design
rIFN-α2A (RO 22-8181) was supplied by Roche products (Dee Why, N.S.W.) as
a freeze dried preparation in ampoules of 3, 9 & 18 x 10^6 u/ml^{-1}(Mu/ml^{-1}).
The preparation was reconstituted immediately before use and given by
s.c. or i.m. injection in the gluteal region. Dose was 3 Mu on days 1-3

then 9 Mu daily from day 4 to week 10. Thereafter dose was 9 Mu or Monday, Wednesday and Friday of each week. Dacarbazine was given by a single i.v. infusion each 21 days as described elsewhere[18] starting at a dose of 200 mg/m^2 and escalating to 400, 800 and 1000 mg/m^2 if tolerated.

All patients had thorough clinical examinations and careful staging of their disease including computerized tomography (CAT scan when appropriate) before initiation of rIFN-α2A treatment. Complete blood cell counts, differential count and platelet counts, blood chemistry, urinalysis, electrocardiogram and chest X-ray were performed prior to commencement of treatment.

All patients were followed up by physical examination at two weekly intervals. Peripheral blood cell counts and biochemistry were repeated each week for 4 weeks, then each 2 weeks to 12 weeks, then 4 weekly thereafter. Chest X-ray and urinalysis was repeated each 4 weeks. Tumour status was reassessed at least once a month by tests pertinent to each patient. Treatment was discontinued if toxicity of grade 3 or 4 developed [16]. After recovery therapy was reinstituted with 25% reduction for grade 3 and 50% for grade 4.

Treatment was discontinued if disease progression noted after 2 months. Therapy could be continued for 6 months in those with no disease progression. In those with complete (CR) and partial (PR) remission treatment was continued for 6 and 12 months respectively.

Criteria of response
These were documented in terms of the WHO criteria as described elsewhere[16]. Complete response: disappearance of all known disease for at least 4 weeks. Partial response: greater than 50% decrease in total tumour size for at least 4 weeks. Stable disease: less than 50% decrease and less than 25% increase for at least 8 weeks. Progressive disease: 25% or more increase in known disease or appearance of new disease.

RESULTS
With one exception (due to early death in the first two days of treatment) all patients were evaluable. Fourteen patients (33%) appeared to have CR (7) or PR (7) of their disease (from the binomial tables 95% confidence limits were 19.08-48.54). Stabilization of disease occurred in 12 patients and in the remaining 17 (39%) disease progressed. As shown in table 1 response rates were higher in those with good performance scores and in male patients.

TABLE 1
RESULTS

		CR (%)	PR (%)	SD (%)	PD (%)
TOTAL		7 (16)	7 (16)	13 (30)	16 (37)
ECOG 0, 1	(36)	6 (17)	7 (19)	11 (31)	12 (33)
ECOG 2	(8)	1 (13)	0	2 (25)	4 (50)
MALE	(27)	5 (19)	6 (22)	8 (30)	7 (25)
FEMALE	(17)	2 (12)	1 (6)	5 (29)	9 (53)

The site of the metastases appeared to be the main determinant of disease. As shown in table 2 the response rate in patients with skin metastases was 50% (95% confidence limits 19–82) and 64% in those with lung ± skin metastases (95% confidence limits 31–89). No responses were recorded in those with metastases solely in lymph nodes. Complete responses were however recorded in 2 patients with visceral (liver and adrenal) disease.

TABLE 2
RESPONSE TO IFN + DTIC ACCORDING TO
SITE OF DISEASE

	% CR + PR	% SD + PD
Skin	50 (5)	50 (5)
Lung ± skin	64 (7)	36 (4)
Lymph nodes ± skin	0	100 (11)
Visceral ± skin	18 (2)	82 (9)

Patient numbers are indicated in brackets.

Figure 1 and 2 indicate clearance of pulmonary and renal metastases and partial clearance of adrenal metastases in one of the patients. Figure 3 illustrates a partial response of pulmonary metastases in a second patient.

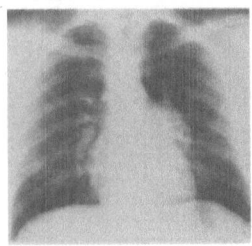 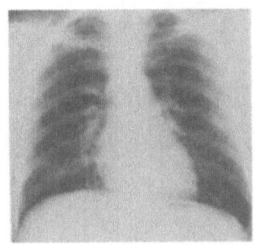

(a)   (b)
Before   After 6 months treatment with
DTIC and rIFN–α2A

Figure 1.   Example of CR of metastasis in L Hilar region.
CR was confirmed by CAT scan.

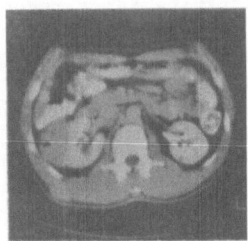 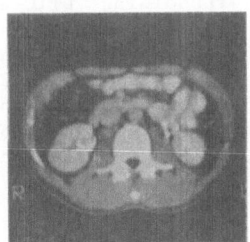 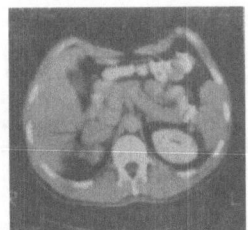

(a)   (b)   (c)
Before   After 6 months treatment   R Adrenal metastasis
with DTIC and rIFN–α2A

Figure 2.   Indicates response of renal metastasis but not adrenal
metastasis to treatment with DTIC and rIFN–α 2A.
Same patient as in figure 1.

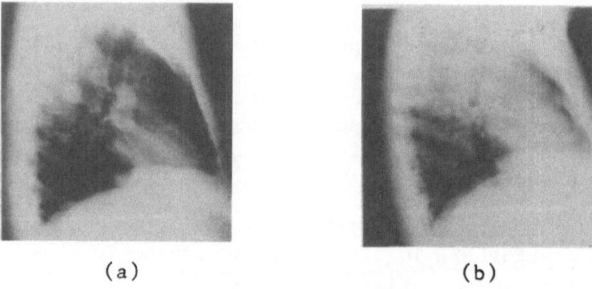

(a)   (b)

Figure 3.   Partial response of pulmonary metastasis after
3 months treatment with DTIC and rIFN-α2A.

Table 3 indicates the response duration to the end of August 1986. With
one exception most responses were continuing at 6 months or longer and in
two instances exceeded one and a half years.   There has been one relapse
in those with CR and two in those with PR.

TABLE 3
RESPONSE DURATION (WKS)

Complete Response				Time to Relapse	
				Old site	New site
Skin	46,	79			
Lung	82,	4,	47		
Liver + skin + LN	57,	24		24	
Partial Response					
Skin	65,	31,	23		65 skin
Lung + skin	28,	38,	31		28 CNS
Lung, Ad, Kid	30				

The most frequent toxicity was nausea and vomiting due to DTIC
administration.   These were alleviated to some extent by the use of
antiemetics but nevertheless this side effect was the most debilitating
to the patients.   Neutropenia ($< 3.0 \times 10^9$/L) was recorded in 15 patients
(36%).   Levels below 2×10^9/L were seen in 4 of the patients but this
was readily corrected by withholding DTIC and restarting at a reduced
dosage.   As recorded in other studies, varying degrees of fever and
muscle aches induced by rIFN-α2A were experienced by most patients at the
commencement of therapy. Fatigue producing 25% or greater reduction in
activity was recorded in 13 patients (30%) and 50% reduction in activity
in 4 (11%).   These effects were most prominent after 3 months of
treatment.   Transient elevations in liver enzymes were recorded in 5
patients.

DISCUSSION
The preliminary results from this study suggest that the combination of
rIFN-α2A and DTIC may provide a significant improvement in treatment of
malignant melanoma than use of either agent alone.   (95% confidence
limits for present study were 19-49% which compares e.g. with those of
9-22% for treatment of 127 patients with DTIC alone in the study by
Costanza et al.,[6].) This appeared particularly evident for patients

with pulmonary and or skin disease and in those with a good performance
status before treatment. Of the 11 patients with pulmonary and or skin
metastases, 7 showed complete or partial responses. The 95% confidence
limit of this response (31–89) appear outside those obtained by treatment
with DTIC alone. e.g. In the study of Einhorn et al.,[9] 18% of 39
patients responded (95% confidence limits 8–34%) which was similar to the
overall response rate in that study.

The response rate of pulmonary metastases to treatment with DTIC alone is
less than the response of metastases at sites in skin or lymph nodes but
higher than that in liver, bone and brain[9,4]. In view of these
considerations it would appear unlikely that the response rate observed
in the present study is due solely to treatment with DTIC and that the
combination of both DTIC and rIFN-α2A was needed to obtain the increase
response rate at these sites. The same may also apply to response of
subcutaneous metastases as the best reported response rate to treatment
with DTIC alone was 28% in 30 patients[9] (95% confidence limits 12–46%).
This was less than the 50% response rate seen in the present study but
confidence limits (19–82) suggest this could occur by chance. The number
of patients with metastases at other sites is as yet too few to determine
whether the response rates in liver or lymph node disease differ
significantly from those seen with DTIC treatment alone.

Further comparison with treatment regimes based on DTIC alone suggest
several other differences. Firstly it is well recognized that the
response to DTIC is higher in females compared to males whereas the
converse appears to be evident in patients treated with both DTIC and
rIFN-α2A. Response in males was 41% compared to 18% in females.
Secondly the duration of responses to treatment with DTIC tends to be
short with a median of 14 weeks[4,6]. Indications from the present study
suggest the duration of response to DTIC and rIFN-α2A may be longer than
this. It is too soon to judge whether survival of patients treated with
rIFN-α2A and DTIC will be superior to those responding to or stabilized
by treatment with DTIC. A much longer follow up period is needed as
patients with non visceral and pulmonary metastases have median survivals
of 7–11 months compared to that of 4.5 months overall[2].

Skin, lymph nodes and lung are the sites for first relapse in
approximately two thirds of patients and the sole manifestation in a
third of patients[2] so that the response rates observed in the present
study (if confirmed) together with the promise of longer response
duration may represent a significant advance in treatment of metastatic
melanoma. The reasons for responses at selective sites is not clear.
Our prejudice is that these sites may facilitate infiltration by
lymphocytes induced by direct effects of IFN-α2A on the lymphocytes or
via effects on antigen expression on melanoma cells. Induction of
changes in the expression of both HLA and melanoma antigens by IFN are
now well documented and may induce immune responses against melanoma
cells[11]. DTIC may complement such effects.

Predominant side effects from the treatment were those of nausea and
vomiting related to DTIC administration and flu like symptoms and fatigue
from rIFN-α2A treatment. Anorexia and weight loss were noted to varying
degrees. Mild neutropenia was detected in approximately 40% of the

patients but was readily reversed by reduction in the dosage of DTIC. In the majority of patients the treatment was well tolerated and allowed continuation of their usual occupation. This together with the relatively simple administration and high response rates compares favourably with more complex treatment regimes where intensive inpatient care is needed[19]. In view of these considerations these present studies suggest that this combination of treatment warrants further analysis in controlled phase III trials.

REFERENCES

1. Amer MH, AL-Sarraf M, Vaitkevicins VK Clinical presentation natural history and prognostic factors in advanced melanoma. Surg. Gynecol. Obstet. 149:168-192, 1979.

2. Balch CM, Soong SJ, Shaw HM et al Prognostic factors in patients with metastatic melanoma at distant sites. In Cutaneous Melanoma, Clinical Management and Treatment Results Worldwide. Eds. Balch CM, Milton GW, Shaw HM and Soon SJ, J.B. Lippincott Co. Philadelphia p.346-352,1985.

3. Carter RD, Krementz ET, Hill GJ et al DTIC (NSL-45388) and combination chemotherapy for melanoma. Studies with DTIC, BCNU, CCNU, Vincristine and hydroxyurea. Canc. Treat. Rep. 60:601,1976.

4. Coates AS and Durant JR Chemotherapy for metastatic melanoma. In Cutaneous Melanoma, Clinical Management and Treatment Results Worldwide. Eds. Balch CM, Milton GW, Shaw HM and Soong SJ, J.B. Lippincott Co. Philadelphia p.275-282, 1985.

5. Comis RL, Carter SK Integration of chemotherapy into combined modality therapy of solid tumors. IV. Malignant melanoma. Cancer Treat. Rep. 1:285, 1974.

6. Costanza ME, Nathanson L, Schoenfeld D Results with methyl-CCNU and DTIC in metastatic melanoma. Cancer 40:1010-1015, 1977.

7. Creagan ET, Ahmann DL, Green SJ Phase II study of recombinant leukocyte A interferon (rIFN-αA) in disseminated malignant melanoma. Cancer 54:2844-2849, 1984.

8. Creagan ET, Ahmann DL, Green SJ Phase II study of "low dose" recombinant leukocyte A interferon (rIFN-α2A) in disseminated malignant melanoma. Am.J.Clin. Oncol. 2:1002-1005,1984.

9. Einhorn LH, Burgess MA, Vallejos C Prognostic correlations and response to treatment in advanced metastatic malignant melanoma. Canc. Res. 34:1995, 1974.

10. Feun LG The natural history of resectable metastatic melanoma. Cancer 50:1656-1663, 1982.

11. Hersey P. The evolving role of recombinant alpha interferons in the treatment of malignancies. Aust. & N.Z. J. of Med. 16:425-437, 1986.

12. Hersey P, Hasic E, MacDonald M et al. Effects of recombinant leukocyte interferon (rIFN-α2A) on tumour growth and immune responses in patients with metastatic melanoma. Br. J. Cancer 51:815-826,1985.

13. Hersey P, MacDonald M, Hall C et al. Immunological effects of recombinant interferon alpha 2A (rIFN-α 2A) in patients with disseminated melanoma. Cancer 57:1292-1300, 1986.

14. Legha SS Interferons in the treatment of malignant melanoma. A review of recent trials. Cancer 57:1675-1677, 1986.

15. Luce JK Chemotherapy of melanoma. Semin. Oncol. 2:179, 1975.

16. Miller AB, Hoogstraten B, Staquet M et al Reporting results of cancer treatment. Cancer 47:207-214, 1981.

17. Namba M, Yamamoto, S, Tanaka H et al. In vitro and in vivo studies on potentiation of cytotoxic effects of anticancer drugs or cobalt 60 gamma ray by interferon on human neoplastic cells. Cancer 54:2262-2267, 1984.

18. Pritchard KI, Quirt IC, Cown DH et al. DTIC therapy in metastatic melanoma: A simplified dose schedule. Canc.Treat.Rep. 64:1123,1980.

19. Rosenberg SD, Lotze MT, Muul LM et al Observations of the systemic administration of autologous lymphokine-activated killer cells and recombinant interleukin-2 to patients with metastatic cancer. N. Engl. J of Med. 313:1485-1492, 1985.

INTRALESIONAL INTERFERON-ALPHA-THERAPY IN ADVANCED MALIGNANT MELANOMA

Peter von Wussow

Dept. of Clinical Immunology, Medical School of Hannover, Hannover, West-Germany

SUMMARY
51 patients with histologically proven metastatic melanoma and at least one skin metastasis were treated intralesionally with interferon-alpha (IFN-α) (26 patients with highly purified natural IFN-α three times 6 Mio. I.U. per week, 25 patients three times 10 Mio. I.U. of a recombinant IFN-α2b (rIFN-2b)). All 51 patients were evaluated both for their overall and their local response to the intralesional IFN-treatment. The overall response consisted of 9 objective remissions with a duration of 2-18 months. 24 (45%) complete and partial local responses were observed. 42 of the 51 patients had at least two different skin lesions: in these patients IFN-injected and non-IFN-injected skin metastases were compared. The difference between the systemic versus the local efficacy was statistically highly significant at a p-level of 0.0004. The results show, that IFN-α has clinically observable anti-tumor activity in malignant melanoma. In addition, it demonstrates that local IFN-α-therapy is significantly superior to systemic therapy.

INTRODUCTION
Until now malignant melanoma belongs to the group of solid tumors which can only marginally be influenced by a systemic treatment. Dacarbacine (DTIC) is still the standard drug for chemotherapy. It induces only in 20% an objective remission and in only maximally 5% a complete response (1,2). IFNs show beside their anti-viral and immunomodulatory properties also anti-proliferative activities against melanoma cell lines in vitro (3,4). In melanoma bearing mice (B16 and Cloudman 91) murine IFN-α,β-preparations demonstrate anti-tumoral activity in vivo (5). In patients in phase I- and phase II-studies natural as well as recombinant IFN-α showed some effectiveness in advanced malignant melanoma (8,9). While IFN-α given i.v. has no anti-tumoral activity, IFN-α given i.m. or s.c. demonstrates to have roughly a 20% rate of complete and partial responses.

In these studies the dose of IFN used ranged from 1 to 50 Mio. I.U. given three times per week (7). However, despite this range a dose response curve could not yet be established. Doses of IFN-α above 50 Mio. I.U. given s.c. are not tolerable due to severe neurological side effects.

In order to study effects of high IFN concentrations on
malignant melanoma metastases we treated patients with
advanced malignant melanoma with intralesional IFN-injections.
If a patient had several lesions we only injected one skin
lesion during the IFN-therapy. Due to this protocol it was
possible to compare a remission rate of both IFN-injected and
non-injected skin lesions in the same patient originating from
the same primary tumor.

INTERFERON-TITERS

IFN-concentrations were measured in the serum. 0, 2, 4, 8 and
20 hrs after intralesional injection of IFN-α plasma and
serum was tested for IFN-activity in an bioassay consisting of
Wish cells and VSV.

PATIENTS

Since 1983 51 patients (21 male, 30 female, 61 year medium age
from 18 to 82 years) with metastatic melanoma have been
treated at the Department of Immunology, Medical School of
Hannover. The patients characteristics and pretreatment data
are presented in table 1. While 11 patients had a radiation
therapy prior to interferon, 24 patients have been pretreated
with chemotherapy consisting of DTIC, platinum and vindesine.
All patients studied had a histologically proven malignant
melanoma. They were in advanced stage II or stage III of their
disease with clear signs of progression. The therapeutic
assessment included blood counts, liver and kidney function
tests, EKG, EEG and apropriate radiological examination for
measuring metastatic lesions. A Karnofsky-index of 60 or more
was required. All patients entering this study had to have at
least one subcutaneous metastases. 26 patients received
natural human leukocyte IFN-α obtained from Bioferon,
Laupheim, Germany. This IFN was produced by buffy coats of
healthy blood donors purified to specific activity of 1×10^7
U/mg protein. This IFN-preparation was injected intra- or
perilesionally three times per week at a dose of 6 Mio. I.U.
In addition 25 patients received rIFN-α 2b from Shering-Plough
administered also intralesionally at a dose of three times 10
Mio. I.U. per week. The patients were examined at monthly
intervals to evaluate treatment. Both the injected lesions and
all other measurable tumor lesions were determined by size. If
progress occurred patients were taken out of the study. If the
disease remained stable or regressed, therapy was continued
until progress occurred. The systemic tumor response was
measured according to the WHO criterias: complete response:
disappearance of all lesions of disease lasting for at least
one month. Partial respone: decrease of more than 50% of the
sum of diameters of all measurable lesions persisting for at
least one month. Progressive disease: increase of more than
25% in the diameter of all measurable lesions or appearence of
any new lesion. In addition, the size of the injected lesion
were measured in order to determine the local efficiency of
this IFN-treatment using the same WHO-criteria adjusted to the
single injected lesion.

RESULTS

A total of 58 patients were entered into the trial. 51
patients are at present evaluable, 26 patients were treated
with the natural IFN- -preparation three times 6 Mio. I.U.
per week, whereas 26 were treated with three times 10 Mio.
I.U. of rIFN- 2b. The two series of patients differed not
significantly in the response rate and in side effects;
therefore the differences in doses and IFN-preparation used
are neglectable and data of all 51 patients were pooled.

Systemic response: Of the 51 patients 3 showed a complete and
6 a partial response. In 5 of the 9 responding patients only
soft tissue metastases were present. In 7 patients the tumor
regression was already observable after 1 month of treatment.
In the remaining 2 patients regression were observable after
1.5 and 2.75 months under therapy, respectively. 16 patients
showed a stable disease in the first month of treatment,
whereas 26 patients had a clear progress during the first 4
weeks of IFN-treatment. The duration of the partial remissions
was between 2 and 6 months. 3 patients who showed a complete
response had no maintenance therapy. Two patients remained in
a complete remission until now after 14 and 20 months,
respectively. One patient relapsed after 6 months. After
reestablishing the IFN-therapy the patient responded again
with a complete disappearance of the tumor. The second
complete response had again a duration of 6 months. After a
second relapse a restarted IFN-therapy proved again to be
effective in the patient. He now showed the third response
after three episodes of IFN-therapy. After the fourth relapse
the tumor did not respond to the reestablished IFN-therapy.
However, increase in the dose to 12 Mio. three times per week
led to a partial response. Four months later the IFN-therapy
had to be reduced due to severe fatigue and apathia. Two weeks
after the dose reduction the tumor showed progression and the
patient died within the next three weeks.

Local remissions: 23 of the 51 injected skin metastases
reduced their size to at least 50% under the therapy. 16 of
the injected skin lesions showed a complete disappearance,
while 7 lesions shrank in size to more than 50%. 26 lesions
showed a stable disease under the injections, whereas two
lesions clearly progressed. During therapy in none of the
lesions observable inflammatory reactions were noted.

Visceral metastases respond to IFN-treatment far less frequent
than soft tissue metastases. Therefore, in order to establish
a comparison of the systemic and local effect of the IFN-
treatment we compared only non-injected and injected skin
lesions for the response (Table III). 42 patients had beside
the injected skin lesion another lesion which was outside the
lymph drainage region of the injected metastases. Since in one
patient always the same skin lesion had been injected with IFN
we were able to compare skin lesions which were every other
day under high IFN-concentrations and skin lesions which were
only under the systemic IFN concentrations as measured in the

blood. While the injected skin lesions showed 14 complete
response, 6 partial responses and 22 stable diseases, the non-
injected skin lesions showed 3 complete responses, 6 partial
responses, 12 stable diseases and 21 progressive diseases.
Therefore 45% of the injected skin lesions responded with a
partial and complete response, whereas only 21% of the non-
injected skin lesions showed a complete and partial response.
This difference was highly significant in an χ^2-test
(p 0.0004). In none of the cases a response of a non-injected
lesion and a non-response of the injected lesion was observed.

TOXICITY

The IFN-α-treatment was well tolerated. 95% of all patients
treated had the typical flue-like symptoms of an IFN-therapy
with fever upto 39.0 to 40°C, headache, chills, and myalgia.
21% had pain at the injection side during and shortly after
the injection. 9% had rather severe headaches over several
hours. 26% of the patients developed apathia and fatigue which
forced to dose reductions. Cardiac arrhythmias and diarrhoea
was not observed. 15% patients showed leukopenia WHO grade I
and 3 showed WHO grade II. Only marginal increases of sGOT and
sGPT were seen with no clinical significance (Table II).

INTERFERON TITERS

In all patients serum levels of IFN have been measured. The
peak levels were obtained at 6 hrs. after injection with a
median maximum titer of 95 U which droped to 22 U after 24
hrs. In contrast, the maximum titers measured in the blood of
the skin lesion: surrounding tissue were observed after 1 hr
with a medium titer of 18.000 U/ml. This level dropped down to
80 U/ml after 24 hrs. Therefore the roughly estimated IFN
tissue concentrations were two logs higher in or at the skin
lesion than the systemic IFN concentrations as measured in the
incubital vains. Interestingly after 48 hrs. no IFN was
measurable, which indicates, that even with local IFN
infections no continuous in IFN-levels can be obtained.

DISCUSSION

We report here on the systemic and local effects of an intralesional IFN-α-treatment in advanced malignant melanoma. The
data presented here clearly show that a local and a systemic
treatment is achieved by intralesional IFN-injections. Three
lines of evidence exist for the systemic effects of IFN-
given intralesionally. First, the pharmakokinetic measured of
the intralesionally administered IFN was quite similar to the
reported kinetic of IFN given subcutaneously (10). Since
conventional systemic IFN-α-therapy is given by the s.c.
route, intralesional injections of IFN leads to the similar
serum titers as the classical IFN-treatment. Second, systemic
side effects were noted. The range of side effects observed in
this trial showed the typical pattern of systemic IFN-
treatment (7,8,9). Third, systemic objective responses of the
melanoma occurred under IFN given intralesionally.

Of the 51 patients treated 3 patients showed a complete and 6
a partial response. This means that 18% of the patients
responded to the treatment. This percent rate is similar to
the rates reported in other trials in malignant melanoma with
IFN given s.c. (7,8,9). The observed local effects of the
intralesional IFN-injections were pain during or shortly after
the injection and local tumor regression. In none of the cases
macroscopically observable inflammatory reactions were noted.
24 of the 51 injected skin lesions showed a partial or
complete response indicating that the local response rate
(45%) was higher than the systemic response (18%).

In order to compare the local and systemic efficacy of this
treatment more accurately we looked at 42 patients who had
beside their IFN-injected skin lesions other skin metastases
which lay outside the lymph drainage area of the IFN-
injection. The difference in the injected and non-injected
lesions were dramatic. While 20 of 42 IFN-injected lesions
regressed in their size to at least 50% (CR + PR), only 9 of
42 non-injected lesions showed a partial or complete response.
This difference - as measured by the adjusted X^2-test - was
statistically highly significant at a p-level 0.0004.

In animals evidence has been accumulated that IFN-α acts both
directly and indirectly through host defense mechanisms on
tumors. The observed clear-cut, highly significant difference
between the local and systemic efficacy, however, suggests
that in malignant melanoma IFN acts directly on the
metastases. Because if IFN acts indirectly in this disease, it
first has to reach and to activate defense cells (for example
macrophages, NK-cells, T-lymphocytes) by entering the blood
sdream. However, these cells will then - once activated -
attack all skin lesions without distinguishing between IFN-
injected and non-injected metastases. Therefore, our clinical
results suggest strongly that IFN-α acts directly on the skin
metastases. Whether within the metastases IFN-α exerts its
effect directly on the tumor cells or through stroma cells
remains to be elucidated.

In conclusion, intralesional IFN-α-treatment has both local
and systemic therapeutic effects with a significant higher
intralesional efficacy. IFN exerts its effect most likely
directly on the metastases. These findings lead to the
consideration that in adjuvant trials in malignant melanoma
IFN-α should be given locally after the resection of the
primary tumor or of regional lymp nodes.

TABLE I:

Patients characteristics of 51 patients
treated with IFN-

treated patients	58
evaluable patients	51
age (mean)	18-81 (61) years
male/female	21/30
pretreatment	
a) wide excision	46
b) lymphadenectomy	41
c) radiation	11
d) chemotherapy	24

TABLE II:

Side effects of the intralesional IFN- -therapy

evalubale patients	58	
flue-like symptoms	55	95%
local inflammation	-	-
pain after IFN-injection	12	21%
headache	5	9%
apathia, fatigue	15	26%
cardiac arrhythmia	0	-
diarrhoe	0	-

Table III:

Local remission of IFN- injected and non-injected skin
metastases in patients with advanced malignant melanoma

evaluable patients	IFN- injected skin metastases	non-injected skin metastases
42	42	42
CR	14	3
PR	6	6
SD	20	12
PD	-	21
PR + CR	45%	21%
SD + PD	55%	79%

_ITERATUR:

1. Pritchard, K.I., Quirt, I.C., Cowan, D.H., Osoba, D., Kutas, G.J.: DTIC-therapy in metastatic malignant melanoma: A simplified dose schedule. Cancer Treat. Rep. 64: 1123-1126

2. Costanzi, J., Fabian, C., Wilson, H., Dixon, D.: Sequential combination chemotherapy for disseminated melanoma. SWOG study. Cancer Treat. Rep. 65: 732-734, 1981

3. Stewart, W.E. II.: The Interferon System. Springer Verlag

4. Salmon, S.E., Durie, B.G.M., Young, L. et al.: Effects of cloned human leucocytes interferons in the human tumor stem cell assay. J. Clin. Oncology I: 217-225, 1983

5. Bart, R.S., Pargio, N.R., Kopf, A.W.: Inhibition of growth of B16 murine malignant melanoma by exogenous IFN. Cancer Res. 40: 614-619, 1980

6. Ratner, L., Norlund, J.J., Lenguel, P.: Interferon as an inhibitor of cell growth. Studies with mouse melanoma cells. Proc. Soc. Exp. Biol. Med. 163: 267-272, 1980

7. Krown, S.E. Burk, M., Kirkwood, J.M. et al.: Human leukocytes interferon (HuLeIF) in malignant melanoma (MM): Preliminary report of the American Society Clinical Trial. Proc. Am. Assoc. Cancer Res. 22: 158, 1981

8. Creagan, E.T., Ahmann, D.L., Green, S.J., Hong, H.J. et al.: Phse II study of low-dose recombinant leukocyte A interferon in disseminated melignant melanoma. J. Clin. Oncol. 2, 1002-1005, 1984

9. Creagan, E.T., Ahmann, D.L., Green, S.J., Hong, H.J. et al.: Phase II study of recombinant leukocyte A interferon (rIFN A) in disseminated malignant melanoma. Cancer 54, 2844-2849, 1984

10. Guttermann, J.U., Fine, S., Quesada, J. et al.: Recombinant leukocyte A interferon. Pharmakokinetics, single-dose tolerance and biologic effects in cancer patients. Ann. Int. Med. 96: 549-556, 1982

COMPARISON OF CLINICAL TOXICITY OF NATURAL ALPHA AND RECOMBINANT GAMMA INTERFERON. RESULTS OF PHASE II TRIALS IN LUNG CANCER

K. MATTSON, L.R. HOLSTI, A. NIIRANEN, S. PYRHÖNEN, M. FÄRKKILÄ
G. HÄRTEL, C-G. STANDERTSKIÖLD-NORDENSTAM, K. CANTELL

1. INTRODUCTION

Interferons (IFN) have been referred to as the prototypic biological response modifiers (BRM). Similarly to corticosteroids they induce a diversity of physiologic changes (1). It is conceivable then that IFNs work by different mechanisms than conventional cytotoxic agents and that their therapeutic optimal dose in cancer therapy not necessarily is equivalent to the maximum tolerated dose. The optimal treatment of cancer in addition involves combined use of various treatment modalities or various drugs. Experimental studies indicate that IFNs can be effectively combined with radiation and chemotherapy (2,3). In addition, IFNs augment the effects of each other and also other biologicals. Such new combined approaches provide possibilities for overcoming resistance of malignant cells. Preliminary evidence from phase I and II trials indicates that qualitatively similar clinical toxicities occur with IFNα, IFNβ and IFNγ (4). In an attempt further to define the clinical spectrum of side effects associated with various types, doses and schedules of IFNs alone and in combined regimens and to select a routine test battery for the monitoring of IFN toxicity we performed systematic large scale tests during four phase II studies in 43 patients with previously untreated lung cancer.

2. PATIENTS AND METHODS

Patient and disease characteristics are shown in Tables 1 and 2.

2.1. IFN regimens

In all studies, IFN was given first as monotherapy, then either in combination with or followed by locoregional radiotherapy (M_0 disease) or chemotherapy (M_1 disease). Twenty-five patients received partially purified human leukocyte IFN (P-IF B+A; sp.act. 1-3 x 10^6 IU/mg protein) (5,6), 3 received highly purified leukocyte IFN (NK2-IFN; sp.act. 1.2 x 10^8 IU/mg protein) obtained by passage of P-IF B+A through a monoclonal antibody affinity chromatography column (NK2 Sepharose, Celltech, England) and 15 were treated with recombinant gamma interferon (rIFNγ from B.coli, > 95 % purity, lyophilized, sp.act. 24 x 10^6 IU/mg protein). IFN doses and schedules as well as radiotherapy programs and chemotherapy regimens were as follows:

TABLE 1. Characteristics of patients with small cell lung
cancer

	TREATED WITH IFNα	
	hihg-dose	low-dose
N	9	5
Male/female	6/3	5/0
Age, years, mean (range)	64 (53-75)	58 (39-70)
Karnofsky 70-100 %	9	5
Weight loss before treatment <10 kg	7	5
$\geq$10 kg	2	0
Histological subtype (WP-L)		
lymphocyte-like 21	2	2
intermediate 22	6	3
21 + 22	1	0
Stage II $T_2N_0M_0$	1	0
III $T_1N_2M_0$	1	1
III $T_2N_2M_0$	6	4
III $T_3N_2M_0$	1	0

TABLE 2. Characteristics of patients with non-small cell lung
cancer

	TREATED WITH	
	rIFNγ	IFNα
N	15	14
Male/female	13/2	9/5
Age, years, median (range)	63 (50-71)	67 (52-71)
Karnofsky $>$ 80 %	12	5
60-70 %	3	9
Diameter of primary tumor $\leq$ 5 cm	9	9
$>$ 5 cm	6	5
Epidermoid	10	9
Adeno	3	3
Large cell	2	2
Stage III M_0	12	9
M_1	3	5

IFNα alone was given either as a high-dose (800 x 10^6
IU/5 D) i.v. or as a low-dose (6 x 10^6 IU/D) i.m. induction
treatment, followed by maintenance treatment (6 x 10^6 IU TIW
or BIW) until tumor progression, to 14 untreated patients with

limited small cell lung cancer (SCLC). If the first site of progression was local or a CNS location, radiotherapy (RT) (55 Gy/20 F/7 wks) was applied and IFN was continued (12 patients). Conventional combination chemotherapy (CT) (CTX 1.2 g/m^2 d 1, VP16 150 mg/m^2 d 1,3,5, VCR 1.2 mg/m^2 d 1+8) was administered (7 patients) only at disease progression outside the thorax.

IFNα alone was administered to 14 patients with non-small cell lung cancer (NSCLC) at a dose of 6 x 10^6 IU/D i.m. 5 days a week for 12 weeks. Responders or those with stable disease were evaluated in a continuation study of IFN + RT (M$_0$, 5 patients; twice daily fractionated 55 Gy/44 F/30 d) or of IFN + CT (M$_1$, 3 patients; 3 cycles cisplatin 90 mg/ m^2 q 4 wks, VDS 3 mg/m^2 every second week for 3 months).

rIFNγ was administered to 15 patients with previously untreated NSCLC as a 4 h i.v. infusion in 5 % dextrose at a dose of 2 mg (48 x 10^6 IU)/m^2 TIW for 12 weeks. Patients with "uncontrolled cardiovascular disease" including uncontrolled hypertension, arrhythmias or a recent ($<$ 3 mo) myocardial infarction were excluded. More or less symptomless patients with optimum medication were therefore not excluded in spite of a history of heart disease. After the completion of 12 wks rIFNγ 6 patients (M$_0$) received RT (twice daily fractionated 55 Gy/ 44 F/30 d) and 4 patients (M$_1$) received CT (3 cycles of cisplatin-VDS as above).

2.2. Evaluation procedures

Patients were evaluated for tumor response by means of a complete restaging including bronchoscopy and CT scans every 6 weeks. Patients who had received at least 4 weeks IFN single treatment were considered evaluable for efficacy. Patients who had received one dose of IFN were considered evaluable for toxicity. WHO guidelines for grading acute and subacute toxicity were used.

Neurotoxicity was evaluated by serial clinical neurological and neurophysiological investigations and psychometric tests as well as by CSF examinations (cell differentials, immunoglobulins, neurotransmitters). Neurophysiological investigations consisted of serial EEG, EMG, visual and auditory evoked brain stem potentials (VEPs and BAEPs) using techniques presented previously (7). Neuropsychological tests included subtests of the Wechsler Memory Scale such as logical memory, digit span and visual reproduction (8).

Cardiotoxicity was evaluated by pretreatment physical examination, chest x-ray with measurement of heart volume and bicycle exercise test with ECG and blood pressure registrations. All investigations were repeated at discontinuation of IFN therapy or as required. When it became evident during the studies that rIFNγ was potentially cardiotoxic the program for monitoring for cardiotoxicity was extended to include ECG recordings and blood pressure measurements before each infusion, halfway through the infusion, at the end of the infusion and on the following day after the infusion.

Radiation pneumonitis induced by IFN-RT was evaluated by serial testing of vital capacity and single-breath diffusing capacity as well as chest x-ray, CT documentation and at the end of the study bronchoalveolar lavage (BAL). The tests were performed before IFN administration, before RT, immediately following RT, and every 3 months thereafter.

Laboratory parameters were monitored by weekly blood controls and by screening of chemistry panels every 4 weeks. Serial serum samples were collected for antibody determinations.

3. RESULTS
3.1.Efficacy
Median total doses of IFN administered in each regimen and tumor responses achieved with IFN monotherapy before the continuation studies with combined treatments are indicated in Table 3.

TABLE 3. Comparison of results of IFN monotherapy in lung cancer

Disease	IFN schedule	Eval. pat. N	Results Response	Duration weeks	Median total dose x 10^6 IU
SCLC	IFNα high-dose	9	3 MR 6 SD	20 - 42 5 - 20	1380
SCLC	IFNα low-dose	5	1 PR 4 SD	12 7 - 17	510
NSCLC	IFNα low-dose	13	13 SD		216
NSCLC	rIFN γ high-dose	10	1 PR, 2 MR 6 SD, 1 PD		3456 (144 mg)

Of 9 patients with SCLC receiving high-dose induction treatment with IFNα 3 achieved minor response for 20, 25 and 42 weeks and 6 achieved stable disease for 5-20 weeks. Seven progressed first at the primary site and were submitted to RT (4 CR, 3 PR). Four patients were administered CT subsequent to IFN-RT. The median survival of 9 patients was 10 months (5-38). Of 5 patients with SCLC receiving low-dose IFNα from the beginning 1 achieved a partial response, duration 12 wks, and 4 achieved stable disease for 7-17 weeks. Five were submitted to RT (3 CR, 2 PR), 4 patients were administered CT subsequent to IFN-RT. The median survival of the 5 patients was 18 months (3-34).

Of 14 patients with NSCLC treated with low-dose IFNα, 13 were evaluable for response. All had stable disease. Of 7 patients who completed the 12 wk course of IFN, 4 had stable and 3 progressive disease. Five patients were subsequently submitted to RT (3 PR, 2 MR) and 3 to CT (1 PR, 2 PD).

Of 15 patients with NSCLC treated with high-dose rIFN γ, 10

were evaluable for response. One had PR, 2 MR, 6 SD and 1 PD. Six patients were subsequently submitted to RT (1 CR, 4 PR, 1 SD) and 4 to CT (1 SD, 3 PD).

3.2. Toxicity

This report focuses on the possible differences between the nature and time pattern of side effects of either natural IFNα or rIFNγ. The adverse effects of IFNs have been grouped as either acute or of later onset and as organ and system toxicities (Table 4).

TABLE 4. Comparison of side effects observed during treatment with natural IFNα or rIFNγ

Side effect		IFNα	rIFNγ
IFN syndrome		+	+
Neurotoxicity		++	−
Cardiotoxicity		−	++
Weight loss		+	−
Myelosuppression			
WBC	↓	++	+
Granulocytes	↓	+	+
Lymphocytes	↑	+	++
Hb	↓	+	−
Platelets	↓	+	−
Change in biochemical parameters			
S-creatinine	↑	+	−
Liver enzymes	↑	+	−
S-albumine	↓	−	++
S-K	↓	+	−
S-Ca	↓	−	++
S-Na	↓	+	−
Interaction with radiotherapy		++	−

Constitutional symptoms. The most characteristic acute toxicity of both IFN types consisted of high temperatures with severe tremor during the peaks, malaise, headache and muscle pain. This so called "flu-like reaction" or the "interferon syndrome" was occasionally very severe when high doses of either IFN type were used. Similar but less severe manifestations which gradually vanished with continuous administration, were common also during low-dose IFNα administration. Premedication with indomethacin was often necessary for patient compliance. The rIFNγ protocol initially did not allow the use of indomethacin for symptomatic treatment. Only when 2 patients dropped out due to severe fever and flu-like symptoms was the protocol amended to allow the necessary and successful use of indomethacin when paracetamol proved ineffective in controlling fever.

Fatigue, weakness, lowered performance, anorexia and weight

loss were dose-limiting with prolonged ($\leq$ 43 weeks) treatment of IFNα . By contrast, we found that these subacute constitutional symptoms were mild or absent in the 15 patients treated with high-dose rIFNγ for $\leq$ 12 weeks.

Severity of the symptoms appeared to increase with increasing intensity of doses and/or increasing frequency of drug administration or increasing duration of treatment. The symptoms became reversible by changing the schedule from daily to TIW or BIW even if the dose (6 x 10^6 IU) remained unchanged. Especially elderly patients experienced progressive profound fatigue after 4-6 weeks of daily IFNα.

Neurotoxicity. With IFNα, especially with the high dose (800 x 10^6 IU), given by continuous i.v. infusion over 5 days, neurotoxicity was clearly dose-limiting and significant neurological, neurophysiological and neuropsychological side effects became apparent (9). The same kind of psychometric and neurophysiological dysfunction also occurred with low-dose i.m. IFNα (Fig. 1), but with a different grade of severity and a different time course: early, after 2-8 days, with high-dose and late, after 3-4 weeks, with low-dose IFNα (10). Side effects of P-IF and NK2 IFNα were indistinguishable.

Clinically the adverse effects of IFNα neurotoxicity consisted of marked behavioral changes, overall mental and motor slowing, loss of appetite and thirst, lethargy and confusion. Psychometric tests showed decrease in the performance of logical memory and of visuoconstructional function, which were accompanied by a diffuse slowing on the EEG of the alpha rhythm with appearance of theta and delta waves predominantly in the frontal lobes and compatible with a diffuse encephalopathy.

The latencies of central conduction times in VEPs and BAEPs increased with increasing cumulative dose of IFNα. After cumulative doses of 400 x 10^6 IU/72 h i.v., deep tendon reflexes were diminished - from clonus to total loss of reflexes - and marked muscle weakness developed. Most elderly patients became bed-bound and were unable to raise their legs in spite of their ambulation three days earlier. Conduction velocities in peripheral nerves did not change (11). This interferon-induced loss of reflexes may be difficult to distinguish from paraneoplastic polyneuropathy. The characteristic interferon-induced syndrome was, however, not associated with sensory deficits.

Psychometric tests and EEGs were useful tools for the characterization, the early detection and the monitoring of this neurotoxicity. Our very preliminary reports suggest that IFN neurotoxicity might be caused by changes in neurotransmitters, IFNs acting as modulators of neurotransmission (12). Only during treatment with high-dose i.v. IFNα was CSF found to contain measurable amounts of IFN. Differential cell counts showed that an increase in the proportion of enlarged lymphocytes in CSF coincided with EEG abnormalities.

By contrast, with rIFNγ at the dose used (up to 144 mg = 3456 x 10^6 IU over 12 weeks) none of our 15 patients developed any abnormalities in spite of systematic and frequent monitoring in logical memory (Fig. 1) and in visuoconstructio-

FIGURE 1

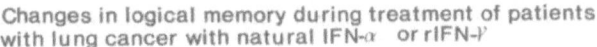

Changes in logical memory during treatment of patients
with lung cancer with natural IFN-α or rIFN-γ

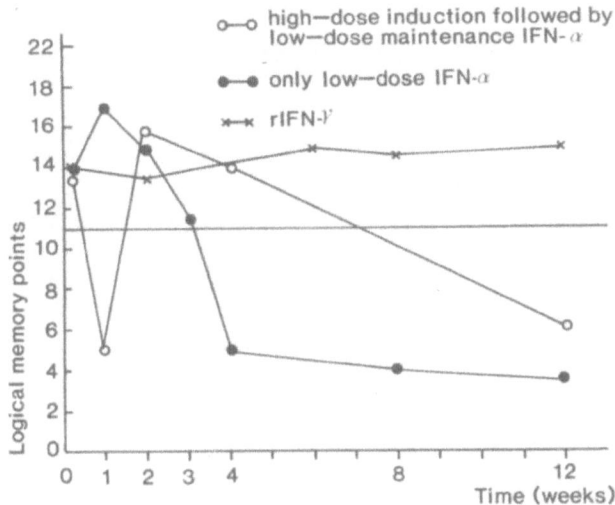

o—o high—dose induction followed by
low—dose maintenance IFN-α

●—● only low—dose IFN-α

×—× rIFN-γ

nal function (Fig. 2 a and b), in EEG recordings, in latencies
of central conduction times and in conduction velocities in
peripheral nerves.

Cardiovascular toxicity. Hypotension defined as drop in sys-
tolic BP ≥ 30 mmHg occurred in 5 and clinically significant
arrhythmias in 8 out of 15 patients treated with rIFNγ.

A 63-year-old woman with a past history of ischaemic heart
disease and coronary artery bypass of 4 years previously discontinued rIFNγ treatment due to AV-block which required
pacemaker inplantation. Occasional right bundle branch block
(RBBB) had been seen on the pretreatment exercise ECG. Two
weeks after the start of rIFNγ 2 mg/m^2 TIW RBBB became constant. A week later treatment was discontinued but within a
week complete AV-block was documented. The patient died of a
myocardial infarction 3 months later.

Atrial fibrillation, irreversible in spite of discontinuation of rIFNγ and symptomatic medical treatment, developed in
one patient with no prior cardiac disease. Cardioversion was
not performed due to medical reasons.

Prolongation of the PQ interval (from 0.08 to 0.22 s) was
reversible on discontinuation of rIFNγ in two previously
healthy male patients.

Two patients developed unifocal and 2 multifocal ventricular

Changes in visuoconstructional ability during treatment with interferon

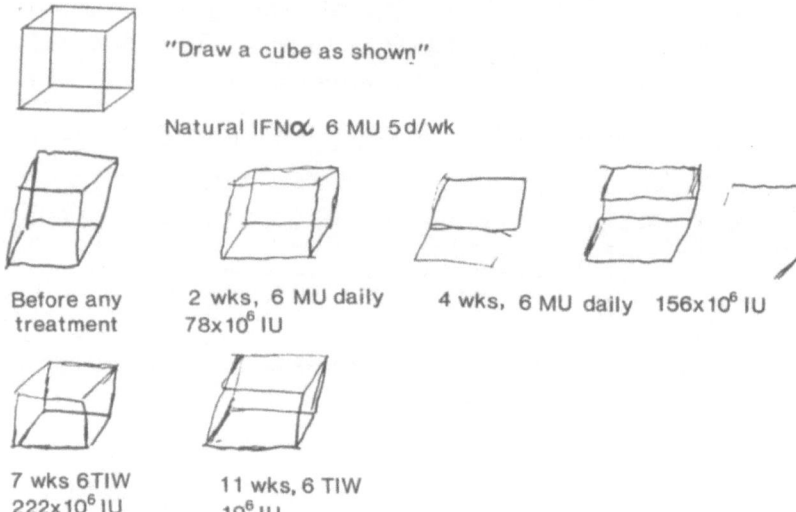

"Draw a cube as shown"

Natural IFNα 6 MU 5d/wk

Before any   2 wks, 6 MU daily   4 wks, 6 MU daily   156x10⁶ IU
treatment   78x10⁶ IU

7 wks 6TIW   11 wks, 6 TIW
222x10⁶ IU   10⁶ IU

Changes in visuoconstructional ability during treatment with rIFN-γ

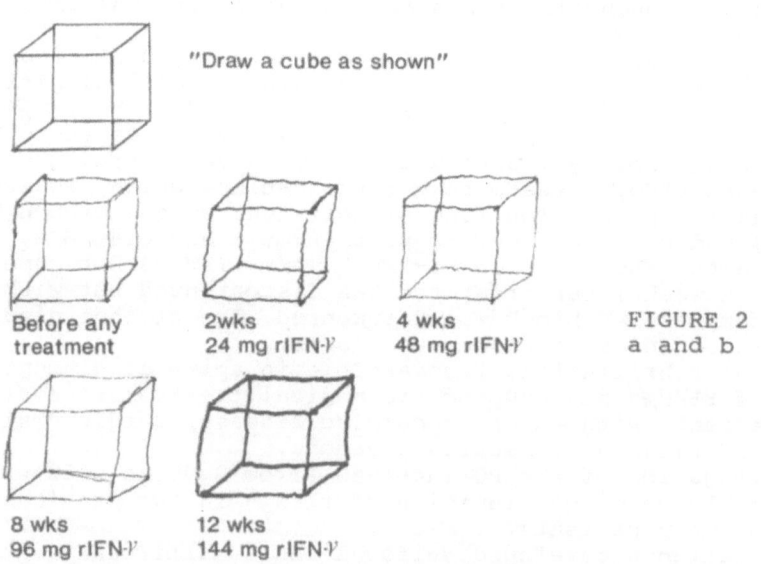

"Draw a cube as shown"

Before any   2wks   4 wks   FIGURE 2
treatment   24 mg rIFN-γ   48 mg rIFN-γ   a and b

8 wks   12 wks
96 mg rIFN-γ   144 mg rIFN-γ

extrasystoles. One of them had a history of occasional unifocal ventricular ectopic beats prior to treatment. rIFN𝛾 had to be discontinued in 2, the other 2 continued rIFN𝛾 plus various antiarrhythmic drugs.

Patients with a prior history of heart disease were excluded from rIFN𝛾 administration as awareness of this risk factor became clear during the study. Quesada (4) reported the development of congestive heart failure following prolonged exposure to IFN𝛼 in 2 patients and 2 patients in whom myocardial infarction complicated the course of IFN𝛼 treatment. Reports in the French lay literature associated IFN administration with myocardial infarction. In vitro, IFN𝛼 alters the response of myocardial fibers (13).

We could never document cardiotoxic events in our 25 patients treated with IFN𝛼. Tachycardia and minor changes in BP seen during IFN𝛼 treatment were related to the acute interferon syndrome.

TABLE 5. Severity of pneumonitis and lung fibrosis following IFN and radiotherapy

	N	Radiological grade of lung reaction		
		Severe	Moderate/mild	None
NSCLC				
IFN 𝛼 low-dose				
1 mo	5	1	2	2
2 mo	4	1	2	1
6 mo	3	1	1	1
9 mo	2	0	1	1
rIFN 𝛾 high-dose				
1 mo	6	0	0	6
2 mo	6	0	1	5
6 mo	6	0	1	5
9 mo	5	0	1	4
SCLC				
IFN 𝛼 high-dose				
2 mo	6	4	1	1
6 mo	3	2	1	0
9 mo	2	2	0	0
IFN 𝛼 low-dose				
2 mo	5	1	2	2
6 mo	4	1	0	3
9 mo	4	1	0	3

Interaction with radiotherapy. The results of recent experimental studies suggest that IFNs potentiate radiation injury (14). The effect of combined IFNα and RT could be evaluated in 11 patients with SCLC (15) and in 5 patients with NSCLC. The effect of RT given after the completion of <12 weeks rIFNγ could be evaluated in 6 patients with NSCLC. The radiological degree of pneumonitis at 2, 6 and 9 months after the completion of RT is shown in Table 5. Our results suggest that both the type and the dose of IFN influence the severity of pneumonitis and its time course. IFNα seemed to enhance radiation effects in a dose-dependent way whereas rIFNγ had no such effects.

Laboratory parameters. Minor, clinically nonsignificant abnormalities were observed (Table 4). Mild IFN -induced leukopenia occurred but did not cause increased incidence of infections or difficulties in simultaneous administration of radio- or chemotherapy. Prolonged therapy with IFN induced mild anemia from which recovery was slow. Similar observations have been reported in the literature (16). Only exceptionally we have observed more severe hematologic complications related to IFN such as thrombocytopenia.

Antibodies. Analysis is currently in progress.

4. CONCLUSIONS

Natural IFNα and rIFNγ induced distinct and qualitatively different patterns of side effects. Dose-limiting side effects of IFNα were fatigue, weight loss and neurotoxicity, while cardiovascular toxicity was dose-limiting for treatment with rIFNγ.

REFERENCES

1. Borden EC: Interferon and cancer: How the promise is being kept. In Gresser I(ed): Interferon V. Orlando, Florida: Academic, p. 43, 1984.
2. Gould MN, Kakria RC, Olson S, Borden E: Radiosensitization of human bronchogenic carcinoma cells by interferon beta. J Interfer Res 4: 123-128, 1984.
3. Mowshowitz SL, Chin-Bow ST, Smith GD: Interferon and cis-DDP: Combination chemotherapy for P388 in CD F1 mice. J Interfer Res 2: 578, 1982.
4. Quesada JR, Talpaz M, Rios A: Clinical toxicity of interferons in cancer patients. A review. J Clin Oncol 4: 234-243, 1986.
5. Cantell K, Hirvonen S, Kauppinen HL, Myllylä G: Production of interferon in human leukocytes from normal donors with the use of Sendai virus. Methods Enzymol 78: 29, 1981.
6. Cantell K, Hirvonen S, Koistinen V: Partial purification of human leukocyte interferon on a large scale. Methods Enzymol 78: 499-505, 1981.
7. Färkkilä M, Iivanainen M. Laukkanen R, Sainio K, Bergström L, Cantell K: Neurological and neurophysiological effects of high dose interferon treatment in amyotrophic lateral sclerosis. In Kirchner H, Schellekens H(eds): The Biology of Interferon System 1984. Amsterdam: Elsevier Science

Publishers, pp. 575-579, 1985.
8. Iivanainen M, Laaksonen R, Niemi ML, Färkkilä M, Mattson K, Niiranen A, Cantell K: Memory and psychomotor impairment following high-dose interferon treatment in amyotrophic lateral sclerosis. Acta Neurol Scand 72: 475-480, 1985.
9. Mattson K, Niiranen A, Iivanainen M, Färkkilä M, Bergström L, Holsti LR, Kauppinen H-L, Cantell K: Neurotoxicity of interferon. Cancer Treat Rep 67: 958-961, 1983.
10. Mattson K, Holsti LR, Niiranen A, Kivisaari L, Iivanainen M, Sovijärvi A, Cantell K: Human leukocyte interferon as part of a combined treatment for previously untreated small cell lung cancer. J Biol Resp Modif 4: 8-17, 1985.
11. Färkkilä M, Niiranen A, Mattson K, Salmi T, Iivanainen M, Cantell K: Neurotoxicity of high- or low-dose natural IFNα and of recombinant IFNγ in patients with lung cancer. In Stewart WE II, Schellekens H(eds): The Biology of Interferon System 1985. Amsterdam: Elsevier Science Publishers, pp. 429-432, 1986.
12. Färkkilä M, Iivanainen M, Härkönen M, Laakso J, Mattson K, Cantell K. Neurotransmitter changes induced by recombinant γ-interferon in patients with lung cancer. J Interferon Res 6: Suppl 1: 46, 1986.
13. Lampidis TJ, Brouty-Boye D: Interferon inhibits cardiac cell function in vitro (41043). Proc Soc Exp Biol Med 166: 181-185, 1981.
14. Dritschilo A, Mossman K, Gray M, Streevalsan T: Potentiation of radiation injury by interferon. Am J Clin Oncol (CCT) 5: 79-82, 1982.
15. Holsti LR, Mattson K, Niiranen A, Standertskiöld-Nordenstam C-G, Stenman S, Sovijärvi A, Cantell K: Enhancement of radiation effects by interferon in the treatment of small cell carcinoma of the lung. Int J Radiol Oncol Biol Phys, accepted for publication, 1986.
16. Krown SE: Interferons and interferon inducers in cancer treatment. Semin Oncol 13: 207-217, 1986.

ANTITUMOR ACTIVITY OF INTERFERON-β AGAINST MALIGNANT GLIOMA IN COMBINATION WITH CHEMOTHERAPEUTIC AGENT OF NITROSOUREA(ACNU)

J.YOSHIDA, K.KATO, T.WAKABAYASHI, H.ENOMOTO, I.INOUE, N.KAGEYAMA

Department of Neurosurgery, Nagoya University School of Medicine

I. INTRODUCTION

Patients with malignant glioma have been regarded as having a very poor prognosis. It has been reported that even an extensive tumor resection resulted without exception in recurrence within a year.(1) Therefore multimodality adjuvant therapy by radiation, chemotherapeutic agents and biological response modifiers is necessary in addition to surgical resection.(2) As chemotherapeutic agents, nitrosoureas are known to be most effective drugs for malignant glioma due to their appropriate pharmacological characteristic enabling them to cross the blood brain barrier.(3) A water soluble nitrosourea ACNU is most frequently used in Japan.(4) On the other hand interferon are recently demonstrated to have antitumor activity against malignant glioma.(5) In the present studies, antitumor effect of combination therapy by IFN and ACNU was evaluated from not only experimental but also clinical points of view.

II. EXPERIMENTAL STUDIES IN VITRO AND IN VIVO

Materials and Methods
Cells; Thirteen human glioma cell lines were used in the present studies. SK-MG-1,-4,-6,-7,-9,-15,-MS,-AO2 were established from the surgical materials of malignant glioma by Pferundschuh et al. in Memorial Sloan Ketering Cancer Institute, U251-MG,-SP,U178-MG by Pontén in University of Uppsala, KNS42 in Kyusyu University, and GL-2 was in our laboratory.
ACNU; (1-(4-amino-2-methyl-5-pirimidinyl)-methyl-3-(2-chloroethyl)-3-nitrosourea hydrochloride was obtained from central research laboratories of Sankyo Company,Japan. HuIFN-β was prepared on the diploid human foreskin cell induced by PolyI:polyC and provided from Toray Industries Inc.Tokyo.
Cell growth inhibition in vitro; Tumor cells were cultured in Eagle's MEM supplemented with 10% fetal calf serum, non-essential amino acid, L-glutamine, and antibiotics. In an attempt to determine cell growth inhibition rate, a certain number of cells(1×10^4)were dispensed into 24 wells of Coastar microtestplate. One day after planting, ACNU of 5 or 10μg/ml and/or HuIFN-β of 1000IU/ml were added to the culture medium. After incubation in 5% CO_2 incubator, culture cells of each well were harvested with 0.25% trypsin and the number of cell was counted serially using a hemocytometer. The %cytotoxicity

was calculated as following formula;(1-cell number of tested well/cell number of control well) X 100.

DNA histogram; For the analysis of DNA histogram, culture cells were harvested with 0.1% trypsin. The cells after washing with PBS, were stained with propidium iodine and analysed with flowcytometry using Cytofluorograf 50H(Ortho Diagnostic System Inc. Mass. USA).

Subcutaneous tumor of nude mice; A human glioma cell line, U251-SP was injected subcutaneously into nude mice and obtained a subcutaneous tumor which was successfully transplanted in our laboratory. For the present experiment, a subcutaneous tumor mass was cut into pieces of about 2mm in diameter. Each fragment was then inoculated into subcutaneous space of abdominal frank using a trocar needle. When each subcutaneous tumor had reached about 6mm in diameter, approximately three weeks later, the nude mice were devided into four groups by randomization. Control group(N=5); This group was given 0.1ml of PBS intraperitoneally three times a week for four weeks. HuIFN-β group(N=6); HuIFN-β of 5×10^5 IU, dissolved in 0.1ml distill water, was administered intraperitoneally three times a week for four weeks. ACNU group(N=5); ACNU of 5mg/kg, dissolved in 0.1ml distill water, was administered intraperitoneally three times a week for two weeks. Combination group(N=8); HuIFN-β of 5×10^5 IU and ACNU of 5mg/kg were administered as same as above. The subcutaneous tumor was measured twice a week for five weeks, and its size was calculated according to the following formula. Tumor size= 1/2 x length X width^2, and the relative size was calculated as follows. (Tested tumor/original tumor) X 100(%). For statistic analysis, student t-test was used to compare tumor size values between control and tested group.

Histological examination; At the end of in vivo experiment, tumor masses in each group were resected and fixed them in 10% formalin for light microscopic observation with hematoxylin and eosin stain.

Results

Cytotoxicity in vitro; The cytotoxicity by HuIFN-β and/or ACNU was classified into three grades. Grade I was less than 90%(one log kill) of %cytotoxicity, grade II was between 90-99%(one-two log kill), and grade III more than 99%(two log kill). The result with 13 glioma cell lines showed additive or synergistic effects by the combination of ACNU and IFN in all cell lines tested. The cytotoxic effect of grade III was seen in two cell lines given ACNU of 5μg/ml, one line treated with IFN of 1000IU alone, and nine treated with the combination, which was much more than that with ACNU of 10μg/ml(Table 1). Two cell lines of SK-MG-6 and U251-SP were received grade I cytotoxicity by ACNU or IFN alone, and grade III cytotoxicity was induced by the combination.

DNA histogram analysed with flowcytometry; Alteration of DNA histogram by ACNU and/or IFN was studied with flowcytometory. Accumulations in the G2M phase with ACNU of 5μg/ml, and in the S phase with IFN of 1000IU/ml were demonstrated, and they were in general diminished and returned to same pattern as those of control within three days. On the other hand marked accumulation in the G1S and S phases at first and following in G2M phase was demonstrated by the combination of both agents and this alteration lasted over a week(Figure 1).

TABLE 1. CYTOTOXICITY OF HuIFN-β AND ACNU AGAINST HUMAN GLIOMA CELL LINES (N=13)

drug %cytotoxicity	ACNU 5µg	IFN-β 1000IU	ACNU+IFN-β 5µg+1000IU	ACNU 10µg
Grade I (<one log kill)	6	6	0	0
Grade II (one-two log kill)	5	6	4	6
Grade III (>two log kill)	2	1	9	7

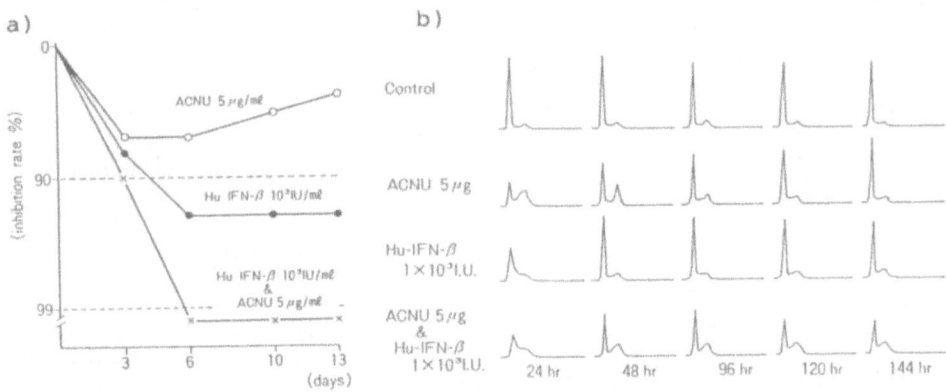

FIGURE 1. %cytotoxicity of a human glioma cell line with HuIFN -β and ACNU (a), and their serial alteration of DNA histogram.

Growth inhibition of subcutaneous tumor in vivo; No nude mice died due to side effect of HuIFN-β or ACNU, and all animals were living without significant loss of body weight at the end of the experiment on 32nd day after starting treatment. In both control and HuIFN-β groups, there was a progressive increase in the tumor size ratio and at the end of experiment, it was nine and seven times bigger in control and HuIFN-β group respectively. In ACNU group, a transient decrease in tumor size ratio was observed from day 13 to day 18. On the other hand, in combination group, there was a gradual and constant decrease after day 11 until the end of experiment, when it was 36%. By statistic analysis, significant difference of tumor growth inhibition in vivo was demonstrated between control, HuIFN-β, or ACNU and combination group(Figure 2).

Histological findings of subcutaneous tumors(Figure 3); Control group: Culture cells of U251-SP produced well circumscribed subcutaneous tumors that grew as solid tumors to about 1.5cm in diameter at 50-70 day. Tumor cells appeared spindle or polyhedral shaped and occasionally had pleomorphic nuclei. Some multinucleated cells were found. The cells were arranged in interdigitating bundles. Numerous mitosis (5-10 mitotic figures per microscopic field of 100 magnifications) and some

small areas of necrosis were also noted. HuIFN-β group: Cell
arrangement, shape of tumor cells and number of mitotic figures were similar to those in control group, although there
was greater variability in the size of nuclei. A marginal
infiltration of lymphocytes were also seen in this group.
ACNU group: There was a large necrotic area in the central
part of tumor. In the peripheral portion, however, tumor
cells and cell arrangement were essentially same as those of
control group. Combination group: A large central necrosis
was seen as well as in ACNU group. Tumor cells in peripheral
portion show degenerative changes with fewer mitotic figures.
Most of these were round and some were giant and multinucleated, whose cytoplasms were distinct and separated each
other.

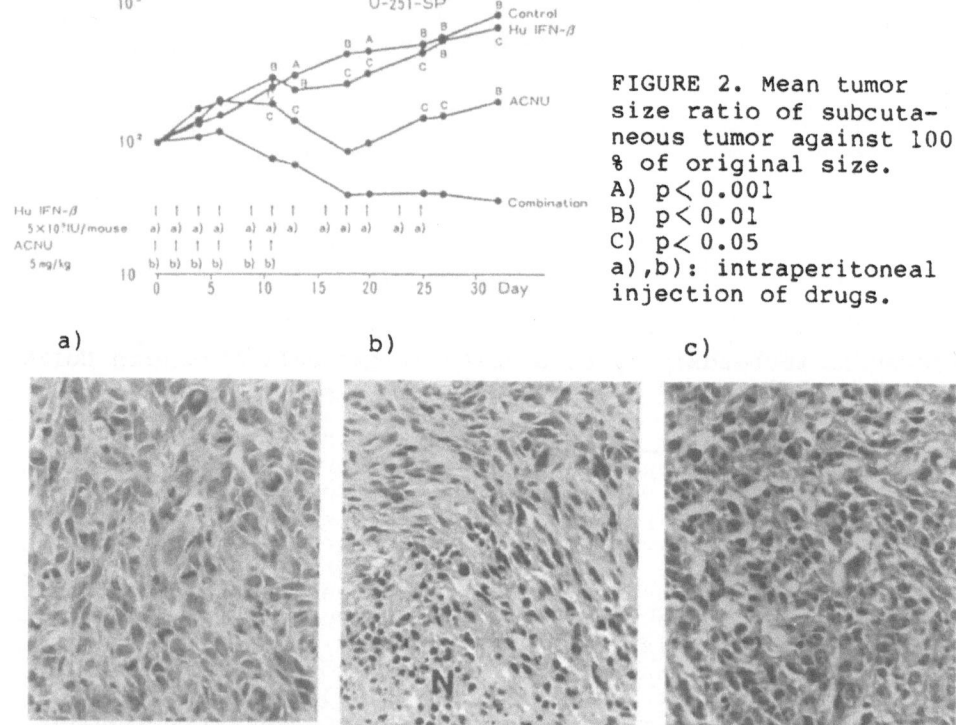

FIGURE 2. Mean tumor
size ratio of subcuta-
neous tumor against 100
% of original size.
A) $p < 0.001$
B) $p < 0.01$
C) $p < 0.05$
a),b): intraperitoneal
injection of drugs.

FIGURE 3. Light micrograph of subcutaneous tumor, a): control
group, b): ACNU group, c): combination group, N: necrosis

III. CLINICAL STUDY

Materials and Methods
Patients selection; Patients were entered to our clinical
study after the histological diagnosis of glioblastoma or
malignant astrocytoma was established except brain-stem glioblastoma, which was diangosed radiologically. Only primary

cases without any preoperative treatment were included and
recurrence cases were excluded. Nagoya University hospital and
seven affiliated hospitals in this trial contributed a total 34
patients and the study covered almost three year period from
October, 1983 through July, 1986.

IFN-ACNU-RADIATION (IAR) therapy; Figure 4 shows the protocol for IAR adjuvant therapy. HuIFN-β was administered postoperatively once a day for 7 days intravenously or intratumor-
ally through Ommaya reservior at the dose of 1-3 million IU.
On day 2, 3mg/kg of ACNU was injected intravenously and from
day 3, radiation was started until 5000-6000rad. The treatment
of IFN and ACNU were repeated two or three times at six weeks
interval if myelosuppression or other side effects did not
developed.

Evaluation of antitumor activity; Efficacy of the IAR therapy was estimated by reduction in tumor size and postdiagnos-
tic survival times. The reduction in tumor size was compared
with our historical control(AR therapy), which was treated
postoperatively with ACNU and radiation but not with IFN. It
was evaluated by comparing tumor size on contrast-enhanced CT
scan between before and after completion of the therapy. The
tumor size was defined as the area of enhanced mass lesion and
the responses were classified into four categories. A complete response(CR) was the disappearance of all tumor rem-
nants for at least 4 weeks. A partial response(PR) denoted a
reduction of 50% or more and a stable condition(ST) was defined as either a 25% decrease or not grater than a 25% increase in tumor size. Progression(PG) denoted an increase of
25% or more.

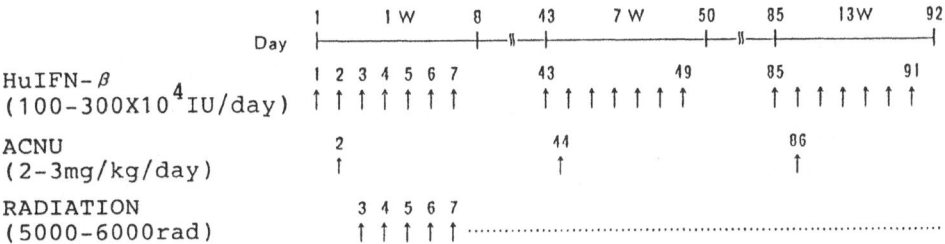

FIGURE 4. Protocol for IAR therapy against malignant glioma

Results

Distribution of patients by age, sex, tumor histology, and
dose of adjuvant therapy were similar in IAR and AR groups and
no statistical significance with regard to these variable was
found in the background investigation. (Table 2)

Reduction in tumor size; Gross total removal of tumor(TR)
was performed in 4 patients and no measruable enhanced mass
lesion was visualized on postoperative CT scans. Residual
tumor was visible in the other 30 patients. With IAR therapy
objective regression of tumor was observed in 18 patients, 9
complete and 9 partial for a response rate of 60%. On the
other hand, in the mode of AR therapy complete and partial
responses were 3 and 7 patients respectively and the response
rate was 35.7%. (Table 3)

Survival time; To compare survival among IAR and AR groups, survival curves were calculated using Kaplan-Meier method and the statistical significance of the survival curves was evaluated with the Cox-Mantel test. The survival rates at one and two years were 74% and 46% in IAR and 70% and 38% in AR group, but the differences were not significant statistically. (Figure 5)

TABLE 2. BACKGROUND INVESTIGATION OF AGE, SEX, HISTOLOGY AND DOSE OF ADJUVANT THERAPY

	SEX M F	AGE ≥40 <40	HISTOLOGY GM MA NV	RADIATION(rad) ≥5000 <5000	ACNU(3mg/kg) 1 2 3-4
IAR GROUP (N=34)	21 13	22 12	20 11 3	26 8	10 18 6
AR GROUP (N=33)	20 13	21 12	22 11 0	26 7	5 12 16
X^2	NS	NS	NS	NS	NS

M:Male, F:Female, GM:Glioblastoma, MA:Malignant astrocytoma, NV:not verity, X^2:chi-square test

TABLE 3. RESPONSE TO ADJUVANT THERAPY ESTIMATED BY REDUCTION IN TUMOR SIZE ON CT SCAN

Response Regimen	TR	CR	PR	ST	PD	Response Rate%
IAR GROUP (N=34)	4	9	9	5	7	60.0
AR GROUP (N=33)	5	3	7	8	10	35.7

TR:Total Removal, CR:Complete Response, PR:Partial Response, ST:Stable, PD:Progression, Response Rate:CR+PR/CR+PR+ST+PD

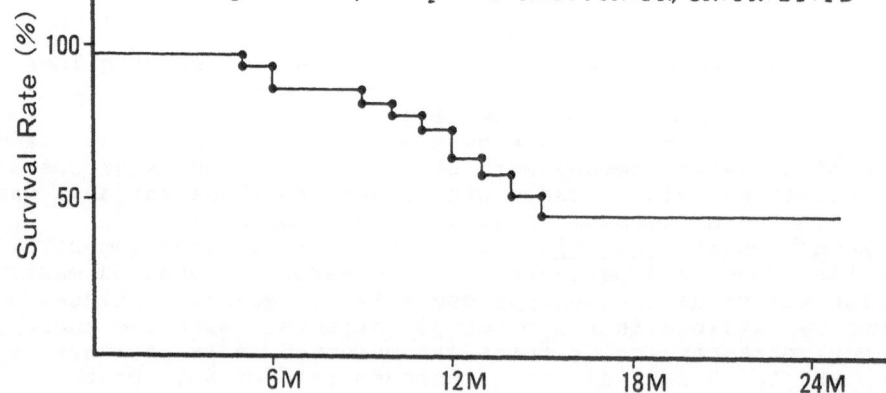

FIGURE 5. Survival curve of patients with malignant glioma treated with IAR therapy.

IV. DISCUSSION

A water soluble nitrosourea derivative, ACNU was reported to
be able to cross the blood-brain barrier and have antitumor
activity against malignant glioma. The activity of ACNU was
dose dependent, then it seems favorable to administer the
agent as much as possible. However, there is a limitation on
amount of the dose because of its persistent and cumulative
bone marrow suppression. Usually 2-3mg/kgw af ACNU is administered intrevenously and it has been reported that the 5μg/g
tissue concentration in tumor could be obtained.(6) Cytotoxicity was tested in vitro by exposing 13 human glioma cell lines
to ACNU at clinical dose of 5μg/ml concentration. Two log
kill cytocidal effect was demonstrated on only two cell lines
and other six cell lines were given less than one log kill
cytostatic effect. These results indicated that ACNU therapy
alone was not fully effective against malignant glioma and it
is necessary to use combination therapy. On the other hand,
experimental studies with IFN support multiple factors for its
antitumor activity. The first factor is as a chemotherapeutic
agent with direct cytocidal or cytostatic effects. The second
one is as a biological response modifier with regulation of
cellular immune system.(7) The third one is as a drug sensi-
tizer with induction of enhanced cytotoxic effect in combination with other chemotherapeutic agents. Human lymphoblastoid
or fibroblast interferon was reported to show synergistic or
additive interaction in vitro and in vivo in combination with
chemotherapeutic agents. Takahashi et al. investigated in
vitro the combined effect of human lymphoblastoid interferon
and various anticancer agents on the growth of human leukemia
cell lines.(8) They reported that the interferon showed a
striking synergistic interaction in combination with acla-
rubicin, cytosine arabinoside or predonine, and also moderately
synergistic with adriamycin or 5-fluorouracil. Similar results
were obtained by Yamamoto et al. using β-type interferon
produced by fibroblast.(9) With in vivo experiment by Chirigos
and Pearson, murine systemic leukemia induced by inoculation of
LSTRA leukemic cells was treated with BCNU and/or interferon.
Combination therapy showed an overall cure rate of 72% over
the 25% cure rate obtained in the group given BCNU alone.(10)
The results of our in vitro study of HuIFN-β show the enhanced
cytotoxic interaction in combination with ACNU on human glioma
cell lines. The interaction was demonstrated on all glioma
cell lines tested including ACNU resistant ones.
The effect was confirmed by in vivo study using human glioma
transplanted nude mice. Reduction in tumor size and histological findings of massive necrosis and tumor cell degeneration were obtained only by the combination therapy. Present
controlled clinical study for postoperative adjuvant therapy
against the patients with malignant glioma, indicated the
importance of interferon therapy especially in combination
with ACNU and radiotherapy. Antitumor activity of IAR therapy
was significantly higher than that of AR therapy from the
evaluation by reduction in tumor size. The response rate was
60% in the former and 35.7% in the latter. However, a signif-

icant increase of survival time could not be obtained at
present in IAR therapy as compared with AR therapy. Almost
half of the patients responded to IAR therapy recurred or took
a progression within a year. From this clinical data, it is
noted that IAR therapy is useful as initial induction therapy
against malignant glioma but it is not enough and some maintenance therapy is necessary after induction of complete or
partial remission.

REFERENCES

1. Walker MD, Green SB, Bayaer DP, et al: Randomized comparison
   of radiotherapy and nitrosourea for the treatment of mlaig-
   nant glioma after surgery. New Eng J Med 303:1323-1329,
   1980
2. Yoshida J, Kobayashi T, Kageyama N.: Multimodality treatment of malignant glioma. Effect of chemotherapy with ACNU
   and immunotherapy with N-CWS. Neurol Med Chirur (Japan) 24:
   19-26, 1984
3. Mori T, Mineura K, Katakura R, et al: A considaration on
   pharmacokinetics of a new water-soluble anti-tumor nitrosourea,ACNU, in patients with malignant brain tumor. Brain
   Nerve (Japan) 31:601-606, 1979
4. Takakura K, Abe H, Tanaka R, et al: Effect of ACNU and
   radiotherapy on malignant glioma. J Neurosurg 64:53-57,
   1986
5. Nagai M.: Clinical trials of human fibroblast interferon
   (HuIFN-β) on malingant brain tumors. J Jpn Soc Cancer Ther
   18:60-68, 1983
6. Saito Y, Nakaya Y, Muraoka K, et al: Chemotherapy of brain
   tumor with nitrosourea derivative(ACNU). Cancer Chemothera
   (Japan) 5:579-794, 1978
7. Gresser I, Maury C, Brouthy-boyé D, et al: Mechanism of the
   antitumor effect of interferon in mice. Nature 239:167-168,
   1972
8. Takahashi I, Oda Y, Lai M, et al: Interaction between human
   lymphoblastoid interferon and chemotherapeutic agents in
   vitro. Acta Med Okayama (Japan) 6:501-504, 1984
9. Yamamoto SH, Tanaka T, Kanmori M, et al: In vitro studies
   on potentiation of cytotoxic effects of anticancer drugs by
   interferon on a human neoplastic cell line(HeLa). Cancer
   Letter 20:131-138, 1983
10.Chirigos MA, Pearson JM.: Cure of murine leukemia with drug
   and interferon treatment. J. Natl Cancer Inst 51:1867-1868,
   1973

INTERFERON ALPHA IN THE TREATMENT OF MALIGNANT DISEASES: SOURCES, DOSES, AND REGIMENS

F. PORZSOLT, W. DIGEL, and H. HEIMPEL

1. INTRODUCTION

Interferon-alpha (IFN-α) has been applied for treatment of several malignancies e.g. hairy cell leukemia (HCL) (7,16,19,28), chronic myelogenous leukemia (CML) (24,25), non-Hodgkin lymphomas (NHL) (2,3), renal cell cancer (RCC) (4,20,27), malignant melanoma (8), and Kaposi's Sarcoma (26). So far, the therapeutic efficacy seems to be more evident in hematologic malignancies than in solid tumors. For example, high response rates (80%) have been shown in HCL and in CML (16,24,25) and lower but relevant response rates (30% or less) have been reported in studies on RCC and melanoma (4,8,20). These results are sufficient to select patients who will benefit very likely from IFN-α therapy. However, one should not draw premature negative conclusions about the lack of efficacy of IFN- if a particular design of a study gives negative results. HCL and RCC are used as examples in this report to provide some evidence that not only IFN doses and treatment regimens but also the source of IFNs may be variables which can considerably influence the results of clinical trials.

2. MATERIALS AND METHODS
2.1 Patients

2.1.1. HCL patients: 14 HCL patients were included in the clinical study. The diagnoses were supported by several tests and were confirmed by bone marrow biopsies in all patients and by spleen histology in 8 patients. The median age was 54 years (36-75 years). The male/female distribution was 13/1. 9 patients received IFN therapy before splenectomy (Sx) and 4 patients had Sx before IFN therapy. One additional patient was splenectomized after two months of primary IFN therapy. Since he had progressive disease seven months after Sx a second IFN therapy was started at this time. This patient was diagnosed as T-HCL.

2.1.2. RCC patients: 92 patients with metastatic RCC were included in a multicenter trial. 82/92 patients were evaluable. The diagnoses were confirmed in 81/83 patients by nephrectomy. Median age was 57 years (21-80 years). The male/female distribution was 58/25.

2.2. Indications for treatment.

HCL patients were treated when they had either progressive or symptomatic disease (e.g. recurrent infections, requirement of platelet or RBC substitution). HCL patients with non-symptomatic stable disease were not treated and are not included in the clinical report. IFN therapy was given only after Sx unless the condition of the patient did not allow or the patient refused surgery. In 4 patients with severe thrombopenia Sx was done after increase of platelets by primary IFN therapy. In RCC, IFN treatment was indicated in all patients with metastatic disease if they had nephrectomy, measureable metastases, and a Karnofsky-index above 60.

2.3. <u>Goals of treatment.</u>

The goal of treatment in HCL was to maintain non-symptomatic stable disease (SD_{ns}). In RCC, complete or partial remission according to usual response criteria should be induced. It was tested by randomization if IFN-α in combination with medroxyprogesterone acetate (MPA) is superior to IFN-α therapy in inducing complete and partial remissions (PR).

2.4. <u>IFN treatment protocols.</u>

2.4.1. <u>HCL.</u> Initially, we used natural IFN-α (Bioferon, Laupheim) and changed to recombinant IFN-α 2C (Boehringer, Ingelheim) because it was available in large amounts. Daily doses and frequency of applications were adjusted to the actual platelet counts as described previously (18). Shortly, starting doses were 0.5 MU daily s.c. when platelet counts were $<15x10^9$/L and were 1.5 MU when platelet counts were $>15x10^9$/L. These daily doses were given 5 x /week and were increased to 3 MU when no substitution of platelets was necessary. The frequencies of applications were reduced from 5/week to 3/week as soon as the platelet counts were $>100x10^9$/L.

2.4.2. <u>RCC.</u> The patients in each of the two randomized study arms (A and B) received 2 MU of recombinant IFN-α 2C (Boehringer, Ingelheim) s.c. 5 x /week for 12 weeks; this dosis was given once per week from week 13 through 48 as far as no progression of the disease was detectable. While patients of arm (A) received no further treatment, patients of arm (B) were treated additionally by MPA 750 mg p.o. daily throughout the study.

2.5. <u>IFN induction in vitro.</u>

For IFN induction either purified PHA (Welcome Laboratories), 10 and 2.5 µg/ml or corynebacterium parvum, 700 and 88 µg/ml or Sendai virus 150 and 15 HAU/ml final concentration was added to one ml of 10^6 mononuclear cells (MNC) of peripheral blood (PB). After 48 hrs incubation the supernatants were harvested and were tested for antiviral activities. Further details have been described previously (17).

2.6. <u>2-5 Oligoadenyl synthetase in mononuclear cells.</u>

The 2-5 AS was measured according to the method described by Revel et al. (25). The enzym containing cytoplasmic fraction was adsorbed to poly I:C cellulose. The activated 2-5 AS synthesizes 2'-5' Oligoadenylates from ATP and α -(^{32}P)-ATP. ^{32}P is separated from not incorporated α -(^{32}P)-ATP by phosphatase treatment. The newly synthetized 2'-5' Adenylate is eluted from an aluminum column and is measured in a ß-counter.

3. RESULTS

3.1. <u>Clinical results in HCL.</u>

3.1.1. <u>Efficacy of platelet adjusted IFN doses.</u> The efficacy of IFN-therapy of HCL can be demonstrated by the decrease of HC's and increase of normal blood cell counts in PB. The PB counts of 15 treatment periods of IFN- therapy are shown in Fig. 1. It is known that HC's are cleared from PB very fast in some patients and more slowly in others. Accordingly, the curve describing the clearance of HC's in Fig. 1 indicates two populations of patients: one population whose HC's are cleared from PB during the first 8 weeks of IFN-therapy represented by the big slope of the curve during this period of treatment. In the second population of patients clearance of HC is not yet completed by week 20 represented by the small slope of the curve from week 8 through 20. In both groups of patients the clearance of HC begins immediately when IFN-therapy is started. The effective increase of platelets in PB begins at week 4, considering initial drops of platelet counts and necessary platelet substitutions in some patients. Comparable reductions of the granulocyte counts and of hemoglobin values were seen during the frist two weeks of treatment but the drop of hemoglobin was frequently compensated by red blood cell transfusions. The

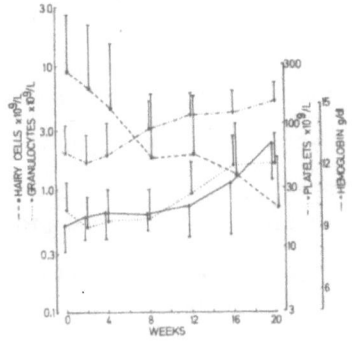

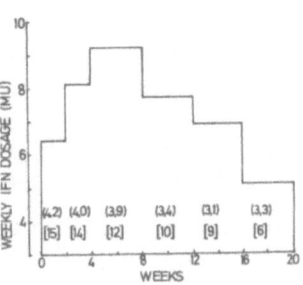

FIG. 1: Kinetics of PB cell counts in HCL during IFN-α therapy. Mean values ± SD from 15 treatment periods. The Hairy Cell counts are based on morphologic differentiation. Normal lower limits for granulocytes: $1.5 \times 10^9/L$, for platelets: $150 \times 10^9/L$, for hemoglobin: 12 g/L.

FIG. 2: Weekly IFN doses in the treatment of HCL. The applications per week are shown in ( ). The numbers of patients are given in [ ].

increase of granulocytes and of hemoglobin begins effectively at the 8th week of IFN-treatment. The decrease of HC's in PB and the increase of platelets, granulocytes and hemoglobin are characterized by the intervals from begin of IFN-therapy to begin of changes (T_0) of blood counts, half-maximal ($T_{1/2}$), and maximal changes (T_{max}) which is usually the lower limit of normal range (Tab. I).

	T_0	$T_{1/2}$	T_{max}
hairy cells	0(13)	4(12)	20(6)
platelets	4(13)	10(11)	20(6)
granulocytes	8(13)	12(10)	16(9)
hemoglobin	8(13)	16(9)	20(6)

TABLE I: Intervals in weeks from begin of IFN-therapy to begin of changes (T_0) of peripheral blood counts, half-maximal ($T_{1/2}$), and maximal changes (T_{max}). The data are based on mean values from (n) courses.

The weekly IFN doses resulting from our dose adjustment strategy are shown in Fig. 2. 3.1.2. Natural versus recombinant IFN-α. Four patients were treated with natural IFN-α for historical reasons. 10 patients received recombinant IFN-α. Comparing both groups, there were no differences in the daily doses, applications per week, and efficacies of treatments characterized by T_0, $T_{1/2}$, and T_{max} (data

not shown). The only difference was the initial HC count which was higher in the group treated by natural IFN-α. These data do not indicate possible differences in the efficiencies of recombinant and natural IFN-α in HCL.

3.2. Experimental results in HCL.

3.2.1. In vitro induction of IFN-α in MNC of HCL. This data confirm and extent our previous results (18) showing that high levels of IFN-gamma but only low levels of or no IFN-alpha can be induced in mononuclear cells from HCL patients. The lack of IFN-alpha inducibility was not restored 1-12 months after IFN-therapy (Tab. II).

			Antiviral activities (U/ml) induced by					
		none	PHA (µg/ml) 10	2.5	c.parvum(µg/ml) 700	88	Sendai(HAU/ml) 150	15
controls:	median (n)	0(5)	300(5)	271(4)	150(5)	565(4)	330(5)	22(4)
	range	0	100-1000	100-585	8-610	310-1000	200-610	8-323
HCL pts: before IFN	median (n)	0(4)	2230(4)	615(4)	3(4)	0(4)	35(4)	5(4)
	range	0-4	102-3230	102-1610	0-15	0-15	10-40	0-8
HCL pts: after IFN	median (n)	0(9)	670(9)	610(5)	0(9)	2(4)	5(9)	0(5)
	range	0-1	56-1900	102-1640	0-33	0-13	0-45	0-26

TABLE II: In vitro induced antiviral activities (U/ml) of peripheral blood MNC from controls and HCL patients before and 1-12 months after IFN-alpha therapy.

3.2.2. 2-5 Oligoadenyl synthetase in MNC of HCL. The base line levels of 2-5 AS of MNC from PB are lower in HCL patients than in controls. After in vitro stimulation with low doses of IFN-α the 2-5 AS levels of both, HCL patients and controls can be increased. In patients with very low base line levels of 2-5 AS only moderate increases are seen after IFN stimulation while in other patients 2-5 AS levels comparable to the base line levels of controls can be induced by IFN in vitro (Fig. 3).

3.3. Clinical results in RCC.

3.3.1. IFN-α + MPA versus IFN-α treatment of RCC. The results of our multicenter study on IFN ± MPA therapy of metastatic RCC will be reported elsewhere in detail. 83 evaluable patients were recruted from 10 centers. Patients of group A (44 pts) and B (39 pts) were comparable as age and sex distributions, Karnofsky indices, TNM classes, primary symptoms, localization of metastases, surgery of metastases, other therapies, and number of drop-out patients are concerned. The results show that minor side effects were equally distributed to both groups. Partial remissions were seen in 3/83 patients after 8, 12, and 24 weeks of treatment. In 45% of patients the treatment had to be discontinued before week 48 mainly because of progressive disease. The survival rates were not different in both groups and were 0.6 after 6 months.

3.3.2. Recombinant versus natural IFN-α in the treatment of RCC. In this part of the report we compared data of our study with results of others. To ensure reliable conclusions only papers which provided sufficient data were considered. So it is shown in Table III that investigators who used alpha-IFN's for treatment of metastatic RCC describe response rates ranging from 0 to 29%. Although it is difficult to compare the various treatment regimens of these studies, it seems that increasing response

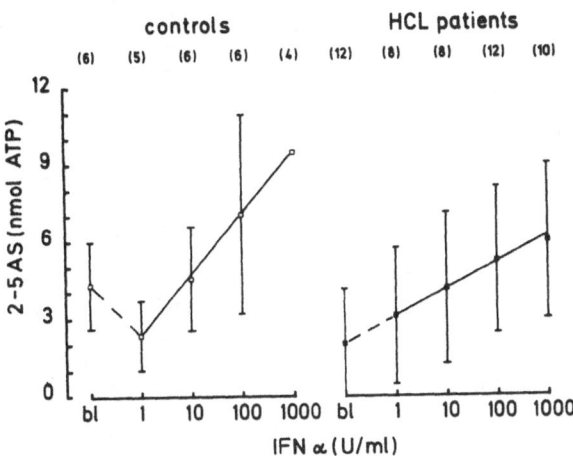

FIG. 3: 2-5 Oligoadenyl synthetase in MNC of 6 controls and 12 HCL patients. Base line values (bl) and values after in vitro stimulation with IFN-α are shown.

rates are seen in the order of recombinant IFN-α (α A, α 2C), human lymphoblastoid IFN-α (HLB), and natural leucocyte IFN-α (leu). Within each of these 3 study groups, recombinant, lymphoblastoid, and leucocyte IFN, high response rates seem to correlate rather with high weekly than high daily IFN doses. The importance of the weekly rather than of daily dose suggests that more than "one hit" is necessary for detectable clinical responses. A remarkable difference seems to exist between applications of high versus low weekly doses of recombinant IFN: whereas no responses were seen using low doses of recombinant IFN such as 10 - 30 MU per week, high response rates comparable to those obtained with leucocyte IFN (29%) were observed when high weekly doses of recombinant IFN-α such as 240 MU per week were applied. It should be mentioned that the high dose protocol caused considerably more side effects than the low dose regimen.

4. DISCUSSION

It's becoming increasingly clear from in vitro studies that several effects on tumor cells such as induction of growth inhibition and/or differentiation or increase of the host's resistance against tumor cells may count for the efficacy of IFN-α in cancer therapy.

Comparing HCL and RCC there are several differences in treatment protocols and in efficacy of treatments. These differences could be indicative for different actions of IFN-α. The necessary doses for efficient treatment are definitely lower in HCL than in RCC. We have described previously (17) and in this paper that the in vitro inducibility of IFN-α in MNC from HCL patients is impaired. It might be assumed in HCL patients that IFN-α production is also impaired in vivo. This assumption is supported by our data on low 2-5 AS levels in MNC from HCL patients. The most likely explanation for reduced IFN-α production is the reduced number of monocytes which is commonly found in PB of HCL patients. The reduced base line values of 2-5 AS can be explained by deficient endogenous IFN-α production. Since monocytes are a major source of IFN-α producing cells we con-

source	Interferon dosage (MU)		eval. pts	CR+PR	ref.
	daily	weekly			
α A	34	240	41	29%	(22)
leu.	3	21	11	27%	(5)
leu.	3	21	19	26%	(24)
leu.	3	21	50	26%	(23)
leu.	3	15	43	16%	(4)
HLB	8.5	25	33	15%	(17)
leu.	3	15	23	13%	(6)**
leu.	1–10	7–70	30	10%	(10)
HLB	3	18–21	18	6%	(15)
HLB	5	15	21	5%	(29)
α 2C	2	10/2	83	4%	(x)
α A	3	21	15	0%	(22)
α A	2x17*	2x17*	15	0%	(16)**

TABLE III: IFN's in the treatment of metastatic RCC. Recombinant alpha IFN'S (αA, α2C) lymphoblastoid IFN-α (HLB), leucocyte IFN-α (leu). *simultaneous application of 17 MU subcutaneously plus intravenously. **Interferon plus Chemotherapy. (x) own study.

sider IFN-α treatment in HCL as substitution of the reduced endogenous IFN-α production. The high IFN doses which are necessary for treatment of RCC do primarily not suggest that a lack of endogenous IFN production has to be substituted.

However, the results of IFN therapy in RCC suggest that there are qualitative differences between recombinant, lymphoblastoid, and leucocyte αIFNs. Unfortunately, we are not aware of sufficient data to titrate carefully the efficacy of lymphoblastoid and leucocyte αIFNs in HCL. So it is possible that the same qualitative differences between sources of αIFNs which have been observed in the treatment of RCC could also exist in the treatment of HCL.

Since natural preparations seem to be superior to recombinant preparations one would have to consider that the used recombinant IFN-preparations contain only one subtype of the IFN-α family. This has also been suggested previously (9). The particular $α_2$ subtype which is available as recombinant IFN-α could be less effective than other subtypes in induction or inhibition of factors which influence proliferation and/or differentiation of HC's. It is also possible that recombinant IFN-$α_2$ acts differently from natural IFN-$α_2$ because of the known differences in two amino acids or because of lack of glycosylation.

Recently, a dose response relationship in the IFN therapy of RCC has been suggested by Kirkwood et al (9) and Krown (10). We confirm this impression as the data in Tab. III indicate that high weekly doses of IFN-α are more important than high daily dosages. This suggests that repeated "IFN-hits" within a week are rather efficient than single "hits" repeated in weekly intervals. It is known from studies on HCL that the frequency of IFN applications obviously influences the expression of IFN receptors. IFN-α receptors in HCL are down-regulated by daily applications of IFN. Since the receptors recover partially after the first "hit" but remain down-regulated after several sequential "hits" (1,12) it seems to be essential to repeat IFN treatment 2 to 3 times per week for keeping the IFN receptors down-regulated. Whether or not down-regulation of IFN receptors will be the key to optimize treatment regimens in HCL and other malignancies remains still open.

Different actions of INF-α in HCL and RCC can also be assumed from the difference in times to response. While the decrease of HC's in PB is detectable immediately after begin of IFN therapy it usually takes several months of treatment to confirm tumor regression in metastatic RCC (15). The tumor load seems to influence the efficacy of IFN in RCC (6) but there is no evidence for such a correlation in HCL. Accordingly, visceral or lung metastases of RCC seem to respond more frequently to IFN than metastases in other organs (22,27). In HCL, there is no obvious difference in responses of HC infiltrations to IFN between different organs. The varying responses of different metastases to IFN therapy cannot yet be explained but may depend on varying degrees of tumor cell differentiation (for review see Sem. Oncology Sept. 1985). Finally, IFN-neutralizing activities have been found in a considerable proportion of sera from RCC patients but only rarely in HCL patients (12). The biologic relevance of these neutralizing activities is absolutely not clear; their presence does obviously not correlate with clinical response failures (20).

In summary, the characteristics of treatment protocols and of clinical responses to IFN-α differ in HCL and RCC in several aspects. Careful analyses of treatment protocols and of IFN responses of other tumors might supplement the basic research in answering the question if different IFN-α-subtypes and/or different IFN-actions mediate the clinical responses in HCL and RCC.

REFERENCES

1 Billard C, Sigaux F, Castaigne S, Valensi F, Flandrin G, Degos L, Falcoff E, Aguet M: Treatment of Hairy Cell Leukemia with Recombinant Alpha-Interferon: II. In Vivo Down-Regulation of Alpha-Interferon Receptors and Tumor Cells. Blood: 67, 821–826, 1986

2 Canellos GP: Interferon in the Treatment of Malignant Lymphoma. Sem Oncol: 12,4 Suppl 5, 25–29, 1985

3 Champlin R, Gale RP, Foon KA, Golde DW: Chronic Leukemias: Oncogenes, Chromosomes, and Advances in Therapy. Ann Int Med 104: 671–688, 1986

4 DeKernion JB, Sarna G, Figlin R, Lindner A, Smith RB: The Treatment of Renal Cell Carcinom with Human Leucocyte Alpha-Interferon. J Urol: 130, 1063–1066, 1983

5 Edsmyr F, Esposti PL, Andersson L, Steineck G, Lagergren C, Strander H: Interferon Therapy in Disseminated Renal Cell Carcinoma. Radiother Oncol: 4, 21–26, 1985

6 Figlin RA, DeKernion JB, Maldazys J, Sarna G: Treatment of Renal Cell Carcinoma with Alpha (Human Leucocyte) Interferon and Vinblastin in Combination: A Phase I-II Trial. Cancer Treat Rep: 69, 263–267, 1985

7 Golomb HM, Ratain MJ, Vardiman JW: Sequential treatment of Hairy Cell Leukemia: A New Role for Interferon. In: DeVita VT, Hellman S, Rosenberg SA (ed): Important Advances in Oncology 1986. Lippincott Co. Philadelphia

8 Kirkwood JM, Ernstoff M: Melanoma: Therapeutic Options with Recombinant Interferons. Sem Oncol: 12,4 Suppl 5, 7–12, 1985

9 Kirkwood JM, Harris JE, Vera R, Sandler S, Fischer DS, Khandekar J, Ernstoff MS, Gordon L, Lutes R, Bonomi P, Lytton B, Cobleigh M, Taylor SJ IV: A Randomized Study of Low and High Doses of Leucocyte Alpha-Interferon in Metastatic Renal Cell Carcinoma: The American Cancer Society Collaborative Trial. Cancer Res: 45, 863–871, 1985

10 Krown SE: Therapeutic Options in Rencal-Cell Carcinoma. Sem Oncol: 12,4, Suppl 5, 13–17, 1985

11 Lau AS, Hannigan GE, Freedman MH, Williams BRG: Regulation of Interferon Receptor Expression in Human Blood Lymphocytes in Vitro and During Interferon Therapy. J Clin Invest: 77, 1632–1638, 1986

12 Linder-Ciccolunghi SN: Improving the Acceptablity and Tolerability of Interferon Therapy in Cancer Patients - A Review. In: Brookes LJ (ed): Recombinant Interferon in Cancer Treatment. Proceedings Satellite Symposium to 3rd European Conference on Clinical Oncology and Cancer Nursing, Stockholm 1985, Professional Postgraduate Services Limited, Macclesfield

13 Marumo K, Murai M, Hayakawa M, Tazaki H: Human Lymphoblastoid Interferon Therapy for Advanced Renal Cell Carcinoma. Urol: 24,6 567-571, 1984

14 Muss HB, Welander C, Caponera M, Reavis K, Cruz JM, Cooper MR, Jackson DV jun, Richards F II, Stuart JJ, Spurr CL, White DR, Zekan PJ, Capizzi RL: Interferon and Doxorubicin in Renal Cell Carcinoma. Cancer Treat Rep: 69, 721-722, 1985

15 Neidhart JA, Gagen MM, Young D, Tuttle R, Melink TJ, Ziccarrelli A, Kisner D: Interferon-Alpha Therapy of Renal Cancer. Cancer Res: 44, 4140-4143, 1984

16 Porzsolt F: Primary Treatment of Hairy Cell Leukemia: Should IFN-therapy Replace Splenectomy? Blut 52: 265-272, 1986

17 Porzsolt F, Janik R, Heil G, Brudler O, Raghavachar Ar, Scholz S, Papendick U, Heimpel H: Deficient IFN-Alpha Production in Hairy Cell Leukemia. Blut 52, 185-190, 1986

18 Porzsolt F, Thomä I, Unsöld M, Wolf W, Obert HJ, Kubanek B, Heimpel H: Platelet-Adjusted IFN Dosage in the Treatment of Advanced Hairy Cell Leukemia. Blut 51, 73-82, 1985

19 Quesada JR, Reuben J, Manning JT, Hersh EM, Gutterman JU: Alpha-Interferon for Induction of Remission in Hairy Cell-Leukemia. N Engl J Med 310: 15-18, 1984

20 Quesada JR, Rios A, Swanson D, Trown P, Gutterman JU: Antitumor Activity of Recombinant-Derived Interferon-Alpha in Metastatic Renal Cell Carcinoma. J Clin Oncol: 3, 1522-1528, 1985

21 Quesada JR, Swanson DA, Gutterman JU: Phase II Study of Interferon-Alpha in Metastatic Renal-Cell Carcinoma: A Progress Report. J Clin Oncol: 3, 1086-1092, 1985

22 Quesada JR, Swanson DA, Trindade A, Gutterman JU: Renal Cell Carcinoma: Antitumor Effects of Leucocyte Interferon. Cancer Res: 43, 940-947, 1983

23 Revel M, Wallach D, Merlin G, Schattner A, Schmidt A, Wolf D, Shulman L, Kimchi A: Methods in Ezymology 79B, 149-161, 1981

24 Talpaz M, Kantarjian HM, McCredie K, Trujillo JM, Keating MJ, Gutterman JU: Hematologic Remission and Cytogenetic Improvement Induced by Recombinant Human Interferon-Alpha A in Chronic Myelogenous Leukemia. N Engl J Med 314: 1065-1069, 1986

25 Talpaz M, Mavligit G, Keating M, Walters RS, Gutterman JU: Human Leucocyte Interferon to Control Thrombocytosis in Chronic Myelogenous Leukemia. An Int Med 99: 789-792, 1983

26 Volberding PA, Mitsuyasu R: Recombinant Interferon-Alpha in the Treatment of Acquired Immune Deficiency Syndrome-Related Kaposi's Sarcoma. Sem Oncol: 12,4, Suppl 5, 2-6, 1986

27 Vugrin D, Hood L, Taylor W, Laszlo J: Phase II Study of Human Lymphoblastoid Interferon in Patients with Advanced Renal Carcinoma. Ca Treat Rep: 69, 817-820, 1985

28 Worman CP, Catovsky D, Bevan PC, Camba L, Joyner M, Green PJ, Williams HJH, Bottomley JM, Gordon-Smith EC, Cawley JC: Interferon is Effective in Hairy Cell-Leukemia. Br J Hematol 60: 759-763, 1985

TREATMENT OF VARIOUS MALIGNANCIES WITH RECOMBINANT IFN-γ.

F. TAKAKU*   and IFN-γ Study Group
*The Third Department of Internal Medicine, Faculty of Medicine, University of Tokyo, Japan

Although most of the clinical experience has been obtained with various types of IFN-α, IFN-γ is reported to induce more pronounced immunomodulation and expected to exert greater antiproliferative activity than IFN-α or β (1).

IFN-γ acts on fibroblasts through receptors which are different from those of IFN-α and β. In vitro studies showed that IFN-γ inhibited the growth of human cancer cell line cells much more effectively than IFN-α and -β (2).

Increase of NK cell activity, activation of tumor cell killing by macrophage, and induction of killer cells in collaboration with interleukin-2 have been reported as showing that IFN-γ exerts its antitumor activity by modulating the immune response against tumors.

In this paper, results of a multi-institutional study on the antitumor effect of recombinant IFN-γ(rIFN-γ) on various human malignancies which has been performed since 1983 will be reported.

MATERIALS AND METHODS

329 patients with various malignancies who were admitted into 57 institutes were studied for phase II trial of rIFN-γ. Human rIFN-γ(S-6810) was produced in E. coli by Biogen SA using mRNA from human spleen lymphocytes (3) and supplied through Shionogi & Co. LTD. It has a specific activity of more than 1×10^7 units/mg·protein with a purity of more than 99%. Eligibility criteria for this study included age over 15 years, life expectancy over 2 months, and performance status under the grade 3. They were also required for normal function of hematopoiesis, liver, kidneys and other requisite laboratory studies. Patients receiving chemotherapy or other treatment within preceding 4 weeks were also omitted from this study. $6-12 \times 10^6$ U/m^2/day of rIFN-γ were administered by one hour drip infusion for at least 4 weeks. In some patients with renal cell carcinoma, 40×10^6 U/m^2/day were administered on days 1-5, and 15-19 as the first and second course by one hour drip infusion (intermittent high dose). Thereafter, the rIFN-γ was given in the same dose on days 29, 31,

33, and 43, 45, 47 as the third and fourth course. Local
administration was performed by injecting 4-12x10^6 U of
rIFN-γ into tumor or an area surrounding tumor every day or
intermittently more than 5 times.

In case of complete or partial response, the interferon
treatment was continued until patients were evaluated as
progressive disease. In case of no change, further treatment
after 4 weeks was decided by each doctor. The therapy was
stopped when the patients were considered as the disease
progressed.

Before the administration of rIFN-γ, prick test was performed on all patients to detect an allergic reaction or
hypersensitivity to this drug beforehand. Informed consent
was given from all patients. Side effects were strictly
checked and administration was stopped in case of necessity
by the decision of each doctor.

Response criteria were defined as complete response (CR),
partial response (PR), minor response (MR), no change (NC),
and progressive disease (PD) according to the classification
of Oye and Shapiro. The similar criteria were applied to
define the response to local administration.

RESULTS

Among 329 cases administered with rIFN-γ (219 systemic and
110 local administration), 237 (142 systemic and 95 local
administration) were eligible for evaluation.

Table 1   Patient Characteristics

| | No. of Patients | | |
| | Interferon Therapy | | |
	Systemic	Local	Total
Total number of patients	219	110	329
Evaluable patients	142	95	237
Male / female	76/66	49/46	125/112
Age in years			
< 39	9	7	16
40 − 49	28	11	39
50 − 59	47	17	64
60 − 69	36	25	61
70 − 79	19	25	44
80 <	3	10	13
Prior therapy			
Surgery	78	35	113
Radiation	26	10	36
Chemotherapy	96	34	130

Treatment was stopped in 15 cases because of side effects. Fifteen cases were dropped from this study because of the death or earlier termination of the treatment. Sixty-two cases were omitted mainly because of the violation of the above described treatment protocols. Patient profile of the 237 eligible cases is shown in Table 1.

Clinical effect of the systemic administration is shown in Table 2. CR was observed in each 1 case of renal cell carcinoma, malignant lymphoma and mycosis fungoides.

Partial response was observed in cases with renal cell carcinoma, head and neck cancer, multiple myeloma, chronic lymphocytic leukemia and mycosis fungoides.

Table 2   Clinical Effect of Recombinant IFN-γ
(I) Systemic Administration

Tumors	No. of Evaluable Patients	Response					Response Rate (%)
		CR	PR	MR	NC	PD	
Renal cell Ca.	61	1	8	1	27	24	9/ 61 (14.8)
Malignant melanoma	8			1	2	5	
Brain tumor	7				3	4	
Hepatic Ca.	5				5		
Ovarian Ca.	5				4	1	
Breast Ca.	4				1	3	
Head and Neck Ca.	3		1		2		1/ 3 (33.3)
Cervical Ca.	3			1	2		
Others	6				3	3	
Multiple myeloma	17		2	3	10	2	2/ 17 (11.8)
Malignant lymphoma	6	1		1	2	2	1/ 6 (16.7)
CLL	5		2	2		1	2/ 5 (40.0)
ATL	5				1	4	
Mycosis fungoides	3	1	1			1	1/ 3 (33.3)
Others	4			1	2	1	
Total	142	2	14	10	64	52	17/ 142 (12.0)

Results of local treatment are shown in Table 3.   CR was observed in cases with Bowen's disease, squamous cell carcinoma, Paget's disease, basal cell carcinoma, mycosis fungoides and malignant lymphoma.   Response rate including CR and PR cases was highest in basal cell carcinoma, followed by Bowen's disease and mycosis fungoides.

Side effects are shown in Table 4.   Fever was observed in most of the cases.   Chills, malaise, anorexia, lethargia, and leukopenia were main side effects.   In patients receiving local administration, similar side effects were observed although the incidence was lower than those in systemic administration. Anaphylactic reactions as well as antibodies against rIFN-γ or E. coil protein were not observed in 70 patients so far studied.

Table 3 Clinical Effect of Recombinant IFN-γ
(Ⅱ) Local Administration

Tumors	No. of Evaluable Patients	Response					Response Rate (%)
		CR	PR	MR	NC	PD	
Malignant melanoma	17		9	4	1	3	9/17 (52.9)
Bowen's disease	14	5	6	3			11/14 (78.6)
SCC	13	1	4	3	4	1	5/13 (38.5)
Mycosis fungoides	11	4	3	1	1	2	7/11 (63.6)
Paget's disease	9	1	4	2	1	1	5/ 9 (55.6)
Malignant lymphoma	7	3			1	3	3/ 7 (42.9)
BCC	6	1	5				6/ 6 (100.0)
Others	8	1	1	2		4	2/ 8 (25.0)
Total	85	16	32	15	8	14	48/25 (56.5)
Bladder Ca.	7		3		3	1	3/ 7 (42.9)
Hepatic Ca.	3				2	1	

Table 4 Toxicity of Recombinant IFN-γ

	No. of Patients and Percent of Total		
	Systemic ⟨171⟩*	Local ⟨96⟩*	Tocal ⟨267⟩*
Constitutional			
Fever 38℃ or over	152 (88.9)	61 (63.5)	213 (79.8)
Chills and shaking	82 (48.0)	16 (16.7)	98 (36.7)
Malaise	76 (44.4)	15 (15.6)	91 (34.1)
Gastrointestinal			
Anorexia	80 (46.8)	12 (12.5)	92 (34.5)
Nausea	30 (17.5)	4 (4.2)	34 (12.7)
Vomitting	14 (8.2)	3 (3.1)	17 (6.4)
Neurological			
Headache	6 (3.5)	3 (3.1)	9 (3.4)
Lethargy	4 (2.3)		4 (1.5)
Muscle and Joint			
Myalgia	4 (2.3)		4 (1.5)
Arthralgia	4 (2.3)		4 (1.5)
Hematological			
Leukopenia	63 (36.8)	17 (17.7)	80 (30.0)
Thrombocytopenia	21 (12.3)	1 (1.0)	22 (8.2)
Anemia	10 (5.8)		10 (3.7)
Hepatic enzyme elevation			
SGOT	36 (21.1)	6 (6.3)	42 (15.7)
SGPT	33 (19.3)	8 (8.3)	41 (15.4)
LDH	17 (9.9)	6 (6.3)	23 (8.6)
Renal			
BUN elevation	9 (5.3)		9 (3.4)
Elevation of serum Cr	5 (2.9)		5 (2.8)
Electrolites			
Decrease of serum Na	6 (3.5)		6 (2.2)

* Number of patients evaluable for toxicities

Ninety-two cases of advanced renal cell carcinoma were treated either by continuous administration of normal daily dose (52 cases) or by intermittent use of high daily dose (40 cases) of rIFN-γ. Sixty-one cases were eligible for evaluation. Thirty-three cases were treated by daily administration of $6-12 \times 10^6$ U/m^2 and other 28 cases were treated by the intermittent high dose as described in Materials and Methods. There was no significant difference between 2 groups in regard to sex ratio, age distribution, clinical stage and prior therapy. Only difference was the performance status. Eight cases of continuous group and 13 cases of intermittent group were in grade 0, respectively. Results of these two treatment protocols are compared in Table 5.

Table 5 Antitumor Effect of Recombinant IFN-γ

Response	No. of Patients	
	Interferon Therapy	
	Intermittent High Doses	Continuous Daily Administration
CR	1 ⎫ (21.4%)	0 ⎫ (9.1%)
PR	5 ⎭	3 ⎭
MR	0	1
NC	10	17
PD	12	12
Total	28	33

Response rate including CR and PR cases was significantly higher in intermittent group as compared to continuous group. Responders in continuous group were given more than 56 days, while those in intermittent group required over 16 days (4 courses) of administration to obtain the good response, suggesting a good response in the patients receiving the administration of longer period.

Percentages of patients showing side effects were higher in intermittent group than in continuous group in all items except fever.

DISCUSSION

The first trial of crude IFNs to cancer patients had been reported as early as 1960. However, extensive clinical trial of IFNs had been impossible until the development of bio-technology to produce rIFNs.

Among IFNs, anti-cancer activity was most extensively studied in IFN-α. This IFN has been reported to be highly active against hairy cell leukemia even in the presence of extensive disease, and effective on patients with chronic myelogenous leukemia, and favorable histology of non-Hodgkin lymphoma. In contrast to the hematologic malignancies, IFN-α had relatively little activity against most solid tumors. Antitumor activity has been reported in malignant melanoma, renal cell carcinoma and Kaposi's sarcoma.

Effectiveness of IFNs on renal cell carcinoma should be noted since this tumor is known to be refractory to various anti-cancer drugs. Response rate of renal cell carcinoma to IFN-α has been summarized as ranging from 0.8% to 20% by a recent review (4).

Twenty-one percent response rate of our patients to inter-mittent high dose treatment seemed to be the best results so far reported on IFNs treatment of renal cell carcinoma. Patients tolerated well to this high dose interferon therapy. Optimum dose for IFN treatment against various malignancies has not yet been settled. In regard to the renal cell car-cinoma, higher response rate by increasing the dose of IFN-α was reported (5). The results obtained in our study also indicated that the response of renal cancer to IFN-γ is dose responsive. It is considered, therefore, that in treating the patients with renal cell carcinoma, higher dose of IFNs could be much more effective.

Effect of local administration of IFN-α and β in cutaneous malignancies has been reported. Local injection of rIFN-α-2a was effective in reducing tumor size in cutaneous T cell lymphoma and mycosis fungoides. Results of our present studies showing the response rate of 52.9% in malignant melanoma, 63.0% in mycosis fungoides, 42.9% in malignant lymphoma were compatible with those of 50% in malignant melanoma by IFN-β and 36 ~ 55.6% in malignant melanoma and 71 ~ 100% in skin malignant lymphoma by IFN-α.

Our results of the high activity of rIFN-γ against basal cell carcinoma of skin which showed 100% (6/6) response rate are worthy of note. This result is considered to reflect the stronger inhibitory activity against squamous cell carcinoma cell line cells of IFN-γ than other IFNs which were observed in studies using cell culture system.

REFERENCES

1. Oldham RK. Biologicals and biological response modifiers:
    New approach to cancer treatment. Cancer Invest 1985;
    3:53 ~ 70

2. Michich E. Biological response modifiers: Their po-
    tential and limitations in cancer therapeutics. Cancer
    Invest 1985;3:71 ~ 83

3. Devos R, Cheroutre H, Fiers W et al. Molecular cloning
    of human immuneinterferon cDNA had its expression in
    eukaryotic cells. Nucl Acid Res 1982;10:2487

4. Neidhart JA. Interferon therapy for the treatment of
    renal cancer. Cancer 1986;57:1696 ~ 1699

5. Vugrin D, Hood L, Taylor W et al. Phase II study of
    human lymphoblastoid interferon in patients with advanced
    renal cell carcinoma. Cancer Treat Rep 1985;69:817 ~ 820

REFERENCES

INTERFERONS AND INTERFERON INACTIVATORS IN AIDS, ARC AND IN CANCER
PATIENTS

Kailash C. Chadha, Julian L. Ambrus and M.G. Ikossi*

Roswell Park Memorial Institute, Buffalo, NY, and *M.D. Anderson Hospital
and Tumor Institute, Houston, TX USA

INTRODUCTION
It is becoming increasingly evident that changes in various parameters of the interferon system and in the levels of prostaglandins and
cyclic nucleotides have a definite role during the health and during the
onset, progression and the final outcome of any disease process. In spite
of ever increasing excitement over the possible clinical application of
interferons, very little is known about the changes and sources of variations in the interferon production capabilities of different individuals.
Heterogeneity of human interferons is now well document from the earlier
work (1) as well as from more recent work based upon DNA recombinant technology (2-4). However, the clinical significance of such a heterogeneity
is not known. Cloned interferons used at present in the treatment of any
type of infections are a product of a single gene. It is quite possible
that expression of different interferon genes varies considerably in different types of infections and malignancies. If that is the case, it may
be necessary to use a custom-tailored interferon preparation for each type
of infection. Such an approach at best appears to be several years away.
In view of lack of such a knowledge, it seems appropriate that serious
attempts should be made towards better understanding of the host's own
interferon system during infection and ways and means of its manipulation
for maximal benefit. In this text, some attempts have been made in this
direction.

RESULTS AND DISCUSSION
For this text, the term 'interferon system' constitutes the determination of the following 3 parameters: (i) the ability of WBCs to synthesize interferon alpha (IFN-α), (ii) the presence or absence of circulating
serum interferons and (iii) the presence or absence of serum interferon
inactivator(s). In the future, this definition of 'interferon system' can
be expanded by including changes in various immune parameters that are
associated with different types of infections.

Interferon system in healthy adults
We have earlier reported (5) that WBCs from normal healthy individuals vary considerably in their ability to synthesize IFN-α when challenged with Sendai virus. They can be divided into two groups: low
producers ($<$ 2,000 IU of IFN-α/10^7 WBCs) and normal producers ($>$ 2,000 IU
of IFN-α/10^7 WBCs). In a normal population, about 5-10% of the individuals
seem to be the low producers. Normal healthy adults do not have any detectable levels of circulating serum interferons ($<$ 5 IU) or interferon
inactivators. It will not be surprising if, in the near future, further
research documents that the low producers are more prone to various types
of infections as compared to the normal producers.

Interferon system in cancer patients

As shown in Table 1, the ability of the WBCs of cancer patients to synthesize IFN-α is considerably reduced as compared to normal healthy adults. Circulating serum interferons were detected in 72% of patients with metastatic neoplastic disease (hepatoma, neuroblastoma, osteosarcoma, and carcinoma of colon, bile, rectum, stomach, etc.) (6). Interferon inactivators were detected in the plasma of 45% of all cancer patients studied (6). Upon successful surgical resection of the tumor, the ability of the WBCs of cancer patients to synthesize IFN-α returns to their normal levels and also serum interferons and interferon inactivators were not detectable (6). However, if the majority of the tumor is not removed, all 3 interferon parameters remained unchanged or even further deteriorated depending upon the nature of post-operative complications (6).

TABLE 1. HuIFN-α production by WBCs of patients with solid tumors

Name	Sex	Age	Diagnosis	HuIFN-α Titer IU/10^7 WBCs
JW	F	57	Hepatoma	39
AB	M	50	Osteosarcoma	50
EG	F	74	Gastric carcinoma	100
BS	F	64	Gastric carcinoma	1050
JB	F	65	Carcinoma of colon	400
LV	F	68	Carcinoma of bile duct	580
BH	F	67	Carcinoma of rectum	1250
BL	F	25	Islet cell carcinoma	980
JK	F	27	Neuroblastoma	1400
JH	M	70	Carcinoma of colon	840
LS	M	65	Carcinoma of stomach	2000
*Mean				790

The mean of HuIFN-α produced by 25 healthy adults was 5100 IU.

Interferon system in patients with AIDS and ARC

Patients diagnosed as having AIDS-related complex (ARC) had no interferon inactivators in their serum, had detectable serum interferon levels (Table 2) and in the majority of cases, their WBCs responded poorly to inducers of IFN-α. As the disease progressed from ARC to full blown AIDS, the response of their WBCs further deteriorated and interferon inactivators appeared in their blood (Table 2). Since no interferon inactivators were detected in ARC patients and in the serum of patients with AIDS-unrelated Kaposi's sarcoma, it seems appropriate to monitor the appearance of "interferon inactivators" as possible markers for the early AIDS diagnosis.

Modulation of interferon production capabilities of WBCs

Tumor cells produce several humoral factors like prostaglandins that we have shown inhibits IFN-α synthesis by the WBC of both low and normal producers (5). A significant enhancement in IFN-α yield was seen when WBCs of low producers were exposed individually to either pentoxifylline or indomethacin prior to or simultaneously during IFN-α induction by Sendai

TABLE 2. Interferon and interferon-inactivator levels in sera from patients with AIDS, ARC and AIDS unrelated Kaposi's sarcoma and healthy adults[1]

Diagnosis	Patient number	Serum IFN-Titer (IU)	IFN-α inactivator(s) titer[2]
AIDS[3]	1	< 5	> 40
	2	< 5	> 40
	50	25	7
	264	< 5	12
	118	< 5	9
	133	32	7
	281	5	8
	37	6	8
	207	13	14
	290	12	5
	265	10	> 40
	1921	18	14
	1922	12	10
	1923	24	8
	1924	21	20
	1925	25	30
ARC[4]	3	36	0
	1664	21	0
	1673	24	0
	1671	45	0
	1887	7	2
	1890	11	4
	1893	11	0
	1895	18	0
AIDS-unrelated Kaposi's sarcoma	4	< 5	0
	5	< 5	0
	6	< 5	0
	1663	7	0
	1684	3	0
	1685	7	0
	1691	< 5	0
	1692	< 5	0
	1681	16	0

[1] A total of 92 normal healthy individuals were tested. None had any interferon or interferon inactivator titers.

[2] Maximal serum dilution that completely eliminated antiviral activity of 25 IU of HuIFN-α.

[3] A total of 52 AIDS patients have so far been examined with identical results.

[4] A total of 20 ARC patients have so far been examined with identical results.

virus (Table 3). Even greater increase in IFN-α yield was seen when PBL of these donors were exposed concurrently to these two drugs (Table 3). However, no increase in IFN-α yield was seen when PBL of normal producers were treated identically. This suggests that perhaps in a clinical situation benefits of interferon treatment can be maximized by combining interferon administration with either of these two drugs in order to minimize the undesirable effects of excessive prostaglandins produced by the tumor cells.

In a separate study, we have shown that treatment of tumor cells with interferon enhances intracellular levels of c-AMP and prostaglandins. Treatment of these cells with indomehtacin significantly reduces the levels of both c-AMP and prostaglandins without having any effect on antiviral and antiproliferative activity of interferons (submitted for publication).

SUMMARY
WBCs of cancer patients have depressed capability to synthesize IFN-α. This reduced ability is restored to normal levels upon complete surgical resection of the neoplasm. Cancer patients often have detectable

TABLE 3. Effect of pentoxifylline and Indomethacin on HuIFN-α production by WBCs of healthy adults

		HuIFN-α yield[1] (IU)			
		Pentoxifylline Conc. (M)			
Donor	Control	10^{-4}	10^{-5}	10^{-6}	10^{-7}
Low producers[2]	235	3,039	1,840	1,103	820
Low producers[3]	400	3,984	2,598	2,179	470
Normal producers[4]	4,650	4,775	5,300	4,710	4,100
		Indomethacin Conc. (M)			
	Control	10^{-5}	10^{-6}	10^{-7}	10^{-8}
Low producers[2]	350	4,550	3,850	2,450	1,400
Low producers[3]	255	3,700	2,300	2,040	780
Normal producers[4]	7,500	8,100	7,350	9,550	8,850
	Control	Pentoxifylline[5] + Indomethacin			
Low producers[2]	580	10,600			
Normal producers[4]	4,500	8,000			

[1] Represents average IFN-α yield from 12 donors in each category.
[2,4] Drug was added at the time of interferon induction.
[3] Drug was added 12 hr prior to interferon induction.
[5] Both drugs were added at the time of interferon induction each at concentrations of 10^{-5}M.

levels of serum interferons and interferon inactivators. Patients with ARC have no detectable serum interferon inactivators. As the disease progresses to full blown AIDS, such inactivators appear in their blood. Monitoring of serum interferon inactivators perhaps can be a useful marker for early diagnosis of AIDS. In in vitro studies, the depressed ability of WBCs of low producers can be enhanced by drugs like indomethacin and pentoxifylline when added to IFN-production media. These drugs are known to suppress intracellular levels of c-AMP and prostaglandins without having any detrimental effect on antiviral or antiproliferative activity of interferons.

REFERENCES

1. Grob PM and Chadha KC: Biochemistry 18:5782, 1979.
2. Nagata S, Mantei N and Weissman C: Nature 287:401, 1981.
3. Goeddel DV, Leung DW, Dull TJ, Gross M, Lawn RM, McCandliss R, Seeburg PH, Ullrich A, Yelverton E and Gray DW: Nature 290:20, 1981.
4. Sehgal PB and Sagar AD: Nature 288:95, 1980.
5. Ikossi MG, Ambrus JL and Chadha KC: Res. Commun. Chem. Pathol. Pharm., in press.
6. Ikossi-O'Connor MG and Chadha KC: J. Surg. Onc. 26:27, 1984.

OVERVIEW:CLINICAL TRIAL WITH NATURAL AND RECOMBINANT INTERFERONS IN CHINA

Hou Yun-te
(Institute of Virology, Chinese Academy of Preventive Medicine, Beijing)

Interferons are a family of proteins which exhibit a number of biological activities such as antiviral, antitumoral and immunomodulatory activities. The antitumoral and antiviral potences of human interferon preparations have been evaluated in patients with malignancies and viral infections since 1970's. But clinical trials for defining the effect of interferons had been restricted by insufficiency of materials. Recent progress of manufacture of various interferons including E.coli produced recombinant interferons has brought enough supplies to investigate their clinical effects. In China, before 1978 most preparations used in clinical trial were crude interferons made from umbilical cord blood leucocytes after induction with New Castle disease virus (NDV) F1 strain. It was demonstrated that the interferon inducing ability of leucocytes from umbilical cord blood is at least 4 times higher than that from adult. Since 1978 there are three types of interferon used in clinical trial in China, namely, human leucocyte, Namalva cell and recombinant interferons. The partial purified leucocyte interferon was prepared from buffy coat as a by-product using modified Cantell method (NDV F1 strain as an inducer instead of Sendai virus) by Chen Du Institute of Biological Product, Guang Chou, Beijing, Shangshai Blood Centers and other institutions. The purity is about 1% ($>10^6$ U per mg protein). The Namalva cell interferon was prepared according to the Finter method and partially purified with a series of chromatography procedures by Shangshai Second Military Medical College, Beijing and Shangshai Institutes of Biological Product. The purity is more than 1% (10^6-10^7 U per mg protein). The human recombinant α A and α D interferons were developed by the Institute of Virology, Beijing. The purity reaches 95%.

About 50 clinical units in China have engaged in clinical trial with different interferon preparations. They have been encouraged by a number of successful results.

Undifferentiated carcinomas of the nasopharynx (NPC) is very common in Southern China; the incidence of NPC reaches 20-30 per 100,000 inhabitants per year. There are more than 19,000 annual cases in China. This cancer is closely associated, whereever observed in the world, with the ubiquitous Epstein Barr herpes virus. Such a virus-associated tumor is an ideal model to test both the antiviral and antitumoral effects of interferon.

In 1980 we adopted human leucocyte IFN (LeIFN) prepared by the Institute of Virology, Chinese Academy of Preventive Medicine, to treat an early case with NPC, female, 47 yr, who had a new growth of 0.5 cm in diameter in her nasopharyngeal region. Biopsy on it revealed a low differentiated squamous carcinoma. We began to treat with LeIFN in dose of $2x10^6$ U, i.m. q.d.. Three days later, the dosage was increased to $4x10^6$ U, i.m.q.d.. Simultaneously $9x10^5$ U was dropped intranasally for local treatment.

Maximal serum IFN titer (124 U per ml) was detected 3hr after intramuscular injection, and low level (26 U per ml) remained 24 hr later. Three weeks later, the lesion regressed to 0.3 cm in diameter and EBV VCA-IgA antibody titer dropped from 1:40 down to 1:10. However, in the fifth week after the initial tratment nasopharyngoscopy revealed an another new growth besides the primary lesion. We continued the treatment till the fifty first day. No regression was detected. We had to shift the treatment with Radio Cobalt 60.

In 1981-1982 an expanded clinical trial had been conducted in China by Dr. Zeng cooperated with Dr. de Thé and Dr. Cantell. Thirty NPC patients (mostly stage II of Ho) were selected and randomized for either IFN treatment followed by Cobalt 60 therapy (IFN group) or radiation therapy alone (radiation group). In the IFN group (15 cases) Cantell's interferon was used at a daily dose of 3×10^6 U for 5 weeks intramuscularly, then 3 times a week for another 5 weeks; after a week of rest, followed by Cobalt 60 therapy administered as 7000 rads to the primary NP tumor, and 6000 rads to the cervical areas with or without lymph node. In the radiation group (15 cases) patients were treated by Cobalt 60 alone. In the IFN group, IFN therapy was discontinued in two cases because of fever or rapid tumor progression. Among the 13 patients who completed their IFN treatment, 5 cases showed noticeable regression of their primary NP tumor for some weeks, 4 cases exhibited no change in the primary tumor mass and 4 cases showed progression. Metastatic cervical lymph nodes were present in 7 patients. These decreased in size in 3 cases, remained stable in one and increased in 3. Comparative evaluation of biopsies taken before and after interferon treatment revealed that after IFN therapy some degenerative process of tumor cells took place in 5 cases and necrosis in I patient. The pathological changes, however, did not correspond to any particular clinical evaluation. IgA-VCA antibodies in serum and saliva were shown to be slightly decreased. These results suggest that human LeIFN exhibits some antitumor effect on NPC.

Partial purified human LeIFN was also applied for treatment of malignant tumors in addition to the NPC, including bile duct carcinoma, adenocarcinoma with bone metastasis, multiple myeloma, lung cancer, primary hepatoma, melanoma, osteosarcoma, nose squamous carcinoma, breast cancer, glioblastoma and so on. Though the preliminary results provide hopeful perspective, any conclusion can not be made because of limited cases tested. As an example, Shi Pengda et al treated 3 cases with primary liver cancer in late stage with i.m. injection of 10^6 U of LeIFN and AraC(100-500 mg) per day for 90 days. The control group consisting of 3 cases received chemotherapy alone. The result showed that IFN seems to improve the general condition of patients, reduce the ascites, restore the liver function, inhibit the rapid growth of the tumor. None of these 3 cases died while the control cases all died during the observation period.

The prevention and treatment of viral hepatitis seems to be the most important in public health in China. The HBV carrier rate is estimated at 8.83% of the population. It means there are about 100 million of cases in China. One believes that HBV is closely associated with hepatoma; there are more than 100,000 annual cases of liver cancer in China.

In 1980 Tong Kuitang et al reported a case of chronic active hepatitis B treated with $0.5-1.0 \times 10^6$ U of LeIFN per day i.m. injection for 27 days. The serum DNA-P dropped from 209 cpm down to 46 cpm; and the liver funtion and the histological finding improved. The DNA-P, however, rose again after termination of the treatment. Feng Tongzhi et al also reported a case of chronic hepatitis B treated with 10^6 U of IFN i.m. injection every

other day for 60 times. The titer of serum HBsAg dropped from 1:4096 down
to 1:128. In addition the swelling liver reduced as well. Ding Zhifeng et
al observed a case of fulminant hepatitis treated with lymphoblastoid IFN,
3×10^6 U per day i.m. Though histological examination showed significant
liver cell regeneration, the patient died due to the kidney failure after
all.

Eight clinical units in different areas of China reported encouraging
results obtained from treatment of total 149 cases with chronic hepatitis
with small dosage of LeIFN ($3-10 \times 10^4$ U per injection), i.m. every other
day for 2-3 months. The titers of HBsAg or HBeAg in IFN group after treatment significantly dropped down as compared with that in control group.

Epidemic hemorrhagic fever is widely prevalent in most areas of China.
There are more than 80,000 annual cases in China. There is a hypothesis of
'immuno-pathogenesis' based on the detection of circulating immune complex
and decrease of the serum C_3 content in patients to explain severe renal
syndrome of this disease. This virus is also shown to be sensitive to the
interferon action in vitro. Authors from three clinical units of different
areas of China reported 88 cases of acute hemorrhagic fever treated with
LeIFN. Some beneficial effects were observed: circulating immune complex
in IFN group was markedly dropped down after treatment as compared with
that in control group. The severe complications in IFN group were also
significantly less than that in control group.

Feg Tongzhi et al reported that 10 cases with viral pneumonia caused
by adenovirus type 3 or 7 were treated with Le IFN for 4-6 days and compared with the control group. The duration for restoration to normal body
temperature averaged 3.75 and 5.0 days, respectively; and for disappearence
of respiratory disturbance, they were 3.8 and 6.5 days, respectively;
rales disappeared until 7.6 and 12.5 days, respectively. Similar results
were obtained by Liang Wenfei et al and Jiang Weishi et al with 64 cases.

Chronic cervicitis, especially its erosive type, is a very common
disease in China; it occurs about 50% among married women. Human papilloma
virus (HPV) and herpes simplex virus (HSV) infections may be related to
its pathogenesis. A double blind controlled study on the efficacy of a
human rIFNαD in treatment of 271 cases with typical cervical erosion was
reported by Dr. Qian Zhiwei et al in this Proceedings. Results obtained
from clinical observation and detection of HPV and HSV indicate that 93.8%
of cases appeared clinically improved and 60% was markedly improved.
Similar results were also obtained by Fang Yilan et al with 263 cases, Gao
Jun et al with 55 cases and Qian Xingguang et al with 76 cases using LeIFN.

Wu Zhijie et al reported 30 cases (31 eyes) with herpetic keratitis
treated with an ointment containing 10^6 U of LeIFN per gram, 3-5 times per
day. Among them 23 eyes were cured for 6.6-26 days as compared with 29.6-
105 days when AraC was used. Similar results were also obtained by Jiang
Wei et al with 39 cases (45 eyes), Zheng Zhenlu et al with 93 cases (95
eyes), Gau Guirong et al with 13 cases (15 eyes), Sun Xiaoqin et al with
33 cases (33 eyes) and Qian Xingguang et al with 49 cases. We demonstrated that Astragalus membranaceus-one of the traditional Chinese materiae
medica in frequent use as toinics-is able to stimulate IFN production not
only in animal model but also in human beings. Placebo controlled field
studies were carried out among 1,137 volunteers. Results obtained indicate
that the incidence of common cold in the IFN treated group was lower than
that of the placebo group, but the efficacy index appeared to be rather
low. The efficacy index was increased in the combined IFN and A.membrana-
ceus treated group as compared with the group treated with IFN alone (2.0-
4.0 against 1.4-1.6). An ointment containing IFN and A.membranaceus also

432

appeared effective in preventing common cold. Results obtained among 587 volunteers indicated that the efficacy index was about 1.9 (P < 0.05). The combined treatment resulted not only in decreased common cold incidence but also in shortening the course of illness; the average course of illness of the patients in the combined treatment group was 2.6 days as compared with 4.6 days in the control group (P<0.05).

The clinical trial with interferon in China as summarized in Tab. 1 and 2.

Tab. 1 Clinical application of IFN in treatment of malignant tumor in China

Malignant tumor	No. tested	Response			
		CR	PR	ST	PG
Nasopharyngeal carcinoma	14	0	6	4	4
Glioblastoma	2	0	0	1	1
Osteosarcoma	2	0	2	0	0
Nose squamous carcinoma	1	0	1	0	0
Breast cancer	6	0	4	0	2
Melanoma	1	0	0	0	1
Hepatoma	3	0	3	0	0
Bile duct carcinoma	1	0	1	0	0
Adenocarcinoma	1	0	1	0	0
Multiple myeloma	1	0	1	0	0
Lung cancer	1	0	1	0	0

CR: complete remission; PR: partial remission;

ST: stable; PG: progression.

Tab. 2 Clinical application of IFN in treatment of viral diseases in
China

Viral disease	No. tested	Response
Chromic hepatitis B	152	Virus markers temporally dropped
Viral pneumonia	74	Clinically improved
Hemorrhagic fever	88	Circulating immune complex dropped down; severe complication decreased
Chronic cervicitis	665	More than 90% of cases clinically improved
Herpetic keratitis	257	More than 80% of cases cured
Common cold	1724	incidence decreased
Japanese B encephalitis	10	No effect
Rotovirus diarrhea	30	Clinically improved
Herpes zoster	100	Clinically improved
Flat wart	50	Clinically improved

INTERFERON PRODUCING CAPACITY IN PATIENTS WITH LIVER DISEASE, DIABETES MELLITUS, CHRONIC DISEASES AND MALIGNANT DISEASES

T.KISHIDA[1,2], M.KITA[1,2], M.YAMAJI[2], E.KANETO[3] and T.TÉRAMATSU[4]
[1]Institut Pasteur de Kyoto, Kyoto, Japan   [2]Kobe Tokiwa
College Research Institute, Kobe, Japan   [3]Kure National
Hospital, Hiroshima,Japan [4]National Hyogo Central Hospital,
Hyogo, Japan

INTRODUCTION

In the past few years we have studied IFN-α producing
capacity in 440 healthy adults and in 110 patients with liver
disease, 86 patients with diabetes mellitus, 54 patients with
malignant diseases, 20 patients with chronic renal failure
and 378 patients with chronic respiratory disease, while we
have examined IFN-γ producing capacity in part of the above
cases. In almost all patients IFN producing capacity showed a
statistically significant decrease: especially titers of
patients with any kind of malignant diseases (tumor, Cancer,
etc.) were low.

The aims of this investigation are:
1) to study the significance of the presence of IFN in
   organisms.
2) to search for the practical clinical use and posology of
   IFN which must be different from the conventional
   chemotherapy and antibiotherapy.

1) is related to my Bio-stabilizer Hypothesis of IFN since
1971 that the normal development, differentiation and the
ultimate health of organisms are always maintained by very
small amount of IFN. Insufficient IFN or no presence of it
leads to fearful results as reported by Gresser et al. after
the administration of anti-IFN, and the subsequent transplant
of the tumor to the nude mouse when IFN-NK system was not
working(1). Minato et al. also described that IFN was deeply

involved in the tumor development and the destiny of the
animal (enhancement of metastasis and death)(2).

As for 2), the investigation was started after we analogiz-
ed from the animal experiments in 1) that even a healthy looking human may have congenital deficiency of IFN producing
capacity or may experience a transient decrease of it due to
various reasons. It would be very important for us to know
what kind of disease would occur as a result of such decrease.
And if we can prevent disease by maintaining the health of a
healthy looking person with low titers, that is to say, by
preserving high levels of IFN-α and IFN-γ in him, the administration of IFN or IFN inducer seems promising. In that case
the clinical use of IFN system must be regarded in the new
concept that IFN is a "Biological Responce Modifier" to be
applied widely in the future. The smaller dose than in the
case of conventional treatment will be enough and the long-
term (one to two years or more) administrations of once or
twice a week will be quite meaningful.

Now the question arises whether a human is born with the
composition which is prone to cancer or chronic virus disease.
Our limited experience has made it evident that patients of
such disease clearly have low titers, however, much longer
time is needed to determine whether this is the cause or the
result of the onset.

MATERIALS AND METHODS
IFN induction

Heparinized blood was taken from healthy donors and patients
and divided into two tubes. Sendai virus was added to one tube
as IFN-α inducer (500HA/ml). Con.A was added to the other tube
as IFN-γ inducer (10μg/ml) after it was diluted four times
with Eagle MEM culture medium. Both tubes were then incubated
at 37°C for 20 hours. The tubes were centrifuged and the
supernatants were stored at -70°C until assayed for IFN.
IFN assay

The IFN activities in culture supernatnats were determined
by a conventional cytopathic effect inhibition assay perfomed
in microtiter plates with FL or WISH cells as the indicator

Table 1.   Monthly variations in HuIFN-α producing
           capacity of normal healthy donors.

Date	IFN titer (IU/ml)			
	M.K.	M.Y.	H.A.	O.Y.
84' 2	7750	3780	5500	5610
3	7300	2380	2150	7210
4	9730	2580	6110	6130
5	8400	3040	6810	8460
6	7740	4120	9120	6820
7	9910	8800	7150	9390
8	5750	5900	8600	6940
9	6220	5550	6810	8520
10	7320	4290	6610	9430
11	4510	5610	6750	5500
12	5990	4320	4390	5210
85' 1	7850	4340	7250	8110
2	5810	3630	5360	7130
3	7270	4630	5400	9310
Mean	7250	4500	6290	7410
±SD	±1520	±1630	±1730	±1480
C.V.	(21%)	(36%)	(28%)	(20%)

cells and vesicular stomatitis virus as the challenge virus(3).

The international reference HuIFN-α (MRC 69/19) and HuIFN-γ
(Gg 23-901-530) were used to calibrate the assay, and the IFN
activities were expressed as international units per ml (IU/ml).

RESULTS

There was not much difference in age, sex and seasons of
IFN-α measurements among healthy donors as I reported in the
previous meeting. The mean value of healthy donors was 5,000
± 3,000 units, and there were only a few cases with less than
2,000 units, as shown in table 1 and 2. On the other hand,

Table 2.   HuIFN-α producing capacity of normal
            healthy donors.

| Age | IFN titer (IU/ml) | | | |
	Male		Female	
10-19	5893 ± 1346	(n=11)	4718 ± 1143	(n= 7)
20-29	5058 ± 2386	(n=18)	4969 ± 1639	(n= 9)
30-39	6275 ± 2663	(n=19)	4002 ± 1916	(n=10)
40-49	5110 ± 2712	(n=43)	4834 ± 1699	(n=21)
50-59	4763 ± 2366	(n=68)	3883 ± 2370	(n=33)
60-69	5367 ± 3280	(n=55)	5320 ± 2185	(n=32)
70-79	4165 ± 2077	(n=37)	4790 ± 2173	(n=39)
80-89	4643 ± 2338	(n= 8)	4139 ± 2980	(n=11)
Total	4938 ± 3127	(n=267)	4808 ± 2989	(n=173)

Table 3.   HuIFN-γ producing capacity of normal
            healthy donors.

| Age | IFN titer (IU/ml) | | | |
	Male		Female	
10-19	ND		ND	
20-29	45 ± 23	(n= 7)	70 ± 66	(n=12)
30-39	61 ± 37	(n=19)	27 ± 17	(n= 5)
40-49	40 ± 25	(n=43)	31 ± 21	(n=18)
50-59	27 ± 18	(n=29)	25 ± 13	(n=16)
60-69	43 ± 32	(n= 5)	25 ± 16	(n= 2)
70-79	33 ± 22	(n= 4)	ND	
Total	45 ± 27	(n=94)	34 ± 31	(n=52)

patients with liver disease such as chronic hepatitis B and
liver cirrhosis possessed medium levels of IFN-α which was
around 3,000±1,500 units. Liver cancer patients had titers
around 1,400±1,200 units and the ones with gastric, pulmonary
and other types of cancer showed titers less than 1,000 units.
Among patients with other chronic diseases, those with diabetes
mellitus and renal failure had low titers of 2,500±1,500,
suggesting their high susceptibility toward cancer and infections diseases (Fig.1).

It is already clear that IFN-α of non-cancerous chronic
diseases such as hepatitis, liver cirrhosis, diabetes and
renal failure decreases significantly to the moderate level.
We are confident that patients with malignant tumors will
show significantly lower titers compared to healthy donors.
From now on we call the decrease of IFN-α producing capacity
"IFN deficiency", and the patients "IFN deficient syndrome"
temporarily.

There was no difference in IFN-γ value by age and sex in
healthy donors. As shown in Table 3 and Fig.1, its value was
around 40 units while patients with chronic renal failure and
liver cirrhosis gave less than 10 units and around 20 units
respectively, which were significantly lower than those of
healthy donors. On the other hand, patients with chronic
hepatitis B and diabetes did not present much difference
compared to healthy donores. Therefore IFN-γ deficiency in
non-cancerous chronic diseases does not correspond to IFN-α
deficiency, and this may suggest some pathological significance. As shown in Fig.1, it is interesting to note both IFN-γ
and IFN-α in liver cancer patients were low (α=1,500,
γ=13).

DISCUSSION AND CONCLUSION
Our investigation has demonstrated that IFN-α and IFN-γ
producing capacities of the total component of blood of healthy
donors remained on a high level regardless of age and sex.
Patients with various viral and non-viral chronic diseases
presented significantly lower moderate titers and patients
with various kinds of cancer gave extremely low values. It is

not known whether IFN producing capacity of such patients was
low before the onset of disease.

Recently we examined a few cases which showed less than
2,000 units of IFN-α producing capacity before the onset or in
the early stage of cancer. We followed one case for a year,
measuring his titers every two months, and found that cancer
was detected when the titer decreased to less than 600 units.
The patient underwent a surgery successfully and his IFN-α
value was returning to that of the healthy adult(unpublished
data). If we could collect more data, we may be able to find
individuals whose susceptibility to virus disease and cancer
is congenitally high, and the detection of cancer in the early

	4887	3191	3364	1442	2447	2218
HuIFN-α	±3072	±1671	±1516	±1229	±1450	±861
HuIFN-γ	41±29	42±26	20±15	13±16	43±21	<10

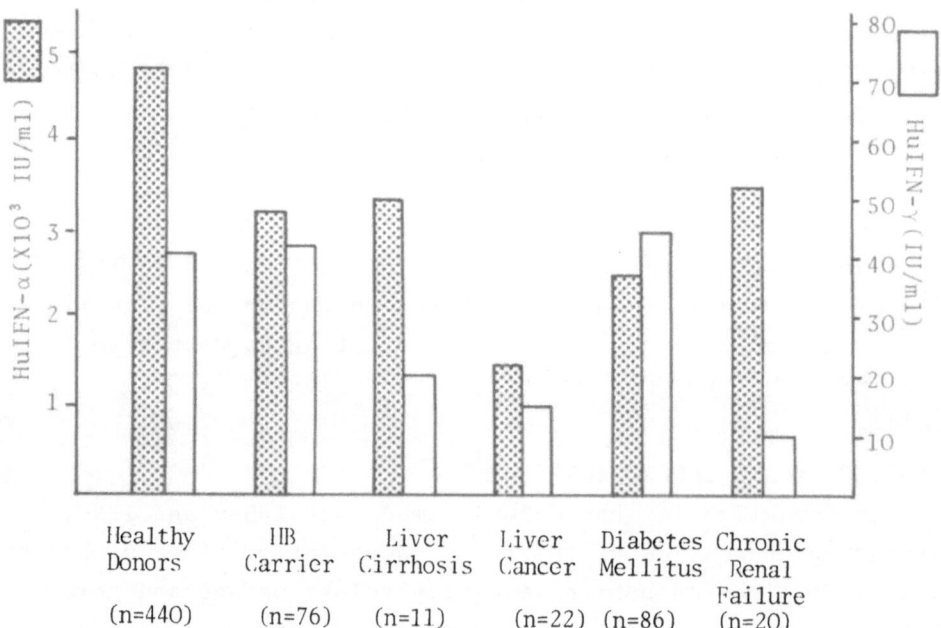

Fig. 1.   Interferon producing capacity in patients with liver diseases,
diabetes mellitus, chronic diseases and malignant disease.

stage of development will be feasible.

We also have studies in progress concerning the relationship between IFN-α and IFN-γ producing capacities and the
natural killer cell activity. So for there is a certain kind
of parallel relationship between IFN-α producing capacity and
NK activity, however, no such link has been found with IFN-γ
producing capacity. The result of this investigation will be
published in the near future after we have collected more
cases.

ACKNOWLEGEMENT

The part of this study was supported by grants from the
Ministry of Health and Welfare through "Cancer Control
Strategy Program". We thank HAYASHIBARA Biochemical Laboratories Incorporated for providing Sendai virus, I. DANJOH for
excellent technical assistance, and Y. KITA for preparation
of the manuscript.

REFERENCES

1. Gresser,I., Tovey,M.G., Bandu,M.T., Maury,C. and Brouty-
   Boye,D. (1976) Role of interferon in the pathogenesis of
   virus diseases in mice as demonstrated by the use of anti-
   interferon serum. I.Rapid evolution of EMC virus infection.
   J. Exp. Med. 144, 1305-1315.

2. Reid,L., Minato,N., Gresser,I., Kadish,A. and Bloom,B.R.
   (1981) Influence of anti-mouse interferon serum on the
   growth and metastasis of tumor cells persistently infected
   with virus and of human prostatic tumors in athymic nude
   mice. Proc. Natl. Acad. Sci. U.S.A. 78, 1171.

3. Imanishi,J., Hoshino,S., Hoshino,A., Oku,T., Kita,M. and
   Kishida,T. (1981) New simple dye-uptake assay for interferon. Biken J. 24, 103-108.

EARLY TREATMENT OF THE COMMON COLD WITH INTRANASAL INTER-
FERON ALPHA-2A (Ro 22-8181): A PRELIMINARY PILOT STUDY

R. Berger[1], M. Just[1], O. Ruuskanen[2], M. Lüdin[1], S.N. Linder-
Ciccolunghi[2]
[1]University Children's Hospital, CH-4058 Basle and [2]Clinical Research
Department, F. Hoffmann-La Roche and Co. Ltd., Switzerland

INTRODUCTION
It is well established that intranasal IFN Alpha-2a is effective in
preventing experimental and natural rhinovirus-induced common cold
(1-3). Seven-day contact prophylaxis is well tolerated and approximately
80% of rhinovirus respiratory virus infections are prevented (4, 5). There
is little information about therapeutic effect of intranasal interferon in
symptomatic patients. Hayden and Gwaltney (6) found only modest effects
on nasal symptoms in experimentally induced rhinovirus infection when
treatment was initiated 28 hours after the virus inoculation.
    In this double-blind, placebo controlled study, we examined the
tolerance and efficacy of early treatment of natural common colds with
intranasal IFN Alpha-2a.

PATIENTS AND METHODS
Study Population
    From November 15, 1985 to February 3, 1986, fifty adults, belonging
to the staff of University Children's Hospital, Basel were admitted to the
study. The patients were asked to enter the study when they felt that
they were developing respiratory tract infection and had experienced at
least one of the following symptoms: sore throat, runny nose, stuffy
nose, cough or sneezing, for less than 24 hours. There were 36 females
and 14 males.
    A physical examination was performed at the pre-study visit. Blood
was taken for haematologic tests (total white blood cell count (WBC) and
differential and erythrocyte sedimentation rate (ESR)). Nasopharyngeal
mucus was aspirated for rapid diagnosis and virus isolation. The patients
received daily personal symptom cards (PSC) to record the signs and
symptoms of the disease twice daily (morning and evening) for the following 5 days. The patients were advised to score their symptoms as
follows: 0 = no symptoms, 1 = mild symptoms, 2 = moderate symptoms, 3 =
severe symptoms. The participants were not allowed to use nasal decon-
gestants or prostaglandin synthetase inhibitors (e.g. aspirin) during the
observation period.
    Re-examinations were performed 6 and 21 days after the first visit. On
day 21 blood was taken for virus serology. Seven patients (4 from IFN
Alpha-2a group and 3 from placebo group) were excluded from the
efficacy analysis due to protocol violations (4 for incomplete or incorrect
completion of PSC, 2 for having treatment assessment performed 3 days
after treatment start and one for starting treatment 2 days after pre-trial
assessment).

Medication and randomisation

By computer-generated code, the patients were randomly assigned to receive either recombinant IFN Alpha-2a or placebo. Patients were instructed to use the nasal spray 4 times daily (2 sprays into each nostril at approx. 6 hour intervals). Each spray contained 0.72 million IU of IFN Alpha-2a in 0.09 ml volume or an equivalent volume of placebo solution. The total daily dose of interferon was 12 million IU. The effect of medication was considered to be evaluable if the patient had used the drug for at least 4 of the 5 days prescribed. The compliance was checked by measuring the amount of liquid in the returned spray bottles.

Virologic studies

Nasopharyngeal secretion specimens (usually 0.5-2.0 ml) were collected by suction through the nostrils with a disposable mucus extractor (Vygon, Ecouen, France). The specimens were stored at -70°C. Rhinovirus isolation (cytopathic effect) was done on HeLa Ohio and MRC-5 cell lines. A double sandwich ELISA (7) was done from culture media to detect adenovirus, influenza A and B virus, parainfluenza virus type 1, 2, and 3 and respiratory syncytial virus antigens. In addition, the original mucus specimens were tested in parallel for these viruses by analogous enzyme immunoassays. Paired blood samples collected on days 1 and 21 were tested for viral antibodies using routine complement-fixation tests.

Antibody to IFN Alpha-2a

The paired serum samples taken at days 1 and 21 were tested by radio-immunoassay for the presence of specific antibodies to IFN Alpha-2a (8).

Statistics

Statistical analysis of total symptom scores was performed using an analysis of covariance with baseline score as covariate. Analysis of overall efficacy assessment (patients' judgement) and tolerability were carried out using Fisher's exact test (2-sided).

RESULTS

Forty-six patients entered the study within 24 hours after the onset of symptoms and signs of acute respiratory tract infection. Four patients entered within 48 hours. The major symptoms reported were stuffy nose (75%), sore throat (62%), and runny noise (36%). Fever was recorded in only one patient. The viral etiology of the acute respiratory tract infection could be confirmed in 37 of 50 patients (74%). Sixteen patients had rhinovirus infection, 8 parainfluenza virus type 1 or 2, 8 influenza A or B virus and 5 respiratory syncytial virus infection.

The overall efficacy assessment by the 43 evaluable patients on day 5 is shown in Table 1.

Forty-eight percent of the patients in interferon group and 8% in the placebo group found the trial medication to be "clearly" effective on day 5 of the study.

The median symptom score for all patients at baseline was 6.0 in both groups. The corresponding numbers at the end of the treatment were 1.0 in the interferon group and 3.0 in the placebo group. There was no significant difference between interferon and placebo in all patients or in patients with laboratory verified rhinovirus infection. The total score summed for all 10 timepoints (5 days) showed superior efficacy in the interferon group (median score 33.0, range 14-85) compared with the total

TABLE 1. Overall efficacy assessment of patients

Group	N	Efficacy Rating		
		Clearly No. (%)	Yes, a little No. (%)	None No. (%)
All patients				
Interferon	21	10 (48)	4 (19)	7 (33)
Placebo	22	4 (18)	11 (50)	7 (32)
Rhinovirus				
Interferon	9	5 (56)	1 (11)	3 (33)
Placebo	4	0 (00)	2 (50)	2 (50)
Non-rhinovirus				
Interferon	12	5 (42)	3 (25)	4 (33)
Placebo	18	4 (22)	9 (50)	5 (28)

score in the placebo group (median score 52.0, range 13-175). However, when the sum of the scores over the 10 timepoints was adjusted for the baseline, this difference was not retained (7.2 in the placebo group and 6.9 in the interferon group) ($p \approx .45$). The imbalance in the size of the rhinovirus subgroups invalidated the comparison between the placebo and interferon groups. However, in the interferon group the rhinovirus subgroup (n = 9) showed a lower median symptom score (24.0) compared with the non-rhinovirus subgroup (n = 12) median symptom score (53.5).

Table 2 shows the number of patients with symptom score decreases by $\geq 50\%$ on days 3 and 5. At the end of the treatment 76% of the patients in the interferon group had a decrease in total symptom score by $\geq 50\%$ as against 52% in the placebo group. The difference was not statistically significant.

TABLE 2. Score reduction $\geq 50\%$ on days 3 and 5

Group	N	Day 3		Day 5	
		n	%	n	%
All patients					
Interferon	21	9	(43)	16	(76)
Placebo	22	8	(36)	12	(55)
Rhinovirus					
Interferon	9	6	(67)	7	(78)
Placebo	4	1	(25)	1	(25)
Non-rhinovirus					
Interferon	12	3	(25)	9	(75)
Placebo	18	7	(39)	11	(61)

There were no differences in the numbers of patients reporting need for acetaminophen use, sleep disturbances or absences from work (data not shown). Bacterial type complications occurred in 3 patients in each treatment group (5 cases of bronchitis and one case of acute otitis media).

There were no withdrawals from the study due to treatment intolerance. Overall, one or more adverse effects were detected in 32% of placebo users and 52% of interferon users (p _ .252). Intranasal adverse effects (Table 3) were the only clinically relevant adverse effects reported i.e. endonasal erosions, nosebleed or blood tinged mucus. None of the adverse effects were severe but 5/6 moderate adverse effects were in the active treatment group.

No patients developed antibodies.

TABLE 3. Adverse experiences during 5 day intranasal interferon or placebo treatment

	Interferon n = 25	Placebo n = 25
Blood-tinged nasal mucus	20%	20%
Nosebleed	12%	4%
Endonasal erosions	12%	4%
Nasal mucosal dryness	12%	12%
Total proportion of patients with one or more intranasal adverse effects	44%	32%

DISCUSSION

Hayden et al (5) and Douglas et al (4) have shown that daily post-contact prophylactic administration of 5 million IU once a day for 7 days prevents common cold in 39 and 41% of contacts compared with the placebo. The efficacies in preventing rhinovirus colds were 78% and 79% respectively. A recent study of Monto et al (9) reported prophylactic efficacy against rhinovirus colds to be 76%. They used 3 million IU daily dose given in two doses over a four week period. In these three studies no prophylactic efficacy was found against parainfluenza, respiratory syncytial or coronavirus infections.

In general short-term prophylactic intranasal interferon is well tolerated. Blood-tinged mucus, nosebleed or nasal mucosal erosions occur in 10-20% of the patients but only require premature discontinuation of therapy in rare cases when treatment is continued for more than one week. Thus rhinovirus infections can be effectively prevented with intranasal interferon but there is no efficacy against other respiratory viruses.

This study demonstrates that early treatment of common cold with 12 million IU daily intranasal IFN Alpha-2a has no clinically relevant effect on the severity of the signs and symptoms of the illness. However, 46% of the patients in the interferon group considered the treatment clearly effective compared to 19% in the placebo group. In the rhinovirus cold sub-group the corresponding numbers were 56% and 0%. It may be that the scoring system used in the study is not sensitive enough to detect the obvious modest therapeutic effect of intranasal interferon. Our findings, however, agree with those of Hayden and Gwaltney (6) who recorded only minimal effect of interferon treatment on the symptoms of established experimental rhinovirus infection.

This study was conducted during the late fall season when many different respiratory viruses exist in the community. We detected the viral etiology in 74% of cases. This exceptionally high number is probably due to the fact that samples were collected during the first 24-48 hours of the illness which is the optimal time. No transport media were used for virus isolation, because samples were collected in the laboratory. Furthermore, new sensitive techniques were used to detect viral antigens in the naso-pharyngeal mucus and in the cell culture media. Although the study started in November, rhinoviruses were the most common viruses found (32%). In addition, parainfluenza virus type 1 and 2, influenza A and B and respiratory syncytial viruses were found. These infections could not be clinically differentiated from rhinovirus infections. Intranasal interferon has neither prophylactic nor therapeutic effect on these latter respiratory virus infections. Thus during the late fall season, specific virus diagnosis to exclude a non-rhinovirus etiology is desirable if interferon is to be used effectively for prophylaxis.

Forty-four percent of the patients in the interferon group and 32% of the patients in the placebo group had local nasal adverse effects (blood tinged mucus, nosebleed and/or nasal mucosal erosions). The figures are higher than earlier reported with lower doses and fewer daily administrations. In this trial, the four times daily drug administration which induces a considerable amount of mechanical irritation to the nasal mucosa was certainly a major etiological factor. This view is supported by the observation of Monto et al (9) who reported that once a day dosing has less side effects than twice a day dosing.

CONCLUSION

In conclusion, the observations of this study suggest that intranasal IFN Alpha-2a early treatment has no clinically relevant therapeutic effect on the severity of (a) respiratory virus infections and (b) laboratory confirmed rhinovirus infections. The slightly increased frequency of non-serious intranasal adverse effects in the interferon group may be improved by reducing the frequency of daily administration.

REFERENCES

1. Scott GM, Phillpotts RJ, Wallace J, Gauci CL, Greiner J and Tyrrell DAJ: Prevention of rhinovirus colds by human interferon alpha-2 from Escherichia coli. Lancet 1982; i:186-189.
2. Herzog C, Berger R, Fernex M, Friesecke K, Haval L, Just M, Dubach UC, Havas L, Just M and Dubach U: Intranasal interferon (rIFN-alpha A, Ro 22-8181) for contact prophylaxis against common cold: a randomized, double-blind and placebo-controlled field study. Antiviral Research 1986; 6:171-176.
3. Intranasal interferon alpha-2 prophylaxis of natural respiratory virus infection. The Journal of Infectious Diseases 1985; 151, No. 46.

4. Douglas RM, Moore BW, Miles HB, Davies LM, Neil RN, Graham MH, Ryan P, Worswick DA and Albrecht JK: Prophylactic efficacy of intranasal alpha-2-interferon against rhinovirus infections in the family setting. New England Journal of Medicine 1986; 314, No. 2:65-70.
5. Hayden FR, Albrecht JK, Kaiser DL and Gwaltney JM: Prevention of natural colds by contact prophylaxis with intranasal alpha-2-interferon. New England Journal of Medicine 1986; 314, No. 2:71-75.
6. Hayden FG and Gwaltney JM: Intranasal interferon alpha-2 treatment of experimental rhinoviral colds. The Journal of Infectious Diseases 1984; 150, No. 2; 174-180.
7. Hennes U, Jucker W, Fischer EA, Krummenacher T, Palleroni AV, Trown PW, Linder-Ciccolunghi S and Rainisio M: Detection of antibodies to recombinant interferon alpha 2a in human serum: Evaluation of three methods. Submitted to J. of Biological Standardization.
8. Halonen P and Meurman O (1982): Radioimmunoassay in diagnostic virology. In: New developments in practical virology. Ed. Howard CR (Alan R. Liss, New York), pp 83-124.
9. Monto AS, Shope TC, Schwartz SA and Albrecht JK: Intranasal interferon alpha-2b for seasonal prophylaxis of respiratory infection. The Journal of Infectious Diseases 1986; 154, No. 1:128-133.

LONG-TERM TOLERANCE OF INTRANASAL RECOMBINANT INTERFERON-β_{SER} IN MAN

FREDERICK G. HAYDEN, M.D., DONALD J. INNES, M.D., STACY E. MILLS, M.D., AND PAUL A. LEVINE, M.D.

1. INTRODUCTION

Intranasal recombinant interferon-alfa2b is protective against natural rhinovirus colds when used for seasonal prophylaxis in open populations (1,2) and for postexposure prophylaxis in the family setting (3,4). However, both natural (5) and recombinant (1,2,6) alfa-interferons (IFN-α) are associated with local nasal intolerance when used for periods longer than one to two weeks. This local toxicity is manifested by complaints of nasal dryness, stuffiness, and blood-tinged mucus and rhinoscopic findings of mucosal bleeding, erosion, and ulceration. Up to one-half of adults given long-term intranasal IFN- α have experienced significant local reactions within four weeks of initiating administration (1,2,5,6). Tolerance studies of rIFN- α have found that over one-half of recipients will develop nasal histopathologic alterations, particularly subepithelial lymphocytic infiltration (7,8). This lymphocytic infiltration can occur as early as four days after initiating exposure and appears to be a precursor to the development of the clinical manifestations of local toxicity (8).

In order to be an effective agent for the seasonal prophylaxis of respiratory viral infections, that is long-term use, an interferon would have to have a significantly improved therapeutic index compared to the recombinant and natural IFN-αs tested to date. One candidate is human beta-interferon (IFN-β), which has a high specific antiviral activity and is active against rhinoviruses in vitro (9,10). IFN- β produced by recombinant DNA technology was initially found to have a 10-fold lower specific activity then that of naturally produced IFN-β. This was postulated to relate to random disulfide bridging among the three cysteine residues within and between the rIFN-β molecules (11). By site-specific mutagenesis of the human IFN-β gene, Mark et al. developed a new molecule designated rIFN-β_{ser} (11). This interferon has a serine substitution for cysteine at amino acid 17 and thus has only one possible intramolecular disulfide bridge. This interferon also differs from natural IFN-β in its lack of glycosylation and lack of the N-terminal methionine residue. The specific antiviral activity of rIFN-β_{ser} is very high ($1-2 \times 10^8$ U/mg protein) and similar to that of fibroblast-derived IFN-β (11,12).

Intitial studies conducted at the MRC Common Cold Unit in Salisbury found that intranasal rIFN-β_{ser} 6 MU/day given in three divided doses was well tolerated during short-term administration (4 days) and that it provided significant protection against experimental rhinovirus colds (13). The overall goal of our studies was to assess the therapeutic index of intranasal rIFN-β_{ser} in healthy adult volunteers. The specific objectives of the trial described in this report were to determine its dose-dependent long-term clinical tolerance and histopathologic effects in uninfected adults.

2. METHODS

The tolerance study was randomized, double-blind and placebo-controlled in design. One hundred and twenty adult volunteers who had no prior interferon exposure participated. These individuals were randomly assigned to receive 3 MU/day (low-dose) or 12 MU/day (high-dose) of rIFN-β_{ser} or placebo once daily for 25 days. The study drug was administered by metered pump nasal spray in a total volume of 0.2 mL/nostil. The study drugs were reconstituted on a daily basis, and neither the interferon nor placebo solutions contained a preservative. The subjects came to the study center once daily for supervised self-administration of the drug. At this time they recorded the presence and severity of various respiratory and systemic symptoms that related to the occurrence of respiratory viral infection or possible interferon toxicity. Symptomatic subjects had nose and throat swabs collected for viral isolation. Nasal exams were performed before and then weekly during the study to look for signs of mucosal irritation. In a total of 35 subjects, punch nasal biopsies were collected before and on the last day or two of dosing for histologic analysis. Immuno-histochemical techniques were used to stain for T and B lymphocytes by previously described techniques (8,14). Routine hematology, blood chemistries and urinalyses were done before and again during the last week of spray administration.

3. RESULTS

3.1 Subjects

Forty subjects were assigned to each of the treatment groups. The majority in all groups were females (73-83% of subjects) whose mean age was about 30 years. A small number of subjects had a prior history of asthma or hay fever, although none was receiving medication for these problems at the time of the study, and the proportion of individuals who continued to smoke during the study was comparable in the three groups (11 high-dose, 14 low-dose, 9 placebo). All subjects except two in the high-dose group completed the 25-day period of drug administration. Both of the dropouts in the high-dose group were unrelated to drug side effects.

3.2 Symptom occurrence

No statistically significant differences were detected between the interferon and placebo groups in the occurrence of nasal or respiratory symptoms during the treatment period. Specifically, the percent of subjects who developed symptoms of nasal intolerance during the entire 25-day treatment period or during the last week of exposure did not differ among the treatment groups (Table 1).

TABLE 1. Symptoms of nasal intolerance during intranasal $rIFN-\beta_{ser}$.

		Percent of subjects							
		Days 0-24			Days 19-24			Moderate or severe		
Symptoms	(MU/day)	12	3	0	12	3	0	12	3	0
Nasal obstruction		39	30	48	11	10	3	10	10	23
Blood in nasal mucus		26	18	20	8	3	3	8	3	3
Nasal dryness		24	18	30	3	3	3	3	0	5
Burning in nose		8	13	10	0	3	3	0	0	3

The complaint of nasal obstruction or stuffiness, which is a common side-effect of IFN-α, was found in a slightly higher proportion of placebo recipients. Blood in nasal mucus was reported by 26% of high-dose interferon recipients but also by 20% of the placebo group. During the last week of dosing only about 10% of subjects in the high-dose group had one or more of these complaints. The percent of subjects who had one of these symptoms of intolerance, which they rated as moderate or marked in severity, also did not differ between the groups. A somewhat smaller portion of the interferon recipients reported moderate or severe nasal obstruction compared to the placebo group (Table 1).

3.3 Nasal exam findings

In contrast, the percent of subjects having abnormal nasal exam findings during the treatment period was higher in the high-dose group compared to the placebo group. During the first two weeks of administration, punctate mucosal bleeding sites were seen about twice as often in the high-dose group (20.5% of subjects) compared to the other two groups (each 10%). Such bleeding typically resolved despite continued drug administration, and minor bleeding sites were often seen in individuals who had no specific nasal symptoms. During the last two weeks of dosing, a significantly higher percentage of high-dose subjects (18.5%) had mucosal bleeding sites detected compared to those in the placebo group (2.5%). During the entire treatment period, 38% of high-dose compared to 12.5% of low-dose or placebo recipients (p<0.05) had punctate bleeding sites detected on one or more exams.

Erosions or ulcerations were seen infrequently in all groups during the first two weeks of the study (2.5% of subjects). By the fourth week of dosing, higher proportions had a detectable abnormalities, but no differences were observed between the groups (20.5% of high-dose subjects, 10% of low-dose, 20% of placebo). Most of these lesions were observed in the anterior septal area of the nose, at sites where the long sprayer tip or perhaps the blast of the spray could have caused direct mechanical trama to the mucosa. However, in addition to the mucosal abnormalities associated with the method of spray administration, the finding of an excess rate of mucosal bleeding suggested that the high-dose interferon group had early signs of mild nasal irritation.

3.4 Histopathologic studies

Evidence of an interferon effect in the high-dose group was also found on histologic analysis of nasal biopsy specimens (Table 2). These samples were evaluated under masked conditions for the presence and degree of stromal lymphocytic infiltration by two pathologists. Compared to preexposure biopsy specimens, the proportion of subjects who had an increased degree of lymphocytic infiltration after drug exposure was 17% of the placebo recipients, compared to 25% of the low-dose and 45% of the high-dose groups.

TABLE 2. Lymphocytic infiltration in nasal biopsies after intranasal rIFN-β_{ser} or placebo for 25 days.

		Mean $\pm$SD score change (pre/post)	
Dose (MU/day)	No. with increase/ total no. (%)	Pathologist 1	Pathologist 2
0	2/12 (17)	0.4$\pm$0.7	0.0$\pm$0.6
3	3/12 (25)	0.4$\pm$0.5	0.3$\pm$0.7
12	5/11 (45)	0.6$\pm$0.8[a]	0.7$\pm$0.9[a]

[a] $p<0.05$, post vs pre, one-tailed sign test.

When the postexposure biopsy samples were compared to the corresponding preexposure ones, the mean change in the degree of lymphocytic infiltration was significant only for the high-dose group (Table 2). In comparison, previous studies of long-term rIFN-α at comparable dosages have found that approximately 60% of interferon recipients will develop increases in lymphocytic infiltration, often of a marked degree (7,8). These changes occur within 10 days and sometimes by 4

days after initiating exposure (8). Thus, the histologic alteration:
observed in the high-dose rIFN-β_{ser} group were qualitatively similar
to but quantitatively less than those observed in previous studies o:
rIFN-α.

3.5 Laboratory studies

Significant laboratory abnormalities were not detected in routine
hematology or chemistry studies. In contrast to earlier studies of
short-term, high-dose (15) or long-term (1) intranasal rIFN-α2b, none
of the rIFN-β_{ser} recipients developed leukopenia.

4. SUMMARY

This study determined that daily nasal sprays of rIFN-β_{ser} at 3
or 12 MU/day for 25 days were not associated with increases in the
frequency or severity of symptoms of nasal intolerance compared to
placebo in healthy adults. Administration of rIFN-β_{ser} 12 MU/day but
not 3 MU/day was associated with a higher frequency of mild mucosal
irritation, specifically punctate mucosal bleeding sites on nasal
exam, and histopathologic evidence of subepithelial lymphocytic infiltration. Compared historically to studies involving long-term
intranasal administration of comparable doses of recombinant or
natural IFN-α, the current study found substantially lower rates of
clinically significant nasal side effects. An earlier study of fibroblast-derived IFN-β found that low doses were ineffective in preventing experimental rhinovirus colds (16). In contrast, studies by
Higgins et al. found that rIFN-β_{ser} 6 MU/day in divided doses significantly protects against experimental colds in the same volunteer
model (13). Together with the tolerance study observations, these
findings suggest intranasal rIFN-β_{ser} may have a more favorable
therapeutic index than previously tested recombinant and leukocyte-
derived IFN-αs. However, the definitive answer to this questions will
depend on the results of field studies of long-term prophylaxis for
naturally acquired rhinovirus infections.

EFERENCES

1. Farr BM, JM Gwaltney Jr, KF Adams, and FG Hayden. Intranasal interferon- α 2 for prevention of natural rhinovirus colds. Antimicrob Agents Chemother 26:31-34, 1984.

2. Douglas RM, JK Albrecht, HB Miles, BW Moore, R Read, DA Worswick, and AJ Woodward. Intranasal interferon-α2 prophylaxis of natural respiratory virus infection. J Infect Dis 151:731-736, 1985.

3. Hayden FG, JK Albrecht, DL Kaiser, JM Gwaltney Jr. Prevention of natural colds by contact prophylaxis with intranasal alpha2-interferon. N Engl J Med 314:71-75, 1986.

4. Douglas RM, BW Moore, HB Miles, LM Davies, NMH Graham, P Ryan, DA Worswick, JK Albrecht. Prophylactic efficacy of intranasal alpha2-interferon against rhinovirus in the family setting. N Engl J Med 314:65-70, 1986.

5. Scott GM, JK Omwubalili, JA Robinson, C Dore, DS Secher, and K Cantell. Tolerance of one-month intranasal interferon. J Med Virol 17:99-106, 1985.

6. Hayden FG, JM Gwaltney Jr, and ME Johns. Prophylactic efficacy and tolerance of low-dose intransal interferon-alpha2 in natural respiratory virus infections. Antiviral Res 5:111-116, 1985.

7. Hayden FG, SE Mills, and ME Johns. Human tolerance and histopathologic effects of long-term administration of intranasal interferon-α2. J Infect Dis 148:914-921, 1983.

8. Winther B, DJ Innes, SE Mills, and FG Hayden. Nasal mucosal responses to recombinant leukocyte A interferon in healthy adults. Presented at the Twenty-fourth Interscience Conference on Antimicrobial Agents and Chemotherapy, Washington, D.C., 1984, abstr. #919.

9. Came PE, TW Schafer, and GH Silver. Sensitivity of rhinoviruses to human leukocyte and fibroblast interferons. J Infect Dis 133:A136-A139, 1976.

10. Harmon MW, SB Greenberg, PE Johnson, and RB Couch. Human nasal epithelial cell culture system: evaluation of response to human interferons. Infect Immun 16:480-485, 1977.

11. Mark DF, SD Lu, AA Creasey, R Yamamoto, and LS Lin. Site-specific mutagenesis of the human fibroblast interferon gene. Proc Natl Acad Sci 81:5662-5666, 1984.

12. Hawkins M, S Horning, M Konrad, S Anderson, K Sielaff, S Rosno, J Schiesel, T Davis, D DeMets, T Merigan and E Borden. Phase I evaluation of a synthetic mutant of β-interferon. Cancer Res 45:5914-5920, 1985.

13. Higgins PG, W Al-Nakib, J Willman, and DAJ Tyrrell. Interferon-β_{ser} as prophylaxis against experimental rhinovirus infection in volunteers. J INF Res 6:153-159, 1986.

14. Winther B, DJ Innes, SE Mills, N Mygind, D Zito, and FG Hayden. Lymphocyte subsets in normal airway mucosa of the human nose. Arch Otolaryngol Head Neck Surg, 1986 (in press).

15. Hayden FG and JM Gwaltney Jr. Intranasal interferon α 2 for prevention of rhinovirus infection and illness. J Infect Dis 148:543-550, 1982.

16. Scott GM, S Reed, T Cartwright, and D Tyrrell. Failure of human fibroblast interferon to protect against rhinovirus infection. Arch Virol 65:135-139, 1980.

THE ROLE OF ALPHA 2 INTERFERON IN COMBINATION WITH ACYCLOVIR IN OPHTHALMIC ZOSTER (an open pilot study)

J.G.M. STRUYK, O.P. VAN BIJSTERVELD
Royal Netherlands Eye Hospital, University of Utrecht, The Netherlands

1. ABSTRACT

Local treatment with acyclovir (ACV) in combination with local and systemic recombinant alpha 2 interferon (IFN) in ophthalmic zoster decreased the incidence of complications as a direct consequence of virus replication in the corneal epithelial cells and Gasserian ganglion respectively i.e.: anterior corneal stroma infiltrates and neurotrophic disturbances of the cornea. The incidence of keratouveitis —believed to be the result of complex mediated immunological reactions— was statistically not different in the two treatment groups. Clinically this complication had a milder course in those patients that received IFN in addition to ACV.

2. INTRODUCTION

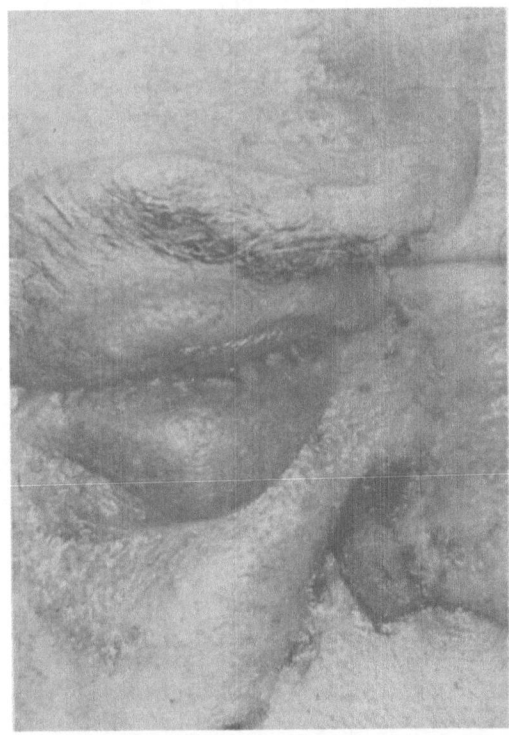

Ophthalmic zoster occurs in about 9 percent of zoster cases (1) and its incidence in the general population is about 0,5% which is low. If ocular complications ensue, —in about half of the cases—, it is a serious disease that can give rise to marked loss in visual acuity or even loss of an eye. In all our patients the eruption started suddenly, preceded some days by neuralgic pains along the first branch of the trigeminal nerve, increasing gradually in intensity. Fever, prostration and lymphadenopathy were constant features. Erythema and oedema because of vasculitis leading to tense and swollen eyelids were sometimes very pronounced (Fig.1).

FIGURE 1. Ophthalmic zoster. Erythema and oedema because of vasculitis of the skin, resulting in tense swollen eyelids.

All ocular structures can be affected in the disease, but the most frequent damage is that caused by virus replication in the corneal epithelial cells, leading to anterior stromal infiltrates (Fig. 2).

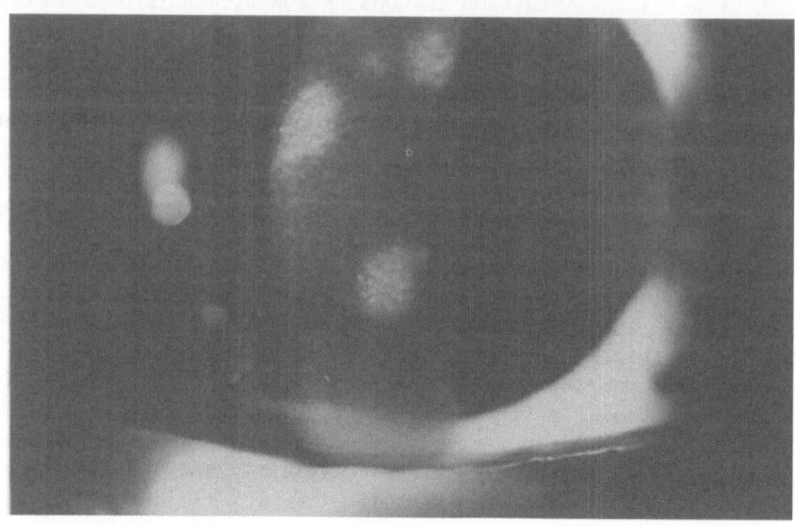

FIGURE 2. Anterior stromal zoster (nummular) keratitis. Soon after the epithelial keratitis, anterior stromal opacities develop, often said to be the first lesion in zoster keratitis.

Virus replication in the Gasserian ganglion sets up an inflammatory reaction which interferes with transmission of nerve impluses but can also cause necrosis of nerve elements. As a result corneal sensitivity decreases and this can lead to neuroparalytic ulcers of the cornea (Fig. 3).

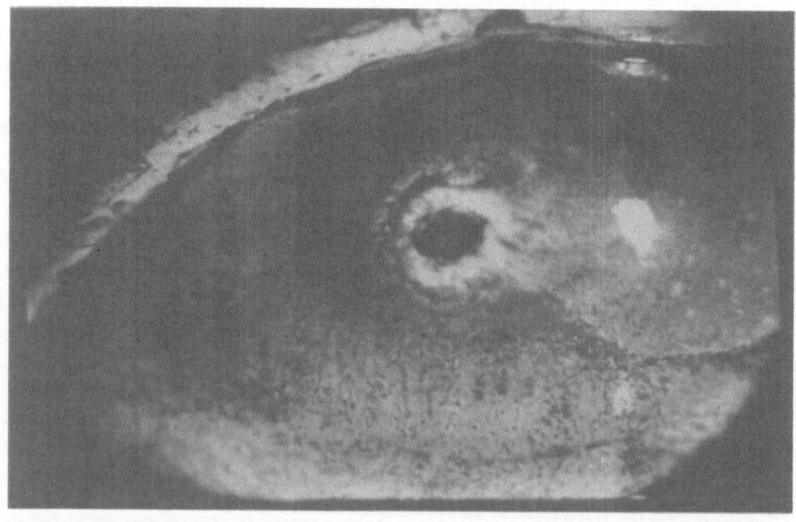

FIGURE 3. Neuroparalytic corneal ulcer after ophthalmic zoster in a patient from the local acyclovir treatment group. This dreaded complication can lead to loss of an eye.

Keratouveitis, another frequent complication in ophthalmic zoster, is caused by complex mediated immunological reactions leading to intense vasculitis of the pars plicata of the ciliary body. This gives rise to exsudation of inflammatory cells and serum constituents in the eye leading to flare, cells, keratitic precipitates (Fig. 4) and even sometimes sterile pus in the anterior chamber. As a result of these inflammatory reactions, deep stromal keratitis develops that can heal with scarring, lipid deposits and vascularisation.

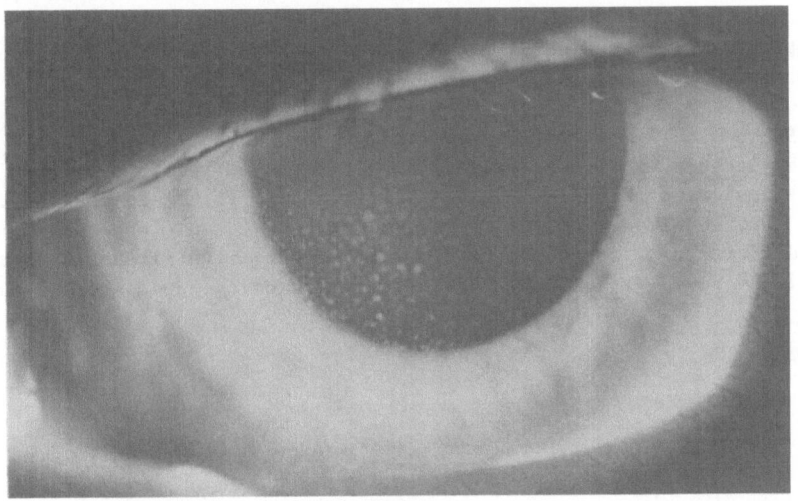

FIGURE 4. Zoster keratouveitis. There is inflammation of the iris and ciliary body with keratitic precipitates.

Since the introduction of selective antiviral agents (2-3), the outlook of zoster eye disease has improved. Local treatment with recombinant alpha 2 interferon in combination with acyclovir has been very effective in the treatment of Herpes Simplex keratitis (4) and this combination we have compared to ACV treatment alone in zoster corneal disease and in zoster keratouveitis.

3. PERSONS AND METHODS

A total of 30 patients were treated and assigned to one of the two treatment groups. Fifteen patients received local ACV alone, as ointment, six times a day, during maximally three weeks. Fifteen other patients received in addition to ACV two IFN eyedrops in total $1,5 \times 10^6$ I.U. per day during 20 days and also 10×10^6 I.U. IFN systemically during 10 days.

All patients had routine ophthalmologic examination with special reference to ocular biomicroscopy with the Haag-Streit slitlamp after vital staining with fluorescein and rose bengal 1% solution. Corneal sensitivity was measured at 20 days and 6 months after the onset of the skin eruption with the aesthesiometer of Cochet and Bonnet (5), using a nylon monofilament with a diameter of 0,12 mm, in 5 different sectors of the cornea. These threshold values of corneal sensitivity were averaged. In this way initial damage and recovery of the sensory function of the cornea could be evaluated.

Patients with serious liver and kidney diseases or with cardial problems or those with malignancies and/or immunosuppressive treatment were excluded from IFN treatment (6). The IFN treatment was stopped when either there was a decline in the number of thrombocytes to less than 100 x 10^9 per L, or a decline in the number of leucocytes to less than 3,5 x 10^9 per L, or a decline in the hemoglobline content of more than 1,5 mmol per L.

Patients were stratified according to the involvement of the nasociliary nerve, age and sex.

4. RESULTS

The incidence of anterior stromal keratitis in the acyclovir-recombinant alpha 2 interferon treated group was 26,6% and that of the group that received acyclovir alone was 46,6%. We could not demonstrate this difference in incidence to be statistically significant in these relatively small groups, but the severity judged by the number of nummular

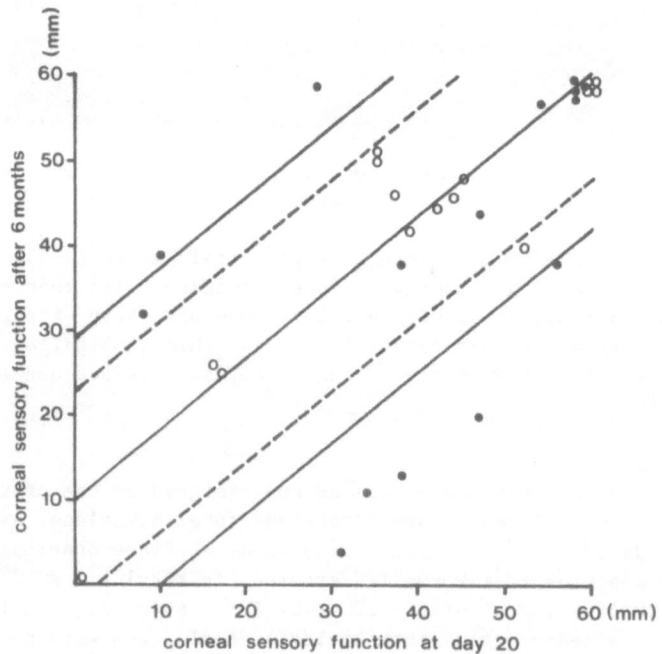

FIGURE 5. Regression of corneal sensitivity at 6 months and at 20 days
after the initial zoster skin eruption in the IFN/ACV treated group (o)
with the 95 (---) and 99 (____) percent confidence limits and in the ACV
treated group (•). There is a significant positive correlation (r=+0.92,
p<0.01) in sensitivity values of the ACV-IFN treated group, but not in the
group that received locally ACV alone. This is due to deterioration of
corneal sensory function in five patients in the ACV treated group.
Measurements were done with a nylon monofilament 0.12 mm diameter and
expressed in mm length of the monofilament at which a touch sensation was
experienced.

anterior stromal infiltrates was significantly reduced in the group that
received both acyclovir and interferon (P<0,05). It would seem therefore,
that the combination therapy as in Herpes simplex, acts synergistically to
reduce virus replication.

A significant positive correlation was found between the corneal sensitivity measured at 6 months and at 20 days after the onset of the skin
eruption for those patients that received the combination therapy but not
for the group that received ACV treatment alone (Fig. 5). This was due to
the fact that in the group that received ACV alone a progressive deterioration at corneal sensory function occurred in 5 cases, two of which developed serious neuroparalytic sequelae. In the ACV-IFN treated group a
moderate deterioration of corneal sensory function occured in one patient.
Recovery in the remaining patients was modest and statistically we could
not demonstrate any difference in recovery rate. The protective effect on
progressive deterioration was most likely the effect of systemic IFN alone
as locally applied ACV cannot reach the Gasserian ganglion in any sufficient concentration.

The incidence of complications caused by virus replication in the corneal epithelial cells and that caused by viral inflammation of the
Gasserian ganglion combined, was significantly reduced in the group that
received acyclovir and recombinant alpha 2 interferon in comparison to the
group that received local acyclovir. With regard to the incidence of
keratouveitis we could not demonstrate a difference in incidence of this
complication between the two treatment groups. Clinically, however, the
course of keratouveitis was milder in those patients that received the
combination therapy. As it is difficult to imagine to have sufficient concentrations of locally applied ACV in the region of the ciliary body, this
effect is most likely due to systemic IFN alone.

5. DISCUSSION

Stratification, particularly with respect to involvement of the nasociliary branches (Hutchinson's sign) is important (7); these branches may
escape but if they are implicated ocular complications are common. However,
the absence of this sign excludes in no way ocular complications. In our
two groups the ACV-IFN treated group had 7 patients with Hutchinson's sign
versus 8 patients in the ACV monotherapy group.

Because of the complications of systemic IFN treatment (6), persons
with certain diseases were excluded from this therapy but not assigned to

the study group that was treated with local ACV. It can be argued therefore, that with respect to the general health condition the two groups were comparable.

As the studied groups were small, statistical evidence for the therapeutic effectiveness in reducing the incidence of the major complications separately could not be demonstrated. The sequelae of direct virus replications in the corneal epithelial cells and in the Gasserian ganglion in combination, however, were statistically significantly reduced in those patients that received ACV and IFN in comparison to those that received ACV alone. Also, the severity of complications was reduced. In keratouveitis, there was a milder clinical course in those patients that received both ACV and IFN. The effectiveness of treatment for anterior stromal infiltrates was the result of the synergistic action of ACV and IFN. The therapeutic effect on ganglionitis and keratouveitis was most likely the result of IFN alone.

5. REFERENCES

1. Hope-Simpson, R.E.: The nature of Herpes Zoster: A longterm study and a new hypothesis. Proc. Roy. Soc. Med. 58: 9-20, 1965.
2. McGill, J.: Topical acyclovir in Herpes Zoster ocular involvement. Br. J. Ophthalmol. 65: 542-5, 1981.
3. McGill, J.; Chapman, C.; Mahakasingam, M.: Acyclovir therapy in Herpes Zoster infection. Trans. Ophthal. Soc. U.K. 103: 111-4, 1983.
4. van Bijsterveld, O.P. and Meurs P.J.: Recombinant alpha-2 interferon and acyclovir therapy for dendritic keratitis. In: The Biology of the Interferon System: H. Kirchner, and H. Schellekens (ed): 569-74, 1984.
5. Cochet, P. and Bonnet, R.: L'esthesiometie Corneenne: Realisation et interpret pratique. Bull. Soc. Ophtalmol. Fr. 61: 541-50, 1961.
6. Quesada, J.R.; Talpaz, M.; Rios, A; Kurzrock, R.; Gutterman, J.U.: Clinical Toxicity of Interferons in Cancer Patients: A Review. Journal of Clinical Oncology 4: 234-43, 1986.
7. Duke-Elder, S.: Viral Kerato-conjunctivitis. Zoster. In: System of Ophthalmology 8 (I): 340-8, 1965. Henry Kingston, London.

PRODUCTION OF A PARTIALLY ACID LABILE ALPHA INTERFERON IN CYTOMEGALOVIRUS PNEUMONITIS.

JANE E. GRUNDY[1], A.M.L. LEVER[1], HEATHER J. MILBURN[2], P. PRATT[1], R.M. DU BOIS[2], AND P.D. GRIFFITHS[1].

Departments of Virology[1] and Thoracic Medicine[2], The Royal Free Hospital, London NW3 2QG, England.

INTRODUCTION

Interstitial pneumonitis is a common complication of organ transplant-ation, especially bone marrow transplantation, and the pathogen most frequently isolated is cytomegalovirus (CMV)[1]. Studies from a murine model using murine CMV (MCMV) have suggested that MCMV interstitial pneumonitis may have an immunopathological basis[2,3]. The ultrastructural pathology of CMV pneumonitis has been reported to be identical to that seen in interstitial lung disease occurring in patients with collagen vascular diseases such as systemic lupus erythrematosus (SLE)[4]. Patients with SLE have been found to have circulating levels of both an acid labile and an acid stable α interferon[5,6] and it has been postulated this may be involved in the immunopathology of SLE[5,7]. MCMV has been shown to induce partially acid labile α interferon after in vivo infection[8,9], and in man this type of interferon is seen in patients with the acquired immunodeficiency syndrome (AIDS)[10,11], who frequently have infections with CMV[12]. We have therefore searched for the presence of acid labile and acid stable interferons in the bronchoalveolar lavage fluid of patients with proven CMV pneumonitis, in order to see whether this might aid in understanding its pathogenesis. In addition, as specific antigenic or mitogenic stimulation of lymphocytes results in the production of γ interferon, we have also looked for the presence of this type of interferon in the lavage fluid. Furthermore, we have analysed whether the presence or absence of such interferons in the lavage fluid has any diagnostic or prognostic value for CMV pneumonitis.

MATERIALS AND METHODS

Patients studied

Thirty patients with 34 episodes of pneumonitis were studied, of whom 16 (20 episodes) had undergone bone marrow transplantation, 5 had undergone renal transplantation, 4 had haematologic malignancies, 3 had fibrosing alveolitis, 1 had sarcoidosis and one was a surgical patient. All patients diagnosed as having CMV pneumonitis were treated with CMV hyperimmune globulin[13], and those with *Pneumocystis carinii* coinfection also received high dose septrin.

Diagnosis of Pneumonitis

All patients had clinical and/or radiological evidence of pneumonitis and underwent fibreoptic bronchoscopy. Bronchoalveolar lavage (BAL) was performed using three 60 ml aliquots of pH 7.4 bicarbonate buffered saline (BBS). Lavage was performed either from lobes with localised abnormality or, in the case of diffuse lung disease, from the lingula or right middle lobe. The aspirated fluid was used for microbiological analysis and an aliquot clarified and stored at -70°C for later interferon assay. The presence of CMV in the lavage was detected by the DEAFF test[14] using CMV specific monoclonal antibodies. Serum samples were taken at the same time as the lavage, stored at -70°C and analysed for interferon.

Assay for Interferon

Interferon α and γ were detected by immunoradiometric assays (Boots-Celltech) based on monoclonal antibodies to the respective interferon types. Where quantity of lavage allowed, samples positive for α interferon were dialysed against a glycine/HCl pH 2.0 buffer for 48 hours, followed by dialysis against PBS, for 24 hours, and then assayed for α interferon. A second aliquot of the same BAL fluid dialysed against PBS for 72 hours was used as a control. Purified lymphoblastoid α interferon (acid stable) or recombinant γ interferon (acid labile) were pH 2.0 treated in parallel.

RESULTS

Alpha interferon was detected in BAL fluid from 3/11 patients with CMV pneumonitis (1 with co-existing *Pneumoncystis carinii*) (Table 1), Alpha interferon was also detected in one patient with sarcoidosis and in two bone transplant recipients, one with *Staphylococcus aureus* and the other with *Aspergillus* (Table 1). The α interferon titres found in BAL fluids and in sera are shown in Table 2.

TABLE 1. The detection of α interferon in the BAL fluid of patients with interstitial pneumonitis.

Patient Group	Diagnosis of Pneumonitis	
	CMV	Non CMV
	No. α Interferon positive/No. tested	No. α Interferon positive/No. tested
Bone Marrow Transplant Recipients	3/8 (3/11 episodes)	2/8 (3/9 episodes)
Renal Transplant Recipients	0/3	0/2
Leukaemia	-	0/4
Fibrosing alveolitis	-	0/3
Sarcoidosis	-	1/1
Surgical	-	0/1

The BAL samples positive for α interferon were analysed for stability at pH 2.0 and the results are shown in Table 3. The α interferon was partially acid labile in two of the three samples. The α interferon from patient no. 5 (see Table 2) with *Staphylocuccus aureus* infection was acid stable (data not shown).

TABLE 2. Titres of α interferon in BAL fluid and serum of patients with pneumonitis.

Patient No.	Diagnosis of pneumonitis	α interferon titre IU/ml	
		BAL fluid	Serum
1	CMV	25.6	5.2
2	CMV	1.7	3.1
3	CMV	1.3	4.7
4	acute Leukaemia	2.0	ND*
5	Staph. aureus	Episode 1 58.0	ND
		Episode 2 7.6	ND

* ND = not done

TABLE 3. Acid lability of α interferon in BAL fluid of patients with CMV pneumonitis.

Patient No.	Interferon titre IU/ml after treatment at	
	pH 7.4	pH 2.0
1	19.0	11.0
2	1.5	1.5
3	4.7	Undetectable

All 3 patients with CMV pneumonitis who had α interferon in the BAL fluid died of their pneumonitis (Table 4). In 10 patients in whom CMV pneumonitis occurred in the absence of detectable interferon, only two died, both after a second episode of pneumonitis. This difference is statistically significant by the X^2 test (P<0.02). The mean time from the onset of symptoms to diagnostic BAL in the α interferon positive patients with CMV pneumonitis was 18.3 days (range 14-27) but was only 5.6 days (range 2-14) in the interferon negative group. Thus interferon was only present in those specimens taken late in the course of CMV pneumonitis.

TABLE 4. The relationship between the clinical outcome of CMV pneumonitis and the detection of α interferon in the BAL fluid.

Interferon detected	Outcome of CMV Pneumonitis	
	Died	Recovered
Yes	3	0
No	2	8

Gamma interferon was not detected in the BAL fluid of any patient with pneumonitis. Gamma interferon (4 IU/ml) was detected in the serum of patient no. 1 (Table 2), who also had α interferon in the serum and BAL fluid, and in two pneumonitis patients with acute leukaemia, one of whom also had a pseudomonas infection.

DISCUSSION

We have found α interferon in the BAL fluid of three patients with CMV pneumonitis. Since the interferon present in the fluid bathing the surface of the lung was substantially diluted by the 180 mls of saline used for the lavage, the interferon concentration in the lung in vivo must have been considerably higher. These three patients also had α interferon in the serum, but when the dilution effect of the lavage is considered, it is likely that the serum concentration was much lower than that at the lung surface, suggesting that the source of α interferon in the BAL fluid was the lung.

Patients with *Pneumocystis carinii* pneumonitis did not have α interferon in the BAL fluid, suggesting that in the patient with co-existing CMV and pneumocystis the virus was the primary interferon stimulus. The presence of α interferon in the BAL fluid of a patient with sarcoidosis is unexplained and we are currently investigating other such patients to see whether or not this is a consistent finding. One patient with two episodes of pneumonitis due to *Staphylococcus aureus* also had α interferon in the BAL fluid.

The detecton of α interferon in the BAL fluid of patients with CMV pneumonitis was associated with a late diagnosis and a poor prognosis. There are several possible explanations for this. Firstly α interferon may only be detected in those patients with a high virus load, which may be seen late in infection and which may not respond to treatment with CMV hyperimmune globulin. On the other hand, in the two patients in whom CMV pneumonitis was fatal and in which α interferon was not detected in the BAL fluid, diagnosis was made within five days of the onset of symptoms, suggesting that fatal outcome and late diagnosis are not closely correlated. However, both these patients died after their second episode of CMV pneumonitis and they may not be representative of the whole group. A second explanation is that there may be a different pathogenesis of CMV pneumonitis between the α interferon positive and negative groups of patients. For example, an overwhelming virus load in the former which stimulates α interferon production and an immunopathological form with a low virus load in the latter. CMV hyperimmune globulin does not neutralise CMV as it occurs in body fluids[15], and therefore would not be expected to be effective in treating pneumonitis of the former type, whereas it may act by blocking recognition by reactive lymphocytes of viral or virally induced antigens on the cell surface in the latter form of CMV pneumonitis. Two different histopathologic patterns of CMV pneumonitis have been described[16].

The α interferon induced in the lung by CMV in patients with pneumonitis was partially acid labile, suggesting it comprised both an acid labile and an acid stable fraction. Acid labile interferon detected by bioassay has been described in the circulation of bone marrow transplant recipeints with CMV infection, where it was assumed to be type because of its acid lability[17]. In our study the use of an immunoradiometric interferon assay which utilizes monoclonal antibodies specific for either α or γ interferon, allowed us to discriminate between the two types and to assay them independently in samples

467

containing both.

As discussed earlier, partially acid labile α/β interferon is
induced by MCMV after in vivo infection[8,9]. Thus the human virus induces
a similar α interferon response to that of murine virus as well as to
that found in patients with some autoimmune diseases[5,6,7] or with AIDS[10,11].
The similarity between the histopathology of CMV pneumonitis and
interstitial lung disease found in patients with autoimmune collagen
vascular diseases[4] was referred to earlier, and the presence of a
similar novel acid labile[5,6] interferon in the two groups is an additional
common feature. Together with the data from the murine model which
suggests that MCMV pneumonitis has an immunopathological basis, these
findings suggest that an aberrant autoimmune type of reaction may
contribute to CMV interstitial pneumonitis in man. The presence of
acid labile α interferon (compared to the lack of γ interferon) may
be involved in the pathogenesis of this condition, but due to the large
dilution of lung secretions in the process of BAL, this may only be
detectable in the BAL fluid of those patients with the most severe
disease.

In conclusion we have demonstrated the presence of partially acid
labile α interferon in the BAL fluid of patients with CMV pneumonitis.
Its presence was found to correlate with both a late diagnosis and
a poor prognosis of CMV pneumonitis.

REFERENCES
1. Meyers, J.D., Flournoy, N., & Thomas, E.D. Non bacterial pneumonia
   after allogeneic marrow transplantation. A review of ten years
   experience. Reviews of Infectious Diseases, 1982, 4, 1119-1132.
2. Shanley, J.D., Pesanti, E.L. & Nugent, K.M. The pathogenesis of
   pneumonitis due to murine cytomegalovirus. Journal of Infectious
   Diseases, 1982, 146, 388-396.
3. Grundy, J.E., Shanley, J.D. & Shearer, G.M. Augmentation of
   graft-versus-host reaction by cytomegalovirus infection resulting
   in interstitial pneumonitis. Transplantation, 1985, 39, 548-553.
4. Hammer, S.P., Winterbauer, R.H., Bockus, D., Remington, F.,
   Scale, G.E. & Meyers, J.D. Endothelial cell damage and
   tubuloreticular structures in interstitial lung disease associated
   with collagen vascular disease and viral pneumonia. American
   Review of Respiratory Disease, 1983, 127, 77-84.
5. Hooks, J.J., Moutsophoulos, H.M. & Notkins, A.L. Circulating
   interferon in human autoimmune diseases. Texas Reports on Biology
   and Medicine, 1981-1982, 41, 164-167.
6. Preble, O.T., Black, R.J., Friedman, R.M., Klippel, J.H. &
   Vilcek, J. Systemic lupus erythematosus. Presence in human serum
   of an unusual acid-labile leukocyte interferon. Science, 1982,
   216, 429-431.
7. Hooks, J.J. & Detrick-Hooks, B. Interferon in autoimmune diseases
   and other immunoregulatory disorder. In Interferon 2 : Interferons
   and the immune system. Editors J. Vilcek & E. De Maeyer, 1984,
   Elsevier Science Publishers B.V. p 165-174.
8. Grundy (Chalmer), J.E., Trapman, J., Allan, J.E., Shellam, G.R. &
   Melief, C.J.M. Evidence for a protective role of interferon in
   resistance to murine cytomegalovirus and its control by non H-2 linked
   genes. Infection and Immunity, 1982, 37, 143-150.

9. Shellam, G.R., Grundy (Chalmer), J.E., Allan, J.E. & Hamett, G.B. Natural resistance to cytomegalovirus and other viruses. In Progress in Immunology Volume 5. Editors Tada, T. & Yamamura, Y., 1983, Academic Press, Japan, p 1209-1217.

10. Destefano, E., Friedman, R.M., Friedman-Kein, A.E., Goedert, J.J., Henriksen, D., Preble, O.T., Sonnabend, J.A. & Vilcek, J. Acid labile human leukocyte interferon in homosexual men with Kaposi's sarcoma and lymphodenopathy. Journal of Infectious Diseases, 1982, 146, 451-455.

11. Eyster, M.E., Goedert, J.J., Poon, M-C. & Preble, O.T. Acid labile alpha interferon: A possible preclinical marker for acquired immune deficiency syndrome in hemophilia. New England Journal of Medicine, 1983, 309, 583-586.

12. Sonnabend, J., Witkin, S.S. & Purtillo, D.T. Acquired Immunodeficiency syndrome, opportunistic infections, and malignancies in male homosexuals. Journal of the American Medical Association, 1983, 249, 2370-2374.

13. Blacklock, H.A., Griffiths, P.D., Stirk, P.R. & Prentice, H.G. Specific hyperimmune globulin for cytomegalovirus pneumonitis. Lancet, 1985, 2, 152-153.

14. Griffiths, P.D., Panjwani, D.D., Stirk, P.R., Ball, M.G., Ganczakowski, M., Blacklock, H.A. & Prentice, H.G. Rapid diagnosis of cytomegalovirus infection in immuno compromised patients by detection of early antigen fluorescent foci. Lancet, 1984, 2, 1242-1245.

15. McKeating, J.A., Griffiths, P.D. & Grundy, J.E. Cytomegalovirus in urine specimens has host microglobulin bound to the viral envelope: A mechanism for evading the host immune response? Journal of General Virology, 1986, (in press).

16. Beschorner, W.E., Hutchins, G.M., Burns, W.H., Saral, R., Tutschka, P.J., Santos, G.W. Cytomegalovirus pneumonia in bone marrow transplant recipients: Miliary and Diffuse patterns . American Review of Respiratory Disease, 1980, 122, 107-113.

17. Rhodes-Feuillette, A., Canivet, M., Devergie, A., Geuckman, E., Mazeron, M-C., Perol, Y. & Peries, J. Circulating interferon after marrow transplant in cytomegalovirus infection. Lancet, 1981, I, 1217.

RANDOMIZED CONTROLLED TRIAL OF A FOUR MONTH COURSE OF
RECOMBINANT HUMAN ALPHA INTERFERON IN PATIENTS WITH CHRONIC
TYPE B HEPATITIS.

A.M. DI BISCEGLIE, J.H. HOOFNAGLE, M. PETERS, K.D. MULLEN,
D.B. JONES, V.K. RUSTGI, M.I. AVIGAN, Y. PARK, J.WAGGONER
and EA JONES. Liver Diseases Section, National Institute of
Diabetes, Digestive and Kidney Diseases, Bethesda, MD., USA.

INTRODUCTION

Chronic infection with hepatitis B virus (HBV) is major
cause of chronic hepatitis, cirrhosis and hepatocellular
carcinoma world-wide (1). No therapy has yet been proven to
be of definite benefit in improving the natural outcome of
this infection. The most promising modality of treatment for
this condition has been prolonged courses of alpha interferon
(IFNα) (2,3). Other agents have either not been successful
in achieving the aims stated above or have resulted in unaccept-
able side effects (4,5). In pilot studies, treatment of
chronic type B hepatitis with IFNα has been shown to result
in a response rate of between 22-40% (6,7). We undertook a
randomized controlled trial of a four month course of recombinant human IFNα (rIFNα) in patients with chronic type B
hepatitis.

PATIENTS AND TREATMENT SCHEDULES

Forty five adult American patients with chronic type B
hepatitis were entered in this controlled trial of rIFNα,
(alfa-2b: Schering Corporation). All of these patients were
known to have hepatitis B surface antigen (HBsAg) in serum
for more than one year, with elevated alanine aminotransferase
(ALT) levels and stable levels of hepatitis B e antigen
(HBeAg), hepatitis B virus DNA (HBV DNA) and HBV-associated
DNA polymerase (DNAp) in serum. Liver histopathologic changes
in these patients were chronic persistent hepatitis in 7,
chronic active hepatitis in 34 and cirrhosis in 4.

The duration of treatment was four months with subsequent
follow up for a further eight months. Patients were randomized
into one of three groups. Group A received 5 million units
(MU) subcutaneously (sc) of rIFNα daily, group B 10 MU sc of
rIFNα on alternate days, and group C received no therapy.
Patients were admitted to hospital for initial evaluation
which included percutaneous liver biopsy. They were taught
to administer rIFNα themselves, and were then seen at weekly
to two weekly intervals for the duration of treatment and
thereafter monthly until one year after randomization, when a
repeat liver biopsy was obtained. Details of the study
protocol were approved by the Clinical Research Subcommittee
of the National Institute of Diabetes, Digestive and Kidney
Diseases and written informed consent was obtained.

METHODS

Serum HBsAg and HBeAg were measured by radioimmunoassays (AUSRIA II, and HBeAg test, Abbott North Chicago). HBV DNA in serum was determined by dot-blot hybridization using a ^{32}P-labeled cloned HBV DNA probe. DNAp in serum was determined by measuring incorporation of [^{3}H]thymidine.

Portions of each biopsy were fixed and stained routinely for histological examination. In selected cases, portions of liver tissue were prepared for DNA extraction and analysis of HBV DNA. Biopsy pairs were tested for HBV DNA by dot-blot and Southern blot analysis in 20 cases. HBV DNA in liver was quantitated by densitometry of dots on autoradiograms and compared to known amounts of cloned HBV DNA spotted on the same membrane. The amount of HBV DNA in liver was expressed as the copy number, or the number of HBV genome equivalents per cell.

RESULTS

TABLE 1. Initial clinical and serum biochemical profile of patients studied .

Group	A	B	C
Number of patients	16	15	14
Mean age (yrs)	37 ± 3*	35 ± 2	37 ± 3
Sex ratio (M:F)	15:1	14:1	13:1
Initial ALT (u/l)	193 ± 41	178 ± 25	197 ± 47
DNAp (cpm/0.2ml)	2753 ± 96	3459 ± 775	2087 ± 473

(*= mean ± SEM)

Serum HBV DNA and DNAp levels decreased within one week of starting rIFNα in all 31 patients. This decrease in DNAp was sustained in only 10 patients and in the remainder, levels returned to pre-existing levels after cessation of therapy. During the four months trial, a sustained loss of HBV DNA and DNAp from serum occurred in 10 of 31 patients. All 10 subsequently became HBeAg negative and 1 of these became HBsAg negative. ALT levels improved with loss of HBeAg. In the same four month period, 1 of 14 controls became negative for HBV DNA, DNAp and HBeAg.

During the follow up period after the four months of therapy, two treated patients became DNAp negative (at 11 and 12 months) and both subsequently became HBeAg negative and serum ALT levels fell to normal. On the other hand in one treated patient who had lost HBeAg with therapy, hepatitis B subsequently reactivated, with reappearance of HBeAg, HBV DNA and DNAp. Among the control subjects, one patient became negative during the year of follow up evaluation.

rIFNα was generally well tolerated. Although minor side effects were common, they were rapidly reversible. The dosage of interferon was decreased in 10 patients because of side effects. When the dosage required reduction, it was

halved and in 4 cases it was halved again. Interferon was
discontinued early in 2 patients, in both cases because of
severe depression. Other side effects of rIFNα included
fever, myalgia, reversible bone marrow suppression, reversible
hair loss and anorexia, weight loss, nausea and vomiting.

TABLE 2. Changes in markers of HBV infection during study
period.

		4 months		12 months	
Group	n	%loss DNAp	%loss HBeAg	%loss DNAp	%loss HBeAg
A	16	25%	12%	31%	19%
B	15	40%	27%	40%	33%
C	14	7%	0%	14%	7%

Figure 1. Shows the serial changes in HBV markers in a patient
who responded to rIFNα. DNAp levels became undetectable within
one month of starting interferon. Shortly afterwards, HBeAg
became negative to be replaced by antibody to e (anti-HBe).
At about the same time HBsAg levels began to decrease and
became negative after three months of treatment. Within one
month of this, serum ALT levels returned to normal or near
normal.

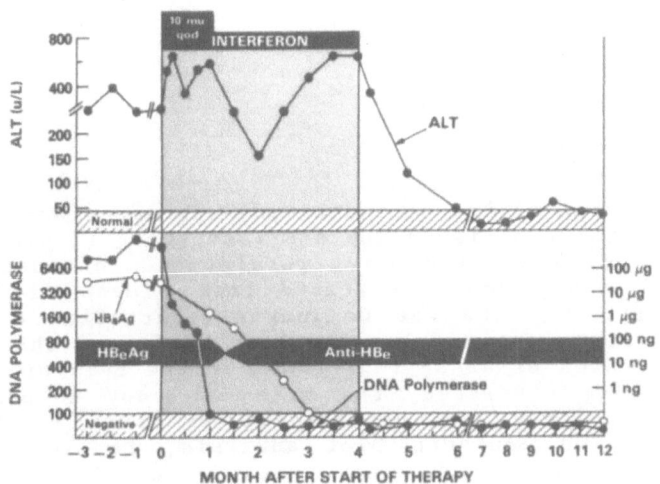

FIGURE 2. This shows serial changes in markers of hepatitis B virus infection in a patient who did not respond to therapy. For the duration of this patient's therapy with 10 million units of rIFNα on alternate days, serum ALT levels declined steadily but did not become normal. Similarly, DNA polymerase levels declined slightly but rebounded to pretreatment levels when therapy was discontinued. HBeAg remained present throughout the study period.

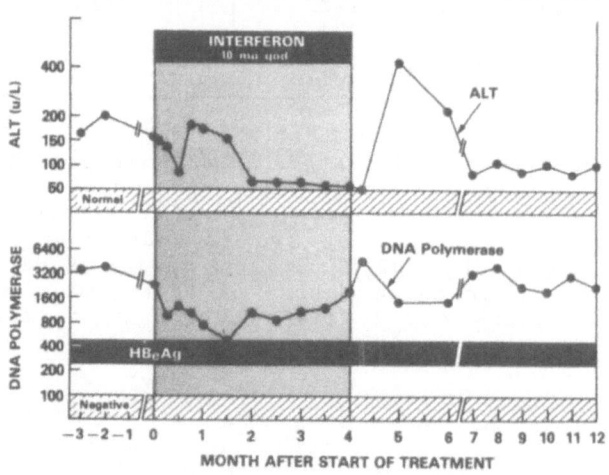

At the time of the second biopsy specimens were available for study from 6 patients who had lost HBeAg (5 treated and 1 control) and 14 in whom HBeAg persisted. In those who lost HBeAg, HBV copy number decreased from a mean of 30 to 1.6 copies/cell. HBV DNA was no longer detectable in specimens from 2 patients one of whom had lost HBsAg and the other who had only traces of HBsAg in serum. In the remaining patients in whom HBeAg persisted, the mean HBV copy number did not change appreciably (22 copies/cell compared to 21 before treatment). On Southern blot analysis, replicative intermediate forms of HBV DNA were still detectable in liver from patients in whom HBeAg persisted in serum, but had apparently disappeared in those patients who lost HBeAg.

Thus at 12 months 35% (11/31) treated patients (Groups A and B) as contrasted to 14% (2/14) control subjects were negative for HBV DNA and DNAp in serum. All of these subjects eventually became negative for HBeAg and had a marked improvement in serum aminotrasferase activities. The difference in seroconversion rate between treated patients and non-treated controls (35% vs. 14%) did not reach statistical significance.

DISCUSSION

This trial was undertaken to evaluate whether a four month course of rIFNα would induce seroconversion from HBeAg to anti-HBe as well as remission in disease activity in patients with chronic type B hepatitis. The findings in this study indicate that such a prolonged course of interferon is associated with the loss of viral replication and a decrease in disease activity in approximately one third of patients. The occurrence of these favorable changes was not significantly greater than that in the control group. Patients were able to administer interferon themselves at home, and prolonged treatment was relatively well tolerated. Fever, myalgias, arthralgias and bone marrow suppression occurred almost universally but tended to be mild and usually did not require discontinuation of therapy.

Interferon was first used in therapy of chronic type B hepatitis almost ten years ago (8). Early studies confirmed a rapid and reproducible reduction in Dane particle-associated viral markers with interferon therapy (9). In many cases however, the levels of these markers returned to pre-treatment values when the interferon was discontinued. It has become apparent that a second and perhaps independent event occurs in some patients after approximately two months of interferon therapy. At this time, HBeAg is cleared from serum and this development is often accompanied by an apparent exacerbation of disease activity, serum ALT activities becoming markedly elevated. A similar sequence of events may occur in non-treated patients (10). These observations suggest that an alteration in immune function may be responsible for cessation of viral replication in some patients.

Randomized, controlled clinical trials of rIFNα therapy in chronic type B hepatitis have become necessary to determine whether rIFNα induces seroconversion from HBeAg to anti-HBe more frequently than the spontaneous rate of such seroconversion. Although the present study demonstrated this to be the case, the difference in response rate between the treated and control groups did not reach statistical significance. In other studies of chronic type B hepatitis interferon treatment was associated with a sustained decrease in viral replication and disease activity in 0% to 46% of cases (11,12). Generally, the numbers of patients studied have been small and many variables have made the studies difficult to compare. Such variables include the use of different types of interferon (recombinant, lymphoblastoid) varying doses and treatment regimens as well as the addition of other agents (e.g. corticosteroids, acyclovir) and different types of patients treated (e.g. male homosexuals, varying ethnic groups). At best though though, interferon as presently used has induced permanent remission of type B hepatitis in no more than 50% of cases (13).

In this study we have shown that although more patients seroconverted from HBeAg to anti-HBe when treated with rIFNα, the difference did not reach stastical significance. This suggests that rIFNα should not be discarded, but that other treatment strategies should be devised. The most promising approach at present is to use alpha interferon in combination with other anti-viral or immunomodulatory drugs.

Our studies of HBV DNA in liver by molecular hybridization demonstrated that when seroconversion from HBeAg to anti-HBe occurs, the viral copy number decreases markedly and in some cases the virus is apparently eliminated. These changes are associated with disappearance of replicative viral DNA forms from the liver, but some patients may be left with residual integrated chromosomal HBV DNA. These findings should encourage attempts at early treatment of chronic type B hepatitis with a view toward shortening the duration of viral replication and thus possibly minimizing the risk of integration of the HBV genome into the host chromosomal DNA. Achievement of this goal may decrease the risk of subsequently developing hepatocellular carcinoma.

REFERENCES

1. Sobelavsky O: HBV as a global problem, in Viral Hepatitis Eds. Vyas GN, Cohen SN and Schmid R. p.347. Philadelphia, The Franklin Institute Press, 1978.
2. Sherlock S and Thomas HC. Treatment of chronic hepatitis due to hepatitis B. Lancet 1985;ii:1343-1346.
3. Gregory PB. Interferon in chronic hepatitis B. (Editorial) Gastroenterology 1986;90:237-240.
4. Hoofnagle JH, Davis GL, Pappas SC, Hanson RG, Peters M, Avigan M, Waggoner JG, Jones EA and Seeff LB. A short course of prednisolone in chronic type B hepatitis. Ann Int Med 1986;104:12-16.
5. Hoofnagle JH, Davis GL, Hanson RG, Pappas SC, Peters MG, Avigan MI, Waggoner JG, Howard R, Jones EA and Straus SE. Treatment of chronic type B hepatitis with multiple ten-day courses of adenine arabinoside monophosphate. J Med Virol 1985;15:121-128.
6. Dooley JS, Davis GL, Peters M, Waggoner JG, Goodman Z and Hoofnagle JH. Pilot study of recombinant human α-interferon for chronic type B hepatitis. Gastroenterology 1986;90:150-157.
7. Dusheiko GM, Di Bisceglie AM, Bowyer S, Sacks E, Ritchie M, Schoub B and Kew MC. Prolonged recombinant leucocyte alpha interferon treatment of chronic hepatitis B. Hepatology 1985;5:556-560.
8. Greenberg HB, Pollard RB, Lutwick LI, Gregory PB, Robinson WS and Merigan TC. Effect of human leukocyte interferon on hepatitis B virus infection in patients with chronic active hepatitis. N Eng J Med 1976;295:517-522.

9. Scullard GH, Andres LL, Greenberg HB, Smith JL, Sawhney WK, Neal EA, Mahal AS, Popper H, Merigan TC, Robinson WS and Gregory PB. Hepatology 1981;1:228-232.
10. Dusheiko GM, Bowyer SM, Paterson A, Song E, Di Bisceglie A and Kew MC. Clinical and serological events accompanying changes in hepatitis B viral replication: case reports. Liver 1985;5:77-83.
11. Omata M, Yokosuka O, Imazeki F and Okuda K. in, Viral Hepatitis and Liver Disease. Eds. Vyas GN, Dienstag JL and Hoofnagle JH, p.670, Grune and Stratton, New York.
12. Lok ASF, Karayiannis P, Brown D, Fowler MJF, Monjardino J, Thomas HC and Sherlock S. Hepatology 1983;3:865-867.
13. Alexander G and Williams R. Antiviral treatment in chronic infection with hepatitis B virus. Br Med J 1986; 292:915-917.

TREATMENT OF CHRONIC HEPATITIS B INFECTION WITH ALPHA (LYMPHOBLASTOID) INTERFERON: RESPONSES IN DIFFERENT SOCIAL AND ETHNIC GROUPS IN COMPARISION TO ADENINE ARABINOSIDE MONOPHOSPHATE.

LEVER A.M.L., CHAN. G., SCULLY L.J., AND THOMAS H.C.

INTRODUCTION

There are approximately 200 million carriers of hepatitis B virus world-wide. This carriage rate is associated with a high incidence of chronic liver disease and a two hundred fold increased incidence of hepat-cellular carcinoma. These patients are those in whom acute infection has not been associated with clearance of the virus and development of immunity, but in whom this mechanism has broken down to produce the chr-. onic carrier state. There are probably many reasons for the chronic carrier state. Approximately 5÷10% of patients infected in adult life in the Western World (1) become chronic carriers whereas 90% of those in the Far East, infected around the time of birth, become carriers.(2). The hepatitis virus is not cytopathic and elimination depends on mobili-sation of host defense mechanisms to lyse virus infected cells and to clear circulating virions (3). Lysis of virus infected cells requires recognition of viral proteins in association with HLA class I antigens on the surface of the cell and it is known that on both uninfected liver and in the chronic carrier state there is very little HLA protein display on the surface of the hepatocyte (4). In the animal model of acute viral hepatitis, this HLA display rises following the detection of interferon. in the circulation and the rise corresponds with biochemical evidence of liver damage.

A reduced ability to produce interferon has been detected in chronic carriers in the Western World (5.,6.), and this may relate to their inability to clear the virus. We have shown that interferon treatment can enhance HLA protein display on the hepatocyte and also increase the level of an interferon activated enzyme, 2-5A synthetase. Untreated, the natural history of chronic hepatitis B carriage is shown in figure 1. Approximately 5-15% of chronic carriers per year will seroconvert from HBeAg to anti-HBe antibody with loss of evidence of viral replication and, following a seroconversion hepatitis which may be quite prolonged, a decrease in liver inflammation. This seroconversion hepatitis may be associated with the development of cirrhosis, particularly if it is severe and prolonged.

TREATMENT WITH ADENINE ARABINOSIDE MONOPHOSPHATE

In a randomised controlled trial of adenine arabinoside monophosphate, given for a four week course, a seroconversion rate of 40% was seen in a study at the Royal Free Hospital, with no change in the control group(7). In a second trial comparing one month of adenine arabinoside to eight weeks of the drug, the seroconversion rate was higher in the patients treated with the shorter course, and this is probably associated with the immunosuppressive effect of ARA-AMP (8). In patients who were treated

(continued........)

478

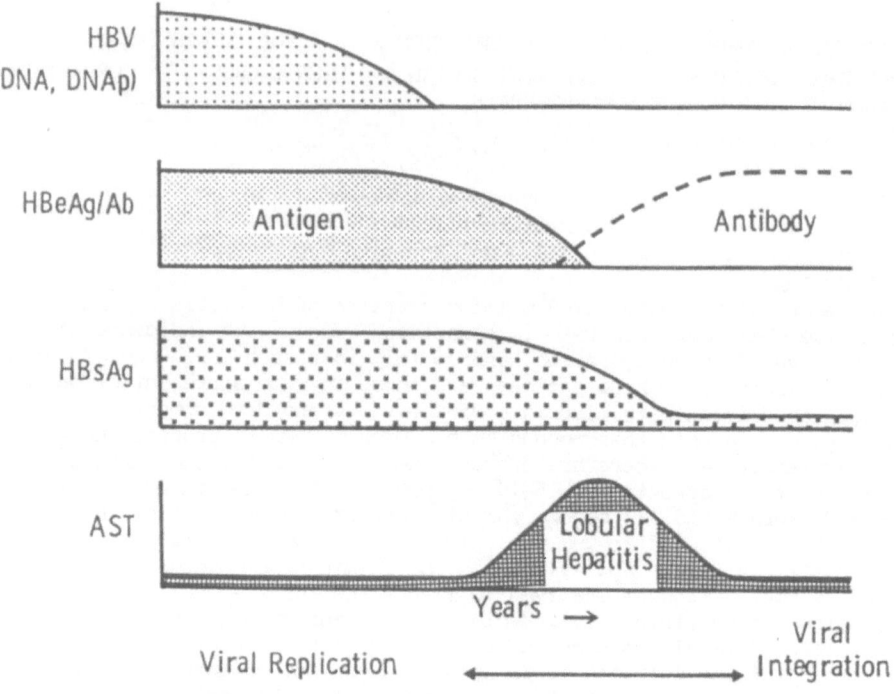

Fig.1. Natural history of chronic hepatitis B infection.
   - HBV DNA, HBeAg, HBsAg detectable in serum during viral replication.
   - Loss of HBV DNA and HBeAg during seroconversion hepatitis.
   - HBsAg detectable at late stage became integrated viral sequences.

successfully, seroconversion occured soon after ARA-AMP was discontinued
presumably when the host immune response had recovered (Figure II). The
longer course had probably produced a more profound immunosuppression from
which the host was less easily able to recover and the virus was able to
establish its mechanisms of chronicity once more. Similar results to these
were reported from France (9), treating predominantly heterosexual patients
whereas (10), in the United States, in spite of an early promising response, all patients relapsed. Following analysis of our results, it appeared that homosexual carriers did not respond at all to ARA-AMP whereas
heterosexual carriers gave a 50-60% response rate (11), this was in the
absence of evidence of HTLV-III infection in the homosexual group. It
would appear that these patients have additional factors causing immunodeficiency.

RESPONSE TO INTERFERON
   A recent study of chronic carriers of hepatitis B virus treated with
lymphoblastoid interferon for three or six months, has shown a seroconver-

(continued......)

-conversion rate of approximately 50% (12). Analysis of the data shows that both homosexuals and heterosexulas have a good response to this treatment, although in the heterosexuals the response rate is probably nearer 70%. Of note, however, is the fact that all the Chinese patients treated in this study, who were presumably infected at birth, none of these responded and alterations in their HBV-DNA, which occurred during treatment, relapsed rapidly on discontinuation of therapy. In successfull treatment, initiation of seroconversion hepatitis was seen (Fig III).

TREATMENT OF CHRONIC HBV INFECTION

ARA-A MONOPHOSPHATE
(INDUCTION/MAINTENANCE FOR 1 MONTH)

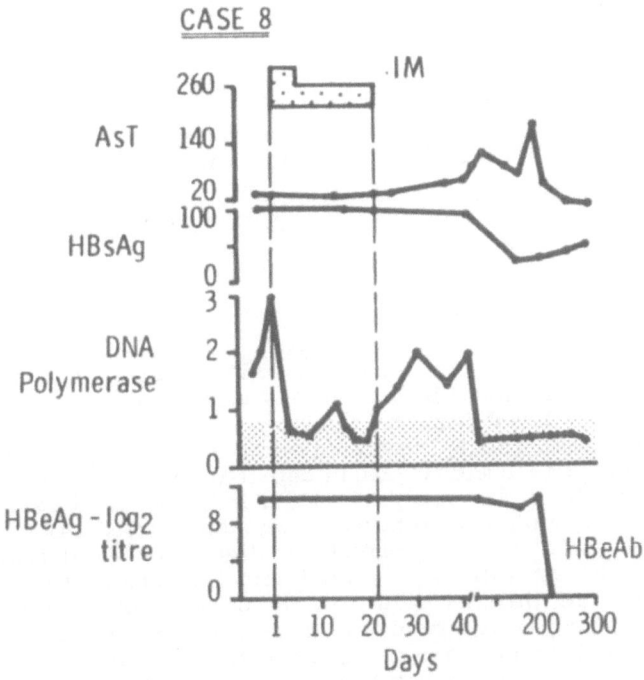

Fig 2: Response to ARA-AMP. Seroconversion hepatitis occurs after treatment has been stopped.

480

with loss of HBeAg between the sixth and eighth week of treatment and development of HBe antibody in half the cases. These patients have now been followed-up for approximately one year and there is no evidence of relapse (Figure IV).

SEROCONVERSION INDUCED BY INTERFERON

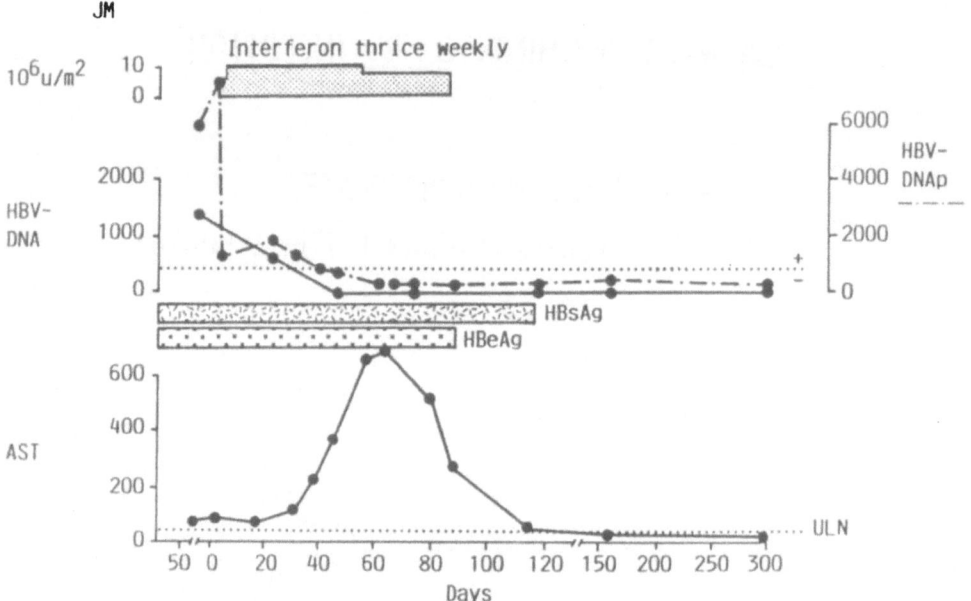

Fig 3. Seroconversion induced by interferon 12 weeks treatment, lobular hepatitis occurs at 10th week followed by loss of HBV markers and normalisation of liver function tests.

In patients treated with interferon who responded, it was noted that the level of serum Beta-2 microglobulin detected in their circulation was higher than those who did not respond indicating that, at a cellular level, interferon was having more effect in the former group (4). We have evidence that interferon can directly effect the display of viral proteins in a chronic carrier, stimulating the production of core protein (nucleocapsid protein) (13) and this may relate to the presence of consensus sequences on the hepatitis genome homologous to sequences found on the human genome upstream from interferon induced genes (14). The presence of these sequences in the hepatitis virus may, by trans-acting effect, reduce the ability of interferon or its second messenger to activate host genes which contribute to the antiviral state induced by interferon (Figure V). Thus in the chronic carrier with high level HBV replication, a reduced efficacy of treatment with interferon would be expected to be seen and these are our findings.

(continued......)

RESPONSE* RATE IN HETEROSEXUAL AND HOMOSEXUAL MALES+

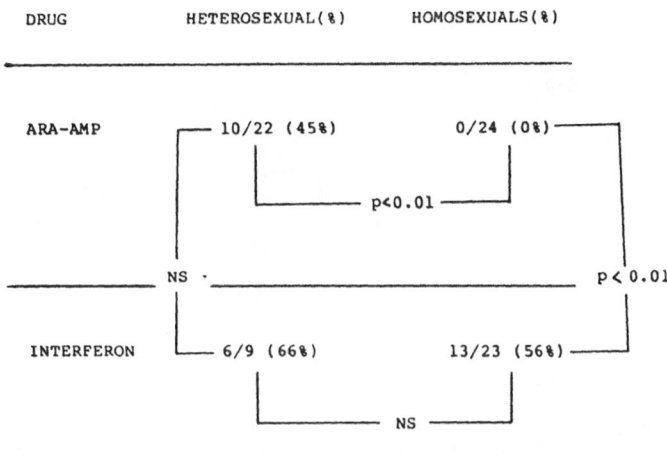

Fig 4: Comparitive response rates of homosexual and heterosexual males to ARA-AMP and lymphoblastoid interferon.

---

* A response is defined as a sustained loss of HBeAg and
HBV-DNA within 1 year of completion of treatment.

+ Results of 2 randomised trials.

POSSIBLE 'TRANS' EFFCT OF IFN CONSENSUS SEQUENCE OF HBV

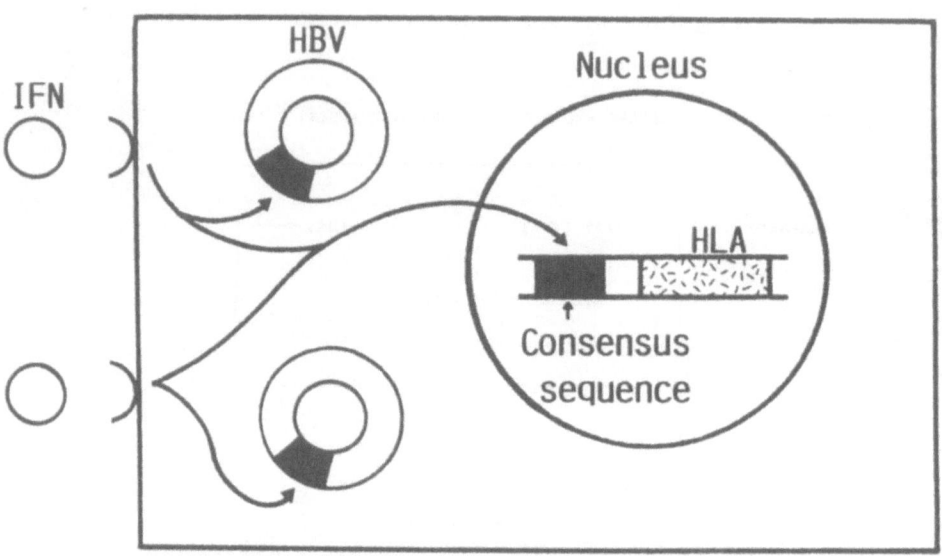

Fig 5: Interferon or second messenger diverted from
host consensus sequences to identical sequences
on HBV genome.

CONCLUSIONS

Interferon appears to be an effective treatment for patients infected after the neotal period in a proportion of whom there may be a defect in interferon production.. Its immunostimulatory effect renders it more effective at treating nomosexual patients than adenine arabinoside monophosphate. In patients infected around the time of birth, there is a high chronicity rate and no evidence of an interferon deficiency . This would not be expected to be as effective a treatment and this is consisttnt with our findings. The presence of consensus sequences on the hepatitis genome, which may be responding to interferon, may make the infected cell less able to be induced into an antiviral statt by the action of exogenous interferon, and in those patients who do not respond to interferon alone, a combination of a powerful virus inhibitor such as ARA-AMP followed by interferon may be the opitmal therapy.

REFERENCES:

1. Violà L.A., Barrison I.G., Coleman J.C., Paradinas F.J., Fluker J.L., Murray-Lyon I.M. Natural History of liver disease in chronic hepatitis B surface antigen carriers: a survey of 100 patients from Great Britain. Lancet 1981: ii; 1156-1159.

2. Beasley R.P., Trepo C., Stevens C.E. and Szmuressw (1977). The e antigen and vertical transmission of hepatitis B surface antigen. American Journal of Epidemiology 105: 94-98.

3. Dudley F., Fox R.A. and Sherlock S. (1972) Cellular immunity and hepatitis- associated Australia anitigen liver disease. Lancet; 723-726.

4. Pignatelli M., Waters J., and Lever A.M.L.,et al (1986a) HLA class 1 antigen on hepatocytes membranes; increased expression during acute hepatitis B and during interferon therapy of hepatitis B. Hepatology (in press).

5. Abb J., Zachoval R., Eisenberg J et al (1985) Production of interferon alpha and interferon gamma by peripheral blood leucocytes from patients with chronic hepatitis B virus infection. Journal of Medical Virology 16;171-176.

6. Lever A.M.L., Ikeda.T., Thomas H.C.,Deficient production of alpha interferon in patients with chronic hepatitis B. Poster IV. 21 l.s.I.R /T N O Meeting 1986.

7. Weller I.V.D., Lok A.S.F., Mindel A et al (1985). A randomised contr-olled trial of adenine arabinoside 5 monophosphate (Ana-AMP) in chro-nic hepatitis B virus infection. Gut 25: 745-751.

8. Weller I.V.D., Bassendine M.F., Craxi A., et al (1982) Successful treatment of HBs and HBeAg positive chronic liver disease: prolonged inhibition of viral replication by highly soluble adenine arabinoside 5'-monophosphate (Ara-AMP). Gut 23: 717-723.

9. Trepo C., Hantz O., Ouzan D. et al (1984) Therapeutic efficacy of Ara-AMP in symptomatic HBe Ag positive CAH: a randomised placibo control study. Hepatology 4: 1055 (abstract).

10. Hoofnagle J.H., Hansor R.G., Minuk G.Y. et al (1984) Randomised contr-olled trial of adenine arabinoside monophosphate for chronic type B hepatitis. Gastroenterology 86: 150-157.

11. Novick D.M., Lok A.S.F., and Thomas H.C., (1984) Diminished responsiv-eness of homosexual men to antiviral therapy for HBsAg-positive chronic liver disease. Journal of Hepatology. 1: 29-35.

12. Scully L.,Thomas H.C. (1985): Antiviral therapy in chronic hepatitis B infection. British Medical Bulletin 41: 374-380.

13. Thomas H.C., Pignatelli M. and Lever A.M.L. (1986) Homology between

(continued......)

484

continued.....

HBV-DNA and a sequence regulating the interferon induced anti-viral
system: a possible mechanism of persistent infection. Journal of Medical
Virology (in press).

14. Friedman R.L. Stark G.R , (1985) interferon induced transcription of
    HLA and Metallothionein genes containing homologous upstream sequences.
    Nature 314: 637-639.

INTERFERON ALFA-n1 (WELLFERON)® IN SEVERE RECURRENT GENITAL WARTS:
REPORT OF A MULTI-STUDY PROGRAM

P.K. WECK, K.F. TROFATTER, R.C. REICHMAN, J.P. MICHA, J. MILLS,
P.H. BRITT AND J.K. WHISNANT
Burroughs Wellcome Co., Research Triangle Park, NC, Duke Medical Center,
Durham, NC, U. Rochester Medical Center, Rochester, NY, U.C. Irvine
Medical Center, Orange, CA, U.C. San Francisco, San Francisco, CA.

INTRODUCTION
   Genital warts, or condylomata acuminata, are benign epithelial
proliferations that occur in both men and women. These warty growths
are associated with types 6 and 11 human papillomaviruses (1,2). The
clinical lesion is usually a rough exophytic wart and in women may be
found on the external genitalia, vagina, perineum and/or anus. The
mucosa of the uterine cervix can also support growth of flat warts. In
men usual sites are the penis, scrotum and/or anus, where both exophytic
and flat warts may be seen. Many cases go undiagnosed unless a magnify-
ing device is used in conjunction with acetic acid to turn the lesion
white.
   The sexual mode of transmission of this disease is well established;
the prevalence of genital warts correlates with sexual activity. These
lesions are highly contagious and up to 70% of the partners of patients
with genital warts have or develop lesions. The incubation period is
from three to nine months, estimated by experimental transmission of
warts to human volunteers. The peak incidence years for females preceed
those for males by about five years. It has been estimated that between
1% and 10% of U.S. women of child-bearing age have genital warts. The
diagnosis may be made by recognition of koilocytes in cervical smears
or by direct colposcopic examination. In 1981, an estimated 946,000
physician consultations were sought for condylomata acuminata (3).
   An estimated one-third of the primary genital warts regress spontaneously within six months; the remaining two-thirds do not. Left
untreated, the warts grow in size and spread to other anatomic locations.
Persistent lesions will cause many patients considerable discomfort,
personal trauma and difficulties in hygiene. Prolonged infection may
precede the development of cervical intraepithelial neoplasia and carcinoma.
   Conventional therapies include topical agents, surgical techniques,
laser therapy and immunotherapy. Podophyllin is the most widely used
caustic agent but it is a known mutagen and teratogen. Surgical removal
is often painful, requires a time period for healing and leaves scars.
Laser vaporization provides immediate relief for the patient but does
not necessarily eradicate the virus infection (4). Recurrences are
known to occur, even though a patient's lesions are no longer detectable either clinically or histologically. Because of the persistent
and recurrent nature of this disease, repeated therapy with several
different modalities is often required.
   Preliminary studies with human leukocyte interferon administered
topically or intramuscularly demonstrated the potential of this type of
therapy (5,6). These trials reported complete eradication of warts in

80 - 90% of patients treated with crude leukocyte interferon. These early results and responses in another HPV disease, juvenile laryngeal papillomatosis, prompted additional trials. Systemic, intralesional and topical administration of interferons have been used for genital warts.

Based on the success of early trials, a progression of controlled trials in severe recurrent genital warts has been conducted using interferon alfa-n1, Wellferon®. The goal of this program has been to identify tolerable and effective doses for severe recurrent or persistent disease. Proper evaluation of disease response has been accomplished by objective measurements at predetermined time intervals. The number, extent and anatomical location of lesions have been recorded. Interferon-associated side effects and toxicities have been evaluated by a standard grading system across different trials. This series of studies has established that interferon alfa-n1 administered systemically at biologically active low doses ($1 MU/m^2$) is effective in treating severe genital warts.

PATIENTS AND METHODS

All patients treated in this series of trials had severe recurrent disease as judged by failure of multiple previous treatments. Histologic confirmation of condylomata acuminata was required. Patients were excluded from study if they had any of the following: renal dysfunction (serum creatinine >1.4 mg/dL), anemia (hemoglobin <10 gm/dL), thrombocytopenia (platelets <100,000/mm^3), leukopenia (WBC <4000/mm^3), clinical or laboratory coagulopathy, hepatic dysfunction, preexisting immunodeficiency, or malignant degeneration in biopsied tissue. Pregnant women or women not practicing adequate birth control were excluded. Informed patient consent was the responsibility of the investigator and was approved by the institutonal review board of each participating center.

Genital warts were documented by recording the bidimensional measurements and numbers of lesions on a schematic diagram of the external genitalia. Photographs were made initially and at each subsequent evaluation. Measurements of condylomatous lesions were made every two weeks during interferon therapy and at monthly intervals for the remainder of the study. Protocol management decisions were based on global disease assessments according to the following definitions: complete response = total clearance of warts, partial response = >50% clearance or reduction in size of warts, or no response = <50% reduction in sum of products of bidimensional measurements. Overall disease response was determined as the percent of baseline lesion measurements.

Routine hematologic evaluation including platelet counts and relevant chemistries were performed at entry. Hematologic, hepatic and renal tests were repeated weekly for the first two to four weeks and every two weeks while the patient received interferon. Sera were obtained before interferon injections, on day one, and post-injection on designated days throughout the study. Serum specimens were assayed for interferon activity in a microtiter system standardized against the NIH human leukocyte standard (G023-901-527).

Symptomatic side effects including fever, chills, fatigue, malaise, nausea/vomiting, and headache were recorded daily by the patients. These and other noted side effects were graded as absent (0), mild (1), moderate (2) or severe (3). Standardized definitions were employed in all trials. This uniform grading scale has been described in detail and is available upon request. Clinical laboratory values and side effects were evaluated at regular intervals during interferon dosing and dose modifications made on the basis of severity grading.

Interferon alfa-n1 (Wellferon®) is a preparation of highly purifed alpha interferons derived from the human lymphoblastoid cell line, Namalwa, following stimulation with Sendai virus (7). This interferon contains at least 16 alpha interferon subytpes and has a specific activity of 1-2 x 10^8 IU/mg protein prior to formulation with human serum albumin. Interferon administration was by the intramuscular or subcutaneous route at doses of 1, 3, 5, or 10 MU/m^2 depending on a particular protocol design. Dosing was usually initiated on a daily basis followed by thrice weekly; however, this was dictated by trial design.

RESULTS

Efficacy Trials
    Three independent trials were conducted to determine a safe, tolerable and effective dose/schedule for the adminstration of Wellferon. These are outlined in Table 1 below.

TABLE 1. Efficacy Trials of Wellferon® in Severe Recurrent Genital Warts

Study	Dose	N	Response (% Patients)		
			Complete	>50% Clear	<50%
Pilot	5 MU/m^2	16	25	69	6
Dose Comparison					
Females	1 MU/m^2	28	21	46	33
Females	3 MU/m^2	43	39	30	31
Males	3 MU/m^2	20	20	30	50
Schedule Comparison (Males)					
5 days/wk x 6	3 MU/m^2	7	43	29	28
Once weekly	10 MU/m^2	10	0	50	50

A pilot study confirmed the reported responsiveness of gential warts to interferon therapy (8). Eligible patients had to have failed at least 10 previous treatments for their condylomata acuminata. Disease measurements at entry ranged from 0.6 to 160.0 cm^2. Wellferon® was administered intramuscularly at 5 MU/m^2/day for 28 days followed by three times weekly for two weeks. At the end of this primary treatment period, 4 of 16 evaluable patients had complete clearance and 11 had >50% but <100% clearance. The one patient classified as a non-responder had a 27% reduction in disease area by week 6. These results demonstrated the activity of Wellferon in genital warts. However, once the dose of 5 MU/m^2 proved too high for this patient population, all patients required at least one level of dose reduction during study. Doses were reduced because of leucopenia and/or intolerable side effects. Statistically significant decreases in mean white counts, lymphocytes, neutrophils and platelets were noted each week. Mean SGOT values were increased during the first six weeks. These laboratory values returned toward normal during the three times per week dosing. The most common interferon-associated side effects were fever, chills, fatigue, headache and nausea/vomiting. Virtually all patients had some degree of chills, fever and/or fatigue on initial dosing. Fever and chills subsided markedly during the first week on study with <10% of patients reporting these as mild by week three. Reductions in doses did not result in loss of response. Overall response evaluations at the end of study showed that 11 of 16 (69%) cleared completely, four (25%) had partial clearance and one (6%) was classified as a non-responder.

Dose Comparison Study: 1 versus 3 MU/m^2

Based on the pilot study results, a multi-center dose comparison trial was designed to test a lower dose in patients with recurrent disease (9). This trial randomized female patients to high or low dose given daily for 14 days then three times weekly for one month. Two additional groups were treated under this protocol. A group of females were treated at 3 MU/m^2 and the protocol was extended to male patients at a dose of 3 MU/m^2. The demographic information for this study is presented below in Table 2. Disease duration, lesion areas and prior therapy are characteristic of resistant/recurrent patients whose natural course does not include spontaneous remission. In general, the two dosage groups were similar in age, lesion area, disease duration and previous therapies.

TABLE 2. Wellferon in Genital Warts: 1 versus 3 MU/m^2 Dose Comparison

Patient Demographics		1 MU/m^2 (n=36)	3 MU/m^2 (n=77)
Sex	Female	83%	64%
	Male	17%	36%
Age (Median in years)		27.9	29.5
Race	White	86%	87%
	Black	11%	10%
	Other	3%	3%
Years from diagnosis	Median	0.9	0.9
	Range	0.2-4.2	0.1-33.0
Lesion area (cm^2)	Median	8.2	11.0
Previous Therapies	Podophyllin	92%	74%
	Cryotherapy	50%	38%
	Cautery	22%	12%
	Surgery	11%	21%
	Laser	3%	6%

There were 114 patients entered on the dose comparison study and 113 were considered evaluable for safety and tolerance of Wellferon®. Fifty-one female patients were rigidly randomized to dose comparison and formed the basis of the analysis. Responses following the primary six-week course for 28 treated at 1 MU/m^2 and 23 treated at 3 MU/m^2 are shown in Table 3. Comparison of measurements to baseline areas demonstrated highly significant decreases at weeks 3, 5 and 7 for both groups. Overall responses for both groups were similar, also.

TABLE 3. Responses to 1 versus 3 MU/m^2 Wellferon: Randomized Females

Disease Assessment	1 MU/m^2 (n=28)	3 MU/m^2 (n=23)
Lesion Areas by Week		
Baseline	1.70 cm^2	1.77 cm^2
Week 3	0.38 (<.001)	0.43 (<.001)
Week 5	0.36 (.001)	0.29 (.002)
Week 7	0.35 (<.001)	0.19 (.004)
Overall Response (Week 7)		
Complete	21%	39%
>50% Clearance	46%	30%
<50% Clearance	29%	9%
Progression	4%	22%

Comparison of efficacy for the two groups was examined by Wilcoxon Rank Sum Test for absolute changes and percent changes from baseline. Although the differences between the two groups favored the higher dose, there were no statistically significant differences (p-values ranged from 0.194 to 0.626). This comparison was extended by use of cumulative response curves. These curves illustrate the cumulative proportion of patients achieving the specified level of disease reduction at the different evaluation points. The data indicated that 3 MU/m^2 induced a more rapid response but the overall response rates after six weeks were not different.

Dose reductions, interruptions or withdrawals from study were associated with the higher dose of 3 MU/m^2. No dose reductions for laboratory alterations or symptomatic side effects were required for patients treated at 1 MU/m^2. A total of nine patients on the higher dose had dose modifications; five for leukopenia, one for thrombocytopenia and three for side effects. One patient at the low and eight patients at the high-dose withdrew from the study. The majority of these were due to intolerance of interferon induced side effects.

Dose-related differences were noted for the incidence of side effects and changes in laboratory parameters for the two groups. Patients at 3 MU/m^2 had higher overall incidences of fever, chills, fatigue/malaise, headache and nausea/vomiting. Scoring for the composite symptom "flu-syndrome" showed a statistically significant difference during week one. The most frequently observed laboratory alterations included decreases in total white counts, lymphocytes and platelets and increases in SGOT and SGPT values. These were observed at both doses.

Maximum changes occurred during the first two weeks, were transient, and returned towards normal over Weeks 3 to 7.

Schedule Comparison in Male Patients

The purpose of this study was to assess the response of resistant/recurrent anogenital warts in males given 6 weeks of therapy with Wellferon administered at different doses and schedules (10). Patients were randomized to receive interferon intramuscularly at 3 MU/m^2 daily, 5 days/week, for 6 weeks or 10 MU/m^2 once weekly for 6 weeks. Of 17 randomized patients, 7 were treated at the low daily dose and 10 at the high weekly dose. Three patients on daily therapy experienced complete clearance whereas none at 10 MU/m^2 cleared. Two patients on daily and five on weekly had $\geq 50\%$ clearance of lesions after 6 weeks of therapy. Three patients on the weekly schedule did not respond. Three patients (1 daily and 2 weekly) withdrew from study due to severe fatigue/malaise and were not evaluable. All patients receiving weekly injections reported fever, chills and severe fatigue. The overall incidence of the six most common side effects were remarkably higher for the 10 MU/m^2 group. This confirmed the efficacy of systemic interferon alfa-n1 for perianal condylomata acuminata. Patient intolerance of 10 MU/m^2 weekly makes this an unlikely treatment regimen for this disease.

CONCLUSIONS/FUTURE TRIALS

The data obtained during the primary six week treatment periods in these studies agree with other reports for interferons in the treatment of genital warts. Systemic administration of Wellferon at doses of 1 or 3 MU/m^2 is active in reducing lesion areas and is associated with the usual biological side effects of interferons. Unlike previous trials, we attempted to treat a defined patient population with severe disease. All patients had a history of resistant/recurrent disease. This type of disease does not undergo spontaneous regression. It can persist over many years and is proven recalcitrant to conventional treatment modalities. Low dose Wellferon therapy is an alternative therapy for such patients. Systemic administration of this potent biological agent, as opposed to intralesional, appears to have certain advantages in treating patients with severe disease. The pain associated with multiple injections of highly keratinized genital warts can be prohibitive. The series of trials described above has resulted in the following conclusions:

- 5MU/m^2 dosing gave significant clinical responses in 15 of 16 patients treated in a pilot study.
- symptomatic side effects and laboratory alterations made doses of 5 MU/m^2 unaccepatable.
- doses of 1 or 3 MU/m^2 resulted in similar overall response rates; however, clearance occurred sooner at 3 MU/m^2.
- interferon side effects are dose-related; 1 MU/m^2 doses are tolerated well and are effective.
- 10 MU/m^2 once weekly is not as effective or well tolerated as lower doses and probably not a feasible regimen.
- responses in males appear to be less frequent than in females.

On-going and future trials with Wellferon in resistant condylomata acuminata will focus on combination with other agents to improve response rates or modify symptomatic side effects.

REFERENCES

1. Gissmann L and zur Hausen H: Partial characterization of viral DNA from human genital warts. Int J Cancer 25:605, 1980.
2. Gissmann L, DeVilliers EM, zur Hausen H: Analysis of human genital warts and other genital tumors for human papillomavirus type 6 DNA. Int J Cancer 29:143, 1982.
3. Condyloma acuminatum - United States, 1966-1981. Morbid Mortal Weekly Rep 32:306, 1983. 4.Ferenczy A, Mitao M, Nagai N, Silverstein S, Crum CP: Latent papillomavirus and recurring genital warts. N Engl J Med 313:784, 1985
5. Ikic D, Bosnic N, Smerdel S, Jusic D, Soos E, Delimar N: Double-blind clinical study with human leukocyte interferon in the therapy of condylomata acuminata. Proc Symp Clin Use Interferon, Zagreb, 1975.
6. Ikic D, Orescanin J, Krusic Z: Preliminary study of the effect of human leukocyte interferon on condylomata acuminata in women. Proc Symp Clin Use Interferon, Zagreb, 1975.
7. Finter NB and Fantes KH: The purity and safety of interferons prepared for clinical use: the case of lymphoblastoid interferon. Gresser I(ed): Interferon. London: Academic Press, 1980.
8. Gall SA, Hughes CE, Mounts P, et al.: Efficacy of human lymphoblastoid interferon in the therapy of resistant condylomata acuminata. Obstet Gynecol 67:643, 1986.
9. Reichman RC, Micha J, Bonnez W, Trofatter K, Weck P, Gall S: A multicenter study of interferon alfa-n1 (Wellferon) treatment for refractory condyloma acuminatum. Proc Intersci Conf Antimicrobial Agents Chemother, 1985.
10. Gottlieb A, Crass R, Mills J: Human interferon alfa-n1 for treatment of severe perianal and rectal condylomata acuminata. Proc Intersci Conf Antimicrobial Agents Chemother, 1986.

ACTIVITY OF INTRALESIONAL INTERFERON ALFA-2B IN VIRAL AND MALIGNANT SKIN DISEASES

KENNETH SMILES, EDWIN PEETS, DANIEL TANNER, & EUGENE TAYLOR

Schering Plough Corp., Kenilworth, N.J., USA 07033

1. INTRODUCTION

Interferon alfa-2b (Intron A®) is a highly purified single molecular species of alpha interferon and represents one of the more than twenty that exist naturally in man (1). It is produced using recombinant DNA techniques in the bacterium E. coli and is purified to a specific activity of approximately 2×10^8 IU/mg of protein. In common with the other alpha interferons, it has potent antiviral and antiproliferative properties when tested in vitro using human cell lines and viruses (2-4). Additionally, it causes the activation of natural killer cells (5), can restore cytotoxicity to functionally deficient monocytes (5), and like other alpha interferons would be expected to enhance antigen expression on cell surfaces (6). These properties suggested that interferon alfa-2b might have clinical utility in treating viral and malignant diseases in man. The ability to achieve high local concentrations in the skin by intralesional injection with comparatively little systemic exposure suggested further that malignant and premalignant skin diseases and those of a viral etiology might be especially amenable to treatment with interferon alfa-2b.

2. GENITAL WARTS

Genital warts are a common sexually transmitted epidermal infection caused by the human papilloma virus. The incidence, severity, and seriousness of sequelae from this disease, especially genital carcinomas, appear to be increasing proportionally due to a lack of specific and effective therapies (7). Podophyllin and destructive therapies are not uniformly effective and their efficacy is accompanied by high rates of recurrence, possibly due to virus being present in the normal appearing skin around the wart margins (8).

Interferon alfa-2b was investigated as a treatment for genital warts in a series of similarly conducted, randomized, double blind, placebo controlled, parallel group studies involving over 400 patients. Patients had from one to five genital warts treated simultaneously, with all treated warts on each patient receiving either interferon alfa-2b or isotonic placebo injections. The warts were injected three times per week for three consecutive weeks, and then followed for at least an additional 9-12 weeks. Ninety-one patients with genital warts provided useful efficacy data in the study shown in Fig. 1. A single wart per patient was treated with either placebo, 10^5 or 10^6 IU of interferon alfa-2b. 10^6 IU was clearly effective in reducing mean lesion area, whereas 10^5 IU produced results similar to placebo. Fifty-three percent of the warts treated with 10^6 IU of interferon alfa-2b were cleared, while only 14% were cleared by placebo and, 19% with 10^5 IU. In a second study, 257 patients provided efficacy data, with 125 receiving 10^6 IU into each of one to three warts/patient and 132 receiving placebo injections. As shown in Fig. 2, 10^6 IU was clearly effective in reducing wart area. In this study, 43% of the patients receiving interferon alfa-2b had complete clearing, and an additional 26% had marked (>75%) improvement. Most patients responded optimally to treatment within one to two months after therapy. Sixty-seven percent of the placebo-treated patients had exacerbation, remained unchanged or had only slight improvement. In the third study, 70 patients provided the efficacy data on treating five warts

494

simultaneously. The patients received 10⁶ IU or placebo into each wart. Interferon alfa-2b was clearly effective in reducing wart volume (Fig. 3). At the end of treatment warts that received interferon alfa-2b had an average 61% decrease in volume, while placebo-treated warts had increased in volume 49%. In these patients with more severe disease, 51% of the warts treated with interferon cleared completely and an additional 19% improved markedly (>75%). Sixty-three percent of the placebo treated patients in this study had no more than an overall slight improvement in their warts.

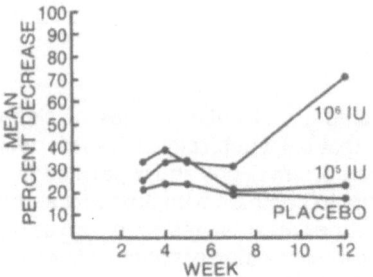

FIGURE 1. Decrease in wart area following treatment with 10⁶ IU, 10⁵ IU of interferon alfa-2b or placebo.

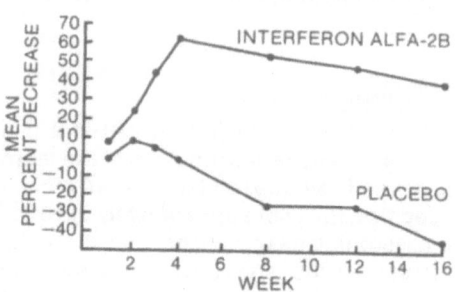

FIGURE 2. Decrease in wart area treating 1-3 warts with 10⁶ IU of interferon alfa-2b.

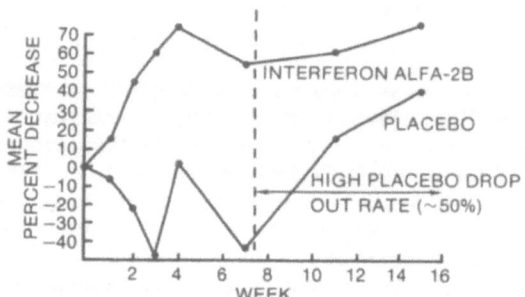

FIGURE 3. Decrease in wart volume treating 5 warts simultaneously with 10⁶ IU of interferon alfa-2b.

During these studies, many patients whose lesions didn't respond to the initial treatments (placebo or interferon) received interferon during a second, unblinded, open labeled phase. Neither prior placebo nor prior interferon treatment appeared to affect response to the open label interferon, since response rates were nearly identical following either blinded or open label treatment. Follow-up data beyond 12 weeks has been gathered. In those patients who could still be contacted and whose warts were cleared by interferon treatment, 70%-80% have so far remained free of disease. No evidence was found that sex, location, or prior therapy significantly altered the response to interferon. Newer and smaller warts tended to respond more favorably; however, this trend was not statistically significant (p>.05).

3. ACTINIC KERATOSES

Actinic keratoses are unsightly erythematous and scaling areas of dysplastic keratino-cytes within the epidermis that are caused by excessive and chronic exposure to ultraviolet light. Left untreated, these lesions can progress to invasive squamous cell carcinoma and may occasionally metastasize. Two double-blind, placebo controlled studies examined the dose required and number of doses required, respectively, to effectively treat these lesions with interferon alfa-2b.

In the first study, sixteen patients each had three lesions treated. Four patients were each randomly assigned to have all three of their lesions treated with either 5x10⁵ IU, 1x10⁵ IU, 1x10⁴ IU, or placebo. Treatments consisted of 0.10 cm³ isotonic injections into the base of each lesion three times per week for three consecutive weeks. The three treated and one to four untreated keratoses were then evaluated weekly during treatment and one month post-treatment. The results at the end of the study are shown in Table 1. A clear dose response was apparent one month after treatment. Of the twelve keratoses injected with the highest dose of 5x10⁵ IU, eleven were cured and one exhibited marked (>75%) improvement. None of the twelve placebo treated lesions cleared, nor was any effect evident on uninjected lesions in any dose group. The lower doses produced intermediate effects.

Table 1.— Effect of Nine Injections of a_2-Interferon at Variable Doses on Actinic Keratoses

Treatment Group	Cleared, No. (%)	Improvement, No. (%)*			No Change No. (%)
		Marked (75)	Moderate (50)	Mild (25)	
5x10⁵ IU per injection	11 (92)	1 (8)	. . .	. . .	. . .
1x10⁵ IU per injection	5 (42)	6 (50)	1 (8)	. . .	. . .
1x10⁴ IU per injection	7 (58)	1 (8)	2 (17)	. . .	2 (17)
Placebo	. . .	2 (17)	4 (33)	2 (17)	4 (33)

*Global assessment by examiner based on reduction of size, scale and erythema.

Using the most effective dose of 5x10⁵ IU, a second study was conducted to determine the number and spacing of the doses required to effect lesion clearing. Forty-five patients were enrolled in an unbalanced block design study consisting of five different dosing schedules. Within each dosing schedule three patients were treated with placebo and five to seven received interferon alfa-2b. The five schedules were: (1) a single dose; (2) three doses in one week; (3) three doses a week apart; (4) six doses over two weeks and (5) nine doses over three weeks. One patient failed to receive the proper number of injections and was excluded from the analysis. Evaluations were identical to the initial study except that uninjected lesions were not specifically evaluated and the timing of evaluations was altered slightly to allow for consistently timed follow-ups to two months in the post-treatment phase. The results at the last evaluation are shown in Table 2. In this study, six injections over two weeks produced clearing in 14 of 15 lesions and marked improvement in the remaining lesion. Thirteen percent of the placebo treated lesions resolved in this study. The unexpected lower efficacy in the nine injection group (67% cleared or markedly improved) compared to the six injection group in this study remains a paradox. Nevertheless, it is apparent that six doses are necessary to achieve a reliable therapeutic effect.

Table 2.—Effect of Different Dosing Schedules of Intralesional a_2-Interferon (IFN) (5×10^5) on Actinic Keratoses

Treatment Group			Improvement, No. (%)*			
No. of Injections/ wk	Total No. of Lesions	Cleared, No. (%)	Marked (75)	Moderate (50)	Mild (25)	No Change, No. (%)
1/0 IFN	18	2 (11)	5 (28)	4 (22)	3 (17)	4 (22)
3/1 IFN	21	10 (48)	6 (29)	2 (10)	1 (5)	2 (10)
3/3 IFN	18	9 (50)	3 (17)	2 (11)	4 (22)	. . .
6/2 IFN	15	14 (93)	1 (7)	. . .	. . .	. . .
9/3 IFN	15	7 (47)	3 (20)	3 (20)	2 (13)	. . .
All Groups Placebo	45	6 (13)	9 (20)	14 (31)	6 (13)	10 (22)

*Global assessment by examiner based on reduction of size, scale, and erythema.

4. MYCOSIS FUNGOIDES

Mycosis fungoides (MF) is a cutaneous T cell lymphoma that is usually slowly progressive and clinically restricted to the skin for many years. Atypical T cells with helper cell markers proliferate in the skin. Although the etiology of the disease is unknown, evidence exists suggesting that it may be a systemic disease of a single clone of malignant T cells, (9, 10) and/or the result of an abnormal immune response to chronic antigenic stimulation (11) or infection with a retrovirus (12, 13). Two controlled, double-blind studies evaluated the effects of interferon alfa-2b injected into cutaneous plaques of early MF.

The first study was of an unbalanced block design in two groups of patients employing three test sites within disease plaques on each patient. Nine patients in Group I were randomly assigned to have one site treated intralesionally with 2×10^6 IU of interferon alfa-2b three times per week for four weeks, a second site treated with a placebo ointment twice daily for four weeks, while the third site remained untreated. Three patients in Group II were randomly assigned to have placebo injections into one site, betamethasone dipropionate ointment (0.05%) applied to the second site, while the third site remained untreated. The test sites and the patient's overall disease status were evaluated throughout treatment and for a month following treatment. Post-treatment biopsies were performed at the test sites three days following the end of treatment. Clinical improvement and statistically significant histological improvement (p<.05) were seen at the interferon treated sites (Fig. 4 and Table 3).

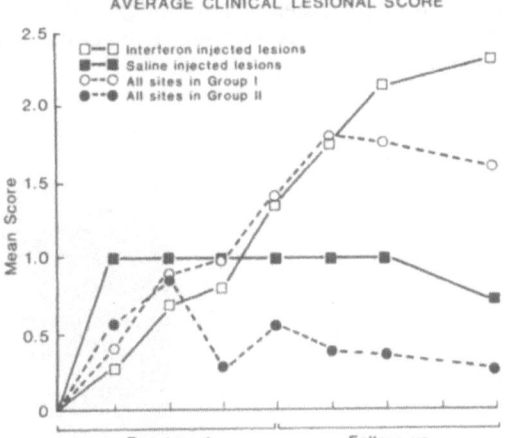

FIGURE 4. Average clinical lesion score (0 = no change, 4 = cleared) over the course of the study.

	Table 3		
	Biopsy Scores		
	Treatment	N	Average Biopsy Score
Group I	Interferon alfa-2b	9	1.89
	Placebo ointment	9	0.89
	Untreated	9	0.67
Group II	Placebo Injection	3	0.00
	Betamethasone dipropionate ointment	3	0.33
	Untreated	3	0.00

*0 = unchanged from pretreatment; 4 = normal skin histology

The second study utilized five test sites within plaques on each of six patients and was of a balanced, paired design. Each of two sites on each patient were randomly assigned to receive 1x10⁶ IU of interferon alfa-2b three times per week for four consecutive weeks. Two sites received isotonic placebo injections and a single site was untreated. These patients were evaluated in the same manner as the first study except that the post-treatment biopsy in one interferon and one placebo treated site was delayed until four weeks following therapy. By one month after therapy, 10 of 12 interferon treated sites had clinically cleared areas of 2-3cm diameter around the injection site while the remaining two showed marked improvement of a similar sized area. With the exception of a single patient, the placebo and

498

untreated sites showed very little change. In all cases, clearing was preceded by a mild inflammatory reaction which upon resolution left clinically normal appearing skin. The results are shown in Fig. 5. Fig. 6 shows an interferon treated site prior to therapy and after one month of additional follow-up. These cleared sites have remained clear during extended follow-up which in some cases is now longer than six months. Histologic examination of the clinically clear areas revealed substantial improvement but with continued presence of disease.

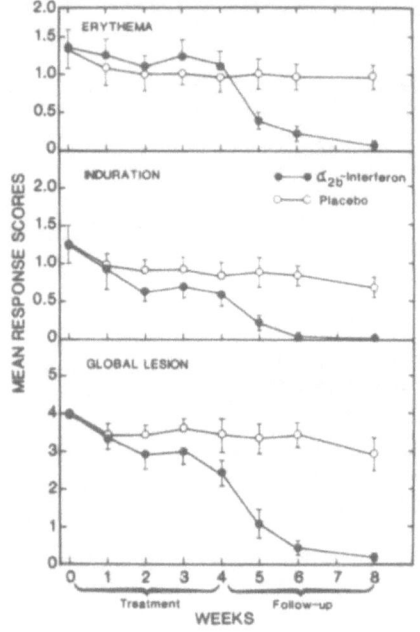

FIGURE 5. Mean response scores during the course of the study (erythema and induration: 0 = none, 2 = moderate; global: 0 = cleared, 5 = exacerbation).

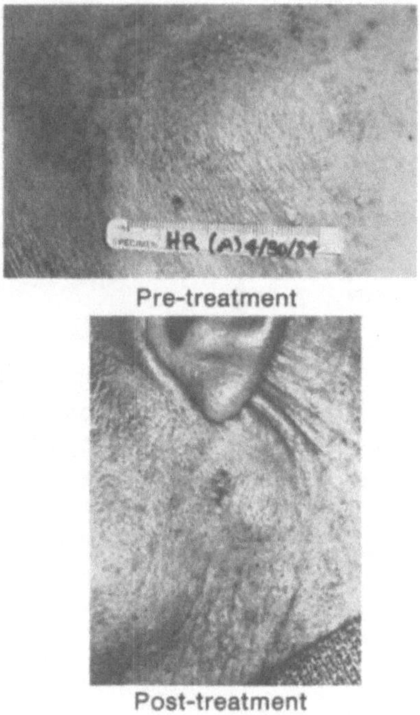

Pre-treatment

Post-treatment

FIGURE 6. Photographs of an interferon alfa-2b treated site prior to and four weeks after therapy.

5. BASAL CELL CARCINOMA

Basal cell carcinoma (BCC) is the most common malignancy in man. Approximately 400,000 cases of this sun-induced tumor are diagnosed each year. Although it is usually a rather nonaggressive tumor that metastasizes only rarely, it can be locally quite destructive. The surgical and chemosurgical procedures most often used produce reliable cures (14 - 16) but can leave an undesirable cosmetic and/or functional deficit. Avoidance of surgery and the possibility of better cosmesis and/or function, suggested that a medicinal therapy for BCC might have significant advantages.

An open, uncontrolled study was conducted in eight patients presenting with either superficial or noduloulcerative BCC (biopsy proven). A single lesion/patient was treated with 1.5×10^6 IU of interferon alfa-2b three times per week for three consecutive weeks. Injections were made into the base and substance of each lesion using an isotonic solution of interferon with a volume of $0.15 cm^3$. Clinical evaluations continued for two months post-treatment. Mild inflammatory changes associated with treatment that resolved slowly during the post-treatment phase were noted. At the end of the follow-up period the entire treated area was excised and histologically examined. Two slides with multiple serial sections were initially evaluated, and to insure that the entire specimen was free of even microscopic foci of BCC, ten more slides with additional deep serial sections were examined. No evidence of BCC was found in any section from any of the eight patients, indicating complete cure of the treated cancer in all cases. The cosmetic result was quite impressive as can be seen in the pre- and eight week post-treatment photographs shown in Fig. 7. At eight weeks post-treatment there was histological evidence of microscopic scar formation and of a resolving chronic inflammatory process.

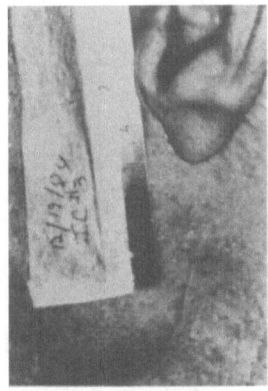   

Pre-treatment   Post-treatment

FIGURE 7.   Photographs of a basal cell carcinoma pre-treatment and eight weeks after treatment with interferon alfa-2b.

6.   TOLERANCE AND SIDE EFFECTS

The most common side effect of interferon alfa-2b in these studies was a constitutional malaise and constellation of symptoms suggestive of a flu-like illness. The incidence and severity of side effects were dose related. In order of decreasing incidence the major signs and symptoms occuring in more than 34% of the patients receiving 1 to 5×10^6 IU were fever, headache, myalgia, and chills. Also occuring in more than 5% of the patients were nausea, fatigue, malaise, dizziness and arthralgia. The incidence and severity of adverse effects tended to diminish with continued treatment as can be seen in Table 4. During the first week of treatment, for example, 46% had fevers which had decreased to 11% during week three. These decreases may be partially explained by concomitant acetaminophen useage. These flu-like symptoms tended to start several hours after treatment and last for less than 12 hours. Pain and other local effects upon injection (i.e. bleeding, itching, purning) were a common finding but were not related to interferon since the incidence rates between interferon and placebo did not differ significantly (22% vs 19%). Local effects, flu-like symptoms, and other adverse experiences resulted in less than 5% of the patients discontinuing treatment. Laboratory abnormalities consisted nearly entirely of decreases in WBC values and occasional elevations of liver enzymes. These changes tended to reverse rapidly following the end of treatment. Although inconvenient to the patient, the side effects noted were not medically serious, were transient, and only rarely caused an interruption of therapy.

Table 4
Adverse Experiences of ≥ 5%
Incidence During and After Treatment
With 1.0-5.0 Million IU Interferon Alfa-2b

PERCENT OF PATIENTS

	OVERALL N-204	WEEK 1 N-204	WEEK 2 N-202	WEEK 3 N-187	WEEK 4 N-186
FATIGUE	15	7	8	9	2
FEVER	52	46	17	11	3
MALAISE	13	7	6	3	0
CHILLS	34	31	10	5	2
CONFUSION	5	5	1	2	0
DIZZINESS	11	9	2	2	0
HEADACHE	47	36	18	11	3
NAUSEA	16	11	5	3	1
ARTHRALGIA	9	6	3	3	1
MYALGIA	37	29	15	12	3

7. CONCLUSIONS AND DISCUSSION

Interferon alfa-2b has shown substantial therapeutic activity in a variety of malignant, premalignant, and virally caused skin diseases when injected locally. In situations where surgical treatments are ill-advised due to medical, psychological, and/or cosmetic result considerations, intralesional interferon alfa-2b might prove to be an advantageous medicinal alternative. Recombinant DNA technology has brought the cost of interferon down to a point that in some situations intralesional interferon alfa-2b may be more cost effective than standard cancer surgery, the use of more exotic surgical methods (e.g. Mohs chemosurgery and laser surgery), and the repetitive use of medical or surgical methods where recurrence rates are high (e.g. genital warts).

The clinical events associated with successful intralesional interferon therapy are quite unlike those associated with standard chemotherapeutic agents and may reflect interferon's mode of action. When interferon works locally, therapy is very often associated with a clinically evident mild inflammatory reaction. Resolution of this inflammation up to one or two months after therapy, is accompanied by cures of long duration. In diseases where we have found no therapeutic activity, such as psoriasis, there is no initial inflammatory event. It is interesting to speculate that from among the in vitro effects associated with interferon, its ability to cause antigen expression may be central to its therapeutic effect. Activation of antigen on abnormal cells (malignant, premalignant, or virally infected) could cause immune recognition that would initiate a destructive immunological reaction against these cells as well as provide for continued local immune surveillance.

REFERENCES

1. Hiscott, J., Cantell, K., and Weissmann, C.: Differential Expression of Human Interferon Genes. Nucleic Acids Res. 12:3727-3746, 1984.
2. Schwartz, J., Lieberman, M., Miller, G.H., et al: Antiviral Activity of Alpha-2 Interferon (Sch 30500) In-vitro and In-vivo. Proc. 24 Intersci. Conf. Antimicrob. Ag. Chemother. 311, 1984.
3. Schwartz, J., Nagabhushan, T.L., Leibowitz, P., et al: Alpha-2 Interferon (Sch 30500) Activity Against RNA and DNA Viruses In-vitro and In-vivo. Antiviral Res. Abstr. 1:142, 1984.
4. Schwartz, J., Smith, S., Siegal, M., et al: In-vitro Cell Antiproliferative Activity of Sch 30500, Alpha-2 Interferon. Blood 64:80a, 1985.
5. Hirsch, R.L., and Johnson, K.P.: The Effect of Recombinant Alpha$_2$- Interferon on Defective Natural Killer Cell Activity in Multiple Sclerosis. Neurology 35:597-600, 1985.
6. Kelly, V.E., Fiers, W., and Strom, T.B.: Cloned Human Interferon-γ, But Not Interferon-β or -α, Induces Expression of HLA-DR Determinants By Fetal Monocytes and Myeloid Leukemic Cell Lines. J. Immunol. 132:240-243, 1984.
7. Bashi, S.A.: Cyrotherapy Versus Podophyllin in the Treatment of Genital Warts. Internat. J. Dermatol. 24:535-536, 1985.
8. Ferenczy, A., Mitao, M., Nagai, N., et al: Latent Papillomavirus and Recurring Genital Warts. New England J. of Med. 313:784-788, 1985.
9. Edelson, R.L.: Cutaneous T Cell Lymphoma: Mycosis Fungoides, Sézary Syndrome and Other Variants. J. Amer. Acad. Derm. 2:89-106, 1980.
10. Edelson, R.L., Cutaneous T Cell Lymphoma in Fitzpatrick, T.B., Eisen, A.Z. and Wolff, K. (eds): Dermatology in General Medicine Update. New York: McGraw Hill Book Co., 1982.
11. Greene, M.H.: Mycosis Fungoides: Epidemiologic Observations. Cancer Treat. Rep. 63:597-606, 1979.
12. Poiesz, B.J., Ruscetti, F.W., Gazdar, A.F., et al: Detection and Isolation of Type C Retrovirus Particles from Fresh and Cultured Lymphocytes of a Patient with Cutaneous T-Cell Lymphoma. Proc. National Acad. Sci. USA. 77:7415-7419, 1980.
13. Poiesz, B.J., Ruscetti, F.W., Reitz, M.S., et al: Isolation of a New Type C Retrovirus (HTLV) in Primary and Uncultured Cells of a Patient with Sézary T Cell Leukemia. Nature 294:268, 1981.
14. Moschella, S., and Hurley, H.: Dermatology. 2nd Edition. Philadelphia: W.B. Saunders Co., 1985, p. 1556, 1567.
15. Mohs, F.E.: Chemosurgery: Microscopically Controlled Surgery for Skin Cancer. Springfield, IL: Charles C. Thomas Publishers, 1978, p. 154.
16. Swanson, N.: Mohs Surgery (Review). Arch. Dermatol. 119:761-773, 1983.

IDIOPATHIC MIXED CRYOGLOBULINEMIA: A NEW THERAPEUTIC APPROACH WITH ALPHA
INTERFERON.

M. CASATO, A. AFELTRA, D. CACCAVO, L. ERCOLI, M. SIENA, F. VACCARO,
L. BONOMO.

CLIN. MED. III^ UNIV. "LA SAPIENZA" OF ROME - ITALY.

INTRODUCTION

Idiopathic mixed cryoglobulinemia (I.M.C.) sustained by diffuse lympho-
proliferation is a disorder of unknown cause characterized clinically by ar
thralgias, purpura, weakness, glomerulonephritis and generalized vasculitis
(Meltzer et Al., 1966). The diagnosis is based on finding cryoprecipitable
serum proteins generally consisting of policlonal IgG and either policlonal
or monoclonal IgM with rheumatoid factor or antigammaglobulin activity. The
treatment of cryoglobulinemia is directed toward minimizing the signs and
symptoms due to the presence of cryoglobulins and consists of drug therapy,
immunosuppressive agents and plasmapheresis. Neither plasmapheresis nor im-
munosuppresors achieve definite successful results in this disease. On the
other hand the good results reported with IFN in lymphoproliferative disea
ses suggested us to try it in 2 patients affected with I.M.C.

PATIENTS

Case n° 1 is a 46 year old man.
Past history included in 1973 recurring abdominal pain with weakness and
diarrhea, till June 1975. In 1974 an itching erythema appeared on both
arms and legs. One year later, in 1975, purpuric lesions appeared on both
legs and disappeared spontaneously in 10 days, but reappeared two months
later extending to the abdomen with arthralgias (wrist and ankles),myalgias,
pretibial edema and weakness. Tests for rheumatoid factors and PCR were
strongly positive while antinuclear antibodies were absent. The ESR was 11
mm. Cryoglobulins were present consisting of homogeneous IgMk and heterogeneous IgG. A bone marrow examination showed a mild plasmocytosis.
The p. was treated with c.steroids and chlorambucil for alternate periods
till April 1980 when necrosis of the right femoral head appeared. In July
1981 cryocrit was 13%, renal failure was manifest as proteinuria, micro-
haematuria and a mild increase of creatinine (1.87 mg/dl) and necrotic purpura and two malleolar ulcers were present. In June 1983, when he first
came to our observation, the major clinical and laboratory findings were:
arterial hypertension (200/110 mmHg), necrotic purpura, painful swelling of
the ankles, two sopramalleolar ulcers and brownish hyperpigmentation of
the skin of both legs, intermittent arthralgias with no redness and swelling, cold sensitivity. Cryocrit was 13%, Hb 10.5g%, ESR 38 mm., positive
test for rheumatoid factor (1/640), C3 was 81 mg/dl (n.r. 70-176) and C4

3mg/dl(n.r. 22-38), hepatitis B virus markers were all negative.
Disease was well controlled with low doses of c.steroids,NSAID and cytotoxic drugs during two years. In July 1985 cryocrit rose to 30% and creatinine to 5.0 mg/dl. Hb was 6.5 g%. There were weakness,anorexy and a few ulcers on both malleolar regions. He was treated with high doses of c.ste-
roids, cytotoxic drugs, plasmapheresis and blood transfusions. In two weeks
we observed an improvement of both clinical and laboratory findings that
lasted for 5 months.
In December 1985 general conditions severely deteriorated, with severe ane
mia (Hb 5 g%), cryocrit 33%, creatinine 5.5 mg%. Necrotic purpura, ulcers
on both legs, significant proteinuria and macrohaematuria were present. The
immunological studies showed a reduction of the T lymphocytes number and a
low NK activity. The single radial immunodiffusion carried out at 37°C show
ed high levels of IgM (515 mg/dl; n.r. 45-250) and a slight decrease of IgG
(588 mg/dl; n.r. 700-1600). The serum electrophoresis tested at 37°C showed
an IgMk monoclonal component.

Case n° 2 is a 39 year old man.
This p. history began in March 1985 with urticaria and purpuric lesions extended to both legs and to the lower part of the abdomen enhanced by cold
and often triggered by standing or substained efforts. Laboratory examinations showed: ESR 35 mm., mild proteinuria, positive test for rheumatoid
factor and lymphocytosis (3.500/cmm). Further studies showed the presence
in serum of cryoprecipitable immunoglobulins. Cryocrit was 10%. The patient
was diagnosed as type II idiopathic mixed cryoglobulinemia. He was treated
with c.steroids for 3 months. In October 1985 the serum aspartate aminotran
sferase and the serum glutamic piruvic transaminase were 78 and 85 respectively. Cryocrit was 50% and the patient complained of paresthesias to both
legs. C.steroid therapy was continued till March 1986 when he came to our
hospital. The clinical examaination showed purpura on both arms and legs
and on the lower part of the abdomen, hepatomegaly, transient loss of con-
sciousness (referred) and paresthesias. Cryocrit was 70%, lymphocyte count
4.600/cmm and platelet count 550.000/cmm. ESR was 35 mm. There was no anemia and urine, creatinine and BUN were normal. Membrane phenotype of B lymphocytes was IgMk. C3 complement component was normal while C4 was not evaluable.

MATERIALS AND METHODS
Both patients have been treated with 3×10^6 I.U./daily of recombinant
human leukocyte interferon rIFNα-A (Hoffmann LaRoche), subcutaneously for
the first week, and with 6×10^6 I.U./daily in the following weeks. Venous
blood was collected from fasting patients in warmed tubes (Vacutainer)
without anticoagulant and allowed to clot at 37°C. The serum, after centrifugation at 37°C, was incubated at 4°C for 5-7 days. The cryocrit was evaluated by centrifugation of the serum in a hematocrit tube at 4°C or by
comparing the serum protein level before and after cryoprecipitation. The
cryoprecipitate was purified by 4 centrifugations and washing at 4°C, the
first 2 being followed by redissolution of the cryoprecipitate at 37°C

(Brouet et. Al., 1974). The cryoprecipitate was characterized by electrophoresis and by immunoelectrophoretic studies of the purified cryoglobulins
against antiserum to whole human serum and monospecific antiserums to α,
μ, γ, k, and λ chains. Serum immunoglobulin and C3/C4 complement component levels have been measured by the radial immunodiffusion technique at
37°C. Peripheral blood mononuclear cells were isolated by Ficoll-Hypaque
density gradient centrifugation at 37°C. B lymphocytes surface Igs were
identified by direct immunofluorescence using FITC-conjugated F(ab')$_2$ fragments of goat anti-human IgM, IgG, IgA, k and burro anti-human Ig λ. T lym
phocytes surface markers were evaluated by the following monoclonal antibodies (Leu series - Becton Dickinson): anti Leu3a and anti Leu2a to identify
the helper/inducer and suppressor/cytotoxic subsets, anti Leu7 and anti Leu
11a to evaluate the natural and antibody dependent cytotoxity effector cells.
Anti Leu M3 and HLA-DR monoclonal antibodies has been used to evaluate the
percentage of peripheral monocytes and HLA-DR$^+$ cells. The NK activity was e-
valuated in 4-hours assay with ^{51}Cr-labeled K562 target cells.

RESULTS
Case n. 1 - After 3 weeks of treatment cryocrit decreased to 12% and creati
nine to 3.8 mg%. After 6 weeks macrohaematuria disappeared and general conditions improved very much. The immuno- and biochemical studies showed a
further decrease of cryocrit to 8% and of creatinine to 2.7 mg%. Hb rose to
10.6 g% (fig. 1). Cellulose acetate electrophoresis and immunoelectrophore-
sis of serum separated at 37°C showed the disappearance of the seric monoclonal component. B lymphocyte surface Igs showed an increase of the percentage of SmIgG parallel to a decrease of IgM positive cells. The total
number of T lymphocytes increased from 611/cmm to 1.500/cmm as well as the
absolute number of Leu7, Leu 11 positive cells and NK activity (28.8% before
treatment; 41% after 6 weeks) (tab. 1). The evaluation of immunoglobulins
carried out at 37°C showed an increase of IgG and a remarkable reduction of
IgM almost to normal values (fig. 2). No modification of C3 and C4 complement component has been observed. After 16 weeks cryocrit was still stable
to 8% and membrane immunofluorescence study showed a normalization of k/λ
ratio of B lymphocytes surface immunoglobulins. In the 16th week we tried to
increase the doses of interferon at 9 x 10^6 I.U./daily to obtain a further
reduction of cryocrit. In 3 weeks collateral effects such as amnesia and
slight degree of confusion occurred without further decrease of cryoglobulins, so we had to reduce the IFN dosage to 3 x10^6 I.U. on alternate days.
The patient has now been on this dosage for 5 months and we still haven't
observed yet either an increase of cryocrit or any modification of clinical
and laboratory parameters.

Case n. 2 - A lowering of cryocrit to 43% was observed in the 4th week of
treatment (fig. 3). The immunological characterization of lymphocytes carried out during the 6th week showed a reduction of the total lymphocyte num
ber and an increase of the percentage of T lymphocytes. K/λ ratio of B lym
phocytes surface immunoglobulins as well as the percentage of IgG, IgA, and
IgM positive cells reverted to normal values (tab. 2). Impairment of liver

CASE N.1

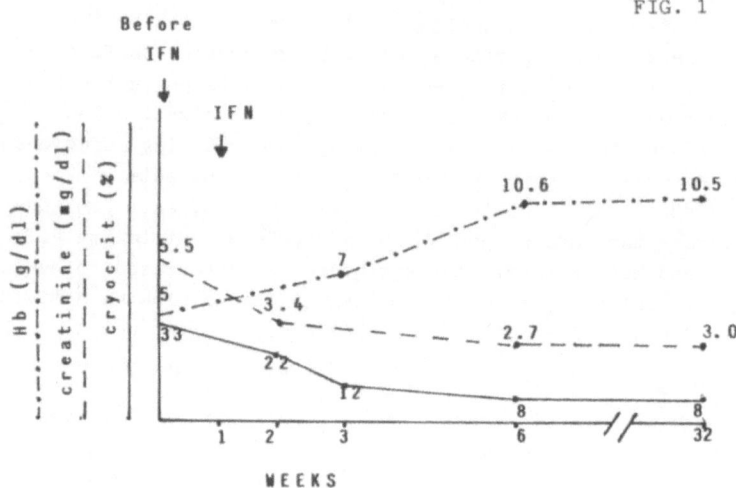

FIG. 1

CASE N.1

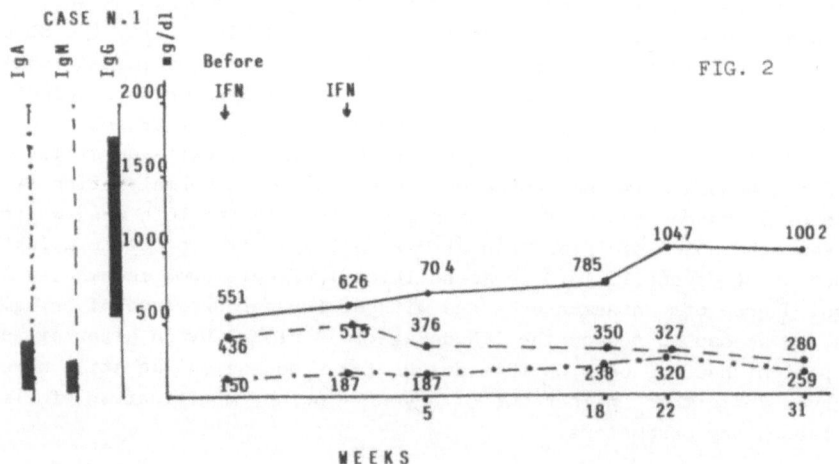

FIG. 2

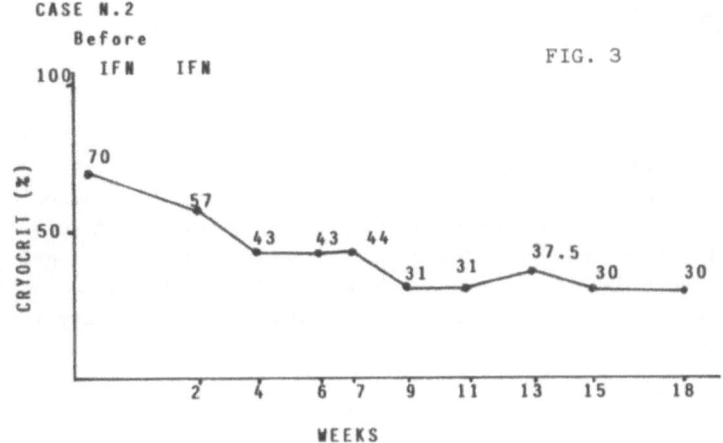

FIG. 3

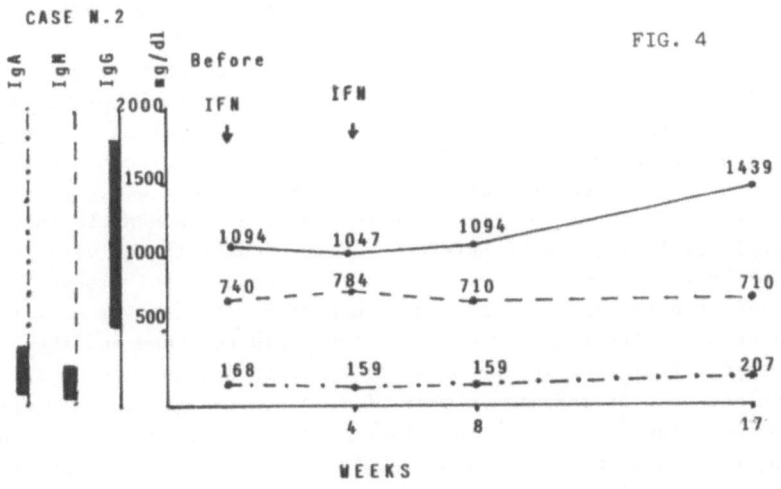

FIG. 4

enzymes and purpuric manifestations were no longer evident. By the 10th week of treatment cryocrit further decreased to 31% and it is still stable at this value. No modification of seric immunoglobulins (fig. 4) and of the complement component has been observed. The patient is now in his 18th week of therapy and no other modifications of immunological parameters have been observed.

COLLATERAL EFFECTS

In the first patient the only collateral effects were fever up to 38°C and nausea during the first 2 weeks of treatment and a transient increase of liver enzymes by the 8th week that disappeared spontaneously. Amnesia and confusion appeared when the dosage of interferon was increased to 9 x 10^6 I.U./daily but disappeared in a few days when we reduced the dosage. In the second patient, besides fever and nausea during the first 2 weeks of treatment, amnesia and confusion appeared in the 15th week without any increase of the dosage.

DISCUSSION

The mechanism by which interferon acts is still unclear. In this preliminary study, recombinant human interferon was effective in inducing clinical and immunological improvement in 2 patients resistant to other conventional therapies (Casato et Al., 1986). It is not clear, however, which of the many biological properties of interferon is responsible for the reduction of circulating cryoglobulins. The following immunological actions can be important:
1) the effects on the monocytes - macrophage functions (Nathan et Al.,1983) could facilitate the elimination of circulating cryoglobulins;
2) the enhancement of functions of T cells, NK cells and K cells (Lindahl et Al., 1972; Saksela et Al., 1981) could restore the physiological control on B cells, that normally produce little quantity of cryoglobulins;
3) the surface phenotype modifications and the normalization of serum immu noglobulins observed in both our patients could be a direct consequence of the action of interferon on B cells maturation (Sidman et Al., 1984).
Additionally, if we consider a viral infection as a pathogenetic factor in mixed cryoglobulinemia, the interferon mediated inhibition of viral replication could have an important role in the treatment of the disease. In the same way IMC as a benign neoplastic disorder of B cells can benefit from the antiproliferative effects of interferon. The rapid response and the low doses requested (it will be interesting to try lower dosage) are probably linked to the presence of the high number of membrane receptors on cryoimmunoglobulins producing B cells. On the other hand the presence of anti-interferon antibodies could in our opinion explain the absence of a further reduction of cryocrit in the second patient thus requesting a longer therapy (Brouty-Boyè et Al., 1981). Further studies are obviously needed to assess the real effectiveness of interferon in IMC but there preliminary data seem to be encouraging.

	before treatment	IFN after 6 weeks	IFN after 16 weeks
CRYOCRIT	33%	8%	8%
MONOCLONAL COMPONENT	IgMk	--	--
LEUKOCYTES/cmm	6.500	8.100	9.000
LYMPHOCYTES/cmm	1.300	3.000	3.240
T LYMPHOCYTES/cmm	611	1.500	1.600
Leu 3a$^+$/2a$^+$	1.8	1.5	1.4
Leu 7$^+$	10%	30%	15%
Leu 11a$^+$	9%	5%	6%
B LYMPHOCYTES (s-Ig$^+$)	9% (117/cmm)	6% (192/cmm)	9% (288/cmm)
s-k$^+$	7%	4%	6%
s-λ^+	2%	1%	3%
s-IgM$^+$	8%	1.5%	2%
s-IgG$^+$	0.5%	4.5%	6%
s-IgA$^+$	0.5%	1%	0.5%
MONOCYTES (Leu M3$^+$)	5%	3%	10%
HLA-DR$^+$ cells	17%	12%	18%
NK activity	28.8%	41%	42%

TAB. 1

	before treatment	IFN after 6 weeks	IFN after 12 weeks
CRYOCRIT	70%	44%	31%
MONOCLONAL COMPONENT	IgMk	IgMk	IgMk
LEUKOCYTES/cmm	8.500	6.700	6.100
LYMPHOCYTES/cmm	4.600	2.546	1.830
T LYMPHOCYTES/cmm	2.700	1.527	1.281
Leu 3a$^+$/2a$^+$	1.4	2.5	2.5
Leu 7$^+$	19%	17%	14%
Leu 11a$^+$	3%	2.5%	3%
B LYMPHOCYTES (s-Ig$^+$)	25%(1150/cmm)	18% (458/cmm)	18% (329/cmm)
s-k$^+$	21%	12%	12%
s-λ^+	4%	6%	6%
s-IgM$^+$	22%	10%	12%
s-IgG$^+$	2%	8%	5%
s-IgA$^+$	1%	0.5%	1%
MONOCYTES (Leu M3$^+$)	14%	10%	13%
HLA-DR$^+$ cells	38%	22%	18%
NK activity	36%	43.4%	45%

TAB. 2

REFERENCES

1. Brouet JC, Clauvel JP, Danon F, Klein M, Seligman M: Biologic and clinical significance of cryoglobulins. Am. J. Med.; 57: 775-88, 1974.
2. Brouty-Boye D, Gresser I.: Studies on the reversibility of the transformed and neoplastic phenotype. I. Progressive reversion of the phenotype of x-ray transformed C3H/10 T ½ cells under prolonged treatment with interferon. Int.J. Cancer 28, 165-173, 1981.
3. Casato M, Afeltra A, Caccavo D, Bonomo L: Interferon treatment in Idiopathic Mixed Cryoglobulinemia. First I.U.I.S. Conference on Clinical Immunology, Toronto, July 5-6, 1986.
4. Huang Kun-Yen: Modulation of Phagocytosis. Texas Reports on Biology and Medicine, vol. 41, 1981-82.
5. Lindahl P, Leary P, Gresser I: Enhancement by interferon of the specific cytotoxicity of sensitized lymphocytes. Proc. Natl. Acad. Sci., USA 69, 721-725, 1972.
6. Meltzer M, Franklin EC, Elias K, McKluskey RT, Cooper N: Cryoglobulinemia. A clinical and laboratory study. Am. J. Med. 40:837-56, 1966.
7. Nathan CF, Murray HW, Wiebe ME, Rubin BY: Identification of interferon ɣ as the lymhokine that activates human macrophage oxidative metabolism and antimicrobial activity. J. Exp. Med. 158, 670-689, 1983.
8. Saksela E: Interferon and Natural killer cells. In Interferon 3 (ed. I Gresser), vol. 3, pp. 45-63. New York and London Academic Press, 1981.
9. Sidman CL, Marshall JD, Shultz LD, Gray WP, Johnson HH: Interferon is one of several direct B cell-maturing lymphokines. Nature (London) 309, 801-804, 1984.

TREATMENT OF RHEUMATOID ARTHRITIS (RA) WITH INTERFERON-GAMMA
(IFN-gamma): RESULTS OF CLINICAL TRIALS IN GERMANY

HANS-J. OBERT
Bioferon GmbH & Co., D-7958 Laupheim, FR Germany

INTRODUCTION
Before the beginning of 1986, the scientific community
would have flatly turned down any clinical protocol that
envisaged treating rheumatic disorders with IFN-gamma. Its
properties of activating T-cells, B-cells and macrophages
and inducing the release of interleukin I and the expression
of class II antigens would have made it a far more likely
candidate for aggravating the inflammatory processes in a
rheumatic limb (1-5). IFN-gamma was also known to be
associated with pronounced side-effects of the type known to
occur with all interferons (6-8).
Since November 1983 we have treated some 120 patients
suffering from rheumatoid arthritis (RA) with preparations
containing natural T-lymphocyte-derived IFN-gamma or
recombinant IFN-gamma (9-13). The following results
represent the state of research as of end of June 1986.

PATIENTS, METHODS AND MATERIAL
Patients
In all studies, patients were entered if they suffered from
definite RA according to ARA criteria, irrespective of the
duration and severity of their disease. They were required
to have an erythrocyte sedimentation rate of not less than
25 mm in 1 hour, and to have received one or more types of
treatment with disease-modifying substances without a
satisfactory response. The onset of their disease had to be
between the ages of 16 and 60, and they had to be aged
between 18 and 65. All patients gave informed consent to
treatment.
Methods
Seven phase II studies were carried out to determine
dosages, mode of administration and tolerance. In these
studies, response was assessed on the basis of rest pain,
pain on movement, and mobility. In a two-armed open
controlled phase II study (14) total daily doses of 50 or
100 µg IFN-gamma were given subcutaneously. In this study
the effect of therapy was monitored using the
internationally accepted semiquantitative parameters: the
Ritchie and Lansbury indices, morning stiffness, morning
pain, rest pain and pain on movement, grip strength and
mobility. The methods used were in accordance with the
recommendations of the European League Against Rheumatism
(EULAR). According to the protocol, a response to treatment

was defined as an improvement of the Ritchie articular index or the Lansbury index by at least 25 % within 20 days. The study lasted 20 days.

The same criteria were used in a multicentre randomised double-blind placebo-controlled phase III trial lasting 28 days, with the aim of confirming or refuting the findings of the open study.

A long-term phase II study of unlimited duration has been running for the past 25 months, and another long-term study lasting 12 months is following on the phase II clinical trial and will be completed in December 1986.

MATERIAL

Three different preparations containing IFN-gamma have been used to treat RA.

1. A preparation derived from human lymphocytes by mitogenic stimulation, and purified by state-of-the-art techniques. This preparation contained 0.1×10^6 I.U. IFN-gamma, 0.5×10^6 I.U. interleukin 2, granulocyte colony-stimulating factor, lymphotoxin and macrophage migration-inhibiting factor (15).

2. A human-lymphocyte-derived preparation highly purified for IFN-gamma, with a specific activity of 1×10^7 I.U. per mg protein (15).

Both of these preparations were produced in collaboration with the German Red Cross, Blood Donor Centre, Springe.

3. Human IFN-gamma produced by genetic engineering, with a specific activity of $>2 \times 10^7$ I.U. per mg protein. This IFN-gamma was produced by cloning IFN-gamma cDNA in an expression plasmid (16), followed by expression of the gene product in Escherichia coli (17). This preparation was produced in collaboration with Biogen S.A., Geneva.

RESULTS

Table 1 gives an overview about 7 pilot studies on 71 patients. There is no difference between intracutaneous, subcutaneous and intramuscular injection of the preparation. Single doses of between 10 and 100 µg IFN-gamma are well tolerated. There was no relation to body weight or body surface area in responding patients; and there was no clear difference between the responses to 10 µg and 100 µg of recombinant IFN-gamma.

After having completed the pilot trials one phase II clinical study was performed. This study comprised two arms, with patients receiving either 50 µg IFN-gamma once daily or 50 µg twice daily by subcutaneous injection. Patients were classified as responders if they either showed a 30 % improvement in the Ritchie articular index within 20 days, or a better than 15 % improvement in the Ritchie index together with at least a 25 % improvement in the Lansbury index.

The results showed that 71 % of the patients receiving 50 µg daily and 54 % of those receiving 100 µg daily fell in the "responder" group.

Of the 49 patients entered in the study, 9 were classified

TABLE 1: PILOT-STUDIES WITH NATURAL OR RECOMBINANT INTERFERON-GAMMA IN RHEUMATOID ARTHRITIS

Study	Subst.	Route	Dosage $(10^6$ I.U.; rsp. µg)	Appl. /week	No. Pts.	Results +	0	-
N	nIFN-γ	i.c.	0.1	1-3	14	14		
A 1	nIFN-γ	s.c.	0.5-1	7	6	6		
A 2	nIFN-γ	s.c.	1	5	10	4	6	
K	rIFN-γ	i.m.	10-50	1-3	18	13	5	
W	rIFN-γ	s.c.	20-500	2-7	6	4	2	
H	rIFN-γ	s.c.	100	7	2	1		1
O	rIFN-γ	s.c.	10-100	3	15	6	9	
Σ total					71	48	22	1

TABLE 2: CHARACTERISTICS OF PATIENTS IN PHASE II CLINICAL STUDY (SUBSTUDY A AND B)

	A	B		A	B
Sex male	5	3	Previous		
female	9	23	medication		
			NSAR	10	14
AGE y (median)	56	47	Gold	10	18
Weight kg (median)	67	57	Steroids	2	12
			Penicillamine	6	19
Disease since y (median)	4.5	9.0	Cytostatics	0	11

TABLE 3: SIDE EFFECTS OF RECOMBINANT IFN-GAMMA IN THE TREATMENT OF RA IN 107 PATIENTS

Pyrexia >38 °C	18 %	Lokal irritation	4 %
Fatigue	7 %	Chills	3 %
Sweating	7 %	Myalgia	3 %
Headache	4 %	Nausea, Vomiting	2 %
Euphoria, Confusion	4 %	Leucocytopenia	2 %

as drop-outs. One patient was entered in error: he had
penicillin allergy, which was an exclusion criterion. Three
patients did not complete 20 days' treatment, two because of
side-effects and one because of ineffectiveness of treatment. Five patients received gradually increasing
doses, i.e. they deviated from the protocol. Of the 8
drop-out patients, 5 would have been classified as
responders as defined above.

Table 2 shows the characteristic of the patients in the
phase II study. All patients had been previously treated
without success; stabilised previous drug therapy was maintained but not altered during the trial. There was no significant difference between the results in the two arms
(treated with 50 µg or 100 µg daily), so the two arms were
evaluated together.

Figure 1 shows the time-courses of the Ritchie articular
pain index and joint index according to Lansbury for responding and non-responding patients. In non-responders, pain
did not change over the 20 days of the study, while
responders showed a 50 % improvement.

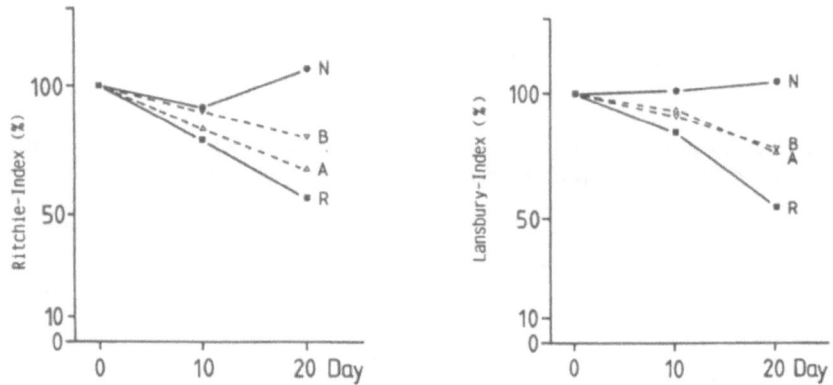

FIGURE 1) Ritchie index (left) and Lansbury index (right)
under treatment with IFN-γ (R = responder; N =
non-responder; A rsp. B = all patients of substudy A rsp. B)

The number of pain-free joints showed a rise over the 20 day
period (Fig. 2). The duration of morning stiffness initially
fell in both responders and non-responders; the improvement
continued in the responders, with an overall fall from 2
hours to 1 hour (Fig. 2). The number of patients with little
or no morning pain increased notably during the 20 days.
Non-responders also showed a less marked tendency in this
direction. The same pattern was seen with rest pain and pain
on movement (Fig. 3). After 20 days there was a larger
number of patients who had only mild pain or none at all.
Mobility of the patients as well as general aspects of the
disease improved markedly in responders, while
non-responders showed no change in this respect (Fig. 4).

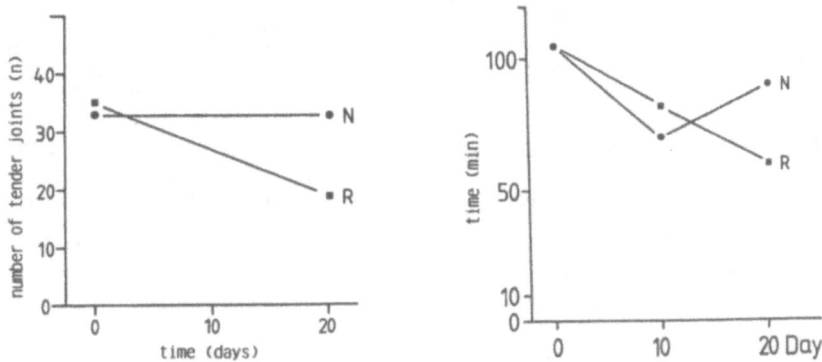

FIGURE 2) Number of tender joints (left) and morning stiffness (right) in responding (R) and non-responding (N) patients with RA under treatment with IFN-γ

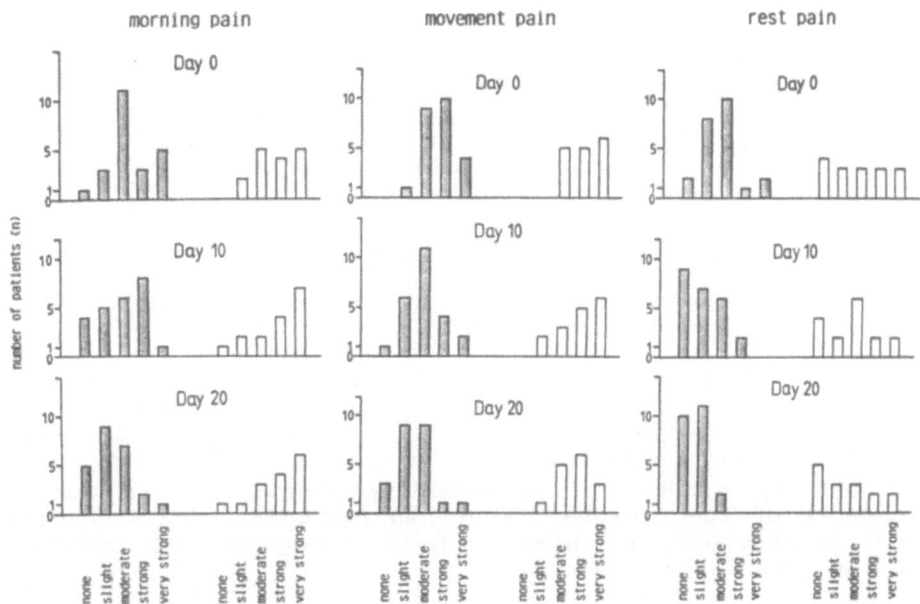

FIGURE 3) Graduation of morning pain (left), pain on movement (middle) and rest pain (right) in responders (dark columns) and non-responders (white columns) during 20 days treatment of RA with IFN-γ

The laboratory tests yielded the unexpected finding that both responders and non-responders showed the same pattern. A striking fact was that the patients in the non-responder group entered the study with worse laboratory values than those in the responder group. Certain laboratory values showed a shift towards the middle of the normal range, representing an improvement in these parameters. The leucocyte count became normal in non-responders and showed a slight fall in responders. No change was observed in platelet counts, in alkaline phosphatase, SGPT or SGOT. Anaemia persisted unchanged.

IFN-gamma is well tolerated. The predominant side-effect is pyrexia of up to 39 °C (Tab. 3). All the side-effects are mild and transient. Side-effects begin about 5-8 hours after injection.

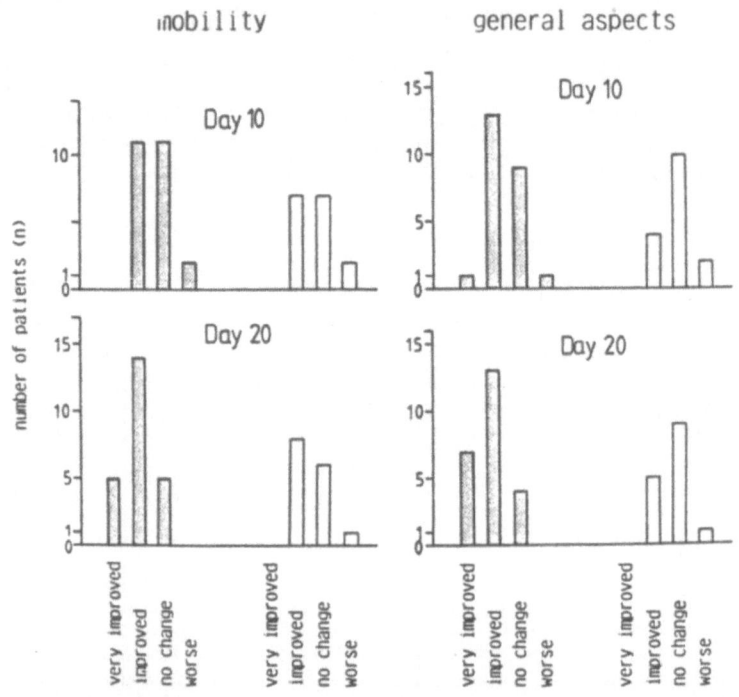

FIGURE 4) Graduation of mobility (left) and of general aspects of the disease (right) in responders (dark columns) and non-responders (white columns) during 20 days treatment of RA with IFN-γ

The results of the phase II study are to be tested in a multicentre double-blind placebo-controlled trial. The duration of the trial is 28 days for each patient. 82 patients have been entered into the study. The criteria of response or non-response are the same as in the phase II study. The trial is not yet completed.

21 patients from the phase II clinical study and 24 patients from our other studies received long-term maintenance treatment with IFN-gamma.

Of these 45 patients, 39 have been receiving successful

treatment for between 4 and 24 months.

It is a striking observation that out of the 8 patients receiving 100 µg IFN-gamma twice or three times weekly, only five have shown persistent improvement on long-term treatment, while out of the 34 remaining patients, receiving between 10 and 50 µg once or twice weekly, 31 have a sustained improvement over long periods. Figure 5 shows the time-course of the Ritchie index in patients out of the phase II clinical trial.

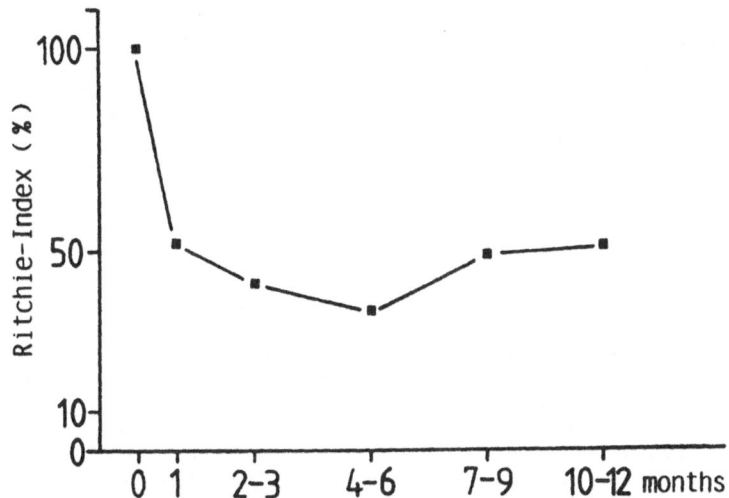

FIGURE 5) Ritchie index of patients with RA during long time therapy with IFN-γ

On the basis of the results in the 120 patients with RA now studied, who have been treated with IFN-gamma within or outside these trials, we recommend a two-tier treatment schedule, divided into initial and maintenance therapy.

The standard dose should be 50 µg IFN-gamma per injection. In case of side-effects, the dose may be reduced to 25 or 10 µg. During maintenance treatment, the dosage frequency and if necessary the dose should be reduced.

Apart from this general guideline, it should also be kept in mind that the optimum dosage frequency also appears to depend on exogenous and endogenous factors. The weather, occupational stresses or intercurrent infections may necessitate a temporary increase in the dosage frequency. This is made easier by the fact that injections of IFN-gamma can be given on an outpatient basis, and the preparation is generally self-administered by the patient under constant guidance and supervision by his doctor.

REFERENCES

1. Vilcek J, Kelker HC, Le J, Yip YK: Structure and function of human interferon-gamma. In: Ford RJ, Maizel AL (eds), Mediators in cell growth and differentiation. New York: Raven press, 1985.
2. Trinchieri G, Perussia B: Immune interferon: a pleiotropic lymphokine with multiple effects. Immunol. Today 6: 131-136, 1985.
3. Leibson HJ, Gefter M, Zlotnik A, Marrack P, Kappler JW: Role of gamma-interferon in antibody-producing responses. Nature 309: 799-801, 1984.
4. Boraschi DS, Censini S, Tagliabue A: Interferon-gamma reduces macrophage suppressive activity by inhibiting prostaglandin E_2 release and inducing interleukin 1 production. J. Immunol. 133: 764-768, 1984.
5. Talal N: Interleukins, interferon and rheumatic disease. Clin. Rheum. Dis. 11: 633-644, 1985.
6. Niederle N, Kurschel E, Schütte J, Krischke W, Schmidt CG, Gerlis L, Gutzwiller E: Phase I study of recombinant human interferon gamma in patients with advanced cancer. In Kirchner H, Schellekens H (eds), The Biology of the Interferon System 1984. Amsterdam, New York, Oxford: Elsevier, 1985.
7. van der Burg M, Edelstein M, Gerlis L, Liang CM, Hirschi M, Dawson A: Recombinant interferon-gamma (Immuneron): results of a phase I trial in patients with cancer. J. Biol. Resp. Mod. 4: 264-272, 1985.
8. Kurzrock R, Rosenblum MG, Sherwin SA, Rios A, Talpaz M, Quesada JR, Gutterman JU: Pharmacokinetics, single-dose tolerance, and biological activity of recombinant gamma-interferon in cancer patients. Cancer Res. 45: 2866-2872, 1985.
9. Obert HJ, Hofschneider PH: Interferon bei chronischer Polyarthritis. Dtsch. Med. Wschr. 110, 1766-1769, 1985.
10. Zilly A, Obert HJ: Therapie der chronischen Polyarthritis. Münch. Med. Wschr. 128: 87-89, 1986.
11. Müller-Faßbender H, Link F: Behandlung der chronischen Polyarthritis mit Gamma-Interferon, EULAR-Kongreß, Wien (in press).
12. Seitz M, Manz G, Franke M: Einsatz von rekombinantem Human-Interferon Gamma bei Patienten mit rheumatoider Arthritis. Z. Rheumatol. 45: 93-99, 1986.
13. Böni A, Hofschneider PH, Obert HJ: Chronische Polyarthritis. Erste Ergebnisse mit Interferon-gamma. Inform. Arzt (in press).
14. Lemmel EM, Franke M, Gaus W, Hartl PW, Hofschneider PH, Machalke M, Miehlke K, Wilms P, Obert HJ: Recombinant human gamma interferon (IF) in the treatment of rheumatoid arthritis (RA): a multicenter phase II trial. Abstract Volume of the 50th Annual Meeting of ARA, New Orleans, USA, 1986 (S.79).
15. Georgiades JA, Baron S, Fleischmann WR jun., Langford M, Weigent DA, Stanton GJ: Immune interferon. In: Came PE, Carter WA (eds), Interferons and Their Applications. Berlin, New York, Tokyo: Springer-Verlag 1984, pp. 305-337.
16. Taya Y, Devos R, Tavernier J, Cheroutre H, Engler G, Fiers W: Cloning and structure of the human immune interferon-gamma chromosomal gene. The EMBO J. 1: 953-958 (1982).
17. Simons G, Remaut E, Allet B, Devos R, Fiers W: High level expression of human interferon gamma in Escherichia coli under control of the PL promotor of bacteriophage lambda. Gene 28: 55-64 (1984).

INTERFERON IN HUMAN BREAST CYSTIC DISEASE

M. MUSCETTOLA*[1], G. GRASSO**, W.R. GIOFFRE'º, F. FRANCHI ºº, G. MONTE-FRANCESCOº

Istituti di Fisiologia Generale*, Anatomia Umana Normale **, Clinica Ostetrica e Ginecologicaºº, Farmacologiaº e Centro Interdipartimentale di Senologia, Università di Siena, Italy.

1. INTRODUCTION

The adequate growth and development of the human breast is dependent on several hormones among which estrogens, progesterone and prolactin are of great importance (1-3). Estradiol (E_2) and progesterone (P) are partly antagonistic and partly synergistic for the normal development and function of the breast, therefore the equilibrium between E_2 and P is necessary to assure tissue eutrophy. There is general agreement that an unbalanced E_2-P ratio may cause benign breast disease in premenopausal women (4, 5). The most common benign breast pathology is gross cystic disease (GCD) (6) characterized by cysts containing fluid secreted by breast epithelial cells. Although gross cysts are not precancerous per se, many studies have documented increased risk of the development of breast cancer among women with GCD (6, 7). Cyst initiation and growth mechanisms are not yet clear, and to elucidate some of these mechanisms breast cyst fluid (BCF) has been extensively studied. BCF has been found to contain immunoglobulins (8), hormones (9) and tumor-associated antigens (10).

Since it seems that a complete regulatory circuit exists between the immune and neuroendocrine systems (11) and serum estrogen and progesterone concentrations have been shown to be significantly affected during the administration of HuIFN-α to healthy women (12) with normal menstrual cycles we decided to evaluate the eventual presence of IFN in the plasma and breast cyst fluid of GCD patients.

2. MATERIALS AND METHODS

Thirty-six premenopausal women (aged 23-50 years), classified as having GCD (6), agreed to participate in the study. BCF samples were collected by needle aspiration during luteal phase. Venous blood samples (10 ml) of thirty control women and of patients were collected in sterile heparinized tubes immediately before aspiration of breast secretion. Plasma and BCF samples were kept at -20°C until assayed for total proteins (13), hormones and antiviral activity (AA). Hormonal assays were performed using specific radioimmunoassay (RIA) methods. E_2 and P were measured after plasma and BCF extraction with diethylether after purification with celite column chromatography (14). AA was assayed in microtiter plates according to Langford et al. (15) using WISH cells and vesicular stomatitis

---

1 - To whom reprint requests should be addressed.

virus (Indiana strain, 70 plaque-forming units per well) as the challenge virus. Every sample was titrated at least twice in duplicate. Results were finally expressed in terms of reference standards (MRC Research Standard B, 69/19, for HuIFN-α (obtained from National Institutes for Biological Standards and Control, Holly Hill, London and IRP for HuIFN-γ obtained by courtesy of NIAID, NIH, Bethesda, Md). The MRC of HuIFN-α (with a defined potency of 3.69897 Log_{10} IU/vial) and the IRP of HuIFN-γ (G - 023 - 901 - 530 with a defined potency of 3.6 Log_{10} IU/vial) when reconstituted in 1.0 ml of sterile distilled water, had a geometric mean titer respectively of 3.98001 Log_{10} IU/ml (SD=0.044; N=30) and of 3.398875 Log_{10} IU/ml (SD=0.274; N=24). All titers are reported in IU/ml. Characterization of AA was carried out according to standard procedures. Neutralization was performed using specific antisera against HuIFN-α, HuIFN-β (obtained from NIAID) and HuIFN-γ (preparation kindly provided by prof. J. Vilček). The residual activity was determined as above. Comparison between individual values was performed according to the t test for paired samples.

3. RESULTS

Our data show that the plasma of GCD patients contained significantly higher IFN levels (P < 0.001) than controls (Table I). IFN was also found in BCF in

Table I

Quantities of some components present in plasma and breast cyst fluid (BCF) obtained from healthy women and gross cystic disease (GCD) patients.

Samples	T. proteins (mg/ml) mean+SD	IFN (IU/ml) mean+SD	E_2 (pg/ml) mean+SE	P (ng/ml) mean+SE
Control plasma* (n=30)	71.9+3.6	8.3+ 3.4[a]	175.3+23.5	10.8+1.3
GCD plasma (n=36)	74.7+5.1	25.4+ 9.0[b]	68.3+ 6.1	6.3+1.9
BCF (n=36)	35.6+7.3	42.2+17.7[c]	64.6+15.2	13.5+5.2

(*) Controls were normal subjects of similar age and menstrual phase. n=number of women; b vs a: P < 0.001; c vs b: P < 0.01.

significantly higher concentrations (P < 0.01) than in the plasma of the same patients. IFN produced was partly acid-labile (after pH 2 treatment) and was not neutralized by antiserum to HuIFN-β but was neutralized by antisera to HuIFN-α and HuIFN-γ . It is concluded that IFNs found in plasma and BCF of GCD patients are of alpha and gamma types. The patients examined had different ranges of IFN types and the percentages of IFNs are presented in Table II. Total proteins found in breast cyst fluids were lower than in plasma. Plasmatic E_2 and P levels were significantly lower than in controls. On the other hand no significant difference was found in the concentrations of estradiol in the plasma and BCF of GCD patients (Table I).

Table II
Range (expressed in %) of IFN types found in plasma and breast cyst fluid (BCF) of gross cystic disease (GCD) patients

Samples	IFN-α range (%)	IFN-γ range (%)
GCD plasma	25 - 50	50 - 75
BCF	10 - 15	85 - 90

4. DISCUSSION

Human benign breast cystic disease has been the subject of several endocrinological studies yet little data are available on its immunological features (8). Our results demonstrate that significant levels of IFN-α and IFN-γ are present in plasma and breast cyst fluid of all the GCD patients examined. Recent studies report the presence of IFN in the blood and secretions of patients with different chronic disease, especially autoimmune diseases (16-19). However, to our knowledge, this is the first report indicating the presence of IFN in a chronic disease believed to be due to hormonal dysfunction. What is the mechanism responsible for IFN production? Has IFN a role in the pathology of benign breast cystic disease or is it to be considered just a marker of the disease? Some investigators found increased estrogen (ER) and progesterone (PR) receptor levels in benign breast disease (20); in addition other authors recently reported that IFN-α augments the expression of ER in human breast cancer tissue in vitro (21); finally Kauppila et al. (12) demonstrated a significant decrease in circulating E_2 and P concentrations during the administration of HuIFN-α to women having normal cycles. Consequently on the basis of our findings, could endogenous IFN be responsible for increased ER expression and decreased E_2 and P levels in GCD patients? Treatment of human breast cancer cells (CG-5) with all three types of IFN resulted in suppression of the proliferative stimulus of estradiol (22). Could IFNs present in GCD patients be protecting against the occurrence of breast cancer? Are the higher IFN levels in BCF the expression of local immune reactivity with synthesis in situ? Or are they due to transport from vascular compartment? Further studies will be necessary to clarify these questions.

522

ACKNOWLEDGEMENTS
This work was supported by funds 60% from MPI Rome. We are particularly grateful to Prof. J. Vilček for the gift of antiserum to HuIFN-γ. We thank Dr. L. Lodi and M. Fedeli for their skilled technical assistance.

REFERENCES
1. Bassler R: The morphology of hormone-induced structural change in the female breast. Curr. Top. Pathol. 53: 1-89, 1970.
2. Ceriani RL, Contesso GP, Nataf BM: Hormone requirement for growth and diffentiation of the human mammary gland in organ culture. Cancer Res. 32:2190 2196, 1972.
3. Dilley WG, Kister SJ: In vitro stimulation of human breast tissue by human prolactin. J. Natl. Cancer Inst. 55:35-36, 1975.
4. Golinger RC: Hormone and the pathophysiology of fibrocystic mastopathy. Surg. Gynecol. Obstet. 146:273-285, 1978.
5. Vorherr H: Fibrocystic breast disease: Pathophysiology, pathomorphology, clinical picture, and management. Am. J. Obstet. Gynecol. 154:161-179, 1986.
6. Haagensen CD, Bodian C, Haagensen DE, Jr: Breast carcinoma: risk and detection. Philadelphia: WB Saunders, 1981.
7. Harrington E, Lesnick G: The association between gross cysts of the breast and breast cancer. Breast 7:13-17, 1981.
8. Agrimonti F, Cavallo R, Frairia R, Violino PL, Borretta G, Angeli A: Immunoglobulin levels in breast cyst fluid: apparent connection with intracystic transcortin (CBG) binding activity. Boll. Ist. sieroter. Milan 61:44-50,1982.
9. Bradlow HL, Rosenfeld RS, Kream J., Fleisher M, O'Connor J, Schwartz MK: Steroid hormone accumulation in human breast cyst fluid. Cancer Res. 41:105-107, 1981.
10. Golinger RC: Hormone and the pathophysiology of fibrocystic mastopathy. Surg. Gynecol. Obstet. 146:273-285, 1978.
11. Blalock JE, Harbour-McMenamin D, Smith EM: Peptide hormones shared by the neuroendocrine and immunologic systems. J. Immunol. 135:858s-861s, 1985.
12. Kauppila A, Cantell K, Jänne O, Kokko E, Vihko R: Serum sex steroid and peptide hormone concentrations, and endometrial estrogen and progestin receptor levels during administration of human leukocyte interferon. Int. J. Cancer 29:291-294, 1982.
13. Gornall AG, Bardawill CJ, David MM: Determination of serum proteins by means of the biuret reaction. J. Biol. Chem. 117:751-766, 1949.
14. Facchinetti F, Genazzani AR: Critical assessment of celite column chromatography in a multiple steroid radioimmunoassay system. J. Natl. Med. All. Sci. 22:61-63, 1978.
15. Langford MP, Weigent DA, Stanton GJ, Baron S: Virus plaque-reduction assay for interferon: microplaque and regular macroplaque reduction assay. In Pestka S (ed): Methods in enzymology, vol. 78, part A, pp. 339-346. New York: Academic Press, 1981.
16. Hooks JJ, Moutsopoulos HM, Geis SA, Stahl NJ, Decker JL, Notkins AL: Immune interferon in the circulation of patients with autoimmune disease. N.

Engl. J. Med. 301:5-8, 1979.

17. Ohno S, Kato F, Matsuda H, Fuji N, Minagawa T: Detection of gamma interferon in the sera of patients with Behcet's disease. Infect. Immun. 36:202-208, 1982.

18. Preble OT, Black RJ, Friedman RM, Klippel JH, Vilček J: Systemic lupus erythematosus: presence in human serum of an unusual acid-labile leucocyte interferon. Science 216:429-431, 1982.

19. Rottoli L, Rottoli P, Muscettola M, Grasso G, Bocci V: Presence of acid-labile interferon in the serum of sarcoidosis patients. Acta virol. 29:91, 1985.

20. Rolland PH, Martin PM, Serment H: Steroid hormone receptors and prostaglandin synthesis in fibrocystic breast disease. In: Angeli A, Bradlow HL, Dogliotti L (eds): Endocrinology of Cystic Breast Disease, pp. 203-209. New York: Raven Press, 1983.

21. Dimitrov NV, Meyer CJ, Strander H, Einhorn S, Cantell K: Interferon as a modifier of estrogen receptors. Ann. Clin. Lab. Sci. 14:32-39, 1984.

22. Iacobelli S, Natoli C, Arnò E, Chirci-D'Afile E, Toppi A, Sbarigia G, Gaggini C: An antiestrogenic action of interferons in human breast cancer cells. Anticancer Res. 6:400, 1986.

EFFICACY OF INTRATHECALLY ADMINISTERED NATURAL HUMAN FIBROBLAST INTERFERON AS TREATMENT OF MULTIPLE SCLEROSIS

L. Jacobs, A.M. Salazar, R. Herndon, P.A. Reese, A. Freeman, R. Josefowicz, A. Cuetter, F. Hussain, W.A. Smith, R. Ekes, J.A. O'Malley, Buffalo, N.Y., Rochester, N.Y., Washington D.C.

INTRODUCTION

In 1981 we reported the results of an open, preliminary study of 20 patients suggesting that intrathecally (IT) administered natural human fibroblast interferon (IFN-B) reduced the neurologic exacerbations of patients with multiple sclerosis (MS).[1,2] We have subsequently conducted a randomized, double-blinded, placebo controlled, two-year, multicenter study to determine definitively whether this treatment is effective in MS.[3] The participating centers for the clinical aspects of the study were the Dent Neurologic Institute, Buffalo, N.Y.; Walter Reed Army Medical Center, Washington, D.C. and the University of Rochester Medical Center, Washington, D.C. The statistics center was the Department of Biomathematics, Roswell Park Memorial Institute, Buffalo, N.Y. Because of its potential importance, this study was monitored throughout its course by a committee of experts appointed by the National Institutes of Health.

MATERIALS AND METHODS

We included in this study 69 patients with definite MS.[4-7] All patients had exacerbating-remitting disease and relatively high pre-study exacerbation rates (minimum of 0.6/year). The patients underwent a complete neurologic examination at the beginning of the study, and the severity of their symptoms and signs were scored according to a modified Kurtzke method.[1,2,8] Their pre-study exacerbation rates were determined by dividing their total number of exacerbations (standard definition) prior to the study by the duration of disease up to the time of randomization and expressed as exacerbations/year. The patients were then randomly assigned (biased coin) to an IFN-B recipient or a control group using a stratification based on pre-study rate (ie less than two exacerbations/year; two or more exacerbations/year).[3] Exacerbation rates during the study were determined by dividing the number of exacerbations occurring during the study by the length of time on the study. The effect of IT administered IFN-B on exacerbation rate was measured by a one-tailed t statistic.

The randomization yielded recipient (n = 34) and control (n = 35) populations with no significant differences in important pre-study clinical parameters (ie age, sex, disease duration, modified Kurtzke disability status or functional group scores). In particular, there was no difference in the pre-study exacerbation rates of the two groups (recipients 1.79, controls 1.98 exacerbations/year; means).

Patients were re-examined on a scheduled basis during the study or whenever they felt exacerbations might be occurring. These examinations were performed by an "examining" physician at each center who was blinded as to what treatment the patients had received. He/she identified new exacerbations occurring during the study and changes in the patients' clinical status (modified Kurtzke method) and also made an overall assessment of the patients' clinical status (improved, unchanged, worse) compared with the beginning of the study. At the end of the study (two years), the on-study exacerbation rates for each patient was determined by dividing the number of exacerbations that had occurred during the study by the duration of the study (exacerbations/year). Changes in clinical status occurring during the study were measured by the modified Kurtzke method and the overall assessment method.[1,2,8]

The IFN-B used in this study was produced by superinduction of human fibroblast cells at Roswell Park Memorial Institute, Buffalo, and had a specific activity of 1×10^7 interferon reference units (IRU) per mg of protein.[1,2] This was the same type of IFN-B as used in the preliminary study and had passed the safety and toxicity tests of the Federal Drug Administration (FDA #1325).

The IFN-B was administered to the recipients by serial lumbar punctures (LPs) performed once per week for the first month and then once per month for the next five months of the study (ie 9 LPs during the first 6 months of the study). The dosage of IFN-B administered was 1×10^6 IRU at the time of each LP. Controls received placebo treatments according to the same schedule. Their treatments consisted of false LPs in which the routine LP rocedure (ie positioning, draping, skin cleansing, local anesthetic) were followed, but the needle was advanced only into the subcutaneous tissues where 5 ml of sterile water was injected. Recipients took Indomethacin capsules, 25-50 mg. every six hours for 24 hours after their treatments and controls took placebo Indomethacin capsules according to the same schedule. Indomethacin, so administered, is known to reduce the side effects of IT administered IFN, and this facilitated blinding of the patients.[9] The treatments were performed by a "treating" physician and the serial clinical examinations were performed by a separate "examining" physician at each center. Patients did not discuss the side effects of treatments with the "examining" physician, nor did he/she question them in that regard. Thus, double-blinding of this study was accomplished by the following: a) Use of separate "examining" and "treating" physicians at each center, b) Instructing patients not to discuss side effects of treatments with "examining" physicians, c) Performance of false LPs in the controls, and d) Administration of small doses of Indomethacin or placebo Indomethacin to recipients and controls respectively.[3]

RESULTS

The figure shows the pre-study and on-study exacerbation

rates for the two groups of patients. The mean pre-study rates of the recipients and controls were nearly identical (recipients 1.79, controls 1.98 exacerbations/year), but the recipients' mean rate during the study (0.76/years) was significantly less than that of the controls (1.48/year), (p < .001). The rates decreased in both groups during the study, but the reduction in the recipients' rate (1.02 exacerbations/year) was significantly more than that of the controls (0.51 exacerbations/year), (p < .04). The greater reduction in rate of recipients compared to controls was consistently observed at all three participating centers. In addition, correlation analysis revealed a strong dependency of exacerbation rate during the study upon the pre-study exacerbation rate in the controls (r = 0.51, p < .001), but this dependency was not observed in the IFN-B recipients (r = .02, p = .45).

Clinically 76.5% of the recipients and 60% of the controls were improved or unchanged and 23.5% of the recipients and 40% of the controls were worsened at the end of the study compared with their status when it began. The degree of worsening was greatest in the controls (mean modified Kurtzke score increases; controls 0.8; recipients 0.3). However, this trend for recipients to be clinically stable or better and controls to be clinically worse was not statistically significant (Chi-square p = .23).

The side effects of treatments (headache, nausea, myalgia, lethargy) occurred with equal frequency in recipients and controls, which may be partially attributed to use of Indomethacin.[9] Low grade fevers (mean 100.7°F) occurred more frequently in recipients (75%) than controls (31%), (Chi-square p < .001). However, this occurrence did appear to break the double-blinding. Questionnaries completed by the patients and "examining" physicians confirmed that both groups were blinded (p = .66 and .07 respectively), and that the majority of patients believed they had received IFN-B (79.3% of recipients, 63.6% of controls).

The recipients developed pleocytosis and CSF protein elevations during treatment. The maximum pleocytosis (range 12-566/cu mm, mean 122/cu mm) and protein elevation (range 24-160/mg/dL, mean 59 mg/dL) occurred during the first five weeks of treatment. By the time of the last treatment, the mean pleocytosis was reduced to 25/cu mm (range 2-188/cu mm) and the mean CSF protein was 46.0 mg/dL (range 15-160 mg/dL). By the time of the last LP at the end of the study (2 years) the mean pleocytosis was 16/cu mm (range 1-61/cu mm) and the mean protein was 45 mg/dL (range 15-160 mg/dL).

DISCUSSION

This study definitively demonstrated that IT administered IFN-B reduces the exacerbations of MS patients with exacerbating-remitting disease. The IFN-B was generally well tolerated and the side effects of treatment seemed to be clearly acceptable for the benefits achieved. The present study confirmed the findings of the preliminary one which first suggested that IT administered IFN-B was of benefit in

528

MS.(1,2)

How IFN-B exerted a beneficial effect in these patients is speculative. Mechanisms of IFN's actions are complex and incompletely understood.(10,11) However, there have been three recent studies which clearly demonstrate that IFN-B acts in a prophylactic and suppressant fashion against the expression of experimental allergic encephalomyelitis, an animal model of human MS.(12-14) The IFN-B was most effective and exerted its beneficial effect at the lowest doses when it was administered directly into the CSF.

We will continue our follow-up of the patients in this study. Follow-up observations of the IFN-B recipients in the preliminary study demonstrated a prophylactic effect against exacerbations that had persisted for 4.4 to 5.3 years without retreatment.(15-17)

We wish to emphasize that the treatment schedule used in this study may not be the optimum one. In the preliminary study, patients underwent thirteen treatments in six months; these were reduced to nine in the present study and both treatment regimens significantly reduced exacerbation rates. It is possible that even fewer treatments with lower doses of IFN-B during six months might provide similar degrees of prophylaxis with less side effects.

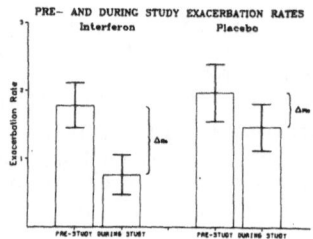

FIGURE - Exacerbation rates before and during the study in 34 IFN-B recipients and 35 control patients with MS. Mean rates (with 2 sd ranges) are shown. The mean re-study rates of the two groups were nearly identical (recipients 1.79, controls 1.98 exacerbations/year), but the recipients' rate during the study (0.76/year) was significantly less than that of the controls (1.48/year), p < .001). The change (reduction) in the recipients' rates ($\triangle$ IFN) during the study was significantly greater than that of the controls who received placebo ($\triangle$ Pla), (p< .04). See text fur further explanation.

REFERENCES

1. Jacobs L, O'Malley J, Freeman A, et al: Intrathecal interferon reduces exacerbations of multiple sclerosis. Science 214:1026-1028, 1981.
2. Jacobs L, O'Malley J, Freeman A, et al: Intrathecal

interferon in multiple sclerosis. Arch Neurol 39:609-615, 1982.

3. Jacobs L, O'Malley JA, Freeman A, Reese P: Intrathecal interferon as treatment of multile sclerosis. A planned multicenter study. Arch Neurol 40:683-686, 1983.

4. Rose AS, Kuzma JW, Kurtzke JF, et al: Cooperative study in the evaluation of therapy in multiple sclerosis: ACTH vs. placebo in acute exacerbations, preliminary report. Neurology 18:1-10, 1968.

5. McDonald WI: What is multiple sclerosis? Clinical criteria for diagnosis, in Davison AN, Humphrey JH, Liversedge LA, et al (eds): Multiple Sclerosis Research, New York, Elsevier North-Holland Inc., 1975, p. 5.

6. Brown JR, Beebe GW, Kurtzke JR, et al: The design of clinical studies to assess therapeutic efficacy in multiple sclerosis. Neurology 29:3-23, 1979.

7. Poser CM, Paty DW, Scheinberg L, et al: New diagnostic criteria for multiple sclerosis: Guidelines for research protocols. Ann Neurol. 13:227-231, 1983.

8. Kutzke JF: Further notes on disability evaluation in multiple sclerosis, with scale modifications. Neurology 15:654-661, 1965.

9. Mora JS, Kao KP, Munsat TL: Indomethacin reduces the side effects of intrathecal interferon. N Engl J Med 310:126-127, 1984.

10. Stewart WE II: The Interferon System. New York, Springer Verlag, 1979.

11. Salazar AM, Gibbs CJ, Gajudsek DC, et al: Clinical use of interferons: Central nervous system disorders, in Came P, Carter WA (Eds): Handbook of Experimental Pharmacology. Berlin, Springer-Verlag, vol. 71, 1983, pp. 472-497.

12. Abreu SL: Suppression of experimental allergic encephalomyelitis by interferon. Immunol Commun 11, 1-7, 1982.

13. Abreu SL, Tondreau J, Levine S, et al: Inhibition of passive localized allergic encephalomyelitis by interferon. Int Arch Allergy Appl Immun 72:30-33, 1983.

14. Hertz F, Deghenghi R: Effect of rat and beta human interferon on hyperacute experimental allergic encephalomyelitis in rats. Agents and Actions 16:397-403, 1985.

15. Jacobs L, O'Malley JA, Freeman A, Ekes R, Reese PA: Intrathecal interferon in the treatment of multiple sclerosis: Patient follow-up. Arch Neurol 42:841-847, 1985.

16. Jacobs L, Salazar AM, Herndon R, Reese PA: Intrathecal interferon in the treatment of multiple sclerosis, in Interferons in the Treatment of Neurologic Disorders, RA Smith (ed): Marcel Dekker, New York, 1986 (In press).

17. Jacobs L, Salazar AM, O'Malley JA: The use of interferon in the treatment of certain diseases of the central nervous system, in The Interferon System, A Current Review, S Baron (ed): University of Texas Press, Galveston, 1986 (In press).

A STUDY ON THE EFFICACY OF RECOMBINANT HUMANαD TYPE INTERFERON (rIFN-αD) IN TREATMENT OF CHRONIC CERVICITIS

Qian Zhiwei,[1] Li Yuying,[2] Yang Xinke,[2] Mao Shujuan[1] and Hou Yunte[2]

ABSTRACT

A double blind controlled study on the efficacy of human rIFN-αD in treatment of 271 cases with typical cervical erosion was carried out. 93.8% of cases appeared clinically improved and 60% was markedly improved. The clinical theurapeutic efficacy was shown to be roughly paralleled with the negative conversion of HSV isolation after treatment of human rIFN-αD. 35 DNA samples isolated from erosive cervices were hybridized with $[\alpha^{32}P]$ labelled pHPVs, among them HPV-16 sequence was found in 18 cases (51.4%); HPV-18-in 5 cases (14.3%); HPV-11-in 2 cases (5.7%); and HPV-6B-in 9 cases (25.7%). Obtained results suggested that HSV and HPV are closely related to the pathogenesis of cervical erosion and this type of chronic cervicitis appears sensitive to human rIFN-αD.

INTRODUCTION

Chronic cervicitis, especially its erosive type, is a very common disease in China; it occurs about 50% among married women (I). Some viral infections, such as human papilloma virus (HPV) and herpes simplex virus (HSV) infections, may be related to its pathogenesis (2-11). A recombinant human interferon αD (rIFN-αD) has been developed in the Institute of Virology, Chinese Academy of Preventive Medicine. In this paper a study on the efficacy of recombinant humanαD type interferon (rIFN-αD) in treatment of chronic cervicitis was carried out.

MATERIALS AND METHODS

SUBJECTS

296 patients with typical cervical erosion were studied in gynecology clinic, Beijing Tian Tan Hospital (mean age 35yr., range 25-56yr.). Among them 90 cases were diagnosed as a mild type, in which the erosive area was less than 1/3 of total cervical surface, as measured by dividers; 48 cases-as a severe type, in which the erosive area more than 2/3; and 158 cases-as a moderate type, in which the size of the erosive area was between the mild and severe types. Another 76 women with clinically normal cervices were taken as a control.

INTERFERON TREATMENT REGIMENS

271 cases were at random divided into five groups and matched by age and erosive grade:

1) Beijing Tian Tan Hospital, Beijing, China
2) Institute of Virology, Chinese Academy of Preventive Medicine, Beijing, China.

Group I: 146 cases (43 with mild type, 78-moderate and 25-severe). A rIFN-αD preparation with the specific activity of 2.8×10^8 units per mg protein was locally administrated by pasting with two layers of gauze soaked up the solution containing 2,5 µg of drug, twice per week for three weeks.

Group II: 22 cases (5 with mild type, 14-moderate and 3-severe). A rIFN-αD preparation was locally administrated by pasting at the dose of 1.25 µg (0.75×10^6 units), also twice per week for three weeks.

Group III: 30 cases (11 with mild type, 9-moderate and 10-severe). A natural leucocyte interferon with the specific activity of 10^6 units per mg protein preparation was locally administrated by pasting at the dose of 2.5 µg as indicated in group I.

Group IV: 20 cases (5 with mild type, 13-moderate and 2-severe). An interferon free preparation preparaed by removing interferon using YOK column (CELLTECH) from the same drug used in group I. Was administrated locally by pasting with the same regimen as indicated in group I and II.

Group V: 53 cases (17 with mild type, 33-moderate and 3-severe). Fuyuanling was locally administrated by spray at the dose of 0.5g, also twice per week for three weeks. This drug contains sulfanilamide and diethyl stilbestrol; it has been routinely used in clinic to treat chronic cervicitis and vaginitis in China and was taken here as a control drug.

The theurapeutic reponse was evaluated as follows:

Completely cured cases were defined as the disappearance of a measurable lesion and the stainable by iodine; markedly improved: an over one grade reduction of the erosive area; improved: a less than one grade reduction of the erosive area; unchanged: no detectable reduction of the erosive area; and progressed: a detectable enlargement of the erosive area.

ISOLATION AND IDENTIFICATION OF HSV

Before and after treatment with IFN the cervical swabs were taken from each patient and inoculated on the monolayer cell culture prepared from suckling rabbit kidney. Isolated cytopathic viruses were identified by neutralization test using rabbit antiserum against HSV-2 on rabbit kidney cell monolayers.

HYBRIDIZATION TEST FOR THE DETECTION OF HPV

38 biopsy specimens were collected from women referred to the Gynecology Clinic of Beijing Tian Tan Hospital. Biopsies for histopathological examination were treated by routine methods and evaluated by the pathology staff of Tian Tan Hospital. 24 biopsy specimens were shown to be cervicitis; 7-metaplasia; 4-hyperplasia; 2-invasive carcinoma and 1-mild dysplasia. Specimen DNAs were extracted routinely. Plasmids containing HPV-16, HPV-18, HPV-11 and HPV-6B DNA, respectively, were kindly provided Prof. H. Wolf (Pettenkoffer Institute, Muenchen) pHPV were labelled with [α^{32}P] dCTP by nick translation to specific activities approximately 1×10^8 cpm/µg DNA. Hybridization tests were performed under stringent condition as described by Gissmann et al(9).

RESULTS

RESPONSE OF PATIENTS WITH CERVICAL EROSION TO HUMAN rIFN-αD TREATMENT

As indicated in Tab. 1., the total theurapeutic response rate of 146 cases with chronic cervicitis (erosive type) to the human rIFN-αD reached 93.8% (group I); 60% of cases in this group appeared markedly improved, as compared with 63.3% in group III when a natural leucocyte interferon was used (P>0.8) and only 37.7% in group for Fuyuanling (P<0.01). The markedly improved rate in group II dropped down to 31.8% when the dose of rIFN-αD was reduced half. In the interferon free group (group IV), however, only two cases among 20 subjects responded to the mask drug in this double blind study. The theurapeutic efficacy of 92 cases was followed up six months after termination of the treatment. All cases in interferon groups (group I, II, III) appeared unchanged or even continuously improved except one which had pregnancy just after termination of the treatment.

Tab. 1 Therapeutic responses of 271 cases with cervical erosion to rIFN-αD, nIFN-αco and Fuyuanling after one course of treatment

Therapeutic response	Group				
	I rIFN-αD	II rIFN-αD	III nIFN-αco	IV IFN-free prep.	V Fuyuanling
Total cases	146	22	30	20	53
complete cured No.	16	11	10	0	2
%	11.0	4.5	33.3	0	3.8
markedly improved No.	71	6	9	0	18
%*	60.0	31.8	63.3	0	37.7
improved No.	50	13	9	2	27
%*	93.8	90.9	93.3	10.0	88.7
unchanged No.	9	2	2	16	6
%	6.2	9.1	6.7	80.0	11.3
progressed No.	0	0	0	2	0
%	0	0	0	10.0	0

* Cumulative rate

On the other side, 12 among 33 subjects in group V recurred during six months after termination of the treatment when Fuyuanling was used (Tab.2).

Tab. 2 Follow up efficacy of rIFN-αD, nIFN-αco and Fuyuanling in treatment of the cervical erosion six months after termination of the treatment

Drug	Total cases	Follow up efficacy					
		Continuously improved		Unchanged		Recurred	
		No.	%	No.	%	No.	%
rIFN-αD	44	25	56.8	18	40.9	1*	2.3
nIFN-αco	15	7	46.7	8	53.3	0	0
Fuyuanling	33	10	30.3	11	33.3	12	36.4

* One case had pregnancy after termination of the treatment

No any side effect was observed during the course of local treatment with natural or recombinant interferon preparation.

Analysed the relationship between clinical efficacy and HSV isolation rate, it was demonstrated that 35% of clinically improved cases appeared virus negative conversion after treatment; 57% of cases-negative unchanged and 5% of cases-positive unchanged before and after treatment; only 2.9% of cases-positive conversion after treatment. (Tab. 3)

Tab. 3 Relationship between the clinical efficacy of human rIFN-αD in the treatment of cervical erosion and HSV isolation rate from erosive cervices

Clinical efficacy or HSV isolation before treatment	Case	HSV isolation before or after treatment			
		Negative conversion	Negative unchanged	Positive unchanged	Positive conversion
Cured	21	11	8	2	0
Markedly improved	70	20	45	3	2
Improved	49	18	27	2	2
Unchanged	2	0	1	1	0
HSV positive before treatment	56	49		7	
HSV negative before treatment	85		81		4

HSV ISOLATION RATE FROM NORMAL AND EROSIVE CERVICES BEFORE AND AFTER TREATMENT WITH rIFN-αD

Result in the Tab. 4 indicated that HSV isolation rate from erosive cervices (30.8%) was 11.8 times higher than that from normal (2.6%). After one course of human rIFN-αD treatment the virus isolation rate markedly dropped down from 33.2% to 8.6%.

Tab. 4 HSV isolation rate from cervices before and after rIFN-αD treatment

Sample	Case	Isolate	HSV isolation rate (%)
Normal cervices	76	2	2.6
Erosive cervices	240	74	30.8
before treat.	208	69	33.2
HurIFN-αD after treat.	175	15	8.6

HYBRIDIZATION OF DNAs ISOLATED FROM EROSIVE CERVICES WITH HPV DNA PROBES

The distribution of HPV sequenses in biopsies obtained from the patients is shown in Tab. 5. Among 35 non-neoplastic lesions (chronic cervicitis, metaplasia and hyperplasia), HPV-16 sequence was found in 18 cases (51.4%); HPV-18-in 5 cases (14.3%); HPV-II-in 2 cases (5.7%) and HPV-6B-in 9 cases (25.7%). 8 of 18 HPV 16 positive specimens were demonstrated to be coinfected with another or other two viral types. One case with mild dysplasia was shown to be HPV-16 positive, wherease one of two cases with invasive carcinoma appeared coinfected with HPV-16, 18 and 6B. (Fig. 1)

Tab. 5 Distribution of HPV sequences in cervical biopsy specimens

Histologic diagnosis	Total no.	Positive number (%) hybridized with HPV DNA probes			
		HPV-16	HPV-18	HPV-11	HPV-6B
Chronic cervicitis:					
Simple	24	12	4	2	6
Metaplasia	7	3	0	0	1
Hyperplasia	4	3	1	0	2
Mild dysplasia	1	1	0	0	0
Invasive carcinoma	2	1	1	0	1
Total	38	20	6	2	10

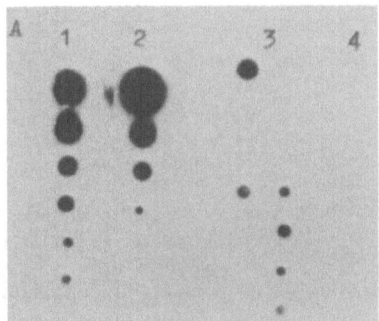 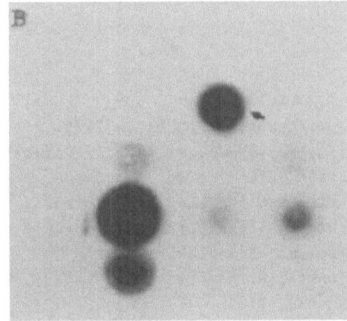

Fig. 1 Autoradiography of hybridization of DNA samples extracted from erosive cervices with [α^{32}P] labelled HPV-16, 11 and 6B DNA
A: 1) Positive control: HPV-6B 10ng-1pg;
   2) Positive control: HPV-11 10ng-1pg;
   3) Samples hybridized with HPV-6B;
   4) Samples hybridized with HPV-11;
B: Samples hybridized with HPV-16; arrow indicates the positive control (HPV-16, 10ng)

DISCUSSION

Result obtained from virus isolation from 491 cervical swabs indicated that HSV isolation rate from erosive cervices was about 12 times higher than that from normal and significantly dropped down after one course of interferon treatment. All virus isolates were characterized as HSV-2 by neutralization test using HSV 2 antiserum and restriction mapping of HSV-2 DNA extracted from several isolates. The clinical efficacy was also shown to be roughly paralleled with negative conversion of HSV isolation after treatment with interferon. These data suggested that HSV infection may be closely related to the pathogenesis of cervical erosion.

The human papillomaviruses are a remarkably heterogenous group encompassing at least 40 distinct types. Among them HPV-16 and HPV-18 are present in a high percentage of cervical, vulval and penile cancer biopsies (8, 10). Scholl et al (1985) reported that 3 biopsies showing squamous metaplasias or mild chronic cervicitis were all negative for HPV-16 DNA sequence using Southern blotting method (12). Prakash et al (1985) detected HPV-16 sequence in two of 17 biopsies taken from cases with chronic cervicitis (5). Wickenden et al (1985) indicated that HPV positive results were obtained in 7% of patients with normal cervices (11). We presented here that about 50% of cases with cervical erosion was demonstrated to be HPV-16 DNA positive using dotting hybridization method. These differences may be attributed to the different geographic regions or different hybridization methods used.

Different types of HPV are classified on the basis of cross hybridization in liquid phase under condition of high stringency. Under stringent condition we found that as much as 44% of biopsies from chronic cervicitis may contain multiple virus types; it seems a coinfection of different types of HPV rather than a cross hybridization with different DNA probes. Of course, the frequency of occurrence of multiple HPV strains in biopsy samples is an unsettled issue at present. L. Gissman indicated that about 15% of condyloma and cervical intraepithelial neoplasia may contain multiple virus strains (13).

In conclusion, we demonstrated a close association between the HSV and HPV infections and chronic cervicitis (erosive type) and this type of common disease is shown to be a good theurapeutic response to human rIFN-αD.

REFERENCES

1. Su Yingkuan, Jiang Sen: (1976) Practive Gynecology. Shan Dong People Press. P.172
2. Hill, EC (1980). in "Current obstetric and Gynecology, Diagnosis and Treatment" ed. by Benson, R.C. 3rd ed. Lange Med. Press p.188
3. Rawls, W.E and Campione Piccardo, J. (1981). In: A.J. Nahmias, W.R. Dowdle and R.E. Schinazi (eds). The human herpes viruses: an interdisciplinary perspective, pp.137-152, Elsevier, New York.
4. Frenkel, N., Roizman, B., Cassal, E. and Nahmias, A. (1972). Proc. Nat. Acad. Sci(Wash) 69, 3784-3789
5. Prakash, S.S., Reeves, W.C., Sisson, G.R., Brenes, M., Godoy, J., Bacchetti, S., de Britton, R.C. and Rawls, W.E. (1985). Int. J. Cancer 35, 51-57.
6. Galloway, D.A. and McDougall, J.K. (1983). Nature (Lond,) 302, 21-24.
7. Park, M., Kitchner, H.C., and Macnab, J.C.M., (1983). Europ. mol. Biol. Org. J. 2,1029-1034.
8. Boshart, M., Gissmann, L., Ikenberg, H., Kleinheing, A., Scheurlen, W. and zur Hausen, H. (1984). Europ. mol. Biol. Org. J. 3, 1151-1157.
9. Gissmann, L., Wolnik, L., Ikenberg, H., Koldovsky, U. and zur Hausen, H. (1983). Proc. nat. Acad. Sci USA 80, 560-563.
10. Durst, M., Gissmann, L, Ikenberg, H. and zur Hausen, H. (1983). Proc. nat. Acad. Sci. USA 80,3812-3815.
11. Wickenden, C., Steele, A., Malcolm A.D.B. and Coleman, D.V. (1985). The Lancet 1,65-67.
12. Scholl, S.M., Kingsley Pillers, E.M., Robinson, R.E. and Farrell, P.J. (1985). Int. J. Cancer 35, 215-218.
13. Crum, C.P., Mitao, M., Levine, R.U. and Silverstein, S. (1985). J. Virol. 54,675-681.

PRESENCE AND POSSIBLE ROLE OF INTERFERONS IN PSORIASIS

M. DEGRÉ, J.K. LIVDEN, J.R. BJERKE, G. HAUKENES and R. MATRE
Dept of Virology, Natl Inst. Public Health, Kapt. W. Wilhelmsen
og Frues Bakteriologiske Institutt, Natl Hospital, University
of Oslo and Depts. of Dermatology and Microbiology & Immunology, University of Bergen, Norway

INTRODUCTION
The etiology of psoriasis is yet to be solved. However, there
is a large body of evidence pointing to involvement of immune
mechanisms in the pathogenesis of this disease. Thus, several
authors reported presence of auto-antibodies against different
structures in the skin, as stratum cornea antigen, basal nuclei
and proliferating cells in psoriatic lesions. Impaired skin
reaction of delayed-hypersensisivity type indicated that cell
mediated immunity functions are also involved. Finally, several
groups of investigators reported a large variety of alterations
in mononuclear cell parameters. A depletion of T cells in
peripheral blood and suppressor cell deffects were observed by
some authors, but not by others. Altered monocyte/macrophage
activity, measured by clearence of ^{51}CR-labelled erythrocytes,
increased chemotaxis and elevated phagocytic and bactericidal
activities, are also common findings.

There is a general agreement that a cellular infiltrate is
present in the dermis already during the early phases of
development of psoriatic lesions. The composition of cellular
infiltrate may vary during different phases, but macrophages/
monocytes and lymphocytes seem to be present in large numbers.
During the initial phase of the present study we have determined the distribution of the infiltrating mononuclear cells by in
situ identification (1). T lymphocytes and macrophages were the
predominant inflammatory cells, and only few B lymphocytes were
detected. Most macrophages were found during clinically active
periods of the disease. Most T cells seem to be of helper/indu-
cer type, since they reacted with OKT3 and OKT4 monoclonal
antibodies, while suppressor T cells were present in minor
quantities. The majority of the T lymphocytes in the skin
lesions expressed Ia-like antigens. This indicated that they
were activated, and are involved in cell-mediated immunity
reactions. Another property of activated T lymphocytes is
production of immune or gamma interferon (IFN-α). Therefore
we decided to test for presence of IFN in the local lesions and
in the serum of psoriatic patients.

MATERIALS AND METHODS

Collection of blister fluids and sera.
Prior to collection of the first samples all local treatment

was interrupted for at least 3 days, in most cases more than
one week. Suction blisters were produced using the Dermovac
suction blister device as described by Kiistala (2) at 175 Hg
below atmospheric pressure. When suction blisters were generated in psoriatic lesions, considerable leakage occurred through
the fragile blister roofs into the cup. This fluid proved to be
sterile, and was pooled with the fluid obtained within the
blisters. Suction blisters were also generated in normal
appearing skin, at least 10 cm away from the lesions and in
skin from normal control persons. From each patient 0.5-1.5 ml
of blister fluid was obtained. Serum samples were drawn at the
same time. All samples were kept at -70°C until testing.

Interferon assay.
Interferon activity was tested by an infectivity inhibition
microtest (3), employing human embryonic fibroblast (HEL) cells
in their fifth to fiftenth passage or A 549 cells and Vesicular
stomatitis virus (VSV). The antiviral activity was calculated
as the reciprocal of the highest dilution preventing development of cytopathic effect in 50% of the wells at the time when
the control virus-infected cells were completely destroyed,
usually after 3 days of incubation. One unit of the standard
preparation G-023-901-527 was equivalent to 1-1.3 units in our
system.

To characterize the antiviral activity representative positive
samples were treated at pH 2 for 2 days, tested on heterologous
Marine Darby bovine kidney (MDBK) cells and were neutralized
with specific anti-human IFN-α (obtained from the Antiviral
Program of the National Institute of Allergy and Infectious
Diseases, Bethesda, Md).

Electron microscopy.
Blister fluids were centrifuged at 100 000 x g for 1 hour. The
pellets were applied to formicon grids and stained by 2%
potassium phosphotungstate, pH 6.6, and examined in a Hitachi
HU 12 A electron microscope at 60 000 magnification.

Treatment of patients.
Groups of patients were given Goeckerman therapy (UV-B radiation and 3% coal tar ointment) 3-6 times a week. Each patient
received an average of 24.6 treatments (range 10-42). A Waldman
UV 1000 treatment cabin (Waldman Co, Villingen-Schwenningen,
West Germany) equipped with 26 Sylvania UVG fluorescent
sunlamps, was used as UV-B source. At 310 nm Waldman 1000 has
an effect of 1.5- 1.8 mW/cm^2. The dose was increased at regular
intervals to obtain
a mild erythema.

RESULTS

In the initial series (4) 9 out of 13 patients with diagnosed
psoriasis vulgaris had significant antiviral activity in the
blister fluids from lesional skin. The titers were moderate,
less than 100 units in all samples. No activity was found from

unaffected skin samples, and only 3 patients had low interferon activity in their serum. Blister fluids from 4 healthy control persons were also negative.

The antiviral activity was characterized as interferon by the usual criteria. The activity was strongly, but not completely reduced by treatment at pH 2 and also neutralized to a large extent by anti-IFN-α. It was concluded that the activity was due mostly to IFN-γ, and to some extent to IFN-α. Furthermore, the difference between the activities in blister fluids and in serum indicated that the interferon was produced locally in the skin.

These findings were confirmed in several series including more patients. The presence of interferon activity in blister fluids and serum from 24 consecutively tested psoriatic patients is shown in TABLE 1.

TABLE 1. Interferon activity in blister fluids and serum from 24 patients with psoriasis vulgaris

	No of patients	IFN U/ml mean ± SD	Per cent positive
Blister fluids			
Lesion	24	33.5 ± 23.9	75
Unaffected	15	11.7 ± 10.2	13
Sera	24	21.0 ± 12.9	42

positive blister fluids and sera were characterized by their sensitivity to pH 2 and neutralization by anti-IFN-α. As shown in TABLE 2, in 7 of the 10 blister fluids tested the IFN titres were significantly reduced (more than 50%) after incubation at pH 2 for 2 days. Part of this reduction could be demonstrated also when the samples were tested in MDBK cells, indicating that pH 2 treatment eliminated some IFN-α activity. Interferon activity in 6 of the 11 blister fluids was reduced by incubation with antiIFN-α. The activity in the serum had similar properties.

TABLE 2. Effect of pH 2 and anti-IFN-α on the IFN activity in sera and blister fluids from patients with psoriasis

		No. of samples with IFN activity			
Treatment	Samples	No.	Reduced	Unchanged	Increased
pH 2	Serum	13	8	4	1
	Blister fluid	10	7	2	1
Anti-IFN-α	Serum	10	4	6	0
	Blister fluid	11	6	5	0

Because it has been suggested recently that viral agents may be involved in the etiology of psoriasis (5,6,7) we have looked for virus particles in the blister fluids by electron-microscopy (8). In 4 out of 8 blister fluids from psoriatic lesions we found particles that were virus like. The diameter of the particles was 100 nm with envelope-like structure and surface projections, like retrovirus particles. Similar particles were found in psoriatic lesions by Dalen and his coworkers (6). They also demonstrated reverse transcriptase activity in connection with psoriatic lesions. The presence of virus particles may explain the production of IFN-α.

Since the implication of these findings is that interferon activity is a measurable parameter in the diseased skin we decided to follow up a group of patients, especially during periods of active treatment. The first samples were taken at least one week after interruption of the local treatment. Also in this group higher levels of IFN were found in blister fluids from lesional skin than from unaffected skin and serum (TABLE 3). New samples were obtained after the patients have received Goeckerman therapy 3-6 times a week in average 24.6 treatments (range 10-42). The patients had higher IFN levels both in blister fluids and specially in the serum compared to the initial samples, indicating systemic production of IFN. Characterization experiments did not clarify whether the composition of different types of IFNs is altered during therapy, but it seems most probable that both IFN-γ and IFN-α is produced. Two patients had decreasing levels of IFN. No paramethers differentiated these patients from the other ones.

TABLE 3. Interferon in blister fluids and sera from patients with psoriasis vulgaris, untreated and after Goeckerman therapy

	No. of patients	Interferon (U/ml) (median $\pm$ SEM)	Percentage positive
Blister fluid			
lesional skin			
untreated	35	35.0 $\pm$ 6.6	77
after treatment	20	50.5 $\pm$ 13.6	90
unaffected skin			
untreated	21	10.0 $\pm$ 3.8	33
after treatment	18	26.5 $\pm$ 8.1	72
Serum			
untreated	34	25.0 $\pm$ 5.8	56
after treatment	22	89.0 $\pm$ 18.2	91

DISCUSSION

Our data show that low but consistent levels of interferons are present in suction blister fluids from psoriatic lesions, and to a lesser extent in the serum of psoriatic patients. The latter finding was confirmed by Diezel et al.(9). The nature of interferon activity seems to be partly of gamma type, which is mostly produced by activated T cells. Our data also indicate that the blister fluid IFN is produced locally in the psoriatic lesions. It is in line with the findings of large numbers of infiltrating T cells in the psoriatic lesions and since HLA-DR antigens were detected on these cells they seem to be activated. The presence of IFN-γ lends support to the notion of an ongoing immunological activity as a part of the development of psoriatic lesions(10).

The presence of the retrovirus particles may possibly explain the production of IFN-α. However, the origin of acid-labile leukocyte interferon is not clear. It has been detected in other immunological abnormities as systemic lupus erythematosus (11,12) and in AIDS (13,14). Incidentally, the possible role of the acid labile interferon induced by the retrovirus, Human immunodefficiency virus (HIV) in the pathogenesis of AIDS gives interesting associations.

The finding of elevated interferon levels during Goeckerman therapy was not surprising. Similar data were reported by Diezel et al. (9). UV radiation can induce IFN production also in in vitro systems (15). Preliminary data (not published) from our laboratories indicate that UV-B radiation of healthy subjects induces interferon production.

The significance of interferons in the pathogenesis of psoriasis is not clear. It is known that interferons do contribute to the disease activity in auto-immune disease in New Zealand mice (16), and manifestation of acute lymphocytic choriomeningitis virus disease in suckling mice (17).Interferons may play a role in systemic lupus erythematosus in man. It is tempting to speculate whether the elevated levels of interferons have a mediator role in the beneficial effect of UV radiation (9) and Goeckerman therapy. On the other hand, exacerbation of psoriasis was seen during IFN-α treatment of malignant diseases (18). More data are needed, especially to correlate disease activity to interferon levels and types both in local lesions and circulating, to further clarify the possible role of interferons.

REFERENCES

1. Bjerke JR, Krogh HC & Matre R: Characterization of mononuclear cell infiltrates in psoriatic lesions. J. Invest. Dermatol. 71:340, 1978.
2. Kiistala U: Suction blister device for separation of viable epidermis from dermis. J. Invest. Dermatol. 50:129, 1968.
3. Dahl H & Degré M: A micro assay for mouse and human interferon. Acta path. microbiol. scand. Sect B. 80:863, 1972.
4. Bjerke JR, Livden JK, Degré M & Matre R: Interferon in suction blister fluid from psoriatic lesions. Brit. J. Dermatol. 108:295, 1983.
5. Guilhou JJ, Vannereau H, Theunynck D, Bergoin M & Blanchard JM: Virologica studies on psoriatic lymphocytes. In:Psoria-

542

sis. Proceedings of the Third International Symposium. Ed.
E.M. Farber & A.J. Cox. Grune & Stratton, New York, 1982,
p.251.
6. Dalen AB, Hellgren L, Iversen OJ & Vincent J: A virus-like
particle associated with psoriasis. Acta path. microbiol.
immunol. scand. Sect. B. 91:221, 1983.
7. Stanbridge EJ: A possible viral etiology of psoriasis. In:
Psoriasis. Proceeding of the Third International Symposium.
Ed: E.M. Farber & A.J. Cox. Grunne & Stratton, New York,
1982, p.67.
8. Bjerke JR, Degré M, Haukenes G, Krogh HK, Livden JK & Matre
R.: T cells, interferons and retrovirus-like particles in
psoriatic lesions. Acta dermato-vener. (Stockholm) suppl
113: 29, 1984.
9. Diezel W, Waschke SR & Sönnichsen N: Detection of interferon
in sera of patients with psoriasis, and its enhancement by
PUVA treatment. Brit. J. Dermatol. 109:549, 1983.
10. Epstein WL: Immunologic factors in psoriasis. In: Psoriasis.
Proceedings of the International Symposium. Ed. E.M. Farber
& A.J. Cox. Stanford University Press, Stanford. 1971,
p.297.
11. Hooks JJ, Moutsopoulos HM, Geis SH, Stahl NI, Decker JL &
Notkins AL: Immune interferon in the circulation of
patients with autoimmune disease. New Engl. J. Med. 301:5,
1979.
12. Preble OT, Black RJ, Friedman RM, Klippel JH & Vilcek J:
Systemic lupus erythematosus: Presence in human serum of
an unusual acid-labile interferon. Science 216:429, 1982.
13. Eyster ME, Goedert JJ, Poon MC & Preble, OT: Acid-labile
interferon. A possible preclinical marker for the aquired
immunodefficiency syndrome in hemophilia. New Engl. J. Med.
309:583, 1983.
14. DeStefano E, Friedman RM, Friedman-Kien AE, Goedert JJ,
Henriksen D, Preble OT, Sonnabend JA & Vilcek J: Acid-labile
human leukocyte interferon in homosexual men with Kaposi's
sarcoma and lymphadenopathy. J. Infect. Dis. 146:451, 1982.
15. Lindner-Frimmel SJ: Enhanced production of human interferon
by u.v. irradiated cells. J. Gen. Virol. 25:147, 1974.
16. Engleman EG, Sonnenfeld G, Dauphinee M, Greenspan JS, Talal
N, McDevitt HO & Merigan TC: Treatment of NZB/NZW F_1 hybrid
mice with Mycobacterium bovis strain BCG or type II interferon preparations accelerates autoimmune disease. Arthritis
Rheumatism 24:1396, 1981.
17. Riviere Y, Gresser I, Guillon JC, Bandu MT, Ronco P, Morel-
Maroger L & Verroust P: Severity of lymphocytic choriomeningitis virus disease in different strains of suckling mice
correlates with increasing amounts of endogenous interferon.
J. Exp. Med. 152:633, 1980.
18. Quesada JR & Gutterman JU: Psoriasis and alpha-interferon.
Lancet 1:1466, 1986.

THE COMPONENTS OF THE INTERFERON SYSTEM IN SLE AND PSORIASIS

L. BORECKÝ, V. LACKOVIČ, D. KOČIŠKOVÁ, Z. ŠUJANOVÁ,
N. FUCHSBERGER, J. PASTOREK, J. ROVENSKÝ, Ľ. SINKA
Institute of Virology, Slovak Academy of Sciences, Bratislava, Research Institute of Rheumatology, Piešťany, Dermatological Clinic, Bratislava, Czechoslovakia

1. INTRODUCTION

Several reports indicate that, in addition to disturbances found in the specific immune reactions, also the broader interferon /IFN/ system, including the NK-cells and IL-2 etc, shows misregulation in autoimmune disorders /Hooks et al., 1981, Waschke and Diezel, 1984, Rovenský et al., 1984, Lange et al., 1984 etc./.
In this study, the properties of IFN and a recently detected IFN inhibitor /"anti-IFN"/, as well as the activity of NK-cells and their response to stimulation with IFN were studied in a longitudinal study in systemic lupus erythematosus /SLE/ and psoriatic patients. These two diseases were chosen as representatives of an organ-non-specific /SLE/ and an organ-specific /psoriasis/ type of disorder /Roitt, 1986/.

2. MATERIALS AND METHODS

2.1. The patients included in the SLE group /70/ fulfilled four or more criteria for diagnosis of the disease /Cohen et al., 1971/. The diagnosis of psoriasis /75 patients/ was based on criteria emploied in the Dermatological Clinic in Bratislava.
2.2. Interferon assay was based on the inhibition of cytopathic effect of vesicular stomatitis virus /VSV/ in human diploid embryonic cells /HEF/. Only titers above 8 units per ml were considered significant. Each assay included a reference IFN preparation.
2.3. Interferon neutralization assay was performed by mixing 16 units per ml of a standardized human leukocyte IFN alpha /or gamma/ with 0.1 ml of patients serum previously heated to 56°C for 30 minutes. After 30 minutes of incubation at 37°C, the remaining IFN activity of the mixture was determined.
2.4. NK-cell cytotoxicity was determined in peripheral blood monocytes /PBMC/ separated from the whole blood by the Ficoll Hypaque technique. Non-adherent mononuclears were suspended in RPMI-medium with 10% calf serum and antibiotics at 4×10^6 cells per ml density. K-562 cells were used as target cells at an effector to target cell ratio 20:1 /Lotzova et al., 1979/.
2.5. Polyacrylamide gel electrophoresis of serum was made according to Laemmli /1970/ while the method of Towbin et al., /1979/ was utilized in Western-blotting procedure.

3. RESULTS

3.1. The IFN found in about 15% of autoimmune patients shows a tendency to disappearance and an unpredictable reappearance during treatment. The titers /with a geometric mean of 13.2/ per ml/ in 11 psoriatic patients were nearly 3 times lower than in 10 SLE patients /36.5 per ml/.

TABLE 1: APPEARANCE AND DISAPPEARANCE OF IFN-α AND ANTI-IFN α
IN SERA OF 4 SLE PATIENTS

| TEST NO. | PATIENTS | | | | | | | |
| | G.A. | | M.E. | | M.M. | | A.R. | |
	IFN	A-IFN	IFN	A-IFN	IFN	A-IFN	IFN	A-IFN
1	16	–	<8	+	<8	–	<8	+
2	<8	+	<8	–	16	–	<8	–
3	<8	–	16	–	<8	+	16	–
4	32	–			16	–	<8	+
IN YEARS	1980 – 1982		1981 – 1982		1983 – 1986		1985 – 1986	

A-IFN+: NEUTRALIZING ACTIVITY PRESENT

The phenomenon of disappearance may explain the broad variation in findings of IFN in sera of SLE patients reported by various authors.

3.2. The IFN found in autoimmune patients is of alpha type, acid /pH2/ and thermolabile /56°C/30 min./.
It was neutralized both by polyclonal and monoclonal anti-IFN alpha sera and showed an about equal antiviral activity on MDBK cells. It was inactive on mouse and chicken cells. /Table 2/.

3.3. Interestingly, an IFN population of similar /acid labile etc./ properties could be separated from the acid stable IFN produced in leukocytes from healthy human blood donors in vitro. By Sephadex G 100 chromatography, it was found in a fraction with approximately 100 000 m.w. /Fig. 1, Curve A/. By rechromatography, this fraction proved to be a mixture of an inducer component /not shown-removable by anti-NDV serum, see Styk et al., 1973/, an acid stable component /which appears as Curve B by rechromatography/, and, an acid labile component /which resembles the SLE-IFN/. The source of the latter is unknown.

3.4. The reason for the temporary disappearance of IFN from the sera of SLE patients is not known. However, an "anti-IFN" was found in 15% of SLE and 20.3% of psoriatic sera in the IFN-free intervals. The anti-IFN, in contradistinction to SLE or psoriatic IFN, is thermostable, has a m.w. between 10 000 and 30 000 and has no proteolytic activity. It could not be found in IFN-positive sera /Table 1 and Fig. 1 and 2/. This suggests that anti-IFN may participate on disappearance of IFN during the disease.

TABLE 2: COMPARISON OF AUTOIMMUNE IFN WITH TWO COMPONENTS FOUND
IN "NORMAL" LEUKOCYTE IFN PREPARATION

IFN SOURCE	AVERAGE TITER ON			TITER ON HEF AFTER TREATMENT WITH ANTISERA		
	HEF	RANGE	MDBK	pH2	antiα	antiϞ
PATIENTS:						
SLE SERA /5/	35.2	/16-64/	35.2	<8	<8	16-64
PSORIATIC SERA /4/	24.0	/16-32/	24.0	<8	<8	16-32
CONTROLS:						
LEUKOC. IFN-ALPHA						
ACIDLABILE FRACTION	10^x		10	<5	<5	10
DETTO						
ACIDSTABLE FRACTION	128^x		128	<8	<8	128
IFN-GAMMA	256^x		NT	<8	256	<8

HEF: HUMAN EMBRYONIC FIBROBLASTS, MDBK: BOVINE CELL LINE, NT: NOT
TESTED, x: UNITS USED FOR TITRATION ON HEF CELLS

FIG. 1: COMPLEXITY OF PEAK A OBTAINED AFTER GEL FILTRATION OF HUMAN
LEUKOCYTE IFN ALPHA /SEPHADEX G-100/

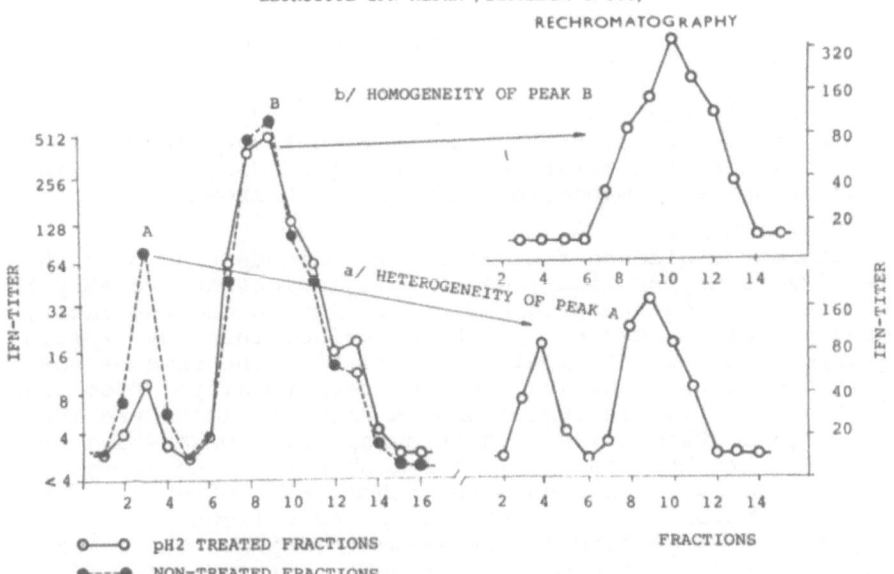

o——o pH2 TREATED FRACTIONS

●----● NON-TREATED FRACTIONS

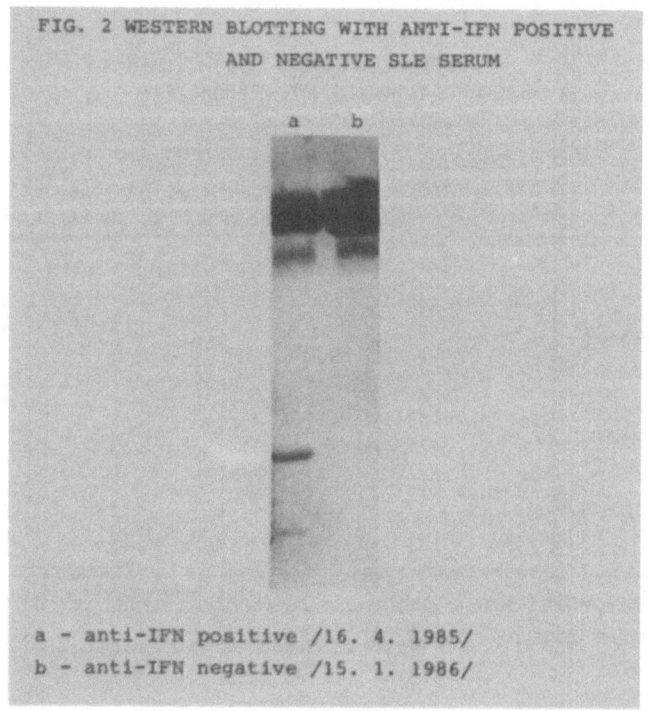

FIG. 2 WESTERN BLOTTING WITH ANTI-IFN POSITIVE
AND NEGATIVE SLE SERUM

a - anti-IFN positive /16. 4. 1985/
b - anti-IFN negative /15. 1. 1986/

3.5. In accordance with the prevailing feeling, the role of
IFN in the pathogenesis of the autoimmune diseases seems
rather negative: a/ Its presence in the serum diminished the
reactivity of NK-cell to exogenous IFN /tolerance?/, b/ IFN
potentiated also the blocking effect on NK-cells of an unknown serum factor found in SLE sera /Fig. 3/. This may
explain the depressed NK-cell activity observed in the majority of patients. A similar factor inhibited also the multiplication of spleen lymphocytes in vitro /not shown/.

4. CONCLUSION
The autoimmune disease are, as a rule, accompanied by disorders of various regulatory mechanisms reflecting probably the
interplay between the progression of disease and reparatory
processes set in movement in the organism. This may explain
the appearance and disappearance of IFN in the same person
during treatment. Such IFN does not seem to differ from other
acid and thermolabile IFNs found in "normal" leukocyte IFN
preparations /Chadha 1984 and this paper/. However, its presence seems to be pathognomic for many autoimmune diseases
and may negatively influence the function of other regulatory
cells and factors /such as NK, IL-2/. As follows from Fig. 3,
the presence of autoimmune IFN seems to depress the reactivity of NK-cells /most probably through a tolerance-inducing
effect/ and, in view of the reported cell-inclusion provoking

FIG. 3: EFFECT OF CONTINUOUS PRESENCE OF IFN IN THE
SERA ON ACTIVITY OF NK-CELLS AND MODULATION
OF NK-CELL ACTIVITY BY A BLOCKING FACTOR

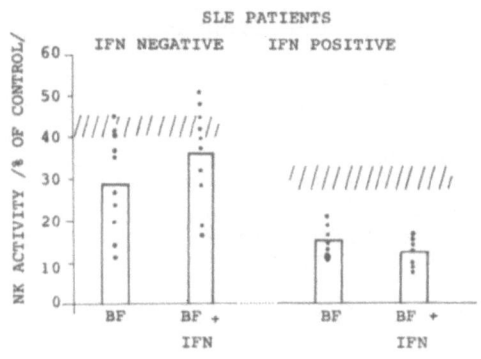

SHADED AREA: BASELINE NK ACTIVITY. OPEN COLUMNS:
MEANS. BF: AUTOLOGOUS SERUM /6 %/, IFN: HUMAN
LEUKOCYTE INTERFERON /500 U PER 0.2 ml/

and 2-5′adenylate synthetase inducing effect of IFN, its presence may explain also the ineffective attempts to cure SLE
with IFN /Rich and Owens, 1984, Preble et al., 1983/. However,
the stimuli for production of autoimmune interferon remain unknown as do the stimuli for release of other factors which
appear in these patients and which are unsufficiently studied.
Among them, the "blocking-factor" seems to inhibit the NK-
cells and, another one, the multiplication of spleen lymphocytes counteracting the IL-2 action /not shown/. They are
presumably not identical with IFN since, in IFN-negative patients, the depressed NK-activity could be overcome by IFN.
The "anti-IFN" appears to be a polypeptid of approximately
10 000 to 30 000 m.w. It may contribute to the temporary
disappearance of IFN from the sera of patients /Rovenský et
al., 1984/.

REFERENCES
1. Chadha KC: Acid labile human leukocyte interferon. In:
   The Interferon System, pp 35-42, Eds.: F. Dianzani, G.B.
   Rossi, Raven Press, N. York, 1985
2. Cohen AS, Reynolds WC, Franklin EC, Kulka MW, Shulman LE,
   Wallace SE: Preliminary criteria for the classification of
   systemic lupus erythematosus. Bull. Rheum. Dis. 21, 643-
   648, 1971
3. Hooks JJ, Moutsopoulos HM, Notkins AL: Circulating interferon in human autoimmune diseases. Tex. Rep. Biol. Med.
   41, 164-168, 1981
4. Lotzova E, McCredie KB, Maroun JA, Dicke KA, Freireich EJ:
   Some studies on natural killer cells in man. Transplant.
   Proc. 11, 1390-1392, 1979

5. Lange A, Havel A, Jacak A, Haas H, Flad HD: Cytotoxic
   and blocking effects of sera from patients with systemic
   lupus erythematosus /SLE/ on NK-cell activity. 6th Europ.
   Immunol. Meeting Interlaken, 3.-8.9. 1984
6. Laemmli UK: Cleavage of structural proteins during the
   assambly of the head of bacteriophage T4. Nature 227, 680-
   685, 1970
7. Preble OT, Rothko K, Klippel JH: Interferon induced 2-5´
   adenylate synthetase in vivo and interferon production in
   vitro by lymphocytes from systemic lupus erythematosus patients with and without circulating interferon. J. Exp.
   Med. 157/6, 2140-2146, 1983
8. Rich SA, Owens TR: Purified recombinant human leukocyte
   interferon IFLRA and IFLRD induce human lupus inclusions
   in raji and daudi cells. J. Interferon. Res. 4/3, 335-345,
   1984
9. Roitt IM: Current theories of autoimmune disorders.
   Sandoz Revue, 1, 27-32, 1986
10. Rovenský J, Lackovič V, Borecký L, Žitňan D, Lukáč J:
    Interferon production by leukocytes obtained from patients
    with systemic lupus erythematosus.in: Physiology and Patho-
    logy of Interferon System, Contr. Oncol. v. 20, pp 280-
    297, Karger, Basel, 1984
11. Styk B, Borecký L, Vetrák J, Fuchsberger N, Lackovič V,
    Hajnická V: Contribution to the problem of purity of in-
    terferon preparations. Acta Virol. 17, 163-166, 1973
12. Towbin H, Staehelin Th, Gordon J: Electrophoretic transfer
    of proteins from polyacrylamide gels to nitrocellulose
    sheets. Procedure and some applications. P.N.A.Sc. /USA/,
    74, 4350-4354, 1979
13. Waschke SR, Diezel W: Interferon in patients with psoria-
    sis. in: Physiology and Pathology of Interferon System,
    Contr. Oncol. v. 20, pp 298-303, Karger, Basel 1984

PHARMACOKINETICS OF [^{123}I]-INTERFERON (IFN) ALPHA-A.

R.A. DIEZ, B. PERDEREAU, M.C. MARTYRE, J. WIETZERBIN, S. CORNAERT*, M. PETER, C. GAUDELET, C. BARBAROUX, T. DORVAL, P. POUILLART, L. GAUCI° and E. FALCOFF. * Commissariat de l'Energie Atomique, ° Hoffmann–La Roche Inc., Bâle, Switzerland and Institut Curie, Biology and Hospital Sections, Paris (France).

1. INTRODUCTION

A major concern in the pharmacological study of human interferons (IFNs) is its species specificity, which makes extrapolation from animal studies difficult. Even in the case of pharmacokinetic studies performed in patients, the data obtained were usually limited to seric concentration-time profiles. To characterize more directly the pharmacokinetic parameters of IFN alpha, we decided to perform the study of IFN concentration in whole blood and serum with the simultaneous determination of its dynamic distribution by scintigraphic counting.

Our results suggest that IFN concentration in serum might to be not representative of the whole blood compartment and that free IFN in circulation is able to bind to mononuclear cells, from which it can be thereafter eluted. The repartition study showed retention of IFN-linked ^{123}I in different organs, including heart and the abdominal compartment. However, the proportion of scintigraphic counts corresponding to IFN or to free iodine remains to be determined.

2. MATERIALS AND METHODS

2.1. Patients. The main clinical characteristics of the 5 disseminated cancer patients studied are summarized in Table 1.

2.2. Treatment. IFN alpha-A (Batch 8413, Specific Activity ≃ 2.02 X 10^8 I.U./ mg protein, Hoffmann-La Roche) was produced as previously described (1). It was coupled to ^{123}I by the Iodogen^R method (2) in the Radiopharmaceuticals Laboratory, Commissariat de l'Energie Atomique, France.

2.2.1. Dose. 15 µg (≃ 3 X 10^6 I.U.) administered by slow intravenous injection (15 min.) performed with an infusion pump (SE 400 Class I, Vial Médical)

2.2.2. Protocol. The evening before the study, lugol was administered to the patients for blocking thyroid captation. Fasting patients received IFN between 1 and 2 P.M. Blood samples were taken just before the injection, at its end and at different times post injection (PI). Scintigraphic counting was

Table 1. <u>Clinical characteristics of the patients</u>

Patient	Primitive	Metastasis	Age Ys	Sex M/F	Race	Height (cm)	Weight Kg	Previous Chemotherapy
1	Breast	Liver Bone	50	F	White	163	70	Yes
2	Breast	Bone	52	F	White	160	56	Yes
3	Ocular Melanoma	Lymph nodes Lung Liver	47	M	White	172	66	Yes
4	Cavum	Lymph nodes Lung ?	21	M	Black	160	59	Yes
5	Cutaneous Melanoma	Liver Bone	60	M	White	174	85	No

performed from the beginning of the injection up to 60 minutes after its end and for 30 min. periods at 3 and 21 h. PI.

2.3. <u>Responses</u>.

2.3.1. <u>Interferon concentration</u>

2.3.1.1. <u>Bioassay</u> in Wish cells against VSV (Indiana strain) as described (3).

2.3.1.2. <u>Gamma emission at 159 keV</u>

2.3.1.2.1. In TCA precipitable material and in PBMC (Multigamma II Counter, LKB). TCA precipitation was previously reported to give comparable results to bioassay determination in mice (4). The technique was slightly modified, since samples were laid and precipitated onto 3M filter paper.

2.3.1.2.2. By external counting, with Maxi Camera II, 400 T Sensor (General Electrics). Numerical treatment of the images was performed with Informatek[R] (Sopha Médical, France).

2.3.2. <u>Characterization of radiolabelled protein in serum</u>, by SDS–PAGE and autoradiography on Kodak XAR–5 film, using Dupont Cronex lightening plus intensifying screens.

2.3.3. <u>Estimation of pharmacokinetic parameters</u> performed by standard techniques on graphical analysis of semilog concentration-time plots.

2.3.4. <u>Determination of surface and internalized interferon</u> was performed by elution with acetic acid (5). In brief, PBMC were obtained from 7 ml heparinized blood samples, drawn at different times, by Ficoll-Paque (Pharmacia) gradient. One million PBMC were washed with PBS-Ca^{++}- Mg^{++} - 10 % FCS and then incubated, in duplicate, with the elution buffer (0.2 M acetic acid in 0.5 M NaCl, pH 2.5) for 5 min. on ice. After 3 min. centrifugation at 200 g (4°C), the pellet and the eluted fraction were considered as internalized and surface IFN respectively. This classical procedure has been shown to reverse IFN binding to its receptor (see 6 for review).

3. RESULTS AND DISCUSSION

The coupling of ^{123}I to IFN alpha-A was accomplished as usually for other iodine isotopes. The estimate of free iodine was about 25 %, as measured by chromatography (not shown). It is less than the usual iodine dose required for thyroid studies and below the safety recommended level. The specific radioactivity showed interpreparation variation, ranging from 3.2 MBq µg protein^{-1} to 7.2 MBq µg protein^{-1}.

The intravenous administration of ^{123}I IFN alpha-A resulted in a predictable pattern of radioactivity in whole blood and serum, as previously shown in rats and rabbits (Perdereau et al., manuscript in preparation). Figure 1 shows the densitometric scanning of serum SDS-PAGE autoradiographies of patient 2. The radioactivity present at the end of the infusion corresponded to a single band of about 20 kD. The next determination (15 min. PI) showed the 20 kD radioactive protein, but also a lower molecular weight band, probably a degradation product. Later samples did not contain as much radioactivity as to give autoradiographic images. Similar results were obtained with 2 additional patients ; in the other 2, autoradiographies were not performed.

The time profiles of blood and serum IFN concentrations of patient 2 measured as TCA-precipitated radioactivity are depicted in Figure 2. They both can be fitted to a biexponential model, though blood levels are systematically higher than serum ones, and its decay is slower. Table 2 presents several pharmacokinetic data from blood compartment calculated from patients 1 to 4, as TCA precipitated radioactivity. Neither blood or serum concentration data from patient 5 could be fitted ; the same was the case for serum data from patient 1. It is not excluded that, at least partially, gamma emission from blood corresponds to free iodine captation or intracellular IFN degradation products.

Bioassay determinations of IFN concentration were always very low. All patients complained of febrile manifestations, including hyperthermia up to

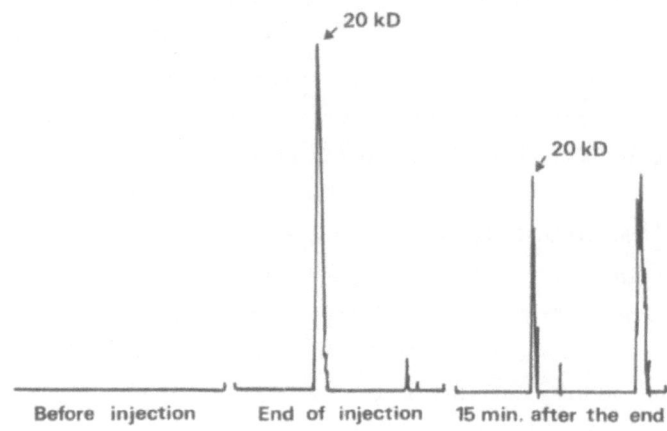

Figure 1. Densitometric scans of autoradiographies corresponding to SDS-PAGE of 50 µl samples of serum.

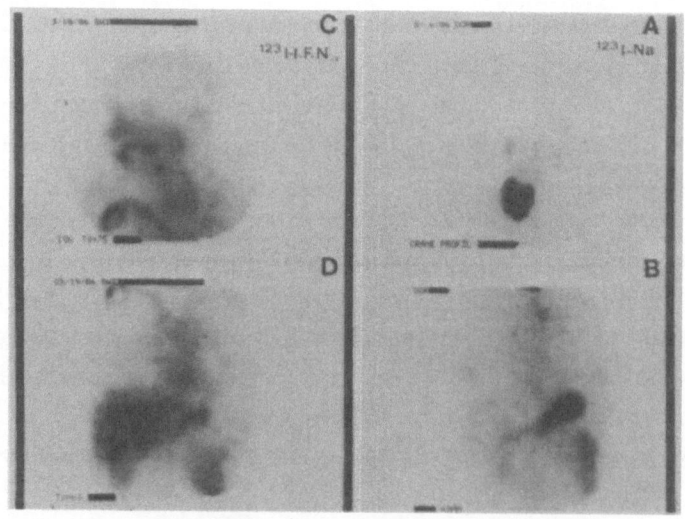

Figure 3. Scintigraphic images obtained in patient 1 (left) after ^{123}I inection and in a control subject (right) undergoing a thyroid captation study with Na ^{123}I. See text for additional explanations.

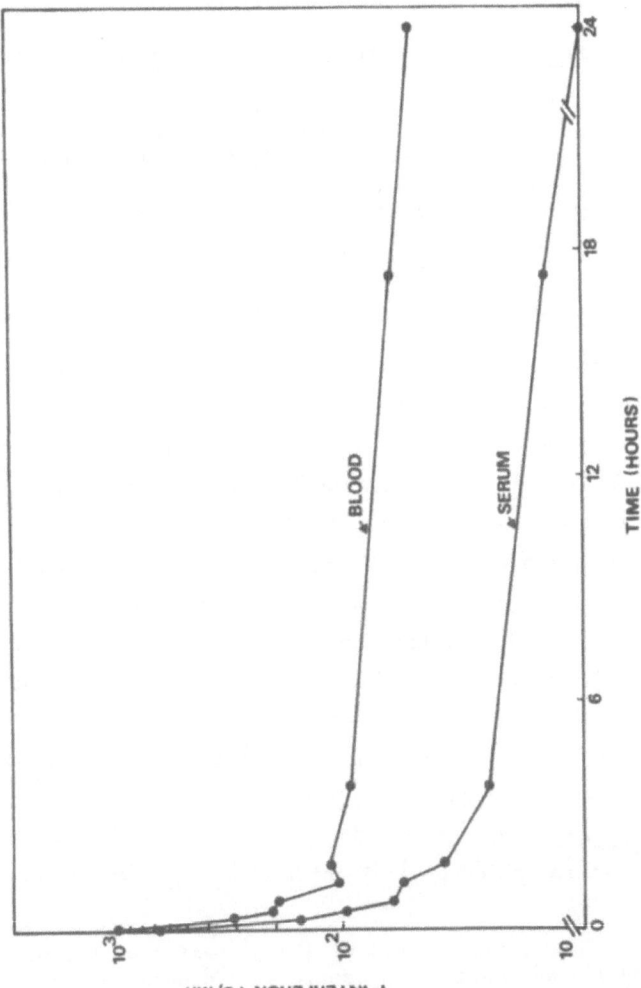

Figure 2. Semilog plots of blood and serum concentration-time profiles of interferon alpha-A after slow intravenous injection.

39°C and in all of them lymphopenia appeared soon after the injection (see below), suggesting that administered IFN was biologically active. It is not clear wether its low activity in the in vitro system is related to the labelling procedure.

Table 2. **Pharmacokinetic parameters of ^{123}I-Interferon alpha-A in whole blood**

Patient	$t1/2\alpha$ (h)	$t1/2\beta$ (h)	α (h^{-1})	β (h^{-1})	Vd ap $(1\ kg^{-1})$	AUC $(IU\ h\ ml^{-1})$	Co $(IU\ ml^{-1})$
1	0.18	9.7	4.12	0.090	0.414	1052	94.5
2	0.70	14.1	0.98	0.049	0.518	1876	91.9
3	0.10	0.5	6.90	1.386	0.255	135	181.8
4	0.13	13.9	5.18	0.050	0.578	1766	87.9

We took advantage of a thyroid study to see Na ^{123}I distribution after i.v. bolus (shown in Figure 3A and 3B). The dose was higher than free iodine in our IFN preparation. Figures 3C and 3D illustrate the scintigraphic study of patient 1. Her countings showed a very similar pattern to the iodine alone study, making it difficult the interpretation of the data.

Very similar images were obtained in the other patients. Individual variations were important but always scintigraphic data behaved as blood and serum pharmacokinetics, i.e. patient 3 had the shortest half-time in any organ tested. Table 3 summarizes the data of accumulation and elimination of ^{123}I IFN in heart, lungs and liver in our 5 patients, as well as of Na ^{123}I in a control subject. As appreciated, the only differential trend is the elimination from heart, which seems faster for ^{123}I IFN.

Table 3. **Scintigraphic determination of ^{123}I-interferon alpha distribution**

Treatment	Effective dose (MBq)	Half-time (min.) of the accumulation phase in			Half-time (h) of clearance from		
		Heart	Lung	Liver	Heart	Lung	Liver
^{123}I-inter-feron (n=5)	66.7* ±10.6	20.4 ±4.4	30.0 ±11.7	27.0 ±7.5	27.4 ±7.2	76.4 ±34.0	18.4 ±1.7
^{123}I-Na (n = 1)	27.75	20.2	–	30.1	47.5	100	24.1

* $\overline{X}$
± SEM

In the case of IFN fate in the kidneys, there was a very important masking effect from ^{123}I accumulation in the stomach and upper gut which hindered valid measurements. The control of TCA precipitable radioactivity in urine was systematically negative, as well as the bioassay.

There was no detectable radioactivity in craneal areas, in agreement with the total absence of TCA-precipitable radioactivity and biological activity in 1 sample of CSF obtained 2 h PI from patient 1. This is also consistent with the reported lack of passage of IFN to the central nervous system at low doses (7).

Many speculations arose about the presence of high ^{123}I counts in oral cavity, nose and paranasal sinuses (Figure 3). At present, no firm explanations can be offered. The main ground for the images to correspond to ^{123}I IFN was the simultaneous detection of TCA-precipitable radioactive material in salive (with a maximum of fifty percent at 90 min. PI, and a rapid decay thereafter, not shown). The IFN activity of salive is under study. However, in the control subject, up to 10 % total salivary radioactivity was TCA precipitable ; its nature remains uncertain.

Table 4. Interferon alpha-A induced lymphocytopenia

Patient	Base-level ($Ly\ mm^{-1}$)	Nadir [$Ly\ mm^{-1}\ (\%°)$]	Time* (min.)
1	884	290 (67)	135
2	1450	497 (66)	244
3	585	249 (57)	120
4	415	238 (45)	123
5	1994	539 (73)	121

* After the beginning of the injection.
° Percentage of decrease.

In order to determine whether blood cells were able to bind ^{123}I-IFN, PBMC were isolated at the same times chosen for the concentration samples. In the course of the study, the 5 patients developed a marked lymphopenia, as previously reported (8), even those whose lymphocyte counts were already low at the beginning of the study (Table 4).

The results of non-eluted interferon per million PBMC at different times are presented in Table 5. After 75 min., neither the ^{123}I counts per 10^6 cells were sufficient to allow us to determine the IFN amount internalized nor the recovery of PBMC to augment their number in the elution procedure. As shown, despite individual variations, there was a time trend to decrease the amount of internalized IFN. The simultaneous evidence of lymphocyte redistribution suggests that lymphocytes leaving the blood compartment were those which had previously bound and internalized IFN. It is not excluded that the lower amounts of non-eluted IFN merely reflects the rapid degradation of internalized IFN, with the delivery of iodine (free or linked to small peptides).

Table 5. **Interferon units bound to 10^6 PBMC (n = 5) estimated as non-eluted gamma emission**

	Time*				
	0	15	30	45	75
$\bar{X}$	0.934	0.256	0.055	0.028	0.007
SEM	0.654	0.128	0.025	0.011	0.007
Range	(0.038-2.878)**	(0.017-0.731)	(0.014-0.142)	(0-0.058)	(0-0.034)

* Minutes after the end of the injection.   ** n = 4.

While heterogenous, in all samples the eluted fraction decreased in the course of the study more rapidly than IFN concentration.

For the samples drawn just at the end of the injection, the eluted percentage of ^{123}I-IFN linked to the cells varied from 18 to 53 % as the in vitro processing time interval increased from 15 min. (n = 2) to 60 min. (n = 4). Normal blood (n = 2) incubated in vitro with added radiolabelled IFN gave comparable results (10 % to 32 %).

We suggest that this approach to the study of the pharmacokinetics of human IFN deserves particular attention. Current studies point to resolve the problems linked to the presence of free iodine in IFN preparations.

We are indebted to Ms Agnes Géhant for the skillful typing of the manuscript.

4. REFERENCES

1. Staehelin T et al. J. Biol. Chem. 256:9750, 1981.

2. Fraker PJ & Speck JC Jr. Biochem. Biophys. Res. Commun. 80:849, 1978.

3. Vaquero C et al. J. interferon Res. 2:217, 1982.

4. Palleroni AV & Bohoslawec O. J. Interferon Res. 4:493, 1984.

5. Wietzerbin J et al. J. Immunol. 136:2451, 1986.

6. Zoon KC & Arnheiter H. Pharmac. Ther. 24:259, 1984.

7. Smith RA et al. Clin. Pharmacol. Ther. 37:85, 1985.

8. Scott GM et al. Antimicrob. Agents Chemother. 23:589-592, 1983.

2'5' A SYNTHETASE ACTIVITY IS INDUCED BY A FACTOR EXCRETED IN HEAT-SHOCKED CELLS. COMPARISON OF THIS FACTOR WITH IFN.

S. CHOUSTERMAN, M.K. CHELBI-ALIX and M.N. THANG.
INSERM U. 245, I.B.P.C., 13 rue P. et M. Curie 75005 PARIS, FRANCE.

1. INTRODUCTION

Heat shock and other stresses, such as treatment with aminoacid analogs or viral infections, have been shown to induce the synthesis of a set of proteins, called heat-shock proteins (hsp), in a wide variety of organisms including bacteria, plants and animals (1).

The 2'5' A system constitutes one of the pathways of the action of interferon (IFN) (2) by leading to the inhibition of protein synthesis, and can for instance inhibit virus replication (3).

We wished to know whether the 2'5' A system plays a role in cell response to heat shock. For this, we measured the 2'5' A synthetase activity in extracts from cells during heat shock and subsequently recovery period.

2. MATERIALS AND METHODS

2.1. Cells and virus

The cells used were Madin-Darby bovine kidney fibroblasts (MDBK) and spontaneously transformed cells from human placental tissue (WISH). MDBK cells were grown in Eagle's medium and WISH cells in Dulbecco's medium, both supplemented with 10 % fetal calf serum. Vesicular stomatitis virus of the Indiana strain (VSV), was grown in the L929 cells. Stock virus samples were usually 10^8 plaque-forming units/ml (PFU/ml).

2.2. Interferon and anti-interferon antibodies

Bovine (β/α) IFN (1.5 10^6 units/mg protein) was produced by challenging primary kidney fibroblasts with Newcastle disease virus (NDV) (4). The bovine (β/α) IFN and antibovine (β/α) IFN antibodies (500 neutralization units/ml) were donated by Dr. La Bonnardière, Thieverval-Grignon, France.

Bovine IFN was titrated on MDBK cells and expressed in relation to the HuLe reference IFN (NIH cat. no. GO23-901-527), which titrates 20,000 international units/ml, and under our conditions (i.e. on MDBK cells) yielded 12,800 units/ml.

2.3. Heat-shock treatment

Cells growing exponentially in Falcon flasks (1.5 10^6 cells) were incubated at 45°C for the period indicated in each experiment. After heat treatment, the culture medium was replaced with fresh medium and incubation continued at 37°C for various periods. At the indicated times, sample were removed for 2'5' A synthetase activity determination in cell extracts, as described in "Assay of 2'5' A synthetase activity". We checked cell viability in MDBK cultures heat-treated for 1 h by trypan blue exclusion. More than 99 % of the cells remained viable throughout the experiment.

2.4. Assay of 2'5' A synthetase activity

MDBK or WISH cells were treated with heat, as indicated above, or with IFN. After different times of incubation at 37°C (see figures), the cells were washed in 140 mM sodium chloride, 3 mM magnesium chloride and 35 mM HEPES pH7.5 and then lysed for 5 min at 0°C in 10 mM potassium chloride, 2 mM magnesium acetate, 7 mM 2-mercaptoethanol, 10 mM HEPES pH7.6 and 0.5 % Nonidet P-40 (5). Cell lysates were then centrifuged for 6 min in an Eppendorf centrifuge and the supernatants were stored until assayed at - 80°C.

558

100 μl of each cell extract were tested for 2'5' A synthetase activity as already described (6).
The protein concentration of the crude extracts was determined according to Spector (7). ATP
conversion into 2'5' A oligoadenylates was calculated. One unit of 2'5' A synthetase activity was
taken to correspond to 1 nmol ATP converted/h at 37°C. Its specific activity was expressed as
units/mg of crude extract protein.

3. RESULTS AND DISCUSSION
3.1. Increase in 2'5' A synthetase activity in MDBK cells and WISH cells after heat shock at 45°C
The basal level of the 2'5' A synthetase activity in MDBK cells is very low (4 units/mg protein).
In contrast, a large enhancement in the enzyme activity was observed when MDBK cells were heat-
treated for various periods of time at 45°C, then transferred to 37°C.

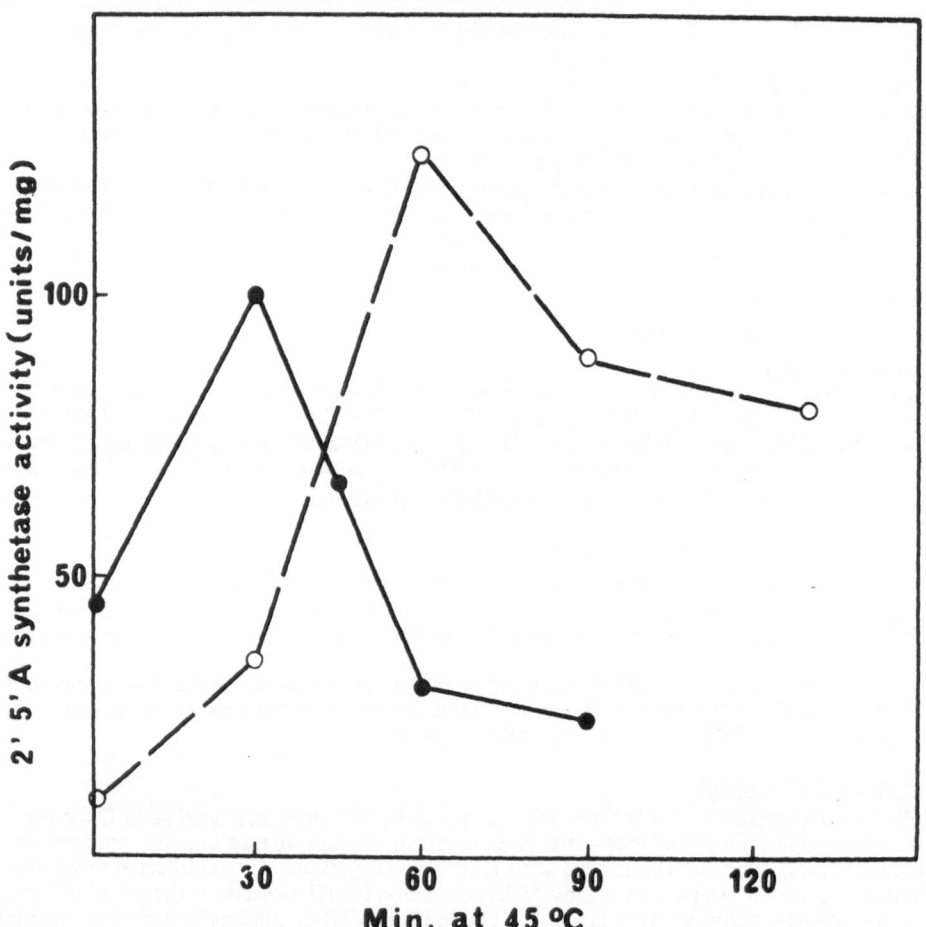

Fig. 1 : Period of heat shock at 45°C required for maximal increase of 2'5' A synthetase activity in
MDBK and WISH cells.

MDBK and WISH cells were incubated at 45°C for different periods of time. After heat treatment,
the medium was replaced with fresh medium before cell transfer to 37°C. After 18 h at 37°C, 2'5' A
synthetase activity was determined in MDBK (O------O) and WISH (●——●)   cell extracts as
described under Methods.

Table I : Effect of different inhibitors on 2'5' A synthetase activity induced after heat shock in MDBK cells.

TREATMENT		2'5' A SYNTHETASE ACTIVITY (units/mg protein)
Untreated cells		4
heat-shocked cells		120
"	+ Actinomycin D (5 μg/ml)	4
"	+ Cycloheximide (5 μg/ml)	5
"	+ Antibovine IFN antibodies	110

MDBK cells were grown exponentially in Falcon flasks (1.5 10^6 cells). Inhibitors or antibovine IFN antibodies (25 neutralizing units/ml) were added 10 min before heat shock and were present throughout the incubations at 45°C and 37°C. Cell extracts and 2'5' A synthetase activity assay were performed as described in Methods.

The maximum increase was obtained when cells were incubated for 1 h at 45°C and then 18 h at 37°C in fresh medium (Fig. 1). Therefore, the subsequent experiments were performed in these conditions. This maximum increase was close to 30 times that of the basal level and corresponded to 120 units/mg protein.This phenomenon was also observed with WISH cells. In this case, the rate of the increase in enzyme activity in response to heat shock was however higher than that of MDBK cells and the maximum was attained earlier. Likewise, the rate of the decline in enzyme activity was also higher (Fig. 1).

Clearly, the 2'5' A synthetase activity induction observed after heat shock is not an early event, since the increase in enzyme activity started only several hours after the beginning of the recovery period at 37°C, and the maximum activity was attained after 18 h recovery (8). It should be added that de novo synthesis of RNA and protein is required for the induction of 2'5' A synthetase activity after heat shock (Table I). Moreover, the kinetics of 2'5' A synthetase induction are different from that observed for the synthesis of the 70 hsp, a prominent hsp found to be highly conserved in most organisms. In a pulse-labeling experiment with ^{35}S-methionine in MDBK cells, we have shown that, as in other eukaryotic cells, a 70 hsp was synthesized during the heat-shock period. Its synthesis was maximum after return to 37°C and then declines rapidly (data not shown).

These results indicate that 70 hsp synthesis precedes the expression of 2'5' A synthetase, so that the latter is obviously not an hsp with an enzymatic activity. The 2'5' A synthetase activity induced after heat shock is therefore a secondary event that occurs relatively late in the recovery period. This is why we assumed that the induction of this activity depends on factor(s) directly induced after heat-shock treatment. These considerations led us to investigate the possible expression of such factor(s) from the heat-shocked cells in their culture medium.

3.2. The culture medium of heat-shocked cells contains a factor(s) capable of inducing 2'5' A synthetase activity in fresh cells.

MDBK cells were heat-treated for 1 h at 45°C and transferred at 37°C for 7 h. The cell-free medium was then separated and tested for its capacity to induce the 2'5' A synthetase activity in fresh cells. The results in figure 2 show the kinetics of induction of enzyme activity in MDBK

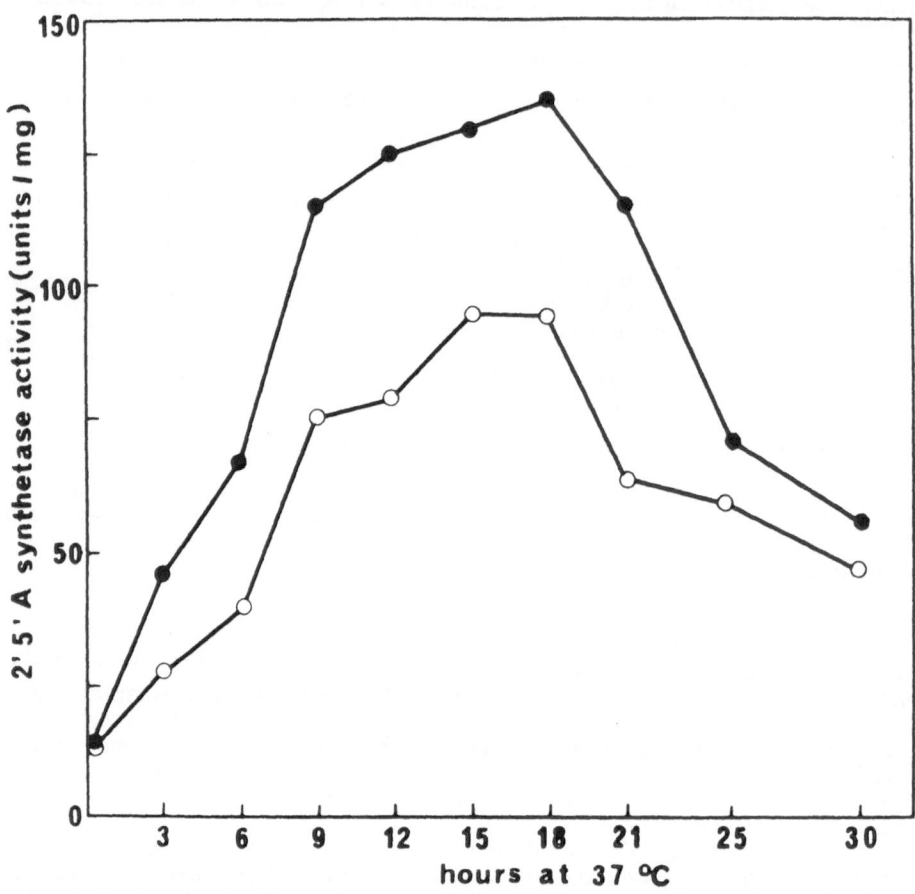

Fig. 2 : Kinetics of induction of the 2'5' A synthetase in MDBK cells by heat-shocked-induced factor (HSIF) or by bovine IFN.

MDBK cells (1.5 10⁶) were heat-treated during 1 h at 45°C, after which the medium was replaced with fresh medium and the culture transferred to 37°C for recovery. After 7 h, the culture medium was assayed for the presence of HSIF. MDBK cells (4 10⁵) were treated with this medium (HSIF) (●———●) or with bovine IFN (O———O). The 2'5' A synthetase activity was determined in cell extracts at the times indicated in the figure.

cells at 37°C by this cell-free medium. The 2'5'A synthetase activity increased for the first 12 h and then reached a plateau (130 units/mg protein). After 18 h, the activity declined but at 30 h still remained 10 fold the basal level. These kinetics resemble those of bovine IFN (Fig. 2). Thus, it is clear that the culture medium of heat-shocked MDBK cells contains a heat-shock-induced factor(s) (HSIF) capable of increasing the 2'5' A synthetase activity in fresh cells.

3.3. Properties of heat-shock-induced factor(s) (HSIF).
To characterize HSIF, we tested the efficiency of the culture medium from heat-treated

Table II : Efficiency of HSIF in different conditions

TREATMENT		2'5' A SYNTHETASE ACTIVITY %
untreated MDBK cells		6
HSIF		100
"	+ Trypsin (30 µg/ml / 1 h)	6
"	+ Ribonuclease A (20 µg/ml / 1 h)	123
"	+ Filtration on G-25 Sephadex column	92
"	+ Heat-stability (56°C / 1h)	43
"	+ Acidic treatment (pH2/22 h)	20
"	+ Antibovine IFN antibodies	93
untreated WISH cells		27
HSIF		100

HSIF (culture medium from heat-shocked cells which have been allowed to recover for 7 h at 37°C) was treated in the following conditions and then used to induce the 2'5' A synthetase activity in fresh MDBK cells during 18 h. The specific activity obtained with HSIF without treatment was taken as 100 %.

MDBK cells (1 h/45°C - 7 h/37°C) after different treatments. From the results shown in TableII, it appears that HSIF behaves as a polypeptide of a molecular weight larger than 5,000. It is rather thermostable but unstable at acid pH. The question then arises of whether HSIF is a heat-shock-induced IFN or not. For this purpose, we assayed the effect of the antibovine IFN antibodies in parallel on HSIF and bovine IFN under conditions in which 250 units of bovine IFN were completely neutralized (Table II). It is clear that the inducing capacity of HSIF remained unaffected whereas that of bovine IFN was abolished. This important observation seems to indicate that HSIF is different from IFN, at least bovine α and β IFN. Besides, it must be pointed out that the increase in 2'5' A synthetase activity induced by IFN in cells treated with HSIF was additive. This additive increase, which roughly corresponded to the induction observed in cells treated with IFN alone, was abolished in the presence of antibovine IFN antibodies. Similarly, we observed an additional increase in the 2'5' A synthetase activity when bovine IFN was added to heat-shocked cells during both heat shock and the recovery period at 37°C (data not shown). Again the additive increase was abolished by the addition of anti-bovine IFN antibodies. On the contrary, the 2'5' A synthetase activity induced after heat shock, in the presence of the same anti-IFN antibodies, remained unaffected (Table I).

Two other types of assay were also performed with HSIF. In the first the antiviral activity of the culture medium containing HSIF was assayed on MDBK cells using VSV as a challenge virus : no detectable antiviral activity was found. Similarly, the heat-shocked cells, infected after 18 h recovery at 37°C, did not develop an antiviral state since the virus titer, expressed as PFU/ml, was the same as that of untreated cells (data not shown). The second type of assay concerned HSIF species-specificity. We did observe that HSIF is also capable of induction capacity in heterologous WISH cells (Table II). It should be mentioned that Taylor et al. (9),

562

working with heat-shocked human B-lymphoblastoid cell lines transformed by Epstein-Barr virus, found elevated levels of antiviral activity due to de novo synthesis of IFN γ ; they also noted that human IFN γ induced by heat shock was species-specific.

Various substances other than IFN can also induce 2'5' A synthetase activity including corticoid hormones (10), butyric acid, DMSO, retinoic acid (11, 12), platelet-derived growth factor (PDGF) (13). It must be emphasized that, in the latter case, 2'5' A synthetase induction is independent of IFN, and Zullo et al. (13) therefore suggested that 2'5' A synthetase and other elements of the IFN system may function as feedback inhibitor of the growth response to PDGF. Similarly, we suggest that 2'5' A synthetase may also function as a feedback regulator in the same response to heat shock.

Accordingly, in heat-shocked cells, allowed to recover at normal temperature, expression of the 2'5' A synthetase gene may constitute part of the rescue mechanism in response to stress, by increasing the turnover of hsp mRNA through the 2'5' A dependent endoribonuclease pathway. Other pathways may be involved in cell recovery through other reactions catalyzed by 2'5' A synthetase (14). It was for instance suggested that in heat-shocked cells, $Ap_4 A$ may have a dual function : to signal stress and terminate the heat shock response (15, 16). We ask ourselves whether the 2'5' A molecules function also as stress-signalling molecules.

REFERENCES

1. Ashburner, M. (ed) : Heat shock : from Bacteria to Man, Cold Spring Harbor Laboratory Press, N.Y., pp 1-7, 1982.
2. Lengyel, P. : Ann. Rev. Biochem. 51, 251-282, 1982.
3. Ball, L.A. In P.D. Boyer (ed) : The enzymes. Vol XV : Nucleic Acids, Part B, New York Academic Press, pp. 281-313, 1982.
4. La Bonnardière, C., De Vaureix, C., L'Haridon, R. and Scherrer, R. : Archiv. Virol. 64, 167-170, 1980.
5. Hovanessian, A.G., La Bonnardière, C. and Falcoff, E. : J. of Interferon Res., 1, 125-135, 1980.
6. Justesen, J., Ferbus, D. and Thang, M.N. : Nucleic Acids Res. 8, 3073-3085, 1980.
7. Spector, T.: Annal. Biochem., 86, 142-146, 1978.
8. Chousterman, S., Chelbi-Alix, M.K., and Thang, M.N. : in Williams, B.R.G. and Silverman, R..H. (eds) : the 2-5 A system, Alan R. Liss Inc. N.Y. pp 67-74, 1985.
9. Taylor, M.W., Long, T., Martinez-Valdez, H., Downing, J. and Zeige, G. : Proc. Natl. Acad. Sci. USA, 81-4033-4036, 1984.
10. Krishnan, I. and Baglioni, C. : Proc. Natl. Acad. Sci. USA, 77, 6506-6510, 1980.
11. Besançon, F. , Bourgeade, M.F., Justesen, J. , Ferbus, D. and Thang, M.N. : Biochem. Biophys. Res. Commun. 103, 16-24, 1981.
12. Bourgeade, M.F. and Besançon, F. : Cancer Res., 44, 5355-5360, 1984.
13. Zullo, J.N., Cochan, B.H., Huang, A.S. and Stiles, C.D. : Cell, 43, 793-800, 1985.
14. Ferbus, D., Justesen,, J. Besançon, F. and Thang, M.N. : Biochem. Biophys. Res. Commun. 100, 847-856, 1981.
15. Bochner, B.R., and Ames, B.N. : Cell, 29, 929-937, 1982.
16. Guedon, G., Sovia, D. , Ebel, J.P., Befort, N. and Remy, P. : EMBO J. 4, 3743-3749, 1985.

ACKNOWLEDGMENTS

We are grateful to Dr. C. La Bonnardière for the gift of bovine (β/α) IFN and antibovine β/α IFN antibodies. We also thank G. Huvet for technical assistance and Dr. S. Conjeevaram for the reading of the manuscript. This work was supported by grants from the Centre National de la Recherche Scientifique (UAC 113), the Institut National de la Santé et de la Recherche Médicale (U. 245) and La Fondation pour la Recherche Médicale.

MAPPING THE ACTION OF INTERFERON ON PRIMATE BRAIN

Richard A. Smith, M.D. and Carlisle P. Landel, Ph.D.,[*] Center for Neurologic Study, San Diego, CA, USA; C.E. Cornelius, M.D., University of California, Davis, CA, USA; Michel Revel, Ph.D., Weizmann Institute, Israel.

INTRODUCTION

Little is known about the effect of interferons on the central nervous system. Presently incurable diseases of the brain with known (SSPE and PML) or postulated (MS and ALS, e.g.) viral etiologies are potentially amenable to treatment with interferon.[1] However, the treatment of neurologic disease with interferon presents special problems because of the action of the blood-brain barrier.[2] Accordingly, the route of administration of interferon may have a significant effect on its distribution. Because titers of interferon in the CSF following systemic administration are relatively low, it might be assumed that little, if any, interferon reaches the nervous system. However, the blood-brain barrier is permeable at three sites: the infundibular recess, the median emminence and the area postreama.[3] As a result, systemically administered interferon may reach high concentrations in some brain sites while failing to do so in others. This may account for the finding that CSF interferon levels may be relatively low (100 IU/ml) at a time when patients exhibit signs of neurotoxicity.[4] In short, CSF interferon titers may not provide a true reflection of what is going on within the brain.

Interferon confers viral resistance upon cells by initiating an intracellular cascade of events which interfere with transcription of the viral template or inhibit translation of viral-specific proteins.[5] One of the events in this cascade is the induction of a number of specific genes, including one coding for a protein known as C56.[6] To examine more directly whether interferon crosses the blood-brain barrier, we determined whether this interferon-induced mRNA was present in rhesus monkey brains after treatment. Specifically, we treated rhesus monkeys either intravenously or intracisternally with interferon, isolated RNA from various tissues, and tested for the presence of C56 mRNA by dot-blot hybridization.

MATERIALS AND METHODS

1. Administration of interferon and tissue isolation. Partially purified alpha interferon (PIF), provided by K. Cantell, Helsinki, was administered to juvenile rhesus monkeys (1.3 kg and 1.75 kg, respectively) following ketamine sedation. An adult animal (9.8 kg) served as a control. One animal, given 7.5×10^6 IU PIF intravenously, was sacrificed 9 hours later and the other, given 1×10^6 IU PIF intracisternally, was posted after 8 hours. These times were chosen on the basis of the known time course of interferon effect. Tissues were removed, snap frozen and stored at $-70°$ C for later study.

[*]Present address: The Salk Institute, San Diego, CA, USA.

2. <u>Preparation of RNA</u>. Frozen tissue was homogenized in ten volumes of 4 M guanidine thiocyanate, 25 mM sodium citrate, 0.5% sarkosyl, 0.1 M mercaptoethanol and 0.1% antifoam A, pH 7.0, and centrifuged at 8000 g for 10 min at 4° C. The supernatant was extracted twice with 1:1 phenol:chloroform and once with chloroform, and was then layered over a 1.2 ml cushion of 5.7 M CsCl, 25 mM sodium acetate, 10 mM EDTA, pH 5.0, in an SW51 centrifuge tube and spun at 35,000 rpm for 18 hours. The RNA pellet was resuspended in RNase-free water, precipitated with ethanol and stored at -70° C until use. Globin mRNA was purchased from Bethesda Research Labs.

3. <u>Preparation of DNA probes</u>. Plasmids containing chicken beta tubulin and human C56 genes were generously provided by J. Chebath, Weizmann Institute, and M. Kirschner, UCSF, respectively.[6,7] These were labeled with ^{32}P by nick translation, resulting in a specific activity of $1 - 5 \times 10^8$ cpm/μg.

4. <u>Hybridization</u>. RNA was resuspended in 20X SSC (SSC = 0.15 M NaCl, 0.015 M sodium citrate), and 40 μg was applied to a nitrocellulose filter using a 96-well vacuum manifold. The filter was air dried and then baked at 80° C for one hour. This dot-blot was then soaked in prehybridization solution (50% formamide, 5X SSC, 50 mM sodium phosphate, 5X Denhardt's solution, 0.1% SDS, and 250 μg/ml sheared, denatured salmon sperm DNA) at 37° C for one hour and then hybridized with prehybidization solution plus 10^6 cpm/ml nick-translated probe overnight at 37° C. The blots were then washed twice at room temperature in 2X SSC, 0.1% SDS, twice for one hour at 50° C in 0.2X SSC, 0.1% SDS, blotted dry, wrapped in saran wrap and autoradiographed at -70° C using Kodak XAR film and a DuPont Cronex intensifying screen.

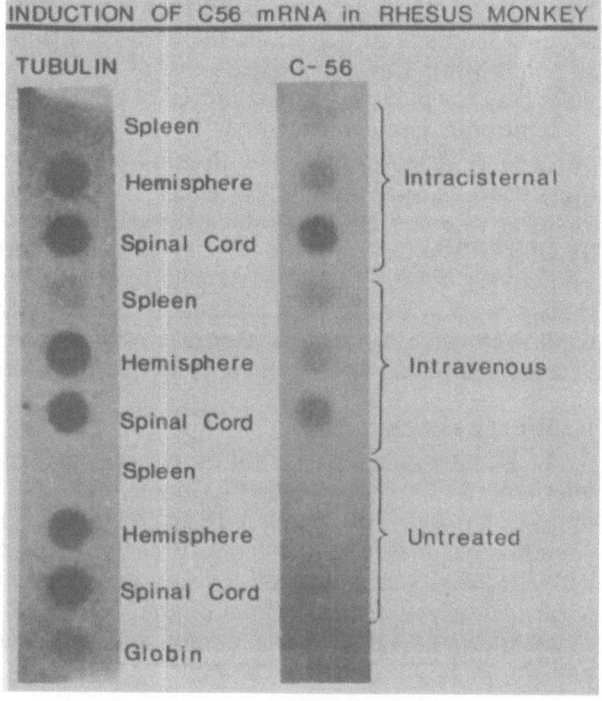

INDUCTION OF C56 mRNA in RHESUS MONKEY

RESULTS

Figure 1 shows dot blots of spleen, cerebral hemisphere, and spinal cord RNAs from the experimental and control animals, as well as globin (control) mRNA, probed with either ^{32}P-labeled C56 or tubulin DNAs. The C56 probe hybridizes with RNA from all tissues in the experimental animals, but not to the RNAs from the control animals or to the globin mRNA. It hybridizes more strongly to the brain RNAs in the intracisternally-injected animal than in the intravenously injected animal, while its binding to the spleen RNAs follows the inverse pattern. The tubulin (control) probe, on the other hand, hybridizes strongly to the brain RNAs from all animals and less strongly to the spleen RNAs. It also hybridizes faintly to the globin mRNA.

DISCUSSION

Interferon appears to have induced C56 mRNA transcription in the brain, spinal cord and spleen. This effect was observed irrespective of the route of interferon administration, since C56 mRNA could be detected in all tissues in the experimental animals but not in the control animal. However, it appears that the route of administration may quantitatively effect the relative amounts of mRNA in the brain and body compartments. The tubulin control demonstrates that sufficient RNA was present in each of the samples to detect this message by hybridization, and that nearly equal amounts of RNA were present in each dot. The relative strengths of the tubulin signals are as expected, since neural tissue produces much more tubulin than spleen. The small amount of non-specific binding of the tubulin probe to the globin mRNA does not alter these conclusions.

This preliminary study suggests that interferon can reach the brain parenchyma after both intravenous and intracisternal injection, and demonstrates for the first time the direct action of interferon on tissues of the CNS. Furthermore, it suggests that while interferon can cross the blood-brain barrier, the route of administration may affect the relative magnitude of the response in the two compartments. Since we have not yet measured the interferon titers in the CSF and plasma of these animals, we do not yet know whether the CSF levels of interferon in the intravenously-injected animal remained low, as would be expected from previous studies. Until this work is carried out, we can only speculate on the reasons for the discrepancy between our results and those demonstrating low CSF titers of interferon after systemic administration. Perhaps these low levels are enough to induce C56 transcription, or perhaps the receptors on brain cells bind interferon so tightly that very little remains free in the CSF. Further studies will resolve this question, and in addition will determine dosage effects and will better define the anatomy and the time course of interferon response in the brain.

REFERENCES

1. Salazar, A.M., Gibbs, C.J., Gajdusek, C.D., and Smith, R.A. 1984. In *Interferons and their applications. Handbook of experimental pharmacology.* (ed. Came, P. and Carter, W.) pp. 471 - 497. Berlin, Springer-Verlag.
2. Smith, R.A., Norris, F., Palmer, D., Bernhardt, L. and Wills, R.J. 1985. *J. Clin. Pharmac. Therap.* 37: 85-88.
3. Brightman, M.W., Prescott, L. and Reese, T.S. 1975. In *Brain endocrine interaction* (ed. Knigge, K.M., Scott, D.D., Kobayashi, H. and Ishii, S.) pp. 146 - 165. Basel, Karger.
4. Smedly, H., Katrak, M., Sikora, K. and Wheeler, T. 1983. *Br. Med.J.* 286: 262-264.

5.   Friedman, R.L., Manly, S.P., McMahon, M., Kerr, I.M. and Stark, G.R.   1984. *Cell* 38: 745-755.
6.   Chebath, J., Merlin, G., Metz, R., Benech, P. and Revel, M. 1983. *Nuc. Acids Res.* 11: 1213-1226.
7.   Valenzuela, P., Quiroga, M., Zaldivar, J., Rutter, W.J., Kirschner, M.W. and Cleveland, D.W. 1981. *Nature* 289: 650-655.

ACKNOWLEDGEMENTS

This work was made possible by the generous support of ALS Health Services, Frances Stockton and The Thagard Foundation. It is dedicated to patients John Stockton, Fred Clark and Charles Bordner, and to Mr. George Thagard, without whom these studies could not have been undertaken.

CHEMICAL AND BIOLOGICAL CHARACTERIZATION OF NATURAL HUMAN LYMPHOBLASTOID INTERFERON ALPHAS

K.C. ZOON, D.L. ZUR NEDDEN, J.C. ENTERLINE, J.F. MANISCHEWITZ, D.R. DYER, R.A. BOYKINS,* J. BEKISZ, AND T.L. GERRARD

Division of Virology and Division of Biochemistry and Biophysics,* Office of Biologics Research and Review, Center for Drugs and Biologics, Food and Drug Administration, 8800 Rockville Pike, Bethesda, MD 20892

INTRODUCTION

The biological and chemical characterization of natural human interferon (IFN) alpha is essential in order to understand the structure and function of this family of related molecules. The identification of multiple species of human IFN-alpha derived from virus-induced lymphoblastoid cells was previously described (1-6). In this study, eighteen species of human IFN alpha (pr1-pr18) derived from Sendai-induced Namalwa cells (Wellferon, Wellcome Research Laboratories) were isolated by sequential monoclonal antibody affinity chromatography using NK2 Sepharose (Celltech), 111/21-Sepharose (7), M2Q-098 Sepharose (Hybritech), and 4F2-Sepharose (Amgen) and high performance liquid chromatography as previously described (3,4), and analyzed for activity in a variety of antiviral and immunological assays. In addition, the apparent molecular weights of the species were determined and several species were analyzed for their carbohydrate content.

RESULTS AND DISCUSSION

The apparent molecular weights of IFN alpha species designated pr1 through pr18 ranged from 17,500 to 22,000 by sodium dodecyl sulfate polyacrylamide gel electrophoresis (SDS PAGE) under nonreducing conditions and 17,500 to 32,000 (pr8) by SDS PAGE under reducing conditions. Two species of IFN-alpha, pr1 and pr2, appear to be glycoproteins based on the presence of neutral and/or amino sugars in these IFN preparations (8). IFN-alpha pr2 was shown to contain glucosamine, galactosamine, fucose, glucose and/or mannose and galactose. Other evidence for the glycoslation of several human IFN-alphas derived from CML and KG-1 cells has been recently reported (9). The lymphoblastoid IFN species exhibited different antiviral specific activities on mouse (1.0×10^3 to 3.2×10^5 units per mg protein) and human (0.2 to 4.1×10^8 units/mg protein) cells, but all exhibited relatively constant antiviral specific activities on bovine cells (1.5-2.9×10^8 units/mg protein).

Several immunomodulatory activities of the IFN species were examined including the induction of interleukin 1 (IL1) secretion by human monocytes, enhancement of natural killer cell activity, enhancement of class I histocompatability antigens and induction of tumoricidal activity. The immunological activities were normalized based on protein and ranked from high to low with the most active species being ++++ and the inactive species being negative (-) (Table I). IL1 secretion by elutriator purified monocytes was quantitated by the murine thymocyte costimulation assay (10). All medium, reagents and IFN were found to be endotoxin-free. The species pr4, pr8, pr9, pr10 and pr12 were very good inducers of IL1 secretion while pr7, pr14, pr16, pr17 and pr18 were poor inducers.

TABLE 1. Immunomodulatory activities of human lymphoblastoid interferon
alphas.[a]

IFN-alpha species	IL1 induction	Enhancement of Natural killer cell activity	Enhancement of HLA-ABC antigens	Induction of monocyte tumoricidal activity
NK2 Column[b]				
pr2	++	++	++++	++
pr3	++	++	++++	+++
pr4	++++	+	+	+/-
pr5	+	+++	+++	+++
pr6	+	++	++	+++
pr7	+/-	++	+++	+
pr8	+++	++	+	+++
111/21 Column				
pr9	+++	++	+	++++
pr10	+++	+	+/-	+
pr11	++	-	+/-	+
pr12	+++	+/-	+/-	+/-
pr13	++	-	+/-	+/-
M2Q-098 Column				
pr14	-	+	+/-	+/-
pr15	+	++	+	+
4F2 Column				
pr16	+/-	++	+	++
pr17	+/-	+++	++	++++
pr18	+/-	+++	+	++
Wellferon	+	++++	++	++++
alpha 2[c]	++	+++	n.d.	++++
gamma[d]	-	++	++++	++++

[a]Activities were normalized based on mg protein.
[b]Data not reported for pr1.
[c]Gift of Schering Corp.
[d]Gift of Biogen Corp.
n.d. Not determined.

The enhancement of natural killer cell activity was assayed using human
peripheral blood mononuclear cells (effectors) and ^{51}[Cr]-labeled K562
cells (targets). IFN concentrations ranged from .002 to 5 ng/ml, and
effector:target ratios of 10:1, 5:1, and 2.5:1 were employed (11).
Species pr5, pr17 and pr18 were good enhancers of NK activity while pr11,
pr12, and pr13 were poor enhancers.

The enhancement of class I histocompatability antigen expression was
measured after a 2 day incubation of HL-60 cells with IFN-alpha species
(12). The cells were stained with mouse monoclonal anti-human HLA-A,B,C
and FITC-goat anti-mouse (Fab')$_2$ and the fluorescence intensity was
quantitated by flow cytometry. IFN species pr2, pr3, pr5 and pr7 were

very good enhancers of class I antigens while pr10, pr11, pr12, pr13 and pr14 were poor enhancers.

The induction of monocyte tumoricidal activity was assayed using elutriator purified monocytes that were pretreated with the various IFN-alpha species for 20 hr before the addition of ^{3}[H]-thymidine-labeled A375 melanoma cells as targets (13). IFN concentrations of 0.033 to 10 ng/ml and effector:target ratios of 20:1, 10:1, and 5:1 were employed. Species pr3, pr5, pr6, pr8, pr9 and pr17 were very good inducers with pr9 and pr17 being essentially equivalent to human IFN-gamma, which has been shown previously to be a potent inducer of monocyte tumoricidal activity (14). Species pr4, pr12, pr13 and pr14 were poor inducers. Thus, IFN-alpha species exhibited immunomodulatory activities which differ among themselves as well as to the starting mixture. In addition, each IFN species showed different degrees of activity for the induction of IL1 secretion by monocytes, enhancement of class I antigen expression, enhancement of natural killer cell activity and induction of monocyte tumoricidal activity.

ACKNOWLEDGEMENTS

We are grateful to Cathy Hobbs for the preparation of this manuscript.

REFERENCES

1. Zoon KC, Smith ME, Bridgen PJ, Zur Nedden D, Anfinsen CB: Proc. Nat. Acad. Sci. USA 76:5601-5605, 1979.
2. Allen G, Fantes KH: Nature (London) 287:408-411, 1980.
3. Zoon KC, Hu R-Q, Zur Nedden D, Nguyen NY: The Biology of the Interferon System 1984 pp. 61-67, Kirchner H, and Schellekens H (eds). Amsterdam: Elsevier Science Publishers B.V., 1985.
4. Zoon KC, Hu R-Q, Zur Nedden D, Gerrard TL, Enterline JC, Boykins RA, Nguyen NY: The Biology of the Interferon System 1985 pp. 55-58, Stewart II WE, and Schellekens H (eds). Amsterdam: Elsevier Science Publishers B.V., 1986.
5. Allen G, Fantes KH, Burke DC, Morser J: J. Gen Virol. 63:207-212, 1982.
6. Yonehara S, Yanase Y, Sano T, Imai M, Nakasawa S, Mori H: J. Biol. Chem. 256:3770-3775, 1981.
7. Arnheiter H, Thomas RM, Leist T, Fountoulakis M, Gutte B: Nature (London) 294:278-280, 1981.
8. Boykins RA, Liu T-Y: J. Biochem. Biophys. Methods 2:71-78, 1980.
9. Labdon JE, Gibson KD, Sun S, Pestka S: Arch. Biochem. Biophys. 232:422-426, 1984.
10. Meltzer MS, Oppenheim JJ: J. Immunol. 118:77-82, 1977.
11. Quinnan Jr GV, Kirmani N, Rook AH, Manischewitz JF, Jackson L, Moreschi G, Santos GW, Saral R, Burns WH: N. Engl. J. Med. 307:7-13, 1982.
12. Kelley VE, Fiers W, Strom TB: J. Immunol 132:240-245, 1984.
13. Gerrard TL, Terz JJ, Kaplan AM: Int. J. Cancer 26:585-593, 1980.
14. Dean R, Virelizier JL: Clin. Exp. Immunol. 51:501-510, 1983.

AUTHOR INDEX

SUBJECT INDEX